# Probability and Bayesian Statistics

W0107327

# Probability and Bayesian Statistics

Edited by
R. Viertl
Technical University of Vienna
Vienna, Austria

Plenum Press • New York and London

Library of Congress Cataloging in Publication Data

International Symposium on Probability and Bayesian Statistics (1986: Innsbruck, Austria)
Probability and Bayesian statistics.

"Based on the proceedings of the International Symposium on Probability and Bayesian Statistics, held September 23–26, 1986, in Innsbruck, Austria"—T.p. verso.
"Dedicated to the memory of Bruno de Finetti"—Pref.
Bibliography: p.
Includes index.
1. Probabilities—Congresses. 2. Bayesian statistical decision theory—Congresses. 3. De Finetti, Bruno—Congresses. I. Viertl, R. (Reinhard) II. De Finetti, Bruno. III. Title.
QA273.A1I58 1986 519.2 87-15298
ISBN-13: 978-1-4612-9050-6 e-ISBN-13: 978-1-4613-1885-9
DOI: 10.1007/ 978-1-4613-1885-9

Based on the proceedings of the International Symposium on Probability and Bayesian Statistics, held September 23–26, 1986, in Innsbruck, Austria

© 1987 Plenum Press, New York
Softcover reprint of the hardcover 1st edition 1987
A Division of Plenum Publishing Corporation
233 Spring Street, New York, N.Y. 10013

All rights reserved

No part of this book may be reproduced, stored in a retrieval system, or transmitted in any form or by any means, electronic, mechanical, photocopying, microfilming, recording, or otherwise, without written permission from the Publisher

To the memory of
Bruno de Finetti (1906–1985)

# PREFACE

This book contains selected and refereed contributions to the "International Symposium on Probability and Bayesian Statistics" which was organized to celebrate the 80th birthday of Professor Bruno de Finetti at his birthplace Innsbruck in Austria. Since Professor de Finetti died in 1985 the symposium was dedicated to the memory of Bruno de Finetti and took place at Igls near Innsbruck from 23 to 26 September 1986. Some of the papers are published especially by the relationship to Bruno de Finetti's scientific work.

The evolution of stochastics shows growing importance of probability as coherent assessment of numerical values as degrees of believe in certain events. This is the basis for Bayesian inference in the sense of modern statistics.

The contributions in this volume cover a broad spectrum ranging from foundations of probability across psychological aspects of formulating subjective probability statements, abstract measure theoretical considerations, contributions to theoretical statistics and stochastic processes, to real applications in economics, reliability and hydrology. Also the question is raised if it is necessary to develop new techniques to model and analyze fuzzy observations in samples.

The articles are arranged in alphabetical order according to the family name of the first author of each paper to avoid a hierarchical ordering of importance of the different topics. Readers interested in special topics can use the index at the end of the book as guide.

The editor wants to thank the referees for their anonymous work. Some of them are also authors in the present volume and their names are not given here. The following scientists who where not participants at the symposium or made no contribution to this Proceedings volume were so glad to act as referees: J.O. Berger, Lafayette, D. Blackwell, Berkeley, L.D. Broemeling, Arlington, L. Crisma, Trieste, I. Csiszar, Budapest, M. Deistler, Wien, P. Diaconis, Stanford, L.E. Dubins, Berkeley, R. Dutter, Wien, W. Ettl, Wien, T.L. Fine, Ithaca, D. Fürst, Roma, P. Hackl, Wien, W. Jammernegg, Graz, A. Kandel, Tallahassee, F. Konecny, Wien, D.V. Lindley, Somerset, M. Luptacik, Wien, G. Marinell, Innsbruck, B. Natvig, Oslo, T. Postelnicu, Bucuresti, H. Rauch, Wien, P. Revesz, Wien, M. Schemper, Wien, K.D. Schmidt, Mannheim, A.F.M. Smith, Nottingham, F. Spizzichino, Roma, H. Stadler, Wien, H. Strasser, Bayreuth, S. Weber, Mainz, G.A. Whitmore, Montreal.

It is the intention of this volume to make Bayesian ideas available for a broader audience and to present different recent developments in probability and statistics. I want to thank PLENUM for publishing this volume in short time which makes it possible to produce an up to date contribution and especially Ms. M. Carter for her kind advice.

R. Viertl

CONTENTS

## STOCHASTIC LINEAR PROGRAMMING WITH RECOURSE UNDER PARTIAL INFORMATION

Peter Abel

Fegro
Praunheimer Str. 5-11
D-6236 Eschborn/Ost, F.R.G.

### 1. INTRODUCTION

Stochastic programming models with random variables with only incompletely known distributions were up to now comparatively seldom analysed, although an attempt to declare probability distribution not always gives a satisfactory description of factors of influence in a decision model: "In any specific problem the selection of a definite probability distribution is made on the basis of a number of factors, such as the sequence of past demands, judgements about trends, etc. For various reasons, however, these factors may be insufficient to estimate the future distribution. As an example, the sample size of the past demands may be quite small, or we have reason to suspect that the future demand will come from a distribution which differs from that governing past history in an unpredictable way" (Scarf, 1958).

One of the reasons for the rarely use of stochastic programming models with only incomplete informations about the distribution of the states in practise may be that often no numerical methods are available for the computation of the optimal solutions of such a model.

In this paper we present a stochastic linear programming model with partial informations about the probability distribution of the random coefficients and have a special look at the possibilities of the numerical computation of all optimal solutions of the presented model.

### 2. THE PROBLEM

The stochastic linear programming model with recourse (two-stage model) with known common probability distribution of the random coefficients is of the form (see Dantzig, 1955)

$$\inf E_P\{c'(\omega)x + \inf\{q'(\omega)y \mid W(\omega)y = b(\omega) - A(\omega)x, \; y \geq 0\}\} \quad (1)$$

$$\text{subject to} \quad x \in X$$

with random variable $(A,b,c,W,q): \Omega \to \mathbb{R}^{mn+m+n+mn'+n'}$ and probability distribution P.

For every $\omega \in \Omega$ is $A(\omega) \in \mathbb{R}^{mn}$ a constant (m,n)-matrix,

$b(\omega) \in \mathbb{R}^{m}$ a constant m-vector,

$c(\omega) \in \mathbb{R}^{n}$ a constant n-vector,

$W(\omega) \in \mathbb{R}^{mn'}$ a constant (m,n')-matrix and

$q(\omega) \in \mathbb{R}^{n'}$ a constant n'-vector.

$X \subseteq \mathbb{R}^n_+$.

If P is not exactly known, but there is merely an information $P \in \Theta$, where $\Theta$ is a set of probability distributions, we have to find an optimal $x_0 \in X$ as solution of the problem

$$\text{"minimize"} \quad E_P\{c'(\omega)x + \min\{q'(\omega)y \mid W(\omega)y = b(\omega) - A(\omega)x, \; y \geq 0\}\} \tag{2}$$

$$\text{subject to} \quad x \in X, \quad P \in \Theta.$$

We suppose that for every $P \in \Theta$ (1) has a finite solution (see Abel, 1984 p. 49f), $X := \{x^{(i)} \mid 1 \leq i \leq M\}$ and $\Omega := \{\omega_1, \omega_2, \ldots, \omega_N\}$ are finite sets and $\Theta$ is a Linear Partial Information (LPI) (see Kofler et al., 1980)

$$\Theta := \{p \in \mathbb{R}^N_+ \mid \sum_{j=1}^{N} p_j = 1, \; Gp \geq h\}$$

with constant matrix G and constant vector h and $p_j := P\{\omega_j\}$ for $1 \leq j \leq N$.

A LPI we get for example, when the probabilities can be put in any order, e.g.

$$p_1 \leq p_2 \leq \ldots \leq p_N.$$

For finding an optimal decision about x we use the minEmax- (or $\Theta$-minimax-) criterion: $x_0 \in X$ is accepted as optimal solution of (2), if

$$\max_{P \in \Theta} E_P\{c'(\omega)x_0 + \min\{q'(\omega)y \mid W(\omega)y = b(\omega) - A(\omega)x_0, \; y \geq 0\}\} = \min_{x \in X} \max_{P \in \Theta} E_P\{c'(\omega)x + \min\{q'(\omega)y \mid W(\omega)y = b(\omega) - A(\omega)x, \; y \geq 0\}\}. \tag{3}$$

For every $x^{(i)} \in X$ and $p^{(j)} \in \Theta$ $(i=1,2,\ldots,M; \; j=1,2,\ldots,N)$ we define

$$v_{ij} := c'(\omega_j)x^{(i)} + \min\{q'(\omega_j)y \mid W(\omega_j)y = b(\omega_j) - A(\omega_j)x^{(i)}, \; y \geq 0\} \tag{4}$$

and have instead of (3)

$$\max\{\sum_{j=1}^{N} v_{i_0 j} p_j \mid p \in \Theta\} = \min_{1 \leq i \leq M} \max\{\sum_{j=1}^{N} v_{ij} p_j \mid p \in \Theta\} =: m_\Theta. \tag{5}$$

Every $x^{(i_0)} \in X$, which fulfills (5), is a minEmax-optimal decision about x for problem (2).

For the computation of a $v_{ij}$ we must solve a linear program of the form (4). When only the coefficients of vector b(.) are random, the computation of the $v_{ij}$'s can be simplified essentially by using methods of the parametric linear programming. We also have simplifications in the computation, when the recourse matrix W(.) contains no random coefficients, especially in case of a so called model with simple recourse. If only vector c(.) contains random coefficients, merely one linear program must be solved.

## 3. SOLVING PROBLEM (5)

For the numerical computation of all solutions of problem (5) we use an algorithm of Abel(1984), p. 150f (see also Abel, 1985):

STEP 1: Setting the starting values.

$$k := 1,$$

$$j_0 := i_0 := 1,$$

$$e_0 := \max_{1 \le j \le N} v_{1j} \text{ and}$$

$$r_{0i} := \min_{1 \le j \le N} v_{ij} \text{ for every } i=1,2,\dots,M.$$

STEP 2: Find an optimal solution $p^{(k)}$ of the linear program

$$\max_{p \in \Theta} \sum_{j=1}^{N} v_{i_{k-1}j} p_j =: m_k. \tag{6}$$

As upper bound for $m_\Theta$ we compute

$$e_k := \min\{m_k, e_{k-1}\};$$

$$j_k := \begin{cases} j_{k-1} & \text{if } e_{k-1} < m_k \\ i_{k-1} & \text{if } e_{k-1} \ge m_k \end{cases} \quad \text{and}$$

$$r_{ki} := \max\{r_{k-1,i}, \sum_{i=1}^{N} v_{ij} p_j^{(k)}\}, \qquad i=1,2,\dots,M.$$

STEP 3: Find $i_k \in \{1,2,\dots,M\}$ with

$$r_{ki_k} = \min_{1 \le i \le M} r_{ki} =: s_k,$$

where $s_k$ is a lower bound for $m_\Theta$.

STEP 4: If $e_k = s_k$ then $x^{(i_k)}$ is an optimal solution of (5) and $m_\Theta := e_k$. Otherwise $k \to k+1$ and goto STEP 2.

The number of iteration steps, needed to compute an optimal solution of (5) with this algorithm, is in no case greater than M. In general, clearly less than M iteration steps are necessary (see Abel(1984), p. 163f).

When $\Theta$ is no LPI only in step 2 other numerical methods must be used for solving (6).

If (5) has more than one solution, we first choose the proposed algorithm to compute one solution. Subsequently, the remaining solutions can be computed simply with an analogous Abel(1984) p. 155 modified version of this algorithm.

## 4. ONLY FINITE $\Omega$

In the last section we described a numerical method for the computation of optimal solutions, when the sets X and $\Omega$ are finite. In this section we want to analyse how we can solve (3) numerically, when we assume that X is a bounded convex polyhedral set as solution set of a linear restriction system of the form $\{x \in \mathbb{R}^n \mid Tx \geq t,\ x \geq 0\}$ with constant matrix T and constant vector t. In this case we get all minEmax-optimal solutions of (2) as solutions of the optimization problem

$$\min_{\substack{x \\ Tx \geq t \\ x \geq 0}} \ \max_{\substack{p \\ Gp \geq h \\ p \geq 0}} \ \min_{\substack{y \\ Wy = b - Ax \\ y \geq 0}} \ \sum_{k=1}^{N} p_k (c'(\omega_k)x + q'(\omega_k)y(\omega_k)) \tag{7}$$

where $y := (y'(\omega_1), y'(\omega_2), \ldots, y'(\omega_N))'$,

$$W := \begin{bmatrix} W(\omega_1) & & & & 0 \\ & W(\omega_2) & & & \\ & & \cdot & & \\ & & & \cdot & \\ & & & & \cdot \\ 0 & & & & W(\omega_N) \end{bmatrix},$$

$A := (A'(\omega_1), A'(\omega_2), \ldots, A'(\omega_N))'$ and

$b := (b'(\omega_1), b'(\omega_2), \ldots, b'(\omega_N))'$.

Using a minimax-theorem of Karlin(1959) p. 28f and the duality theorem of the linear programming we get the to (7) equivalent linear program

$$\min -h'u \tag{8}$$

$$\text{subject to} \quad \begin{aligned} Cx + Qy + G'u &\leq 0 \\ Ax + Wy &= b \\ Tx &\geq t \\ x &\geq 0 \\ y &\geq 0 \\ u &\geq 0 \end{aligned}$$

where $C := (c(\omega_1), c(\omega_2), \ldots, c(\omega_N))'$ and

$$Q := \begin{bmatrix} q'(\omega_1) & & & & 0 \\ & q'(\omega_2) & & & \\ & & \cdot & & \\ & & & \cdot & \\ & & & & \cdot \\ 0 & & & & q'(\omega_N) \end{bmatrix}.$$

Under this assumption consequently all minEmax-optimal decisions about x are solutions of linear program (8) and every solution of (8) is a minEmax-optimal decision. For the numerical solving of (8) we recommend not to use the standard Simplex-algorithm, but to take the special structure of the restriction system into consideration and use for example the decomposition of the dual problem as solution method.

## 5. THE GENERAL CASE

An essential assumption in the previous sections was the finiteness of $\Omega$ (and especially in section 4 the presence of a LPI). In this section we now want to renounce this supposition and analyse problem (3) under the assumption that X is a convex polyhedral set, without requiring the finiteness of $\Omega$.

As generalization of the LPI defined in section 2 we have in this case the so called Stochastic Partial Information (SPI) (see Kofler et al., 1980)

$$\Theta = \{P \mid P = \sum_{k=1}^{N} \lambda_k P_k, \ \sum_{k=1}^{N} \lambda_k = 1, \ \lambda_k \geq 0, \ k=1,2,\ldots,N\},$$

where for every $k \in \{1,2,\ldots,N\}$ $P_k$ is a probability measure. When only the coefficients of vector b(.) in (2) are random variables and we have a SPI for everey coefficient, we can compute the minEmax-optimal decisions about x analogous Abel(1984) p. 198f.

When we get a $\Theta$ based on informations about the mean respectively the variance of single coefficients of A(.), b(.), c(.), q(.) and/or W(.) for example in form of fixed upper and lower bounds (so far as mean and variance generally do exist), we have, depending on the respective structure of $\Theta$, for several $\Theta$ deterministic optimization models available, which are equivalent to (3) and solvable with numerical standard methods (see e.g. Dupačová, 1980; Huelsmann,1971, 1972, 1972a; Jagannathan, 1977 and Theodorescu, 1972).

## REFERENCES

Abel, P., 1984, "Stochastische Optimierung bei partieller Information," Hain, Königstein/Taunus.

Abel, P., 1985, Ein numerisches Verfahren zur Bestimmung maxEmin-optimaler Aktionen, Statistische Hefte, 26: 31-41.

Dantzig, G. B., 1955, Linear programming under uncertainty, Management Science, 1: 197-206.

Dupačová, J., 1980, On minimax decision rule in stochastic linear programming, in: "Studies on mathematical programming," A. Prékopa, ed., Akadémiai Kiadó, Budapest.

Huelsmann, J., 1971, Spieltheoretische Ansätze im stochastischen linearen Optimieren, Operations Research Verfahren, 10: 236-243.

Huelsmann, J., 1972, Zweistufiges stochastisches Programmieren bei Unsicherheit über die Verteilung von (A,b,c), Operations Research Verfahren, 13: 236-245.

Huelsmann, J., 1972a, Optimale Entscheidungen im zweistufigen stochastischen Programmieren als Lösung linearer Programme, Operations Research Verfahren, 14: 316-321.

Jagannathan, R., 1977, Minimax procedure for a class of linear programs under uncertainty, Operations Research, 25: 173-177.

Karlin, S., 1959, "Mathematical methods and theory in games, programming and economics I," Addison-Wesley, Reading.

Kofler, E., Menges, G., Fahrion, R., Huschens, S., and Kuß, U., 1980, Stochastische Partielle Information (SPI), Statistische Hefte, 21: 160-167.

Scarf, H., 1958, A min-max solution of an inventory problem, in: "Studies in the mathematical theory of inventory and production," K. J. Arrow, S. Karlin, and H. Scarf, ed., Stanford University Press, Stanford.

Theodorescu, R., 1972, Random programs, Mathematische Operationsforschung und Statistik, 3: 19-47.

# APPLIED GEWR(n,p,q) NORMAL DISCOUNT BAYESIAN MODEL: AN AUSTRIAN ECONOMIC CASE STUDY

M. Akram

Business And Management Department
Bahrain University
Isa Town, Bahrain

The theory of Generalised Exponentially Weighted Regression (GEWR) and dynamic Bayesian models has been given previously by Harrison-Akram(1982), Akram-Harrison(1983) and Akram(1984). This paper breifly reviews some of the main results and applies them to seasonal data concerned with the disposable personal income in Austria. For the selection of an appropriate model a new Stepwise Identification Procedure(SIP) based on a nonparametric measure, called Average String Length(ASL), is used. Both short and long term full forecasts and trends are obtained from a single model using on-line Bayesian learning procedure. The model applied yields optimum forecasts in the senses of minimum mean square error and whiteness of one step ahead forecast errors.

## 1. INTRODUCTION

In economic systems we often encounter time series containing additive coloured noise. For such series Generalised Exponentially Weighted Regression theory has been developed and given by Harrison-Akram(1982), Akram-Harrison (1983) and Akram(1984). In their work, they introduced a wide class of parsimonious dynamic linear models and applied them to data sets from various walks of life. This theory, which is based on linear filtering using an exponentially weighted system and Bayesian formulation, is briefly reviewed in section 2. In section 3 a GEWR Normal Discount Bayesian Model is given. Section 4 introduces on-line Bayesian learning procedure. Section 5 discusses Stepwise Identification Procedure and section 6 describes a particular form of the model used to analyse data concerned with disposable personal income of Austria and presents short and long term forecasts along with trends.

## 2. GENERALISED EXPONENTIALLY WEIGHTED REGRESSION (GEWR)

### 2.1 Definition of GEWR

At time t, for forecasting future outcomes $Y_{t+i}$, assume a local model

$$Y_{t+i} = \underline{f}_i \underline{\theta} + \varepsilon_{t+i}$$

where for an integer i $\underline{f}_i$ are (lxn) row vectors of some known functions of independent variables or constants and $\varepsilon_{t+i}$ is a sequence of coloured noise. By definition, the elements of $\underline{f}_i$ vectors are functions of time, generally described by constants, polynomials and trigonometric functions as in Brown (1962). For time series $\underline{f}_i = \underline{f}\, \underline{G}^i$, where $\underline{G}$ is (nxn) transition matrix of full rank with non zero elements on its main diagonal. The eigenvalues of this matrix determine the form of forecast function. It is assumed that coloured noise arises from ARMA(p,q) noise process

$$\phi_p(B)\, \varepsilon_t = \eta_q(B)\, \delta_t$$

where $\phi_p(B) = \prod_{i=1}^{p} (1 - \phi_i B)$ and $\eta_q(B) = \prod_{i=1}^{q} (1 - \eta_i B)$ are polynomials in B, the backward shift operator, of degrees p and q respectively, the roots of which are assumed to lie outside unit circle; and $\delta_t$ is a white noise sequence of random variables, iid with mean zero and variance V. The coloured noise vector $\underline{\varepsilon}_t' = (\varepsilon_t, \varepsilon_{t-1}, \ldots, \varepsilon_1)$ is such that $\underline{\varepsilon}_t \sim (\underline{0}\ ;\ \underline{P}_t^{-1} V)$, where $\underline{P}_t$ is a (txt) precision matrix at time t.

For a discount factor $0 < \beta < \min|\lambda_i|^2$, where $\lambda_i$ are eigenvalues of $\underline{G}$, defining $\underline{\beta}_t^{\frac{1}{2}} = \text{diag}(1, \beta^{\frac{1}{2}}, \beta, \ldots, \beta^{\frac{1}{2}(t-1)})$ the GEWR estimate $\underline{m}_t$ of $\underline{\theta}$ based on observations $y_t, y_{t-1}, \ldots, y_1$ is that vector value of $\underline{\theta} \in R^n$ which minimizes

$$\underline{\varepsilon}_t' \, \underline{\beta}_t^{\frac{1}{2}} \, \underline{P}_t \, \underline{\beta}_t^{\frac{1}{2}} \, \underline{\varepsilon}_t \, .$$

## 2.2 Recurrence Relations

Defining:

$$\psi(B) = \phi_p(B)/\eta_q(B) = \sum_{i=0}^{\infty} \psi_i B^i \quad ; \quad \psi_0 = 1$$

$$\psi_t(B) = \sum_{i=0}^{t-1} \psi_i B^i$$

$$\underline{u}_t = \underline{f}\, \psi_t\, (\beta^{\frac{1}{2}} \underline{G}^{-1}) = \underline{f} \sum_{i=0}^{t-1} \psi_i (\beta^{\frac{1}{2}} \underline{G}^{-1})^i$$

$$z_t = \psi_t\, (\beta^{\frac{1}{2}} B) y_t = \sum_{i=0}^{t-1} \psi_i\, (\beta^{\frac{1}{2}})^i y_{t-i}$$

$$d_t = \underline{u}_t \underline{G}\, \underline{m}_{t-1}$$

and one step ahead forecast error $e_t = z_t - d_t$, the recurrence relations for computing $\underline{m}_t$ are

$$\underline{m}_t = \underline{G}\, \underline{m}_{t-1} + \underline{A}_t e_t$$

$$\underline{K}_t = \underline{G}\, \underline{Q}_{t-1}^{-1} \underline{G}'/\beta$$

$$\underline{A}_t = \underline{K}_t \underline{u}_t' /(1 + \underline{u}_t \underline{K}_t \underline{u}_t')$$

$$\underline{Q}_t^{-1} = (\underline{I} - \underline{A}_t \underline{u}_t) \underline{K}_t .$$

For these recurrence relations, no matrix inversions are involved, but if required for any intermediate estimate, they can be found from the expression

$$\underline{Q}_t = \underline{u}_t' \underline{u}_t + \beta(\underline{G}^{-1})' \underline{Q}_{t-1} \underline{G}^{-1}.$$

## 2.3 Limiting Results

Under restriction $0 < \beta < \min|\lambda_i^2|$ the following limiting results exist. As $t \to \infty$

i) $\text{Lim } \underline{u}_t = \underline{u}$ , $\text{Lim } \underline{A}_t = \underline{A}$ and $\text{Lim } \underline{Q}_t = \underline{Q}$

ii) $\text{Lim } \{ \prod_{i=1}^{n} (1- \lambda_i B) z_t - \prod_{i=1}^{n} (1- \beta B/\lambda_i) e_t \} = 0.$

The last result is true, irrespective of GEWR is optimal or not. However, if $e_t \sim (0, \sigma_e^2)$ then subject to Box-Jenkins(1976) restrictions on the eigenvalues $\lambda_i$ (i=1,..., n), $Y_t$ has a limiting ARIMA representation. For more detail see Harrison-Akram(1982) and Akram(1984).

# 3. GEWR NORMAL DISCOUNT BAYESIAN MODEL (GNDBM)

## 3.1 Autoregressive Form

An autoregressive form of GNDBM is presented here as in practice this form is most commonly used owing to the fact that quite often ARMA processes with invertible moving average part can be modelled as parsimonious finite AR processes. A GNDBM of order n,p,0, written as GEWR(n,p,0), where n is the degree of polynomial required to represent low frequency component or trend and p is as stated earlier, is defined as follows.

$$Z_t = \underline{u}_t \underline{\theta} + \delta_t \quad ; \quad \delta_t \sim N(0 ; V)$$

where $Z_t$ is the series derived from the original observations $Y_t$ for t=1,2,.. ... , such that

$$Z_t = \begin{cases} \sum_{i=0}^{t-1} \psi_i \beta^{i/2} Y_{t-i} & \text{if } t \leq p \\ \sum_{i=0}^{p} \psi_i \beta^{i/2} Y_{t-i} & \text{if } t > p \end{cases}$$

$$\underline{u}_t = \begin{cases} \underline{f} \sum_{i=0}^{t-1} \psi_i ( \beta^{\frac{1}{2}} \underline{G}^{-1} )^i & \text{if } t \leq p \\ \underline{f} \sum_{i=0}^{p} \psi_i ( \beta^{\frac{1}{2}} \underline{G}^{-1} )^i & \text{if } t > p \end{cases}$$

The AR(p) representation of coloured noise $\varepsilon_t$ is $\psi(B)\ \varepsilon_t = \phi_p(B)\ \varepsilon_t = \delta_t$ . Consequently a GNDBM formulation based upon $Z_t$ is $\{ \underline{u}_t, \underline{G}, V, \beta \}$ which for $t > p$ becomes a constant GNDBM defined over quadruple $\{ \underline{u}, \underline{G}, V, \beta \}$.

## 3.2 Recurrence Relations

For a given prior $( \underline{\theta}_{t-1} \mid D_{t-1} ) \sim N( \underline{m}_{t-1} ; \underline{C}_{t-1} )$, where

$D_{t-1} = y_{t-1}, y_{t-2}, \ldots, y_1$, the posterior $(\theta_t \mid D_t) \sim N(\underline{m}_t ; \underline{C}_t)$, where $D_t = (y_t, D_{t-1})$, is computed through the following recurrence relations.

$$\underline{R}_t = \underline{G}\,\underline{C}_{t-1}\,\underline{G}'/\beta$$

$$\hat{Y}_t = V + \underline{u}_t\,\underline{R}_t\,\underline{u}_t'$$

$$\underline{A}_t = \underline{R}_t\,\underline{u}_t'\,(\hat{Y}_t)^{-1}$$

$$\underline{C}_t = (\underline{I} - \underline{A}_t\,\underline{u}_t)\underline{R}_t$$

$$e_t = z_t - \underline{u}_t\,\underline{G}\,\underline{m}_{t-1}$$

$$\underline{m}_t = \underline{G}\,\underline{m}_{t-1} + \underline{A}_t\,e_t.$$

### 3.3 Forecast Function

k-steps ahead forecast function for the original series $Y_t$ is

$$F_t(k) = \underline{u}_{t+k}\,\underline{G}^k\,\underline{m}_t - \sum_{i=1}^{\ell} \psi_i\,\beta^{i/2}\,x_t(k-i)$$

where $\underline{u}_{t+k}\,\underline{G}^k\,\underline{m}_t = E(Z_{t+k} \mid D_t)$, $\ell = \min(p, t-1)$ and

$$x_t(k-i) = \begin{cases} y_{t+k-i} & \text{if } k \leq i \\ F_t(k-i) & \text{if } k > i \end{cases}$$

### 3.4 Seasonality

In case of seasonal time series $\psi(B)$ for the GNDBM is replaced by $\psi_\zeta(B)$ which is defined as follows.

$$\psi_\zeta(B) = \phi_p(\beta^{\frac{1}{2}}B).\,S_s(B) = \sum_{i=0}^{\zeta} \psi_i B^i = \prod_{i=1}^{\zeta}(1 - \gamma_i B)$$

where $S_s(B)$ is a polynomial in B of degree s for seasonality such that

$$S_s(B) = \sum_{j=0}^{s} S_j\,B^j = \prod_{j=1}^{s}(1 - r\mu_j B)$$

and $S_j$ are real but $\mu_j$ occur in complex conjugate pairs and $0 < r < 1$ is a damping factor. The series $Z_t$ and vector $\underline{u}_t$ are redefined, replacing p by $\zeta = p + s$ and using the coefficients $\psi_i$ of the polynomial $\psi_\zeta(B)$. For more detail see Akram(1984).

## 4. ON - LINE BAYESIAN LEARNING PROCEDURE

For recurrence relations (3.2) if variance V is unknown, then at time t, it is estimated as:

$$\hat{V}_t = L_t/N_t$$

$$L_t = \beta_v\,L_{t-1} + (1 - \underline{u}_t\,\underline{A}_t)d_t$$

$$N_t = \beta_v\,N_{t-1} + 1$$

$$d_t = \min(e_t^2, \;\S\hat{Y}_t)$$

where $0 < \beta_v < 1$ is a discount factor for variance learning and § is a confidence factor, corresponding to distance between some $\sigma$-limits, say $2\sigma$ or $3\sigma$. For example § = 4 for $2\sigma$ limits and § = 6 for $3\sigma$ limits. For most practical situations, $\beta_v$ close to one and § = 4 is recommended. For more detail see Akram(1984).

Comment

If there is no original information in the system then no contribution to the estimate of $V_t$ is made during the first few points. For this period in place of $\hat{V}_t$ the prior estimate $\hat{V}_0$ is used. As a rule of thumb, it is recommended to use $\hat{V}_t$ after n+p+q observations, where n,p and q are as defined earlier, and minimum values of $L_0$ and $N_0$.

## 5. STEPWISE IDENTIFICATION PROCEDURE

In practice the noise process for a GEWR application is well represented by an AR(p) process of order p=1 or p=2. For identification of type of noise and subsequent selection of some appropriate GEWR(n,p,0) model, various approaches may be used, such as, Yule-Walker equations ( Yule(1927) and Walker(1931) ), Autocorrelation, Partial Autocorrelation and Durbin-Watson (1950). However, here a simple nonparameteric procedure, called Stepwise Identification Procedure (SIP), introduced by Akram (1984) is briefly reviewed. This approach is based on Average String Length (ASL), the mean distance between successive peaks or troughs of residuals or one step ahead forecast errors obtained by applying some GEWR(n,p,0) model to the time series of interest. The steps involved in the identification procedure are as follows.

A GNDBM GEWR(n,0,0) is applied to the data with some appropriate values of n,$\underline{f}$, $\beta$ and V or $\beta_v$ if variance is estimated on-line using Bayesian Learning Procedure (4). ASL of one step ahead forecast errors is computed and compared with the theoretical values of ASLs given in Appendix B. This comparison gives us approximate value of $\phi$ , the AR coefficient, which in turn helps us to identify the nature of residuals with respect to colour or whiteness and the suitability of the model applied. For whiteness of residuals the computed value of ASL should not be significantly different from 2 ( a value corresponding to $\phi = 0$ ) at a certain level of significance. We see this by formulating a null hypothesis ASL = 2 ( i.e. the residuals form a white noise sequence ) against some alternative hypothesis, say ASL$\neq$0. The null hypothesis is accepted or rejected according to the critical region bounded by the the critical values

$$2(N+1)/(N \pm Z_\alpha \sqrt{N})$$

where for large N+1 observations $Z_\alpha$ is a standard normal variate value at certain $\alpha$ , the level of significance. Acceptance of null hypothesis ensures whiteness of residuals, whereas, the rejection confirms the presence of colour in the residuals. Subsequently, the whiteness of residuals tell us that the model GEWR(n,0,0) is suitable for the series under study. This gives us green light to go ahead to find forecasts and trends, both short and long term.

In case of rejection of null hypothesis , i,e. the computed value of ASL is significantly different from 2, we look at the AR(1) coefficient corresponding to the computed value of ASL in Appendix B and adopt an approximate value of this coefficient for GEWR(n,1,0) model. The values of n, $\underline{f}$, $\underline{G}$, $\beta$, V and $\beta_v$ are taken as selected before. This new model is applied to the data, residuals are obtained and ASL is computed again as before and checked the whiteness of the residuals. We go on cycling the identification procedure until we see whiteness of the residuals or in turn confirm the suitability of the model applied. In case of unsatisfactory results of GEWR(n,1,0), we move to GEWR(n,2,0) with AR(2) coefficients suggested by the computed ASL values. For moving forward we go on retaining the identified AR coefficients in a successive manner. First we estimate $\phi_1$, then $\phi_2$ and $\phi_3$ and so on.

## 6. CASE STUDY

A quarterly seasonal data set concerned with the disposable personal income in Austria, consisting of 104 observations (1954-79)(appendix A) is analysed by applying GEWR(n,p,0) Normal Discount Bayesian Model (3.1), i.e.

$$Z_t = \underline{u}_t \underline{\theta} + \delta_t \quad ; \quad \delta_t \sim N(0; V)$$

with n = 2. For low frequency or trend, which shows a continually decreasing growth rate following an asymptotic S-shaped growth curve, Gompertz function

$$y_t = a\, b^{\rho^t} \quad ; \quad a > 0\ , \quad 0 < b,\rho < 1$$

is used . In order to establish a link between this function and the GNDBM, a log analogue of this function is used. Following the procedure explained by Akram (1984), for our GEWR model, the following setting is considered for operation.

$$\underline{f} = (1 \quad 1),\ \underline{G} = \text{diag}(1,\ 0.994),\ \beta = 0.98 \quad \text{and} \quad r = 0.99\beta^{\frac{1}{2}}.$$

The dynamic system of the model is initiated by using prior

$$\underline{\theta}_0 \sim N\left[\begin{pmatrix} 6 \\ -3 \end{pmatrix} \begin{pmatrix} 1 & -1.1 \\ -1.1 & 1 \end{pmatrix}\right]$$

On-line variance learning is used by setting $\beta_v = 0.99$, $\hat{V}_0 = 1$ and $N_0 = 5$. For first 2+s+p observations $\hat{V}_0$ and after that $\hat{V}_t$ are used. $Z_t$ and $\underline{u}_t$ are derived by using formulation (3.1).

For quarterly seasonal data, seasonal polynomial (3.4) is considered as follows.

$$S_3(B) = (1 + r^2B^2)(1 + rB).$$

This seasonal polynomial gives us full harmonic representation for the quarterly data under study.

First GEWR(2,0,0) form of the GNDBM is applied by considering $\phi(\beta^{\frac{1}{2}}B)=1$ and $\psi_3(B) = S_3(B)$. One step ahead forecasts along with residuals are obtained using the recurrence relations (3.2). The residuals give us ASL ≅ 17, a figure significantly different from 2 (at 5% level of significance). This reflects inadequacy of model for the data under study. Looking at the table of theoretical ASL values (Appendix B) it is decided to consider $\phi = 0.9$ for onward use.

Following SIP we proceed to GEWR(2,1,0) form of the GNDBM choosing $\phi_1 = 0.9$ and derive $Z_t$ and $\underline{u}_t$ again considering

$$\psi_4(B) = (1 - \phi_1\beta^{\frac{1}{2}}B).S_3(B).$$

This model yields one step ahead forecast errors having ASL ≅ 3. Still a value significantly different from 2. For this approximate value we select $\phi = 0.5$ for onward use. This reflects incapability of the GEWR(2,1,0) to filter whole coloured noise. It has filtered quite a lot but not all.

In the light of information provided by ASL, it is decided to move a step forward and formulate GEWR(2,2,0) form of GNDBM by choosing AR(2) coefficients $\phi_1 = 0.9$ and $\phi_2 = 0.5$. For this selection $Z_t$ and $\underline{u}_t$ are derived again by considering

$$\psi_5(B) = (1 - \phi_1\beta^{\frac{1}{2}}B)(1 - \phi_2\beta^{\frac{1}{2}}B).S_3(B).$$

As usual, one step ahead forecasts are obtained along with residuals. This time ASL computed for the residuals is 2.2, a value not significantly different from 2. It indicates that the residuals form a sequence of white noise, a confirmation of suitability of the model. Other than this, the model GEWR(2,2,0) with the identified AR coefficients yields one step ahead forecasts with less than 0.06% Mean Square Error and 1.5% Mean Average Deviation, lowest in its class of models.

After ensuring the suitability of GEWR(2,2,0) model, all short and long term forecasts along with trends are obtained from this model. One step ahead forecasts along with observations are displayed in fig.a and one step ahead trend is shown in fig.b. Long term forecasts (10 and 20 steps ahead) and trends are displayed in figures c to f.

COMMENT

For analysis of data, log transformation is used in line with the log form of Gompertz function and all forecasts and trends are obtained from a single GEWR(2,2,0) model. The results show that a joint modelling scheme where low frequency (trend), medium frequency (seasonal variations) and high frequency (coloured noise) components are incorporated within same framework, the low frequency is well protected from the high frequency, especially in the long run.

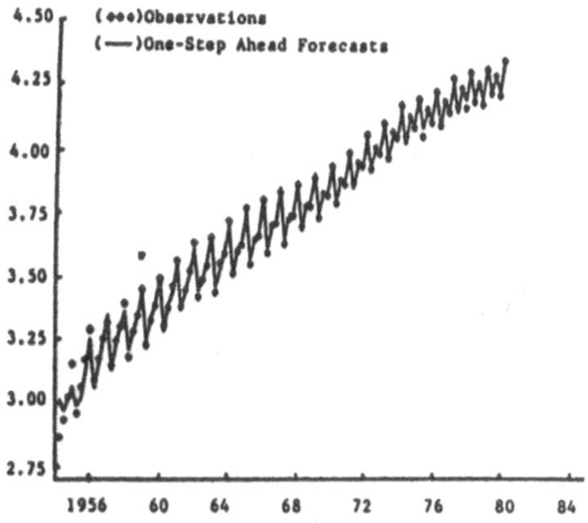


Fig.a

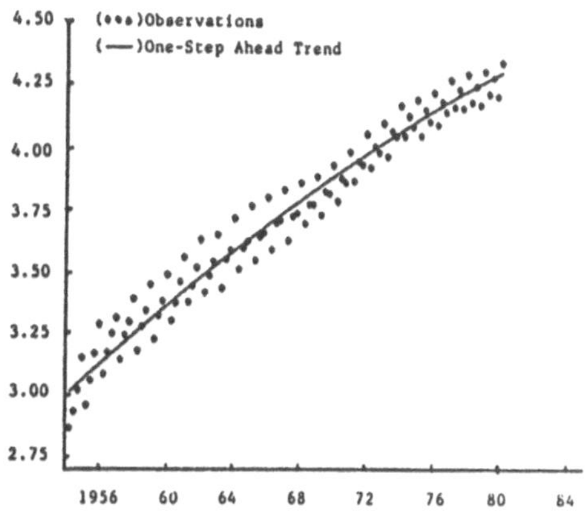


Fig.b

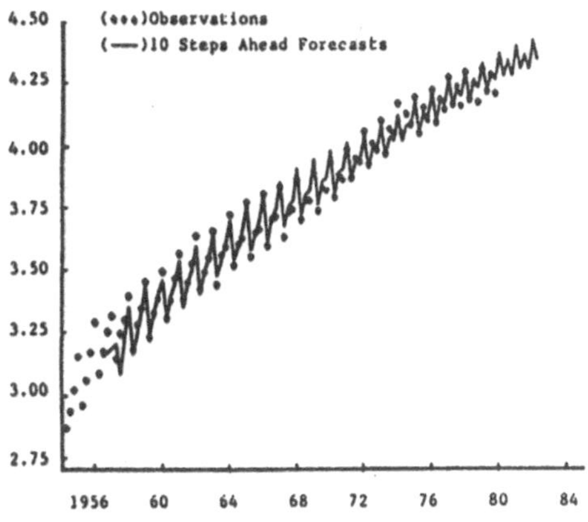


Fig.c

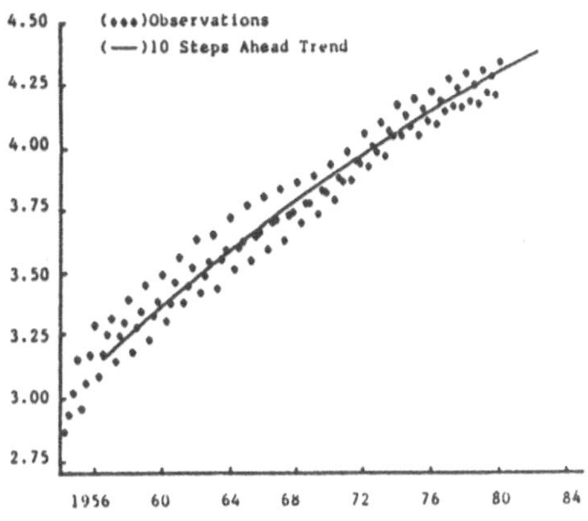


Fig.d

Austrian Disposable Personal Income (1954-79)--Continued next page

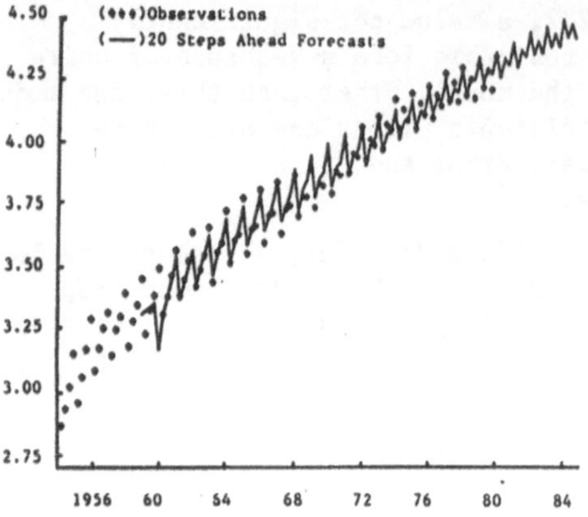


Fig.e

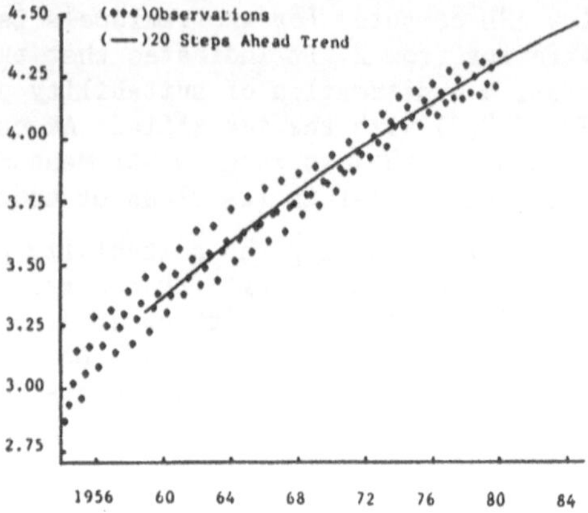


Fig.f

Austrian Disposable Personal Income (1954-79)--Continued

REFERENCES


Akram, M., 1984, Generalised Exponentially Weighted Regression and Dynamic Bayesian Forecasting Models, " Ph.D thesis ", Warwick University, Coventry, U.K.

Akram, M., Harrison, P. J., 1983, Some Applications of the GEWR, " International Time Series Conference, April 11-15, 1983 ", Nottingham, U.K.

Box, G. E. P., Jenkins, G. M., 1976, " Time Series Analysis and Control ", Holden-Day, U.S.A.

Brown, R. G., 1962, " Smoothing, Forecasting and Prediction of Discrete Time Series ", Prentice Hall, U.S.A.

Durbin, J., Watson, G. S., 1950, Testing for Serial Correlation in Least Square Regression, Biometrika, 37, 409-28.

Harrison, P. J., Akram, M., 1982, Generalised Exponentially Weighted Regression and Parsimonious Dynamic Linear Modelling, " Time Series Analysis: Theory and Practice 3 ", Proceedings of the International Time Series Conference, May 24-28, 1982 , North Holland, U.S.A.

Yule, U. G., 1927, On the Methods of Investigating Periodicities in Disturbed Series with special reference to Wolfer's Sunspot numbers, Philosophical Trans., A226, 267-298.

Walker, G., 1931, On Periodicities in Series of Related Terms, Proceedings of Royal Soc., London, U.K. 518-32.


KEY WORDS: Generalised Exponentially Weighted Regression; coloured noise process; Normal Discount Bayesian Model; Average String Length; Stepwise Identification Procedure; on-line Bayesian learning procedure.

APPENDIX A

| Year | 1 | 2 | 3 | 4 |
|---|---|---|---|---|
| 1954 | 17.62 | 18.86 | 20.58 | 23.51 |
| 1955 | 19.31 | 21.42 | 23.93 | 26.91 |
| 1956 | 21.98 | 24.00 | 25.96 | 27.66 |
| 1957 | 23.34 | 25.80 | 27.23 | 29.87 |
| 1958 | 24.19 | 26.70 | 28.47 | 31.65 |
| 1959 | 25.41 | 27.94 | 29.61 | 32.98 |
| 1960 | 27.40 | 29.40 | 32.04 | 35.37 |
| 1961 | 29.52 | 31.51 | 33.94 | 38.06 |
| 1962 | 30.70 | 32.87 | 34.75 | 38.76 |
| 1963 | 31.22 | 35.12 | 36.44 | 41.44 |
| 1964 | 33.72 | 36.77 | 37.71 | 43.61 |
| 1965 | 34.96 | 38.47 | 39.15 | 45.05 |
| 1966 | 36.57 | 40.68 | 41.10 | 46.43 |
| 1967 | 37.88 | 41.79 | 42.36 | 47.74 |
| 1968 | 40.61 | 43.93 | 43.92 | 49.01 |
| 1969 | 42.08 | 46.31 | 45.94 | 51.38 |
| 1970 | 44.53 | 48.65 | 47.99 | 54.08 |
| 1971 | 48.25 | 52.21 | 51.51 | 58.11 |
| 1972 | 50.90 | 55.29 | 54.00 | 60.62 |
| 1973 | 53.14 | 58.74 | 57.69 | 65.01 |
| 1974 | 57.71 | 62.36 | 59.88 | 66.60 |
| 1975 | 57.78 | 63.93 | 61.07 | 68.43 |
| 1976 | 60.37 | 66.05 | 63.47 | 72.19 |
| 1977 | 64.69 | 69.50 | 64.46 | 73.80 |
| 1978 | 66.02 | 70.48 | 65.35 | 74.70 |
| 1979 | 68.33 | 72.94 | 67.74 | 77.31 |

Unit: Billions of Schillings

Source: Austrian Institute of Economic Research, Vienna, Austria.

APPENDIX B

Average String Lengths of Autoregressive Type Coloured Noise

| $\phi$ | ASL | $\phi$ | ASL | $\phi$ | ASL | $\phi$ | ASL |
|---|---|---|---|---|---|---|---|
| -0.999 | 1.01 | -0.50 | 1.50 | 0.00 | 2.00 | 0.50 | 3.00 |
| -0.99 | 1.05 | -0.49 | 1.51 | 0.01 | 2.01 | 0.51 | 3.03 |
| -0.98 | 1.07 | -0.48 | 1.52 | 0.02 | 2.03 | 0.52 | 3.07 |
| -0.97 | 1.08 | -0.47 | 1.52 | 0.03 | 2.04 | 0.53 | 3.10 |
| -0.96 | 1.10 | -0.46 | 1.53 | 0.04 | 2.05 | 0.54 | 3.14 |
| -0.95 | 1.11 | -0.45 | 1.54 | 0.05 | 2.07 | 0.55 | 3.18 |
| -0.94 | 1.12 | -0.44 | 1.55 | 0.06 | 2.08 | 0.56 | 3.22 |
| -0.93 | 1.14 | -0.43 | 1.56 | 0.07 | 2.09 | 0.57 | 3.26 |
| -0.92 | 1.15 | -0.42 | 1.57 | 0.08 | 2.11 | 0.58 | 3.30 |
| -0.91 | 1.16 | -0.41 | 1.58 | 0.09 | 2.12 | 0.59 | 3.34 |
| -0.90 | 1.17 | -0.40 | 1.58 | 0.10 | 2.14 | 0.60 | 3.39 |
| -0.89 | 1.18 | -0.39 | 1.59 | 0.11 | 2.15 | 0.61 | 3.43 |
| -0.88 | 1.19 | -0.38 | 1.60 | 0.12 | 2.17 | 0.62 | 3.48 |
| -0.87 | 1.20 | -0.37 | 1.61 | 0.12 | 2.18 | 0.63 | 3.53 |
| -0.86 | 1.21 | -0.36 | 1.62 | 0.14 | 2.20 | 0.64 | 3.59 |
| -0.85 | 1.21 | -0.35 | 1.63 | 0.15 | 2.21 | 0.65 | 3.64 |
| -0.84 | 1.22 | -0.34 | 1.64 | 0.16 | 2.23 | 0.66 | 3.70 |
| -0.83 | 1.23 | -0.33 | 1.65 | 0.17 | 2.24 | 0.67 | 3.76 |
| -0.82 | 1.24 | -0.32 | 1.66 | 0.18 | 2.26 | 0.68 | 3.82 |
| -0.81 | 1.25 | -0.31 | 1.67 | 0.19 | 2.28 | 0.69 | 3.88 |
| -0.80 | 1.26 | -0.30 | 1.68 | 0.20 | 2.29 | 0.70 | 3.95 |
| -0.79 | 1.27 | -0.29 | 1.68 | 0.21 | 2.31 | 0.71 | 4.02 |
| -0.78 | 1.27 | -0.28 | 1.69 | 0.22 | 2.33 | 0.72 | 4.10 |
| -0.77 | 1.28 | -0.27 | 1.70 | 0.23 | 2.35 | 0.73 | 4.18 |
| -0.76 | 1.29 | -0.26 | 1.71 | 0.24 | 2.36 | 0.74 | 4.26 |
| -0.75 | 1.30 | -0.25 | 1.72 | 0.25 | 2.38 | 0.75 | 4.35 |
| -0.74 | 1.31 | -0.24 | 1.73 | 0.26 | 2.40 | 0.76 | 4.44 |
| -0.73 | 1.32 | -0.23 | 1.74 | 0.27 | 2.42 | 0.77 | 4.54 |
| -0.72 | 1.32 | -0.22 | 1.75 | 0.28 | 2.44 | 0.78 | 4.64 |
| -0.71 | 1.33 | -0.21 | 1.76 | 0.29 | 2.46 | 0.79 | 4.76 |
| -0.70 | 1.34 | -0.20 | 1.77 | 0.30 | 2.48 | 0.80 | 4.88 |
| -0.69 | 1.35 | -0.19 | 1.78 | 0.31 | 2.50 | 0.81 | 5.01 |
| -0.68 | 1.35 | -0.18 | 1.79 | 0.32 | 2.52 | 0.82 | 5.16 |
| -0.67 | 1.36 | -0.17 | 1.80 | 0.33 | 2.54 | 0.83 | 5.31 |
| -0.66 | 1.37 | -0.16 | 1.81 | 0.34 | 2.57 | 0.84 | 5.48 |
| -0.65 | 1.38 | -0.15 | 1.82 | 0.35 | 2.59 | 0.85 | 5.66 |
| -0.64 | 1.39 | -0.14 | 1.84 | 0.36 | 2.61 | 0.86 | 5.87 |
| -0.63 | 1.40 | -0.13 | 1.85 | 0.37 | 2.64 | 0.87 | 6.09 |
| -0.62 | 1.40 | -0.12 | 1.86 | 0.38 | 2.66 | 0.88 | 6.40 |
| -0.61 | 1.41 | -0.11 | 1.87 | 0.39 | 2.68 | 0.89 | 6.64 |
| -0.60 | 1.42 | -0.10 | 1.88 | 0.40 | 2.71 | 0.90 | 6.97 |
| -0.59 | 1.43 | -0.09 | 1.89 | 0.41 | 2.74 | 0.91 | 7.35 |
| -0.58 | 1.44 | -0.08 | 1.90 | 0.42 | 2.76 | 0.92 | 7.80 |
| -0.57 | 1.44 | -0.07 | 1.91 | 0.43 | 2.79 | 0.93 | 8.35 |
| -0.56 | 1.45 | -0.06 | 1.93 | 0.44 | 2.82 | 0.94 | 9.02 |
| -0.55 | 1.46 | -0.05 | 1.94 | 0.45 | 2.85 | 0.95 | 9.89 |
| -0.54 | 1.47 | -0.04 | 1.95 | 0.46 | 2.87 | 0.96 | 11.10 |
| -0.53 | 1.48 | -0.03 | 1.96 | 0.47 | 2.90 | 0.97 | 12.80 |
| -0.52 | 1.48 | -0.02 | 1.97 | 0.48 | 2.94 | 0.98 | 15.68 |
| -0.51 | 1.49 | -0.01 | 1.99 | 0.49 | 2.97 | 0.99 | 22.20 |
| | | | | | | 0.999 | 70.24 |

# USING INFLUENCE DIAGRAMS TO SOLVE A CALIBRATION PROBLEM

R. E. Barlow
University of California
Berkeley, CA 94720

R. W. Mensing and
N. G. Smiriga
Lawrence Livermore National Laboratory
Livermore, CA 94550

## INTRODUCTION

A measuring instrument measures a unit and records an observation y. The non-measurable variable of interest, the "true" measurement, x, of the unit is to be inferred from y, the measurable variable. If $p(y|x)$ is the likelihood of y given x and x has prior $p(x)$, then by Bayes' Theorem

$$p(x|y) \propto p(y|x)p(x).$$

Let $x_o$ and $\sigma_o^2$ be the mean and variance of $p(x)$. We will assess the likelihood, $p(y|x)$, using a linear regression model

$$y = \alpha + \beta(x-x^*) + \epsilon \qquad (1.1)$$

where $x^*$ is specified and a priori $(\alpha,\beta) \perp x \perp \epsilon$ and $\epsilon$ is $N(0,\sigma^2)$ with $\sigma$ specified. (These assumptions could, of course, be relaxed; e.g. $\sigma^2$ unknown, $\epsilon$ dependent on x, etc. However, our assumptions are convenient and sufficiently general to provide conclusions of general interest.) It follows that $p(y|\alpha,\beta,x-x^*)$ is $N(\alpha+\beta(x-x^*),\sigma^2)$.

The "center", $x^*$, of the likelihood model and the prior for x are intertwined. The natural choice for $x^*$ is the mean of the prior for x, namely $x^* = x_o$. This is reasonable since our attention is focused on calculating $p(x|y)$. The line, with $x^* = x_o$, is $y = \alpha + \beta(x-x_o)$ where $\alpha$ and $\beta$ are unknown and of course y cannot be observed without error. Of course, the prior for $(\alpha,\beta)$ depends on $x^* = x_o$ and it is natural to assume that $p(\alpha,\beta|x_o) = p(\alpha|\beta,x_o)p(\beta)$ since only $\alpha$ depends on $x_o$.

Figure 1.1 is an influence diagram describing the logical and statistical dependencies between unknown quantities, decision alternatives and values (losses or utilities). The decision may be an estimate for x given y. If the value or loss is

$$w(d,x) = (d-x)^2$$

then the optimal decision will be the posterior mean for x given y. The next section will discuss influence diagrams in more detail.

*The Calibration Experiment*

The purpose of the calibration experiment is to learn about $(\alpha,\beta)$ so that given a future observation $y_f$ we can reduce our uncertainty about a future "true" measurement $x_f$. To calibrate our measuring instrument, we record n measurements

$$\mathbf{y} = (y_1, y_2, \ldots, y_n)$$

on n units all of whose "true" measurements,

$$\mathbf{x} = (x_1, x_2, \ldots, x_n)$$

are specified before hand. Based on our prior, $p(x_f)$, and our regression model (1.1), our problem is to determine $\mathbf{x} = (x_1, x_2, \ldots, x_n)$ (subject to feasibility constraints) so as to minimize some overall loss function. The experimental design for the calibration experiment is called $\mathbf{x}$.

The following assumptions will be made relative to the calibration experiment.

<u>Assumption 1</u>. The future "true value", $x_f$, is independent of $(\alpha,\beta)$, $\mathbf{x}$, and $\mathbf{y}$. The future observation, $y_f$, is independent of $(\mathbf{x},\mathbf{y})$ given $(\alpha,\beta)$.

<u>Assumption 2</u>. The value function $w(d,x_f)$ is a loss function and depends only on d (the decision regarding $x_f$ taken at the time we observe $y_f$) and the "true value" $x_f$. For example, we are ignoring the cost of performing the experiment.

<u>Assumption 3</u>. The feasible region, R, for the experimental design, $\mathbf{x}$, is bounded. That is, infinite $x_i$ values are not allowed in practice.

Figure 1.2 is an influence diagram representation for our problem. We seek an optimal experimental design subject to $\mathbf{x} \in R$. For a more detailed discussion of this problem and references to other approaches see Chapter 10 of Aitchison and Dunsmore (1980). Hoadley (1970) discusses the calibration inference problem in some detail and points out the difficulties with the maximum likelihood estimator for $x_f$ given an observation $y_f$ and data $[(x_i,y_i), i = 1, 2, \ldots, n]$ from a calibration experiment. Brown (1982) and Brown and Sundberg (1985) extend Hoadley's results using a multivariate

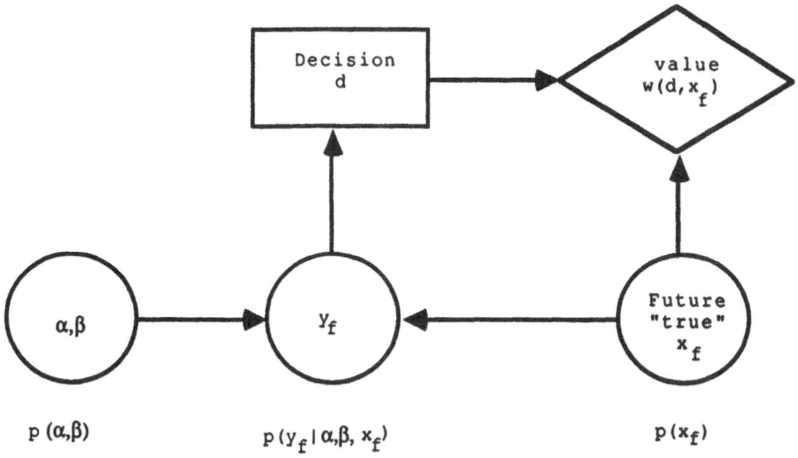


Fig. 1.1 Influence diagram for the inference calibration problem

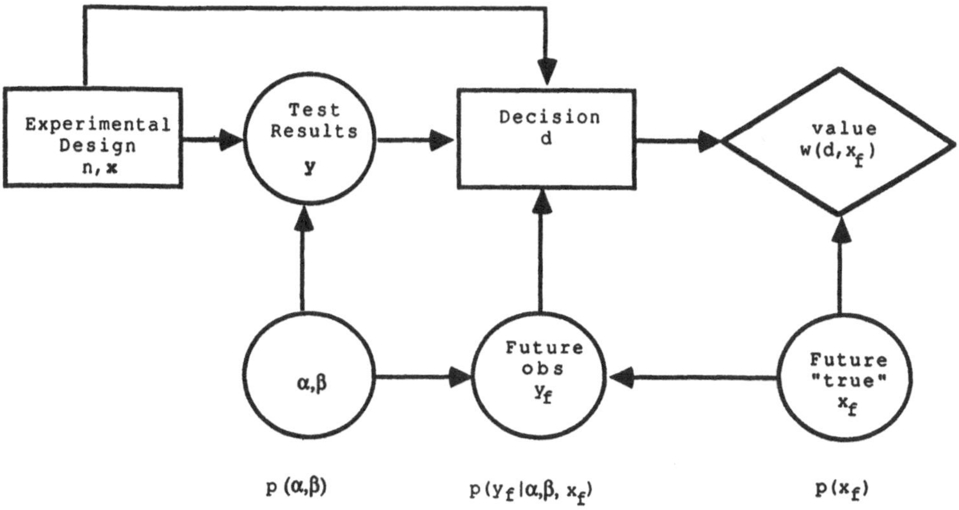


$$y = \alpha + \beta(x - x_0) + \epsilon$$

$$\epsilon \sim N(0, \sigma^2)$$

Fig. 1.2 The calibration experimental design problem

formulation. However, they do not consider the problem of optimal Bayesian experimental design. The definitive reference for Bayesian design for linear regression is Chaloner (1984). The objective of this paper is to discuss the calibration experimental design problem.

*Summary of Results*

Based on the likelihood it is shown that the experimental design may be summarized by

$$n,$$

$$\bar{x} - x_o = \sum_1^n (x_i - x_o)/n,$$

and

$$v_x = \sum_1^n (x_i - x_o)^2$$

where

$$|\bar{x} - x_o| \leq \sqrt{v_x/n}$$

If $\beta$ is known, the optimal experimental design depends only on n and corresponds to taking n as large as possible. The values of x are immaterial. If $\alpha$ is known, the value of the experimental design depends only on $v_x$ and is decreasing in $v_x$ for fixed n. If both $\alpha$ and $\beta$ are unknown, the optimal design can be found by performing a three dimensional search over $(n, \bar{x}-x_o, v_x)$.

## USING INFLUENCE DIAGRAMS

Influence diagrams are discussed by Shachter (1986). He also provides an influence diagram solution algorithm for decision problems.

In Figures 1.1 and 1.2 circles denote random quantities while rectangles denote decision alternatives. Diamonds denote deterministic functions of their immediate input variables. The arrows denote influence. Thus the two input arrows to $y_f$ indicate that $y_f$ depends on both $x_f$ and $(\alpha, \beta)$. In this case the assessed probability function for $y_f$ given $(\alpha, \beta)$ and $x_f$ is $N[\alpha+\beta(x_f-x_o), \sigma^2]$. Each circle node has a weight attached. This weight is a conditional probability function which only depends on <u>immediate</u> predecessor nodes.

The influence diagram is first of all an acyclic directed graph. As such there always exists an ordered list of nodes which preserves the graph ordering. For example, in Figure 1.2 an ordered list is

$$(n, \mathbf{x}) \prec (\alpha, \beta) \prec \mathbf{y} \prec x_f \prec y_f \prec d \prec w.$$

From the ordered list and the weights attached to circle nodes we can calculate a unique representation for the joint probability function corresponding to probability nodes. From Figure 1.2, for random quantities,

$$\alpha,\ \beta,\ \mathbf{y},\ x_f,\ y_f$$

the joint probability function is

$$p(\alpha,\beta)\,p(\mathbf{y}\,|\,\alpha,\beta,n,\mathbf{x})\,p(x_f)\,p(y_f\,|\,\alpha,\beta,x_f)$$

It can be easily verified that if two probability nodes have only output arrows, then they are unconditionally independent. From Figure 1.2 we see that $(\alpha,\beta)$ and $x_f$ are unconditionally independent as required by assumption 1. Since there is no arc connecting $(\mathbf{x},\mathbf{y})$ and $y_f$ it follows that $(\mathbf{x},\mathbf{y}) \perp y_f$ given the status of their immediate predecessor nodes, namely $(\alpha,\beta)$ and $x_f$. But, since $(\mathbf{x},\mathbf{y}) \perp x_f$, it follows that $(\mathbf{x},\mathbf{y}) \perp y_f$ given $(\alpha,\beta)$.

To find the optimal experimental design, we will reduce Figure 1.2 to just two nodes, namely the decision node $(n,\mathbf{x})$ and the value node $w(d,x_f)$. The value node is deterministic; i.e. the value is determined given d and $x_f$. The value node has only input arrows.

The solution algorithm starts with the value node. The nearest decision node in the ordered list is d. Fix the immediate predecessors of node d; namely, $(n,\mathbf{x})$, $\mathbf{y}$ and $y_f$. These denote information available at the time of decision. Next, eliminate all other probabilistic predecessors of the value node; namely, $x_f$ and $(\alpha,\beta)$. This is done by arc reversal and Bayes' Theorem. Figure 1.3 shows the influence diagram after reversing the arc from $(\alpha,\beta)$ to $\mathbf{y}$. Note that the posterior distribution for $(\alpha,\beta)$ now depends on both $(n,\mathbf{x})$ and $\mathbf{y}$.

The next step is to reverse the arc from $(\alpha,\beta)$ to $y_f$. After reversal, node $(\alpha,\beta)$ has only input arcs; i.e. all the information in this node relative to our design problem has been extracted. Hence, at this point node $(\alpha,\beta)$ is deleted leaving the influence diagram of Figure 1.4.

Before the decision node can be eliminated, we must first eliminate node $x_f$ by reversing the arc from $x_f$ to $y_f$ and the arc from $x_f$ to w. Figure 1.5 shows the influence diagram after reversing the arc from $x_f$ to $y_f$. The next step is to reverse the arc from $x_f$ to w. After this reversal, node w has value

$$\int_{-\infty}^{\infty} w(d,x_f)\,p(x_f\,|\,y_f,\mathbf{y},n,\mathbf{x})\,dx_f$$

$$= E_{x_f|y_f,\mathbf{y},n,\mathbf{x}}[w(d,x_f)\,|\,y_f,\mathbf{y},n,\mathbf{x}].$$

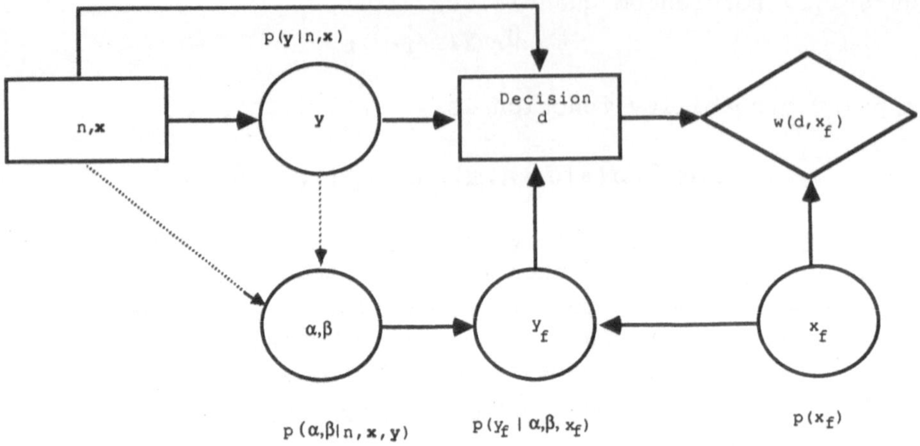


Fig. 1.3 The calibration problem after arc reversal

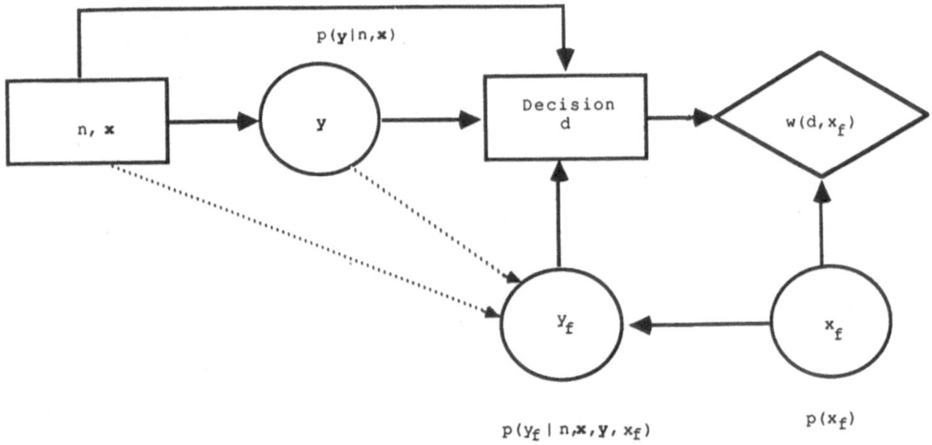


Fig. 1.4 The influence diagram after elimination of node (α, β)

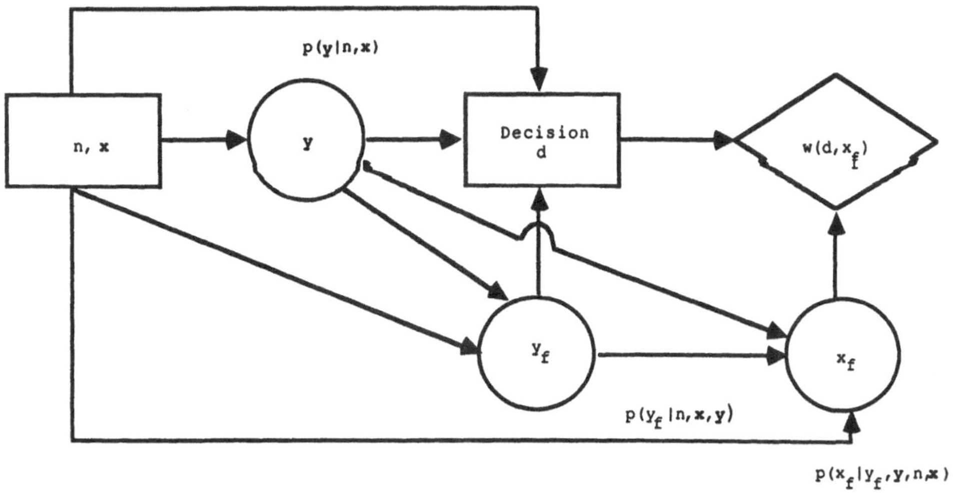


Fig. 1.5 The influence diagram after reversing the arc from $x_f$ to $y_f$

To eliminate the decision node we calculate

$$\operatorname*{Min}_{d} \int_{-\infty}^{\infty} w(d, x_f)\, p(x_f \mid y_f, \mathbf{y}, n, \mathbf{x})\, dx_f$$

where

$$p(x_f \mid y_f, \mathbf{y}, n, \mathbf{x}) \propto p(y_f \mid x_f, \mathbf{y}, n, \mathbf{x})\, p(x_f)$$

and

$$p(y_f \mid x_f, \mathbf{y}, n, \mathbf{x}) = \iint p(y_f \mid \alpha, \beta, x_f)\, p(\alpha, \beta \mid \mathbf{y}, n, \mathbf{x})\, d\alpha d\beta.$$

Finally, we compute

$$W(\mathbf{x}) = E_{\mathbf{y} \mid n, \mathbf{x}}\, E_{y_f \mid \mathbf{y}, n, \mathbf{x}}\, \operatorname*{Min}_{d}\, E_{x_f \mid y_f, \mathbf{y}, n, \mathbf{x}}\, [w(d, x_f) \mid y_f, \mathbf{y}, n, \mathbf{x}] \qquad (2.1)$$

where

$$p(y_f \mid \mathbf{y}, n, \mathbf{x}) = \int p(y_f \mid x_f, \mathbf{y}, n, \mathbf{x})\, p(x_f)\, dx_f.$$

The optimal design is the minimizer of $W(\mathbf{x})$. If we take squared error loss, $(d-x_f)^2$, as our value function then the posterior mean is the minimizer of

$$E_{x_f|y_f,\mathbf{y},n,\mathbf{x}}[(d-x_f)^2|y_f,\mathbf{y},n,\mathbf{x}]$$

and $W(\mathbf{x})$ becomes

$$W(\mathbf{x}) = E_{\mathbf{y}|n,\mathbf{x}}E_{y_f|\mathbf{y},n,\mathbf{x}}\mathrm{Var}(x_f|y_f,\mathbf{y},n,\mathbf{x}).$$

LIKELIHOOD AND THE OPTIMAL EXPERIMENTAL DESIGN

Under the assumption that observation errors, $\{\epsilon_i | i = 1, 2, \ldots, n\}$ are independent $N(0, \sigma^2)$, but without specifying prior distributions, we can determine some of the structure of the optimal experimental design. This can be done using the sufficient statistics for $(\alpha, \beta)$ corresponding to our likelihood model. As noted before, the purpose of the calibration experiment is to learn about $(\alpha, \beta)$. The likelihood for $(\alpha, \beta)$ given the data is

$$L(\alpha,\beta)|\mathrm{Data},x_o) \propto \exp\{-\sum_1^n [y_i - \alpha - \beta(x_i-x_o)]^2/2\sigma^2\}.$$

A priori assume $\alpha \perp \beta \perp \epsilon$ and let $E(\alpha) = a$, $E(\beta) = b$, $\mathrm{Var}(\alpha) = \sigma_a{}^2$, and $\mathrm{Var}(\beta) = \sigma_b{}^2$. Define

$$e_i = y_i - a - b(x_i-x_o)$$

and rewrite

$$y_i - \alpha - \beta(x_i-x_o) = [y_i - a - b(x_i-x_o)] - (\alpha-a) - (\beta-b)(x_i-x_o)$$

$$= e_i - (\alpha-a) - (\beta-b)(x_i-x_o)$$

so that

$$L(\alpha,\beta|\mathrm{Data},x_o)$$

$$\propto \exp\{-[n(\alpha-a)^2 + (\beta-b)^2 \sum(x_i-x_o)^2 -2\sum e_i[(\alpha-a) +$$

$$(\beta-b)(x_i-x_o)] + 2(\alpha-a)(\beta-b)\sum(x_i-x_o)]/2\sigma^2\}. \qquad (3.1)$$

Clearly $n$, $\sum_1^n(x_i-x_o)$, $\sum_1^n(x_i-x_o)^2$, $z_1 = \sum_1^n e_i$ and $z_2 = \sum_1^n e_i(x_i-x_o)$ are sufficient statistics for $(\alpha,\beta)$ since $x_o$, $a$, $b$ and $\sigma$ are specified. It follows that the posterior density for $(\alpha,\beta)$ also depends on the data only through $n$, $\sum_1^n(x_i-x_o)$, $\sum_1^n(x_i-x_o)^2$, $z_1$ and $z_2$.

## Theorem 3.1

$W(\mathbf{x})$ depends on $\mathbf{x}$ only through

$n$,

$$\bar{x} - x_o = \sum_1^n (x_i - x_o)/n$$

and

$$v_x = \sum_1^n (x_i - x_o)^2$$

N. B. This is true for all value functions $w(d, x_f)$ and priors on $(\alpha, \beta)$ and $x_f$. Were $w$ to also depend on $\mathbf{x}$ but only through $n$, $\bar{x}-x_o$ and $v_x$, theorem 3.1 would still hold.

Proof:

The purpose of the calibration experiment is to learn about $(\alpha,\beta)$ so that, in the future, we may make a "good" decision about $x_f$. Since $n$, $\bar{x}-x_o$, $v_x$, $z_1$ and $z_2$ are sufficient statistics for $(\alpha,\beta)$, the test results, $\mathbf{y}$, may be summarized by $z_1$ and $z_2$.

If we examine the derivation of (2.1) carefully, we see that (2.1) can be rewritten as

$$W(\mathbf{x}) = E_{z_1,z_2|n,\bar{x},v_x,x_o} E_{y_f|z_1,z_2,n,\bar{x},v_x,x_o} \operatorname*{Min}_d E_{x_f|y_f,z_1,z_2,n,\bar{x},v_x,x_o} [w(d,x_f)|y_f,z_1,z_2,n,\bar{x},v_x,x_o] \quad (3.2)$$

Hence, from (3.2), we need only show that the joint distribution of $(z_1,z_2)$ depends on $\mathbf{x}$ only through $n$, $\bar{x}-x_o$ and $v_x$. It is easy to show that $(z_1,z_2)$, given $(\alpha,\beta)$, is bivariate normal where $z_1$ given $(\alpha,\beta)$ is

$$N[n(\alpha-a) + (\beta-b)\sum_1^n (x_i-x_o),\ n\sigma^2]$$

and $z_2$, given $(\alpha,\beta)$, is

$$N[(\alpha-a)\sum_1^n (x_i-x_o) + (\beta-b)\sum_1^n (x_i-x_o)^2,\ \sigma^2\sum_1^n (x_i-x_o)^2]$$

while

$$\text{Cov}(z_1, z_2|\alpha,\beta) = \sigma^2\sum_1^n (x_i-x_o)\,. \qquad \text{QED}$$

## Corollary 3.2

If $\sigma_b = 0$, i.e. we are certain that $\beta = b$, then $W(\mathbf{x})$ depends on $\mathbf{x}$ only through $n$. The "levels" $(x_1, x_2, \ldots, x_n)$ are immaterial and we might just as well take

$$x_1 = x_2 = \ldots = x_n = x_o$$

or any other values that we like.

Proof:

If we are certain that $\beta = b$; i.e. $\sigma_b = 0$, then (3.1) becomes

$$L(\alpha|\text{Data},x_o) \propto \exp\{-[n(\alpha-a)^2 - 2\sum_1^n e_i(\alpha-a)]/2\sigma^2\}.$$

Hence n and $z_1 = \sum_1^n e_i = \sum_1^n [y_i - a - b(x_i-x_o)]$ are sufficient for $\alpha$.

Since $z_1$ given $(\alpha,\beta=b)$ is

$$N[n(\alpha-a),\ n\sigma^2]$$

it follows that $W(\mathbf{x})$ depends on $\mathbf{x}$ only through n. QED

<u>Corollary 3.3</u>

If $\alpha$ is known, i.e. $\sigma_a = 0$, then $W(\mathbf{x})$ depends on $\mathbf{x}$ only through $v_x$. Further, for fixed n, $W(\mathbf{x})$ is decreasing in $v_x$.

In this case, $W(\mathbf{x})$ is minimized for those $\mathbf{x}$ belonging to R for which $v_x$ is maximum.

Proof:

If $\sigma_a = 0$, then (3.1) becomes

$$L(\beta|\text{Data},x_o) \propto \exp\{-[(\beta-b)^2\sum_1^n (x_i-x_o)^2 - 2(\beta-b)\sum_1^n e_i(x_i-x_o)]/2\sigma^2\}.$$

Hence $\sum_1^n (x_i-x_o)^2$ and $z_2 = \sum_1^n e_i(x_i-x_o)$ are sufficient for $\beta$. Since $z_2$ given $(\alpha = a, \beta)$ is

$$N[(\beta-b)\sum_1^n (x_i-x_o)^2,\ \sigma^2\sum_1^n (x_i-x_o)^2]$$

it follows that when $\alpha=a$ is known, $W(\mathbf{x})$ depends on $\mathbf{x}$ only through $v_x$.

Suppose

$\sum_1^n (x_i-x_o)^2 < \sum_1^n (x_i'-x_o)^2$. Clearly we can find $x_{n+1}$ such that

$$\sum_1^n (x_i'-x_o)^2 = \sum_1^n (x_i-x_o)^2 + (x_{n+1}-x_o)^2$$

$$= \sum_{1}^{n+1} (x_i - x_o)^2 .$$

By the expected information inequality [see Raiffa and Schlaifer (1961)], the expected value function can only decrease if we perform additional calibration experiments. Hence $W(\mathbf{x})$ is decreasing in $v_x$ for fixed n. QED

<u>Determining the Structure of the Optimal Experimental Design</u>

Since

$$\sum_{1}^{n} (x_i - \bar{x})^2/n \geq 0$$

it follows that

$$\sum_{1}^{n} (x_i - x_o + x_o - \bar{x})^2/n = \sum_{1}^{n} (x_i - x_o)^2/n - (\bar{x} - x_o)^2 \geq 0$$

and

$$|\bar{x} - x_o| \leq \sqrt{v_x/n} .$$

Consequently, the minimization problem with respect to $\mathbf{x}$ can be transformed to a minimization problem with respect to only three variables, namely n, $\bar{x}-x_o$, $v_x$ where

$$|\bar{x} - x_o| \leq \sqrt{v_x/n} .$$

Since $\bar{x}-x_o$ and $v_x$ are symmetric functions of an experimental design $\mathbf{x}$, it follows that, for fixed n, any permutation of the coordinates of an experimental design solution is also a solution (if allowed by the feasibility constraints). Figure 3.1 shows the nature of the possible $(x_1, x_2)$ solutions for $v_x$ fixed and n = 2. The darkened arcs on the circumference show the possible designs for a fixed $v_x$ (up to permutations of coordinates). For fixed $v_x$, possible solutions are traced out by the intersection of the line $\bar{x}-x_o = c$ with the circumference of the circle $(x_1-x_o)^2 + (x_2-x_o)^2 = \sqrt{2v_x/n}$ as c varies from $-\sqrt{v_x/n}$ to $\sqrt{v_x/n}$.

The optimal experimental design $\mathbf{x}$ can, in theory, be found through a three dimensional search over the feasible region R. One strategy would be to fix n and, using a computer, calculate a three dimensional plot of $W(\mathbf{x})$, as given by (3.2), versus $\bar{x}-x_o$ and $v_x$. Figure 3.2 illustrates the 3 dimensional plot for a fixed n. The plot shows the surface of $W(\mathbf{x})$ as a function of $|\bar{x} - x_o| \leq \sqrt{v_x/n}$ .

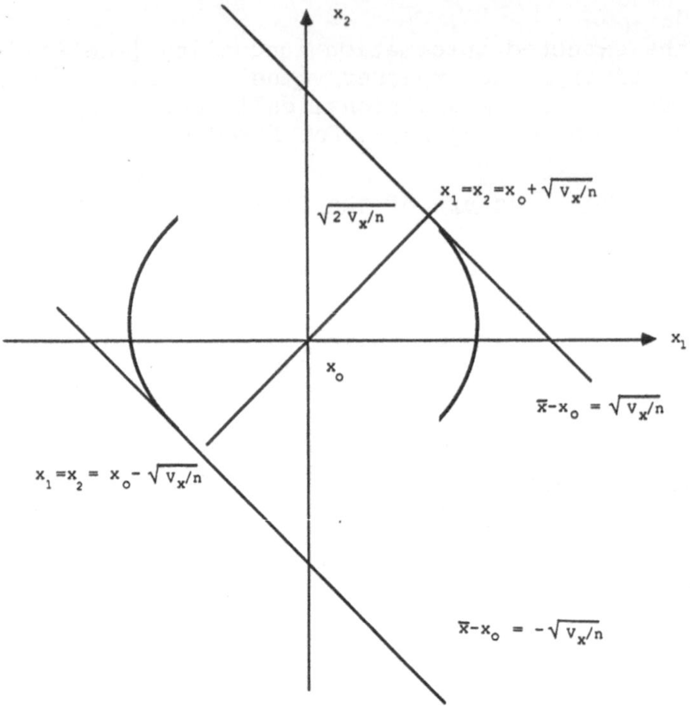


Fig. 3.1 N = 2

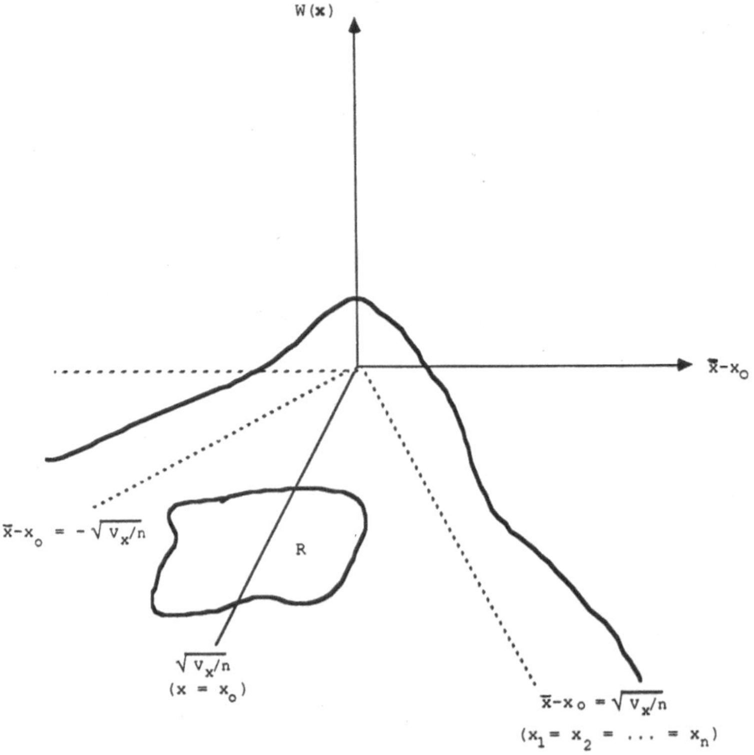


Fig. 3.2 $\sigma_b \neq 0$, $n \geq 2$ and fixed

The case of $x_1 = x_2 = \ldots = x_n = x_o$

Suppose we are uncertain about both $\alpha$ and $\beta$. From (3.1) we see that if $x_1 = x_2 = \ldots = x_n = x_o$, then

$$L(\alpha, \beta | \text{Data}) \propto \exp\{-[n(\alpha-a)^2 - \sum_1^n e_i(\alpha-a)]/2\sigma^2\}$$

so that in this case the data provide no direct information about $\beta$. If, in addition, the prior for $(\alpha,\beta)$ satisfies

$$p(\alpha,\beta | x_o) \propto p(\alpha | x_o)\ p(\beta)$$

i.e. $\alpha$ and $\beta$ are a priori independent given $x_o$, then

$$p(\alpha,\beta) | \text{Data}, x_o) \propto L(\alpha | \text{Data}, x_o)\ p(\alpha | x_o)\ p(\beta)$$

and the posterior marginal for $\beta$ is the same as the prior marginal for $\beta$. Intuitively, if $\beta$ is unknown and $\sigma_b >> \sigma_a$, the experimental design

$$x_1 = x_2 = \ldots = x_n = x_o$$

is a local maximum for the final expected value since values of $x_i$ near $x_o$ will provide information about $\beta$ and hence tend to reduce the final expected value.

Computational Considerations

The calculation of $W(\mathbf{x})$ as expressed in (3.2) assumes that $p(x_f | y_f, z_1, z_2, n, \bar{x}-x_o, v_x)$ and $p(y_f | z_1, z_2, n, \bar{x}-x_o, v_x)$ are available. To obtain these densities, we must first calculate the posterior density for $(\alpha,\beta)$ given n, $\bar{x}-x_o$ and $v_x$. In the case of a bivariate normal prior for $(\alpha,\beta)$, the posterior density will again be bivariate normal and $p(y_f | x_f, z_1, z_2, n, \bar{x}-x_o, v_x)$ will be univariate normal. In a future paper we investigate the computational problems in more detail.

## INFLUENCE DIAGRAMS AND THE SURE THING PRINCIPLE

The Sure Thing Principle [Savage (1954)] asserts that if decision d is preferred to $d^*$ for any possible value of a quantity, say $\alpha$, then d is also preferred to $d^*$ when $\alpha$ is unknown. Suppose n is fixed for our calibration experimental design. By Corollary 3.3 and the Sure Thing Principle we might at first infer that the optimal experimental design corresponds to those $\mathbf{x} \in R$ for which $v_x$ is maximum regardless of whether $\alpha$ is known or unknown. This reasoning is easily seen to be incorrect, since by Corollary 3.2 and the Sure Thing Principle we would also have concluded that the optimal experimental design depends only on n.

The resolution of this seeming contradiction can be seen from the influence diagram, Figure 1.2. There are two decision nodes, say $d_1$ and $d_2$. Hence the correct statement of the Sure Thing Principle would require that a decision pair $(d_1,d_2)$ be preferred to $(d_1^*,d_2^*)$ for any value of $\alpha$. In fact, the decision node corresponding to estimating $x_f$ will depend on $\alpha = a$ when $\alpha$ is known and on $\beta = b$ when $\beta$ is known.


## Acknowledgements

We would like to acknowledge Tony O'Hagan for helpful suggestions and for providing facilities for one of the authors for much of this research in connection with the Bayesian study year at Warwick University. Ross Shachter taught us how to use influence diagrams. Dennis Lindley read our manuscript carefully and provided very useful comments.


## REFERENCES


Aitchison, J. and I. R. Dunsmore (1975). Statistical Prediction Analysis. Cambridge Univ. Press, Cambridge,England. (Paperback edition 1980)

Brown, P. J. (1982). Multivariate Calibration. (With Discussion). J. Roy. Statist. Soc., B 44, 287-321.

Brown, P. J. and Rolf Sundberg (1985). Confidence and Conflict in Multivariate Calibration. Universitet Stockholm Research Report No. 140, August 1985. (Inst. For Forsakringsmatematik och Matematisk Statistik).

Chaloner, K. (1984). Optimal Bayesian Experimental Design for Linear Models. Annals of Statistics, 12, 283-300.

Hoadley, B. (1970). A Bayesian Look at Inverse Linear Regression. J. Amer. Stat. Ass., 65, 356-369.

Raiffa, H. and Schlaifer, R. (1961). Applied Statistical Decision Theory. MIT Press, Cambridge, Mass.

Savage, J. L. (1954). Foundations of Statistics. Wiley, New York, N.Y.

Shachter, R.D. (1986). Evaluating Influence Diagrams. In: Reliability and Quality Control. A. P. Basu, ed. Elsevier Science Publishers (North Holland). pp.321-344.

# RELIABILITY OF A COMPLEX SYSTEM FROM BAYESIAN VIEWPOINT

Asit Basu*
University of Missouri
Columbia, Missouri, U.S.A.

Ghasem Tarmast
Ahwaz University
Iran

## 1. INTRODUCTION

Let X and Y be two random variables with cumulative distribution functions F(x) and G(y) respectively. Let Y be the strength of a component subject to a stress X. Then the component fails if at any moment the applied stress (or load) is greater than its strength or resistance. Reliability of the component is then given by

$$R = P(X < Y) \tag{1.1}$$

The above model has been useful in a number of areas, specially in the structural and aircraft industries. As an example consider the following. A solid propellant rocket engine is successfully fired provided the chamber pressure X generated by ignition stays below the burst pressure Y of the rocket chamber. If X > Y, the engine blows up and the operation is a failure.

From practical considerations it is desirable to draw inference about R and other similar measures. In many situations, the distribution of X (or of both X and Y) will be completely known except possibly for a few unknown parameters and it is desired to obtain parametric solutions. The problems of estimating the reliability functions, both for simple and complex systems, have been considered by many. For a bibliography of available results see Basu (1977a, 1977b, 1981, 1985) and Bhattacharyya and Johnson (1975). However, most results are based on sampling theory approach. Enis and Geisser (1971) and Zacks (1977) have considered the problem from Bayesian point of view. In this paper we consider Bayesian approach for general systems.

A number of complex systems are described in Section 2. Bayesian analysis, assuming noninformative prior and conjugate prior distributions, are given in Sections 3 and 4. Finally a model based on multivariate normal distribution is discussed in Section 5.

*This research was funded by a grant from the Research Council of the Graduate School, University of Missouri-Columbia. Part of this research was carried out while A.P. Basu was visiting the University of Warwick, Coventry, England.

## 2. COMPLEX SYSTEMS

Consider a physical system. A system is called simple if it consists of a single component. Otherwise it is called a complex system. a complex system, consisting of p components, is called a k-out-of-p system if it functions if and only if at least k of these p components functions successfully. Such a system occurs quite naturally in many physical and biomedical models. As an example of a 2-out-of-3 system, consider an airplane which can function satisfactorily if and only if at least two of its three engines are functioning. When $k = p$ (or $k = 1$) we obtain series (or parallel) systems as special cases of k-out-of-p systems. In this section we shall consider three k-out-of-p systems.

First, consider a simple system of strength Y which is subjected to p different stresses $X_1, X_2, \ldots, X_p$. An example of interest is the case where a beam of strength Y is subjected to p different stresses $X_1, X_2, \ldots, X_p$. Let us assume that the $X_i$'s are independently and identically distributed with a common distribution function $F(x) \equiv F(x;\theta_1)$. Let the cdf of Y be $G(y) \equiv G(y,\theta_2)$ and assume that the $X_i$'s and Y are independent. Then the reliability of the system is given by

$$R_1 = \text{Prob}(\text{at least } k \text{ of the } X_i\text{'s} < Y)$$
$$= \sum_{j=k}^{p} \binom{p}{j} \int_{-\infty}^{\infty} [F(y)]^j [1-F(y)]^{p-j} dG(y) . \tag{2.1}$$

Assume that F and G satisfy the Lehmann alternative, that is, assume X and Y tc have proportional failure rates. Let

$$[1 - G(x;\theta_2)] = [1 - F(x;\theta_1)]^{\theta_1/\theta_2} . \tag{2.2}$$

The exponential distributions and the Weibull distributions with common shape parameter satisfy (2.2). From (2.1) and (2.2) we obtain

$$R_1 = \sum_{j=k}^{p} \binom{p}{j} \Gamma(j + 1)\, \Gamma(p + \frac{1}{\lambda} - j)/\{\lambda\, \Gamma(p + \frac{1}{\lambda} + 1)\}$$
$$= \Gamma(p + 1)\, \Gamma(p + \frac{1}{\lambda} + 1 - k) \,/\, \{ \Gamma(p + 1 - k)\, \Gamma(p + \frac{1}{\lambda} + 1)\} \tag{2.3}$$

where $\lambda^{-1} = \theta_1/\theta_2$. The last expression is obtained using the result

$$\binom{a + b - 1}{b - 1} + \binom{a + b - 1}{b} = \binom{a + b}{b}. \tag{2.4}$$

Next a p-component system with strengths $Y_1, Y_2, \ldots, Y_p$ respectively is considered where each component is subjected to the same stress X. As an example, let X denote the flow of current through an electric component assembled from several subcomponents with abilities to accommodate currents $Y_1, Y_2, \ldots, Y_p$. As before X and $Y_i$'s are assumed independent. Let the cdf of X be $F(x)$ and the common cdf of the $Y_i$'s be $G(y)$ where F and G satisfy (2.2). Here the reliability of the system is given by

$$R_2 = P(\text{at least } k \text{ of the } Y_i\text{'s} > X)$$
$$= \sum_{j=k}^{p} \binom{p}{j} \int_{0}^{\infty} [G(x)]^{p-j} [1 - G(x)]^j \, dF(x). \tag{2.5}$$

For Lehmann alternatives

$$dF(x) = \lambda\{1 - G(x)\}^{\lambda-1}\, dG(x). \tag{2.6}$$

In this case, using (2.4), $R_2$ can be shown to be given by

$$R_2 = 1 - \frac{\Gamma(k + \lambda)\ \Gamma(p + 1)}{\Gamma(p + \lambda + 1)\ \Gamma(k)} . \tag{2.7}$$

Finally, consider a more general p-component system where the ith component of strength $Y_i$ is subject to stress (shock) $X_i$, $i = 1,2,\ldots,p$. Assuming as before that $X_i$'s and $Y_i$'s are independent with $X_i \sim F(x)$ and $Y_i \sim G(y)$, the reliability $R_3$ for this k-out-of-p system is given by

$$R_3 = \sum_{j=k}^{p} \binom{p}{j} [P(X < Y)]^j\ [P(X > Y)]^{p-j}. \tag{2.8}$$

In the special case, when $X \sim e(\theta_1)$ and $Y \sim e(\theta_2)$ with cdf's $F(x;\theta_1) = 1 - e^{-x/\theta_1}$, and $g(y;\theta_2) = 1 - e^{-y/\theta_2}$, we have

$$P(X < Y) = \frac{\theta_2}{\theta_1 + \theta_2} = \frac{\lambda}{1 + \lambda} \quad \text{and} \quad P(X > Y) = \frac{\theta_1}{\theta_1 + \theta_2} = \frac{1}{1 + \lambda}. \tag{2.9}$$

In this case (2.8) reduces to

$$R_3 = \sum_{j=k}^{p} \binom{p}{j} \frac{\lambda^j}{(1 + \lambda)^p} . \tag{2.10}$$

## 3. BAYESIAN ESTIMATION BASED ON NONINFORMATIVE PRIORS

In this section we shall consider Bayesian estimation of $R_1$,$R_2$ and $R_3$. Considerable literature exists about the choice of a suitable prior distribution. In Sections 3 and 4 noninformative priors, and conjugate priors are considered.

For simplicity X and Y are assumed to have independent exponential distributions with cdf's

$$\begin{aligned} F(x;\theta_1) &= 1 - e^{-x/\theta_1}, \ x \geq 0,\ \theta_1 > 0, \\ G(y;\theta_2) &= 1 - e^{-y/\theta_2}, \ y \geq 0,\ \theta_2 > 0. \end{aligned} \tag{3.1}$$

A reasonable noninformative prior distribution for $\theta_i$ is given by

$$h(\theta_i) = \frac{1}{\theta_i},\ \theta_i > 0,\ (i = 1,2). \tag{3.2}$$

Let $X_1,X_2,\ldots,X_{n_1}$ and $Y_1,Y_2,\ldots,Y_{n_2}$ be two independent random samples from F and G respectively. Then the maximum likelihood estimator of

$\lambda = \frac{\theta_2}{\theta_1}$ is given by $\hat{\lambda} = \frac{\hat{\theta}_2}{\hat{\theta}_1} = \frac{\bar{y}}{\bar{x}}$ where $\bar{x} = \frac{1}{n_1} \sum_1^{n} x_i$, $\bar{y}$ analogous.

The maximum likelihood estimators of $R_1$,$R_2$, and $R_3$ can be readily obtained by replacing $\theta_1$,$\theta_2$, and $\lambda$ by their respective estimators in (2.3), (2.7) and (2.10).

Lemma 3.1. The posterior density of $\lambda = \theta_2/\theta_1$, based on the noninformative prior distributions (3.2), is given by

$$h(\lambda|\bar{x},\bar{y}) = \frac{\Gamma(n_1+n_2)u^{n_2}}{\Gamma(n_1)\,\Gamma(n_2)} \cdot \frac{\lambda^{n_1-1}}{(\lambda+u)^{n_1+n_2}}, \quad \lambda > 0. \tag{3.3}$$

where $u = n_2\bar{y}/n_1\bar{x}$.

Proof: Straightforward.

The Bayes estimators of $R_1$, $R_2$, and $R_3$ are given by

Theorem 3.1. The Bayesian estimator of $R_1$, using the noninformative priors (3.2), is given by

$$\tilde{R}_1 = \frac{\Gamma(n_1+n_2)\Gamma(p+1)}{\Gamma(n_1)\Gamma(n_2)\Gamma(p+1-k)} \int_0^1 \frac{\Gamma(p+1-k+\frac{1-y}{uy})}{\Gamma(p+1+\frac{1-y}{uy})} y^{n_1-1}(1-y)^{n_2-1}\,dy, \tag{3.4}$$

where $u = n_2\bar{y}/n_1\bar{x}$.

Proof: $\tilde{R}_1 = E(R_1|\text{data})$

$$= \int_0^\infty R_1 h(\lambda|\bar{x},\bar{y})\,d\lambda.$$

Substituting for $R_1$ from (2.3), and integrating out $\lambda$ (3.4) is obtained.

Theorem 3.2. Bayesian estimator of $R_2$, using the noninformative prior (3.2), is given by

$$\tilde{R}_2 = 1 - \frac{\Gamma(n_1+n_2)\Gamma(p+1)}{\Gamma(n_1)\Gamma(n_2)\Gamma(k)} \int_0^1 \frac{\Gamma(k+\frac{uy}{1-y})y^{n_1-1}(1-y)^{n_2-1}}{\Gamma(p+1+\frac{uy}{1-y})}\,dy. \tag{3.5}$$

Proof: Similar to proof of Theorem 3.1.

Theorem 3.3. Bayesian estimator of $R_3$ in (2.10) using the noninformative prior (3.2), is given by

$$\tilde{R}_3 = \sum_{j=k}^{p} \binom{p}{j} \frac{\Gamma(n_1+n_2)u^j}{\Gamma(n_1)\Gamma(n_2)} \int_0^1 \frac{y^{n_1+j-1}(1-y)^{n_2+p-j-1}}{[1+(u-1)y]^{p}}\,dy. \tag{3.6}$$

Proof: Similar to that of Theorem 3.1.

Numerical comparisons of Bayesian and maximum likelihood estimates of $R_1$, $R_2$, and $R_3$ are carried out through simulation. Estimates of the mean square error (MSE) and bias with $n_1 = n_2 = 20$ are obtained from 1000 trials for the k-out-of-3 and k-out-of-4 systems with $\lambda = 1,2,3$, and 4.

The tables 1, 2, and 3 show the estimated bias and MSE. The bias and MSE of both the maximum likelihood and Bayes estimates appear to be nearly equal.

Table 1
The maximum likelihood and Bayesian estimators of the reliability of k-out-of-p system when a simple system with strength Y is subjected to the stresses X1,X2,...,Xp under the assumption that the Xi and Y are independent exponential distributions.

| | | | Bias | | Mean Sq. Error | |
|---|---|---|---|---|---|---|
| k | Lamda | R(k,p) | MLE | Bayesian | Sm | Sb |
| | | | p=3 | | | |
| 1 | 1 | 0.75 | 0.022 | - 0.012 | 0.0011 | 0.0044 |
| 2 | 1 | 0.5 | 0.044 | 0.0033 | 0.0038 | 0.01 |
| 3 | 1 | 0.25 | 0.041 | 0.012 | 0.0039 | 0.0087 |
| 1 | 2 | 0.86 | - 0.006 | - 0.011 | 0.0019 | 0.0021 |
| 2 | 2 | 0.69 | - 0.0018 | - 0.0077 | 0.0063 | 0.0065 |
| 3 | 2 | 0.46 | - 0.00033 | - 0.0019 | 0.01 | 0.01 |
| 1 | 3 | 0.9 | - 0.0044 | - 0.008 | 0.0012 | 0.0013 |
| 2 | 3 | 0.77 | - 0.0084 | - 0.014 | 0.0041 | 0.0041 |
| 3 | 3 | 0.58 | - 0.007 | - 0.011 | 0.0097 | 0.0093 |
| 1 | 4 | 0.92 | - 0.0027 | - 0.0057 | 0.00078 | 0.00082 |
| 2 | 4 | 0.82 | - 0.0039 | - 0.0092 | 0.0031 | 0.0032 |
| 3 | 4 | 0.66 | - 0.0083 | - 0.013 | 0.0084 | 0.0081 |
| | | | p=4 | | | |
| | | | Bias | | Mean Sq. Error | |
| k | Lamda | R(k,p) | MLE | Bayesian | Sm | Sb |
| 1 | 1 | 0.8 | 0.017 | - 0.017 | 0.00067 | 0.0035 |
| 2 | 1 | 0.6 | 0.036 | - 0.0053 | 0.0026 | 0.0085 |
| 3 | 1 | 0.4 | 0.037 | - 0.0078 | 0.0034 | 0.0099 |
| 4 | 1 | 0.2 | 0.039 | 0.015 | 0.004 | 0.0076 |
| 1 | 2 | 0.89 | - 0.0059 | - 0.011 | 0.0013 | 0.0016 |
| 2 | 2 | 0.76 | - 0.0064 | - 0.014 | 0.0045 | 0.0051 |
| 3 | 2 | 0.61 | - 0.0042 | - 0.01 | 0.0092 | 0.0097 |
| 4 | 2 | 0.41 | 0.00064 | 0.00065 | 0.011 | 0.01 |
| 1 | 3 | 0.92 | - 0.0049 | - 0.008 | 0.00069 | 0.00073 |
| 2 | 3 | 0.83 | - 0.0091 | - 0.015 | 0.0033 | 0.0035 |
| 3 | 3 | 0.71 | - 0.01 | - 0.016 | 0.0069 | 0.0068 |
| 4 | 3 | 0.53 | - 0.00045 | - 0.0036 | 0.011 | 0.011 |
| 1 | 4 | 0.94 | - 0.0029 | - 0.0054 | 0.00044 | 0.00047 |
| 2 | 4 | 0.87 | - 0.0079 | - 0.013 | 0.0022 | 0.0023 |
| 3 | 4 | 0.77 | - 0.0074 | - 0.013 | 0.0052 | 0.0052 |
| 4 | 4 | 0.62 | - 0.011 | - 0.016 | 0.0092 | 0.0088 |

Table 2

The maximum likelihood and Bayesian estimators of the reliability of k-out-of-p system when the system with strength Y = [Y1,Y2,...,Yp] is subjected to the stress X under the assumption that the Yi and X are independent exponential distributions.

| | | | | | | |
|---|---|---|---|---|---|---|
| p=3 | | | | | | |
| | | | Bias | | Mean Sq. Error | |
| k | Lamda | R(k,p) | MLE | Eayesian | Sm | Sb |
| 1 | 1 | 0.75 | 0.035 | -0.013 | 0.0022 | 0.0092 |
| 2 | 1 | 0.5 | 0.044 | 0.0076 | 0.0041 | 0.01 |
| 3 | 1 | 0.25 | C.026 | 0.0035 | 0.0018 | 0.0043 |
| 1 | 2 | 0.9 | -0.012 | -0.021 | 0.0031 | 0.0036 |
| 2 | 2 | 0.7 | -0.0061 | -0.012 | 0.0085 | 0.0088 |
| 3 | 2 | 0.4 | 0.0075 | C.0086 | 0.0071 | 0.0071 |
| 1 | 3 | 0.95 | -0.0095 | -0.016 | 0.0013 | 0.0014 |
| 2 | 3 | 0.8 | -0.011 | -0.018 | 0.006 | 0.0058 |
| 3 | 3 | 0.5 | 0 | 0 | 0.007 | 0.0067 |
| 1 | 4 | 0.97 | -0.0057 | -0.011 | 0.00063 | 0.00072 |
| 2 | 4 | 0.86 | -0.0068 | -0.014 | 0.0041 | 0.0041 |
| 3 | 4 | 0.57 | 0.00058 | -0.001 | 0.0067 | 0.0064 |
| p=4 | | | | | | |
| | | | Bias | | Mean Sq. Error | |
| k | Lamda | R(k,p) | MLE | Bayesian | Sm | Sb |
| 1 | 1 | 0.8 | 0.029 | -0.016 | 0.0018 | 0.0071 |
| 2 | 1 | 0.6 | 0.042 | -0.0039 | 0.0039 | 0.011 |
| 3 | 1 | 0.4 | 0.037 | 0.0055 | 0.0034 | 0.0082 |
| 4 | 1 | 0.2 | 0.025 | 0.0096 | 0.0017 | 0.0035 |
| 1 | 2 | 0.93 | -0.01 | -0.02 | 0.002 | 0.0026 |
| 2 | 2 | 0.8 | -0.011 | -0.019 | 0.0073 | 0.0075 |
| 3 | 2 | 0.6 | 0.0039 | 0.0012 | 0.0095 | 0.0095 |
| 4 | 2 | 0.33 | 0.003 | 0.0055 | 0.0054 | 0.0055 |
| 1 | 3 | 0.97 | -0.0073 | -0.0013 | 0.00085 | 0.00098 |
| 2 | 3 | 0.8y | -0.012 | -0.02 | 0.0041 | 0.0041 |
| 3 | 3 | 0.71 | -0.0053 | -0.01 | 0.008 | 0.0076 |
| 4 | 3 | 0.43 | -0.0017 | -0.00017 | 0.0075 | 0.0072 |
| 1 | 4 | 0.99 | -0.0048 | -0.0092 | 0.00026 | 0.00033 |
| 2 | 4 | 0.93 | -0.011 | -0.018 | 0.0024 | 0.0024 |
| 3 | 4 | 0.79 | -0.011 | -0.017 | 0.0078 | 0.0075 |
| 4 | 4 | 0.5 | -0.0021 | -0.0021 | 0.0077 | 0.0073 |

## 4. BAYESIAN ESTIMATION USING CONJUGATE PRIORS

In this section informative priors, which are natural conjugate priors for exponential distributions, are considered. As in Section 3, let the cdf of X and Y be given by (3.1). The prior distribution of $\theta_i$ is assumed to be the inverted gamma distribution with density

$$h(\theta_i) = \frac{\alpha_i^{\nu_i}}{\Gamma(\nu_i)} e^{-\alpha_i/\theta_i} (1/\theta_i)^{\nu_i+1}, \quad \theta_i > 0,\ \nu_i > 0,\ \alpha_i > 0,\ (i=1,2). \tag{4.1}$$

Here the parameters $\alpha_i$ and $\nu_i$ are chosen to reflect prior information.

Lemma 4.1. The posterior density of $\lambda = \theta_2/\theta_1$ is given by

$$h(\lambda|\bar{x},\bar{y}) = \frac{\Gamma(n_1 + n_2 + \nu_1 + \nu_2)}{\Gamma(n_1 + \nu_1)\Gamma(n_2 + \nu_2)} \cdot \frac{u^{n_2+\nu_2} \lambda^{n_1+\nu_1-1}}{(\lambda + u)^{n_1+n_2+\nu_1+\nu_2}}, \tag{4.2}$$

where $u = \dfrac{\alpha_2 + n_2\bar{y}}{\alpha_1 + n_1\bar{x}}$.

Proof: Straightforward.

Using $h(\lambda|\bar{x},\bar{y})$, as in Section 3, the Bayesian estimates of $R_1$, $R_2$ and $R_3$ are obtained as given by the following theorem.

Table 3

The maximum likelihood and Bayesian estimators of the reliability of k-out-of-p system when the system with strength Y = [Y1,Y2,...,Yp] is subjected to the stress $\underline{X}$ = [X1,X2,...,Xp] under the assumption that the Xi and the Yi are independent exponential distributions.

| p=3 | | | | | | |
|---|---|---|---|---|---|---|
| | | | Bias | | Mean Sq. Error | |
| k | Lamda | R(k,p) | MLE | Bayesian | Sm | Sb |
| 1 | 1 | 0.87 | 0.87 | 0.011 | 0.0008 | 0.0008 |
| 2 | 1 | 0.5 | 0.057 | 0.054 | 0.0065 | 0.006 |
| 3 | 1 | 0.12 | 0.032 | 0.041 | 0.0025 | 0.0023 |
| 1 | 2 | 0.96 | - 0.0077 | - 0.014 | 0.0008 | 0.0009 |
| 2 | 2 | 0.74 | - 0.0073 | - 0.016 | 0.0097 | 0.0093 |
| 3 | 2 | 0.3 | 0.0079 | 0.012 | 0.0099 | 0.0093 |
| 1 | 3 | 0.98 | - 0.0048 | - 0.0088 | 0.0003 | 0.0004 |
| 2 | 3 | 0.84 | - 0.013 | - 0.023 | 0.0055 | 0.0054 |
| 3 | 3 | 0.42 | - 0.0031 | - 0.0026 | 0.011 | 0.011 |
| 1 | 4 | 0.99 | - 0.0027 | - 0.0052 | 0.0001 | 0.0002 |
| 2 | 4 | 0.9 | - 0.0076 | - 0.016 | 0.0034 | 0.0035 |
| 3 | 4 | 0.51 | - 0.0059 | - 0.0081 | 0.011 | 0.011 |

| p=4 | | | | | | |
|---|---|---|---|---|---|---|
| | | | Bias | | Mean Sq. Error | |
| k | Lamda | R(k,p) | MLE | Bayesian | Sm | Sb |
| 1 | 1 | 0.94 | 0.011 | 0.002 | 0.0002 | 0.0003 |
| 2 | 1 | 0.94 | 0.011 | 0.0021 | 0.0002 | 0.0003 |
| 3 | 1 | 0.31 | 0.054 | 0.06 | 0.0069 | 0.0062 |
| 4 | 1 | 0.062 | 0.023 | 0.033 | 0.0015 | 0.0015 |
| 1 | 2 | 0.99 | - 0.0044 | - 0.0088 | 0.0002 | 0.0002 |
| 2 | 2 | 0.89 | - 0.015 | - 0.026 | 0.0052 | 0.0053 |
| 3 | 2 | 0.59 | - 0.0074 | - 0.012 | 0.016 | 0.015 |
| 4 | 2 | 0.2 | 0.006 | 0.014 | 0.0079 | 0.0075 |
| 1 | 3 | 1 | - 0.0026 | - 0.0049 | 0.0001 | 0.0001 |
| 2 | 3 | 0.95 | - 0.01 | - 0.019 | 0.0019 | 0.0021 |
| 3 | 3 | 0.74 | - 0.012 | - 0.022 | 0.012 | 0.011 |
| 4 | 3 | 0.32 | 0.0016 | 0.0054 | 0.012 | 0.011 |
| 1 | 4 | 1 | 0.0011 | - 0.0001 | 0 | 0 |
| 2 | 4 | 0.97 | - 0.0091 | - 0.016 | 0.001 | 0.0011 |
| 3 | 4 | 0.82 | - 0.0081 | - 0.019 | 0.0065 | 0.0065 |
| 4 | 4 | 0.41 | - 0.0015 | - 0.0009 | 0.013 | 0.012 |

Theorem 4.1. Bayesian estimators of $R_1, R_2$ and $R_3$, using the conjugate prior distributions (4.1), are given by $\tilde{\tilde{R}}_1, \tilde{\tilde{R}}_2$ and $\tilde{\tilde{R}}_3$ respectively, where

$$\tilde{\tilde{R}}_1 = \frac{\Gamma(n_1+n_2+\nu_1+\nu_2)\Gamma(p+1)}{\Gamma(n_1+\nu_1)\Gamma(n_2+\nu_2)\ \Gamma(p+1-k)} \int_0^1 \frac{\Gamma(p+1-k+\frac{1-y}{uy})}{\Gamma(p+1+\frac{1-y}{uy})} \times y^{n_1+\nu_1-1}(1-y)^{n_2+\nu_2-1}\,dy \tag{4.3}$$

$$1-\tilde{\tilde{R}}_2 = \frac{\Gamma(n_1+n_2+\nu_1+\nu_2)\Gamma(p+1)}{\Gamma(n_1+\nu_1)\Gamma(n_2+\nu_2)\ \Gamma(k)} \int_0^1 \frac{\Gamma(k+\frac{uy}{1-y})}{\Gamma(p+1+\frac{uy}{1-y})} \times y^{n_1+\nu_1-1}(1-y)^{n_2+\nu_2-1}\,dy, \tag{4.4}$$

and

$$\tilde{\tilde{R}}_3 = \sum_{j=k}^{p} \binom{p}{j} \frac{\Gamma(n_1+n_2+\nu_1+\nu_2)}{\Gamma(n_1+\nu_1)\Gamma(n_2+\nu_2)}\, u^j \times \int_0^1 \frac{y^{n_1+\nu_1+j-1}(1-y)^{n_2+\nu_2+p-j-1}}{[1+(u-1)y]^p}\,dy. \tag{4.5}$$

Proof: Starting out with (4.2) instead of (3.3) the results follow immediately from (3.4), (3.5), and (3.6).

## 5. ESTIMATION FOR p-COMPONENT SERIES SYSTEMS

In this section we derive the Bayesian estimator of the reliability of a p-component series system. Let $Y_i$ denote the strength of the i-th component which is subject to stress $X_i$, $i = 1,2,\ldots,p$. Then the reliability of the system is given by

$$R_4 = P(X_i < Y_i,\ i = 1,2,\ldots,p) = P(Z > 0), \tag{5.1}$$

where $X = (X_1,\ldots,X_p)'$, $Y = (Y_1,\ldots,Y_p)'$, $Z = Y - X$. Let $Z_1$, $Z_2$, and $Z_n$, be a random sample. Assume that Z follows the multivariate normal distribution with mean vector $\mu$ and covariance matrix $\Sigma$. We want to obtain the Bayesian estimator of $R_4$. It is well known that the vague prior distribution of $\mu$ and $\Sigma$ is given by

$$p(\underset{\sim}{\mu},\underset{\sim}{\Sigma}^{-1}) = p(\underset{\sim}{\mu})p(\underset{\sim}{\Sigma}^{-1}) \propto |\underset{\sim}{\Sigma}|^{\frac{p+1}{2}}. \tag{5.2}$$

The posterior distribution of $\underset{\sim}{\mu}$ and $\underset{\sim}{\Sigma}^{-1}$ in this case is given by

$$p(\underset{\sim}{\mu},\underset{\sim}{\Sigma}^{-1}|\text{data}) = K|\underset{\sim}{\Sigma}^{-1}|^{\frac{n-p-1}{2}} \times \exp[-\tfrac{1}{2}\operatorname{tr}\underset{\sim}{\Sigma}^{-1}((n-1)\underset{\sim}{S}+n(\bar{\underset{\sim}{Z}}-\underset{\sim}{\mu})(\bar{\underset{\sim}{Z}}-\underset{\sim}{\mu})')], \tag{5.3}$$

where

$$K = \frac{n^{p/2}|(n-1)\underset{\sim}{S}|^{\frac{n-1}{2}}}{2^{\frac{n-p}{2}}\pi^{p(p+1)/4}\prod_{i=1}^{p}\Gamma\left(\frac{1}{2}(n-i)\right)},$$

## REFERENCES


Basu, A. P., 1977a, A generalized Wilcoxon-Mann-Whitney statistic with some applications in reliability, in: "The Theory and Applications of Reliability, Vol. I," C. P. Tsokos and I. N. Shimi, ed., Academic Press, New York.

Basu, A. P., 1977b, Estimate of reliability in the stress-strength model, in: Proc. 22nd Conf. on Design of Experiments in Army Research, Development and Testing, Arlington.

Basu, A. P., 1981, The estimation of P(X<Y) for a distribution useful in life testing, Naval Research Logistics Quarterly, 28:383.

Basu, A. P., Estimating the reliability of complex systems - a survey, in: "The frontiers of modern statistical inference procedures," E. J. Dudewicz, ed., American Science Press, Columbus.

Basu, A. P., and Tarmast, G., 1987, Some inequalities for multivariate distributions with applications, Unpublished research report, Univ. of Missouri, Columbia.

Bhattacharyya, G. K., and Johnson, R. A., 1974, Estimation of reliability in a multicomponent stress-strength model, J. Amer. Stat. Assoc., 69:966.

Bhattacharyya, G. K., and Johnson, R. A., 1975, Stress-Strength Models for System Reliability, in: "Reliability and Fault Tree Analysis," R. E. Barlow, J. B. Fussell, and N. D. Singpurwalla, ed., SIAM, Philadelphia.

Enis, P., and Geisser, S., 1971, Estimation of the probability that Y<X, J. Amer. Stat. Assoc., 66:162.

Zacks, S., 1977, Bayes estimation of the reliability of series and parallel systems of independent exponential components, in: The Theory and Applications of Reliability, Vol. II," C. P. Tsokos and I. N. Shimi, ed., Academic Press, New York.

## INFORMATION IN SELECTION MODELS

M. J. Bayarri
Dep. Estadística e IO
Fac. Matemáticas
Univ. de Valencia
Burjasot. Valencia
Spain

M. H. DeGroot
Dept. of Statistics
Carnegie-Mellon Univ.
Pittsburgh, PA 15213
U.S.A.

### Abstract

The experiment in which a selection sample is drawn from some distribution involving an unknown parameter $\theta$ is compared according to various criteria with the usual experiment in which an unrestricted random sample is drawn from that distribution. The Fisher information is studied for several of these experiments, and conditions under which the experiments are ordered with respect to the criterion of sufficiency or pairwise sufficiency are presented. Standard problems involving selection samples from the normal, binomial, and Poisson distributions are discussed in detail. Some results for general exponential families and for selection models involving bivariate observations are also considered.

### 1. Introduction

In many situations, experimenters are not able to draw a random sample from the population in which they are interested, and statistical models that incorporate the restrictions under which the observations were obtained must be developed. In this paper, we consider problems in which observations are obtained only from certain selected portions of the population, either because experimental conditions make it impossible to obtain data from the whole population or because the experimenter chooses to restrict the observations in this way.

Consider a random variable X that is distributed over a certain population according to the (generalized) density $g(x|\theta)$ and suppose that it is desired to make inferences about the unknown value of the parameter $\theta(\theta \in \Omega)$. The usual statistical analysis assumes that a random sample from $g(x|\theta)$ is obtained. In this paper we will assume, however, that each observation is restricted to lie in a specified subset S of the sample space, so the analysis is based on a random sample from the following density:

$$f(y|\theta) = \frac{g(y|\theta)}{\Pr(X \in S|\theta)} \quad \text{for } y \in S \ , \tag{1.1}$$

and $f(y|\theta) = 0$ otherwise. It is assumed that $\Pr(X \in S|\theta) > 0$ for all $\theta \in \Omega$. The model (1.1) is called a *selection model* or a *truncation model*, and a random sample from (1.1) is called a *selection sample.*

Selection samples occur frequently in practice, and several examples were given in Bayarri and DeGroot (1986a) together with a Bayesian analysis of these models. The name "selection models" in this context is due to Fraser (1952, 1966), although the term "selection" was used in a more general setting by Tukey (1949).

In this paper, our main interest is in comparing the experiment in which a selection sample is obtained with that in which an unrestricted random sample from $g(x|\theta)$ is obtained. In some problems, this comparison is carried out by studying the Fisher information in each type of experiment. In certain cases, stronger results are obtained based on the theory of the comparison of statistical experiments as developed originally by Blackwell (1951, 1953). His method can be described as follows:

Let $E_X = \{X, \mathcal{X} ; g(\cdot|\theta), \theta \in \Omega\}$ denote a statistical experiment in which a random variable or random vector X defined on some sample space $\mathcal{X}$ is to be observed, and the distribution $g(\cdot|\theta)$ of X depends on a parameter $\theta$ taking values in the parameter space $\Omega$. Also, let $E_Y = \{Y, \mathcal{Y}; f(\cdot|\theta), \theta \in \Omega\}$ denote another statistical experiment with the same parameter space $\Omega$. Then the experiment $E_X$ is said to be *sufficient* for the experiment $E_Y$ (denoted $E_X \succsim E_Y$) if there exists a stochastic transformation of X to a random variable Z(X) such that, for each $\theta \in \Omega$, the random variables Z(X) and Y have identical distributions. The relationship $E_X \succsim E_Y$ holds if and only if for every decision problem involving $\theta$ and every prior distribution on $\Omega$, the expected Bayes risk from $E_X$ is not greater than that from $E_Y$.

Some other properties of the relationship $E_X \succsim E_Y$ should be noted. Let $E_X^n$ denote the experiment in which $E_X$ is independently performed n times so that a random sample $X_1, \ldots, X_n$ is obtained, and let $E_Y^n$ be defined analogously. Then $E_X \succsim E_Y$ implies $E_X^n \succsim E_Y^n$ for every value of n.

The experiment $E_X$ is said to be *pairwise sufficient* for the experiment $E_Y$ (denoted $E_X \succsim_2 E_Y$) if for every pair of values $\theta_1$, $\theta_2 \in \Omega$, $E_X$ is sufficient for $E_Y$ when the parameter space is restricted to contain just the two values $\theta_1$ and $\theta_2$. Clearly if $E_X \succsim E_Y$ then $E_X \succsim_2 E_Y$. However, the converse does not necessarily hold.

If $\theta$ is a k-dimensional vector and $\Omega$ is an open subset of $R^k$, we shall let $I_X(\theta)$ and $I_Y(\theta)$ denote the kxk Fisher information matrices for the experiments $E_X$

and $E_Y$ respectively, under the standard regularity conditions. We shall use the notation $E_X \underset{\sim F}{>} E_Y$ whenever $I_X(\theta) - I_Y(\theta)$ is nonnegative definite for all $\theta \epsilon \Omega$. The relationship $E_X \underset{\sim}{>} E_Y$ implies a similar ordering in terms of Fisher information; i.e., if $E_X \underset{\sim}{>} E_Y$ then $E_X \underset{\sim F}{>} E_Y$. However, the converse does not necessarily hold. Moreover, since the Fisher information can be obtained from the Kullback-Leibler information by considering pairs of values of $\theta$ that are arbitrarily close to each other, it can be shown that if $E_X \underset{\sim 2}{>} E_Y$ then $E_X \underset{\sim F}{>} E_Y$. Some of these relations and other properties of the comparison of experiments are described in Stein (1951), Stone (1961), Kullback (1968, pages 26-28), Torgersen (1970, 1972, 1976), Hansen and Torgersen (1974), and Goel and DeGroot (1979). Some new examples and counterexamples will be given in this paper.

In Section 2, the relation $E_X \underset{\sim F}{>} E_Y$ is studied for problems in which X has a normal distribution with unknown mean $\theta$ and Y is restricted to lie in different selection sets, as well as for problems in which X has a binomial or Poisson distribution and Y has the corresponding truncated distribution with the zero class missing. In these cases, when the selection set is the upper tail of the distribution it is found that $E_X \underset{\sim F}{>} E_Y$.

In Section 3, we study the conditions under which a selection sample from an exponential family provides greater Fisher information than an unrestricted random sample. When the selection set is the upper tail, these conditions can be determined by the behavior of the hazard-rate function.

In Section 4, we consider some examples of other types of selection involving bivariate observations in which the selection mechanism restricts the values of one of the variables.

In Section 5, we study the relations $\underset{\sim}{>}$ and $\underset{\sim 2}{>}$ for different experiments involving unrestricted and selection samples from the normal distribution with an unknown mean and known precision. It is shown that $E_X \underset{\sim 2}{>} E_Y$ when the selection set is the upper tail and that $E_Y \underset{\sim}{>} E_X$ when the parameter space contains just two points and the selection set contains both tails of the normal distribution chosen symmetrically with respect to those points.

In Section 6, we study the relations $\underset{\sim}{>}$ and $\underset{\sim 2}{>}$ for the truncated binomial and Poisson distributions with the zero class missing. It is shown that the relation $E_X \underset{\sim}{>} E_Y$ does not hold for either of these distributions, where $E_X$ is the experiment in which an unrestricted random sample is observed and $E_Y$ is the corresponding selection experiment. It is also shown that for the binomial distribution with n = 2, $E_X \underset{\sim 2}{>} E_Y$, thus providing an interesting example in which the parameter space is an open subset of the real line and one experiment is pairwise sufficient but not sufficient for another one.

Due to restrictions of space, most of the results are presented in this paper without any derivation or proof. Full details can be found in Bayarri and DeGroot (1986b).

## 2. Fisher information for selection models

In this section we will study the Fisher information for selection samples from some standard distributions in order to compare this information with that obtained from unrestricted random samples. Here and throughout the paper we shall let $E_X$ denote the experiment in which an observation X is obtained from an unrestricted density $g(\cdot|\theta)$ and let $E_Y$ denote the experiment in which an observation Y is obtained from a selection model for which the density $f(\cdot|\theta)$ is of the form given in (1.1).

We shall begin by considering various selection sets for problems in which X has a normal distribution with unknown mean $\theta$ and known precision which, without loss of generality, we take to be 1. Suppose first that for a specified value of $\tau$, the observation Y is restricted to the set $Y \geq \tau$. Then the p.d.f. of Y is

$$f(y|\theta) = \frac{\phi(y-\theta)}{1-\Phi(\tau-\theta)} \quad \text{for } y \geq \tau , \tag{2.1}$$

where $\phi(\cdot)$ and $\Phi(\cdot)$ denote the standard normal p.d.f. and d.f. respectively. Under the usual regularity conditions, the Fisher information about a real-valued parameter $\theta$ obtained from an arbitrary random variable U with density $h(u|\theta)$ is given by

$$I(\theta) = E\left\{ -\frac{\partial^2}{\partial\theta^2} \log h(U|\theta) \right\} . \tag{2.2}$$

In the example we are considering it is well known that $I_X(\theta) = 1$ for $-\infty < \theta < \infty$. Furthermore, it can be found that

$$I_Y(\theta) = 1 + \frac{1}{[M(\tau-\theta)]^2} [(\tau-\theta)M(\tau-\theta) - 1] , \tag{2.3}$$

where $M(\lambda)$ is Mills' ratio defined by

$$M(\lambda) = \frac{1-\Phi(\lambda)}{\phi(\lambda)} \quad \text{for } -\infty < \lambda < \infty . \tag{2.4}$$

It follows from the properties of $M(\lambda$ that $I_Y(\theta) < 1$. Hence, $E_X \underset{\sim}{>}_F E_Y$ for any selection point $\tau$.

The analysis for a selection sample from the lower tail of the normal distribution is similar. Suppose next that the observation Y is restricted to the set $S = \{y;\ y \leq \tau_1 \text{ or } y \geq \tau_2\}$, where $\tau_1 < \tau_2$ are specified real numbers. Then the p.d.f. of Y is

$$f(y|\theta) = \frac{\phi(y-\theta)}{1-\Phi(\tau_2-\theta)+\Phi(\tau_1-\theta)} \quad \text{for } y \leq \tau_1 \text{ or } y \geq \tau_2 . \tag{2.5}$$

It can be shown that $I_Y(\theta) < I_X(\theta)$ for some values of $\theta$ and that this inequality is reversed for other values of $\theta$, so neither of the relationships $E_X \underset{\sim F}{>} E_Y$ nor $E_Y \underset{\sim F}{>} E_X$ holds.

Finally, suppose that the observation Y is restricted to the interval $\tau_1 \leq Y \leq \tau_2$. The p.d.f. of Y is now

$$f(y|\theta) = \frac{\phi(y-\theta)}{\Phi(\tau_2-\theta)-\Phi(\tau_1-\theta)} \quad \text{for } \tau_1 \leq y \leq \tau_2 . \tag{2.6}$$

For this selection model, it can be shown that $I_X(\theta) \geq I_Y(\theta)$ for all $\theta$. Hence, $E_X \underset{\sim F}{>} E_Y$. It is noteworthy that an unrestricted random sample provides greater Fisher information for all possible values of $\theta$ than a selection sample from any bounded interval irrespective of its location or its length.

Other selection models that have been widely treated in the statistical literature are the truncated binomial and Poisson distributions in which the zero class is missing (David and Johnson, 1952; Irwin, 1959; Cohen, 1960; Dahiya and Gross, 1973; Sanathanan, 1977; Blumenthal and Sanathanan, 1980; and Blumenthal, 1981). It can be shown that for both the binomial and the Poisson distributions an unrestricted random sample provides greater Fisher information than a selection sample with the zero class missing.

## 3. Selection from an exponential family

In this section we will consider the question of whether we gain or lose Fisher information when a selection sample rather than a random sample is obtained from a distribution belonging to an exponential family. We begin by considering an arbitrary density $g(\cdot|\theta)$ indexed by a real-valued parameter $\theta$ lying in an open subset $\Omega$ of the real line and an arbitrary specified selection set S, so the selection model $f(\cdot|\theta)$ is given by (1.1). If we let

$$s(\theta) = \Pr(X \in S|\theta) \tag{3.1}$$

then under the usual regularity conditions

$$I_Y(\theta) = E\left[ -\frac{\partial^2}{\partial\theta^2} \log g(Y|\theta) \right] + \frac{d^2}{d\theta^2} \log s(\theta) . \tag{3.2}$$

Suppose now that the unrestricted model for an observation X is represented by a density $h(\cdot|\omega)$ of the following form:

$$h(x|\omega) = a(x)b(\omega)\exp\{u(x)v(\omega)\} . \tag{3.3}$$

In other words, we are assuming that $h(x|\omega)$ belongs to an exponential family for which the natural parameter is $\theta = v(\omega)$. If we now reparametrize the family in terms of $\theta$, the density of X becomes

$$g(x|\theta) = a(x)c(\theta)\exp\{\theta u(x)\} \tag{3.4}$$

and

$$-\frac{\partial^2}{\partial\theta^2} \log g(x|\theta) = -\frac{d^2}{d\theta^2} \log c(\theta) . \tag{3.5}$$

Since (3.5) is a constant that does not depend on x, it follows that $I_X(\theta)$ as well as the expectation on the right-hand side of (3.2) are given by (3.5). Hence,

$$I_Y(\theta) = I_X(\theta) + \frac{d^2}{d\theta^2} \log s(\theta) , \tag{3.6}$$

so that $I_Y(\theta) \geq I_X(\theta)$ for all $\theta$ if and only if log $s(\theta)$ is convex and $I_Y(\theta) \leq I_X(\theta)$ for all $\theta$ if and only if log $s(\theta)$ is concave.

It is well known that under the usual conditions, since $\theta = v(\omega)$ the Fisher information $I(\theta)$ about $\theta$ and the Fisher information $I^*(\omega)$ about $\omega$ satisfy the following relation for any experiment:

$$I^*(\omega) = I(\theta)\left(\frac{d\omega}{d\theta}\right)^2 . \tag{3.7}$$

Thus, $I^*_Y(\omega) \geq I^*_X(\omega)$ for all values of $\omega$ if and only if $I_Y(\theta) \geq I_X(\theta)$ for all values of $\theta$. In other words, a relation of the form $E_Y \underset{\sim}{>}_F E_X$ is defined unambiguously regardless of the parametrization used. It follows that in order to determine whether the experiments $E_X$ and $E_Y$ are ordered with respect to the relationship $\underset{\sim}{>}_F$, we need only determine whether the function log $s(\theta)$ is convex or concave. In the remainder of this section we will consider selection sets of the form $Y \geq \tau$ so that $s(\theta) = 1 - G(\tau|\theta)$, where $G(\cdot|\theta)$ is the d.f. corresponding to the density $g(\cdot|\theta)$.

Suppose that the distribution $G(\cdot|\theta)$ is absolutely continuous and $\theta$ is either a location parameter (i.e., $G(x|\theta) = G_0(x-\theta)$) or a scale parameter (i.e., $G(x|\theta) = G_0(\theta x)$). Then the convexity or concavity of log $s(\theta)$ can be easily studied in terms of the hazard-rate or failure-rate function

$$r_0(x) = \frac{g_0(x)}{1-G_0(x)} . \tag{3.8}$$

It follows that for both types of families, $E_Y \underset{\sim}{>}_F E_X$ if and only if $r_0(x)$ is a decreasing function of x.

For example, consider again the normal distribution with unknown mean $\theta$ and precision 1, so that $g(x|\theta) = \phi(x-\theta)$. In this problem, $\theta$ is a location parameter and the distribution $G_0$ is the standard normal distribution, for which it is known that the hazard-rate function is increasing. Hence, $E_X \underset{\sim}{>}_F E_Y$.

On the other hand, suppose that the mean of the normal distribution is known to be 0 and the precision $\theta$ is unknown, so that

$$g(x|\theta) = \theta^{1/2} \phi (\theta^{1/2} x) . \tag{3.9}$$

In this case, $\theta$ is the natural parameter and although it is not a scale parameter, it can be shown that $\log s(\theta)$ is convex. Hence, $E_Y \underset{\sim}{>}_F E_X$, which means that in this case a selection sample provides greater Fisher information than a random sample from the whole population.

It should be noted that if $\tau = 0$ in this example, the experiments $E_X$ and $E_Y$ are equivalent not only in the sense that $I_X(\theta) = I_Y(\theta)$ for all values of $\theta$, but also in the sense that both $E_X \underset{\sim}{>} E_Y$ and $E_Y \underset{\sim}{>} E_X$.

As another example, suppose that X has a gamma distribution for which the shape parameter $\alpha$ is known and the scale parameter $\theta$ is unknown; that is

$$g(x|\theta) = \frac{\theta^{\alpha}}{\Gamma(\alpha)} x^{\alpha-1} e^{-\theta x} \quad \text{for } x > 0 . \tag{3.10}$$

For this exponential family, $\theta$ is the natural parameter and, as its name implies, it is a scale parameter. It is known (Barlow and Proschan, 1975, Chapter 3) that a gamma distribution has an increasing hazard-rate function if $\alpha > 1$ and a decreasing hazard-rate function if $0 < \alpha < 1$. Therefore, if $\alpha > 1$, then $E_X \underset{\sim}{>}_F E_Y$, whereas if $\alpha < 1$, then $E_Y \underset{\sim}{>}_F E_X$. Of course, if $\alpha = 1$, the gamma distribution reduces to the exponential distribution for which the hazard-rate function is constant and the experiments $E_X$ and $E_Y$ are equivalent.

## 4. Other types of selection

We will now consider briefly some selection models involving bivariate observations in which the selection mechanism restricts the values of one of the variables. One example that has been extensively discussed in the literature (Blackwell and Girshick, 1954, Chapter 12; Lehmann, 1986, p. 87-88; and DeGroot, 1970, p. 444-445) compares the experiments in which a selection sample can be drawn from one of four different subpopulations.

We will consider a continuous version of this type of problem. Suppose that U and V have a bivariate normal distribution for which the means $\mu_1$ and $\mu_2$ and the variances $\sigma_1^2$ and $\sigma_2^2$ are known, and the correlation $\theta$ is unknown. Without loss of generality, we shall take $\mu_1 = \mu_2 = 0$ and $\sigma_1^2 = \sigma_2^2 = 1$. Consider the following two experiments:

($E$) A random sample of n bivariate observations is drawn from the bivariate normal distribution.

($E_v$) A random sample of n observations is drawn from the conditional distribution of U given V = v. This sample can be regarded as a selection sample from the subpopulation for which V = v.

Let $I(\theta)$ and $I_v(\theta)$ denote the Fisher information in the experiments $E$ and $E_v$, respectively. Then it can be shown that for all $\theta$,

$$\begin{aligned} I(\theta) &> I_v(\theta) \quad \text{if } v^2 < 1 , \\ I(\theta) &< I_v(\theta) \quad \text{if } v^2 > 1 , \\ I(\theta) &= I_v(\theta) \quad \text{if } v^2 = 1 . \end{aligned} \tag{4.1}$$

It is interesting to note, as indicated in (4.1), that the Fisher information obtained from an unrestricted random sample is identical to that obtained from a sample from the conditional distribution of U given V = 1 or given V = -1. It is also interesting to note, as indicated in (4.1), that $E \underset{\sim F}{>} E_v$ whenever $-1 < v < 1$ and $E_v \underset{\sim F}{>} E$ whenever $v < -1$ or $v > 1$. We do not know if these experiments are ordered by the sufficiency relation $\underset{\sim}{>}$.

A general class of selection models that includes these examples is the following: Suppose that the random vector (U, V) has a joint distribution that depends on the parameter $\theta$ and that observations can be obtained only when V lies in some selection set. In some problems, an observation may consist of the pair (U, V), while in others just U is observed. Some examples in econometrics are studied by Heckman (1976), Amemiya (1984) and Little (1985).

## 5. Sufficiency in normal experiments

Consider again the problem discussed at the beginning of Section 2 in which X has a normal distribution with unknown mean $\theta$ and precision 1, and the observation Y is restricted to the set $Y \geq \tau$. It was shown there that $E_X \underset{\sim F}{>} E_Y$. We will now prove the stronger result that $E_X \underset{\sim 2}{>} E_Y$. Because of the comments in Section 1, it is sufficient to consider experiments based on just one observation.

Suppose then that the parameter space contains just two values $\theta_0$ and $\theta_1$. For any experiment $E$ and any value of $\alpha$ $(0 < \alpha < 1)$, let $\beta(\alpha|E)$ denote the probability of a type 2 error when the likelihood ratio test for distinguishing between $\theta_0$ and $\theta_1$ is carried out with the specified probability $\alpha$ of a type 1 error. It was shown by Torgersen (1970, 1976) that for any two experiments $E$ and $E^*$ with the same parameter space $\Omega = \{\theta_0, \theta_1\}$, $E \underset{\sim}{>} E^*$ if and only if $\beta(\alpha|E) \leq \beta(\alpha|E^*)$ for all values of $\alpha$ $(0 < \alpha < 1)$. Hence, in the problem we are now considering, the desired conclusion that $E_X \underset{\sim 2}{>} E_Y$ can be obtained by showing that, for any pair of values $\theta_0$ and $\theta_1$, the function $\beta_X(a)$ is not larger than the function $\beta_Y(a)$ over the interval $0 < a < 1$, where $\beta_X(a) = \beta(a|E_X)$ and $\beta_Y(a) = \beta(a|E_Y)$.

Without loss of generality we will assume that $\theta_0 = 0$ and $\theta_1 > 0$. The likelihood ratio test at level $a$ based on the experiment $E_X$ rejects the hypothesis $H_0$: $\theta = \theta_0$ when $X \geq c_X(a)$, where

$$c_X(a) = \Phi^{-1}(1-a) \ . \tag{5.1}$$

Similarly, the likelihood ratio test at the same level $a$ based on the selection experiment $E_Y$ rejects $H_0$ when $Y \geq c_Y(a)$, where

$$c_Y(a) = \Phi^{-1}[1-a+a\Phi(\tau)] \ . \tag{5.2}$$

For any given value of $a$, let $L_X(a)$ denote the likelihood ratio for the observation X evaluated at $X = c_X(a)$, and let $L_Y(a)$ be defined similarly. In order to establish that $\beta_X(a) \leq \beta_Y(a)$ for $0 < a < 1$, it is sufficient to show that

$$\frac{\beta'_Y(a)}{\beta'_X(a)} = \frac{L_Y(a)}{L_X(a)} = A \exp\{\theta_1[c_Y(a) - c_X(a)]\} \tag{5.3}$$

is an increasing function of $a$. In (5.3), A is a constant not involving $a$. In turn, since $\theta_1 > 0$ it is sufficient to show that $D(a) = c_Y(a) - c_X(a)$ is an increasing function of $a$, which follows from the fact that for any given value of $a$, the function $b/\phi[\Phi^{-1}(ab)]$ is an increasing function of b for $0 < b \leq 1$. Hence, $E_X \underset{\sim}{>}_2 E_Y$.

A similar argument shows that if $E_Z$ is a selection experiment of the same type as $E_Y$ but with a larger selection point $\tau$, then $E_Y \underset{\sim}{>}_2 E_Z$. We do not know whether the experiments $E_X$, $E_Y$, and $E_Z$ are ordered by the sufficiency relation $\underset{\sim}{>}$.

Consider next the problem in which X again has a normal distribution with unknown mean $\theta$ and precision 1, but in which Y is now restricted to lie in the two-tailed selection set $Y \leq \tau_0$ or $Y \geq \tau_1$. Suppose again that the parameter space contains just two values $\theta_0$ and $\theta_1$, and that the selection points $\tau_0$ and $\tau_1$ are symmetrically placed with respect to $\theta_0$ and $\theta_1$ so that $\tau_1 - \theta_1 = \theta_0 - \tau_0$. Without loss of generality we can assume that $\theta_1 = -\theta_0 = \mu > 0$ and $\tau_1 = -\tau_0 = \tau > 0$. We will show that $E_Y \underset{\sim}{>} E_X$ in this particular problem. It should be noted that it is not true that $E_Y \underset{\sim}{>}_2 E_X$ when the parameter space $\Omega$ is the entire real line because, for fixed values of $\tau_0$ and $\tau_1$, when we restrict the parameter space to contain just two points, the relation $E_Y \underset{\sim}{>} E_X$ will not hold for all pairs of values of $\theta$.

To establish that $E_Y \underset{\sim}{>} E_X$, we will again compare the functions $\beta_X(a)$ and $\beta_Y(a)$. In this problem, we must show that $\beta_Y(a) \leq \beta_X(a)$ for $0 < a < 1$. For testing the hypotheses $H_0$: $\theta = -\mu$ and $H_1$: $\theta = \mu$, the likelihood ratio test at level $\alpha$ based on the experiment $E_X$ rejects $H_0$ when $X \geq c_X(\alpha)$, where

$$c_X(\alpha) = \Phi^{-1}(1-\alpha)-\mu \ . \tag{5.4}$$

Similarly, the likelihood ratio test at the same level $\alpha$ based on the selection experiment $E_Y$ rejects $H_0$ when $Y \geq c_Y(\alpha)$. Let

$$B = 2 - \Phi(\tau-\mu) - \Phi(\tau+\mu) \tag{5.5}$$

and let

$$\alpha_0 = \frac{1}{B}\,[1-\Phi(\tau+\theta)] \ . \tag{5.6}$$

Then

$$c_Y(\alpha) = \begin{cases} \Phi^{-1}(1-\alpha B)-\mu & \text{for } \alpha \leq \alpha_0 \, , \\ \Phi^{-1}[(1-\alpha)B]-\mu & \text{for } \alpha > \alpha_0 \, . \end{cases} \tag{5.7}$$

For $\alpha \leq \alpha_0$, the relation $\beta_Y(\alpha) \leq \beta_X(\alpha)$ is equivalent to the relation

$$\frac{1}{B}\,\Phi[2\mu+\Phi^{-1}(\alpha B)] \geq \Phi[2\mu+\Phi^{-1}(\alpha)] \ . \tag{5.8}$$

In turn, (5.8) follows from the fact that the left-hand side is a decreasing function of B over the interval $0 < B \leq 1$. A similar argument applies for $\alpha > \alpha_0$. Thus, $E_Y \underset{\sim}{>} E_X$. Furthermore, if $E_Z$ is a selection experiment of the same type as $E_Y$ but with a larger value of $\tau$, then $E_Z \underset{\sim}{>} E_Y$.

## 6. Sufficiency in binomial and Poisson experiments

In this section we will consider again the truncated binomial distribution with the zero class missing. It was stated in Section 2 that if X has a binomial distribution with parameters n and $\theta$ and Y has this truncated binomial distribution, then $E_X \underset{\sim}{>}_F E_Y$. We will now give a simple argument which shows that the relation $E_X \underset{\sim}{>} E_Y$ does *not* hold.

In order for the relation $E_X \underset{\sim}{>} E_Y$ to hold, there must exist a stochastic transformation $h(y|x)$ such that for all values of $\theta$ in the interval $0 < \theta < 1$,

$$\sum_{x=0}^{n} h(y|x)g(x|\theta) = f(y|\theta) = \frac{1}{1-(1-\theta)^n}\binom{n}{y}\theta^y(1-\theta)^{n-y} \quad \text{for } y = 1,\dots,n \ . \tag{6.1}$$

The stochastic transformation $h(y|x)$ must be a nonnegative function such that

$$\sum_{y=1}^{n} h(y|x) = 1 \qquad \text{for } x = 0,1,\dots,n \ . \tag{6.2}$$

Since for any given value of y and any possible stochastic transformation h, the left-hand side of (6.1) must be a polynomial in $\theta$ whereas the right-hand side, is not a polynomial, it follows immediately that (6.1) cannot be satisfied for all values of $\theta$. Hence, it cannot be true that $E_X \gtrsim E_Y$ . Of course, it also cannot be true that $E_Y \gtrsim E_X$ since $E_X \gtrsim_F E_Y$ and the experiments are not equivalent.

Next, consider the problem in which X has a Poisson distribution with mean $\theta$ and Y has a truncated Poisson distribution with the zero class missing. Again, the following argument shows that the relation $E_X \gtrsim E_Y$ does *not* hold even though $E_X \gtrsim_F E_Y$.

In order to have $E_X \gtrsim E_Y$, there must exist a stochastic transformation $h(y|x)$ defined for x = 0,1,2... and y = 1,2,... such that for all values of $\theta > 0$,

$$\sum_{x=0}^{\infty} \frac{h(y|x)e^{-\theta}\theta^x}{x!} = \frac{e^{-\theta}\theta^y}{y!(1-e^{\theta})} \quad \text{for } y = 1,2,\ldots . \tag{6.3}$$

By expanding $1-e^{-\theta}$ in a Taylor series, it can be shown that no such stochastic transformation can exist.

To conclude this paper we return to the truncated binomial experiment and we will show that when n = 2, $E_X \gtrsim_2 E_Y$. We will do this by explicitly constructing a stochastic transformation that satisfies (6.1) and (6.2).

Suppose then that the parameter space contains just two values $\theta_0$ and $\theta_1$ $(0 < \theta_0 < \theta_1 < 1.)$ When n = 2, Y can just take the values y = 1 and y = 2. Since $h(y|x)$ must be such that $h(2|x) = 1 - h(1|x)$ for x = 0, 1, 2, we just have to find three numbers $h(1|x)$ (x = 0, 1, 2) such that $0 \le h(1|x) \le 1$ and satisfying the two equations

$$h(1|0)(1-\theta_i)^2 + h(1|1)2\theta_i(1-\theta_i) + h(1|2)\theta_i^2 = \frac{2(1-\theta_i)}{2-\theta_i} \quad \text{for } i = 0,\ 1 . \tag{6.4}$$

It can be shown that there are infinitely many solutions of the system (6.4) satisfying the restriction $0 \le h(1|x) \le 1$ for x = 0, 1, 2. One simple solution is obtained by taking $h(1|2) = 0$ and solving (6.4) for $h(1|0)$ and $h(1|1)$. In this way it is found that

$$h(1|0) = 1 - \frac{\theta_0\theta_1}{(2-\theta_0)(2-\theta_1)} ,$$

$$h(1|1) = 1 - \frac{(1-\theta_0)(1-\theta_1)}{(2-\theta_0)(2-\theta_1)} , \tag{6.5}$$

$$h(1|2) = 0 ,$$

provides the desired stochastic transformation. Thus, we have developed an interesting example of experiments in which the parameter space is an open subset of the real line and $E_X \gtrsim_2 E_Y$ but it is not true that $E_X \gtrsim E_Y$.

**Acknowledgments**

This research was carried out in part while M. J. Bayarri was visiting at Carnegie-Mellon University and M. H. DeGroot was visiting at Ohio State University, and was supported by the Spanish Ministry of Education and Science and the Fulbright Association under grant 85-07399 and by the National Science Foundation under grant DMS-8320618. We are deeply indebted to Prem K. Goel for several valuable discussions about this work.

**References**

Amemiya, T. (1984). Tobit models: a survey. Journal of Econometrics **24**, 3-61.

Barlow, R. E., and Proschan, F. (1975). Statistical Theory of Reliability and Life Testing. New York: Holt, Rinehart and Winston.

Bayarri, M. J., and DeGroot, M. H. (1986a). Bayesian analysis of selection models. Technical Report 365, Department of Statistics, Carnegie-Mellon University.

Bayarri, M. J., and DeGroot, M. H. (1986b). Information in selection models. Technical Report 368, Department of Statistics, Carnegie-Mellon University.

Blackwell, D. (1951). Comparison of experiments. Proceedings of the Second Berkeley Symposium on Mathematical Statistics and Probability, 93-102. Berkeley, California: University of California Press.

Blackwell, D. (1953). Equivalent comparison of experiments. Annals of Mathematical Statistics. **24**, 265-272.

Blackwell, D., and Girshick, M. A. (1954). Theory of Games and Statistical Decisions. New York: John Wiley & Sons.

Blumenthal, S. (1981). A survey of estimating distributional parameters and sample sizes from truncated samples. In Statistical Distributions in Scientific Work (C. Taillie, G. P. Patil, and B. Baldessari, eds.), Vol. 5, 75-86. Dordrecht: Reidel.

Blumenthal, S., and Sanathanan, L. P. (1980). Estimation with truncated inverse binomial sampling. Communications in Statistics **A9**, 997-1017.

Cohen, A. G. (1960). Estimation in truncated Poisson distributions when zeros and some ones are missing. Journal of the American Statistical Association **55**, 342-348.

Dahiya, R. G., and Gross, A. J. (1973). Estimating the zero class from a truncated Poisson sample. Journal of the American Statistical Association **68**, 731-733.

David, F. N. and Johnson, N. L. (1952). The truncated Poisson. Biometrics **8**, 275-285.

DeGroot, M. H. (1970). Optimal Statistical Decisions. New York: McGraw-Hill.

Fraser, D. A. S. (1952). Sufficient statistics and selection depending on the parameters. Annals of Mathematical Statistics **23**, 417-425.

Fraser, D. A. S. (1966). Sufficiency for selection models. Sankhyā A **28**, 329-334.

Goel, P. K., and DeGroot, M. H. (1979). Comparison of experiments and information measures. Annals of Statistics **7**, 1066-1077.

Hansen, O. H., and Torgersen, E. N. (1974). Comparison of linear normal experiments. Annals of Statistics **2**, 367-373.

Heckman, J. (1976). The common structure of statistical models of truncation, sample selection, and limited dependent variables and a simple estimator for such models. Annals of Economic and Social Measurement **5**, 475-492.

Irwin, J. O. (1959). On the estimation of the mean of a Poisson distribution from a sample with the zero class missing. Biometrics **15**, 324-326.

Kullback, S. (1968). Information Theory and Statistics. New York: Dover.

Lehmann, E. L. (1986). Testing Statistical Hypotheses, 2nd Edition. New York: John Wiley & Sons.

Little, R. J. A. (1985). A note about models for selectivity bias. Econometrica **53**, 1469-1474.

Sanathanan, L. P. (1977). Estimating the size of a truncated sample. Journal of the American Statistical Association **72**, 669-672.

Stein, C. (1951). Information and the comparison of experiments. Unpublished Report. University of Chicago.

Stone, M. (1961). Non-equivalent comparisons of experiments and their use for experiments involving location parameters. Annals of Mathematical Statistics **32**, 326-332.

Torgersen, E. N. (1970). Comparison of experiments when the parameter space is finite. Z. Wahrscheinlichkeitstheorie und Verw. Gebiete **16**, 219-249.

Torgersen, E. N. (1972). Comparison of translation experiments. Annals of Mathematical Statistics **43**, 1383-1399.

Torgersen, E. N. (1976). Comparison of statistical experiments. Scandinavian Journal of Statistics **3**, 186-208.

Tukey, J. W. (1949). Sufficiency, truncation and selection. Annals of Mathematical Statistics **20**, 309-311.

# APPROXIMATIONS IN STATISTICS FROM A DECISION-THEORETICAL VIEWPOINT

José M. Bernardo

Departamento de Estadística
Universidad de Valencia
Burjassot, Valencia, Spain

SUMMARY

The approximation of the probability density $p(.)$ of a random vector $\mathbf{x} \in \mathbf{X}$ by another (possibly more convenient) probability density $q(.)$ which belongs to a certain class $\mathcal{Q}$ is analyzed as a *decision problem* where the action space is the class $\mathcal{Q}$ of available approximations, the relevant uncertain event is the actual value of the vector $\mathbf{x}$ and the utility function is a *proper scoring rule*. The logarithmic divergence is shown to play a rather special role within this approach. The argument lies entirely within a Bayesian framework.

## 1.APPROXIMATION AS A DECISION PROBLEM

Let $p(.)$ be the probability density of a random vector $\mathbf{x} \in \mathbf{X}$ with respect to some dominating measure, simply denoted $d\mathbf{x}$, and suppose one is interested in approximating $p(.)$ by a density $q(.)$ which belongs to a class $\mathcal{Q}$ of possibly more tractable distributions. For instance, one may wish

- To approximate a complicated probabilistic model $p(\mathbf{x}|\theta)$ by a member $q(\mathbf{x}|\omega)$, $\omega=\omega(\theta)\in\Omega$ of a more tractable family (for example, a multivariate normal)

- To describe prior opinions in a mathematically tractable form (for example, a finite mixture of distributions which are conjugate to some probabilistic model)

- To approximate posterior distributions by distributions which satisfy specific additional restrictions, (for example, *reference* posteriors, or posteriors within a class of easily integrable distributions)

From a Bayesian decision-theoretical viewpoint, the problem posed may be seen as a decision problem where the action space is the class $\mathcal{Q}$ of available approximations, the relevant uncertain event is the particular value of $\mathbf{x}$ which eventually obtains and the loss function represents the loss suffered when $p(.)$, the true distribution of $\mathbf{x}$, is replaced by a member $q(.)$ of the class $\mathcal{Q}$.

Without loss of generality, we may write such (opportunity) loss as the difference

$$l\{q(.),\mathbf{x}\} = u\{p(.),\mathbf{x}\} - u\{q(.),\mathbf{x}\} \tag{1}$$

between the utility $u\{p(.),\mathbf{x}\}$ which obtains if the true distribution is used and the utility $u\{q(.),\mathbf{x}\}$ obtained if $q(.)$ is used instead. Since $u\{q(.),\mathbf{x}\}$ measures the reward, or *score*, attained if if $q(.)$ is predicted and $\mathbf{x}$ obtains, utility functions of the type $u\{q(.),\mathbf{x}\}$ are often referred to as *scoring rules* (see *e.g.* Savage 1971, Lindley 1985).

It is only natural to require that, before $\mathbf{x}$ is actually observed, the *expected* loss of using an approximation $q(.)$, rather than the true distribution $p(.)$, must be *non-negative* and zero if, and only if, $q(.)=p(.)$ almost everywhere. Indeed, it would seem strangely perverse to expect a net gain by using the wrong distribution!. Thus, we assume,

$$l\{q(.)\} = \int_{\mathbf{X}} p(\mathbf{x})\ l\{q(.),\mathbf{x}\}\ d\mathbf{x} \geq 0 \tag{H1}$$

with $l\{q(.)\}=0$ iff $q(.)=p(.)$ a.e. Using (1), this assumption implies that, for all $q(.)$,

$$\int_{\mathbf{X}} u\{q(.),\mathbf{x}\}\ p(\mathbf{x})\ d\mathbf{x} \leq \int_{\mathbf{X}} u\{p(.),\mathbf{x}\}\ p(\mathbf{x})\ d\mathbf{x}$$

which is the definition of a *proper* scoring rule, where the reward is maximized if, and only if, the selected distribution $q(.)$ is equal (a.e.) to the true distribution. Examples of proper scoring rules include

$$u\{q(.),\mathbf{x}\} = A \log q(\mathbf{x}) + B(\mathbf{x}), \quad (\textit{logarithmic})$$

$$u\{q(.),\mathbf{x}\} = A\ \{2q(\mathbf{x}) - |q(.)|_2^2\} + B(\mathbf{x}), \quad (\textit{quadratic})$$

$$u\{q(.),\mathbf{x}\} = \frac{A}{\alpha-1}\left[\left\{\frac{q(\mathbf{x})}{|q(.)|_\alpha}\right\}^{\alpha-1} - 1\right] + B(\mathbf{x}), \quad (\textit{spherical}),$$

where

$$|q(.)|_\alpha = \left\{\int q^\alpha(\mathbf{x})\ d\mathbf{x}\right\}^{1/\alpha}, \quad \alpha > 1$$

is the $L_\alpha$ norm. Those scoring rules are respectively associated to the names of Good(1952); Brier (1950) and de Finetti (1962) and Good (1971). The spherical utility functions contain the logarithmic as their limit as $\alpha \to 1$.

Summing up, the approximation of $p(.)$ by some $q(.)$ in $\mathcal{Q}$ is a decision problem whose optimal solution is to choose that density $q(.)$ in $\mathcal{Q}$ which maximizes

$$\int p(\mathbf{x})\ u\{q(.),\mathbf{x}\}\ d\mathbf{x}$$

where $u\{q(.),\mathbf{x}\}$ is *any proper* scoring rule. We shall now find the conditions under which the appropriate utility function is precisely the *logarithmic* function defined above.

## 2. LOGARITHMIC DIVERGENCE

In a problem of inference, it is often the case that the utility obtained when $q(.)$ has been predicted and $\mathbf{x}$ has been observed only depends on the probability density $q(\mathbf{x})$ attached to the value *actually* observed. Thus, we may further assume,

$$u\{q(.),\mathbf{x}\} = u\{q(\mathbf{x}),\mathbf{x}\}. \qquad \textbf{(H2)}$$

A scoring rule which satisfies (H2) is called a *local* scoring rule.

It should be obvious that (H2) does not carry the same *normative* weight as (H1), but it *does* describe however a large class of interesting situations. Indeed, if one is trying to approximate the probabilistic model $p(\mathbf{x}|\theta)$ which is supposed to describe the behaviour of $\mathbf{x}$ by another model $p(\mathbf{x}|\omega)$, $\omega=\omega(\theta)\in\Omega$ which belongs to some convenient family of distributions, the assumption $u\{q(.),\mathbf{x}\}= u\{q(\mathbf{x}),\mathbf{x}\}$ is nothing but a version of the *likelihood principle*, in that the utility of the prediction $q(.)$ depends on the data $\mathbf{x}$ obtained, but not on the data which could have been obtained but was not.

*Theorem 1.* If $\mathbf{x}$ is a random vector which may take three or more distinct values, a differentiable proper local scoring rule is necessarily of the form

$$u\{q(.),\mathbf{x}\} = A \log q(\mathbf{x}) + B(\mathbf{x}), \quad A>0$$

*Proof.* The discrete version of this result was proved by Good(1952) for the binomial case, mentioned by McCarthy (1956), proved by Aczel and Pfanzagl (1966) and generalized by Savage (1971); a continuous version was stated by Bernardo (1979a).

*Theorem 2.* Let $\mathcal{Q}$ be a class of strictly positive densities on the support of $p(.)$. Under (H1) and (H2), the loss to be expected if $p(.)$ is approximated by a member $q(.)$ of $\mathcal{Q}$ is of the form

$$l\{q(.)\} = A \int p(\mathbf{x}) \log \frac{p(\mathbf{x})}{q(\mathbf{x})}\, d\mathbf{x}, \quad A>0 \qquad (2)$$

Moreover, $l\{q(.)\}$ is (i) non-negative, (ii) invariant under one-to-one transformations of $\mathbf{x}$ and, (iii) additive in the sense that if $\mathbf{x}=(\mathbf{x}_1,\mathbf{x}_2)$, $p(\mathbf{x})=p(\mathbf{x}_1)p(\mathbf{x}_2)$ and $q(\mathbf{x})=q(\mathbf{x}_1)q(\mathbf{x}_2)$, then $l\{q(\mathbf{x})\}=l\{q(\mathbf{x}_1)\}+l\{q(\mathbf{x}_2)\}$.

*Proof.* By Theorem 1, assumptions (H1) and (H2) imply that $u\{q(.),\mathbf{x}\} = A \log q(\mathbf{x}) + B(\mathbf{x})$ and the required expression then obtains from substitution into (H1). But (2) is the well-studied *logarithmic divergence* of $q(.)$ from $p(.)$, which is known to have the stated properties (see, e.g. Kullback, 1959).

Theorem 2 implies that the utility function used has an attractive information theoretical interpretation; indeed, with the definition of information provided by Shannon (1948) and further discussed within a statistical context by Lindley (1956) and Good (1966), the expected loss (2) may be regarded as the *amount of information* which has been lost in the approximation.

Theorem 2 also implies that whenever a local utility function is appropriate, the expected loss of an approximation to $p(.)$ which gave probability zero to non-null sets under $p(.)$ would be infinite. Thus, we obtain as a corollary another version of Lindley's *Cromwell's rule*: never approximate by probability zero something which is not logically impossible!.

We have established that, under (H1) and (H2) the optimal approximation of $p(.)$ by some $q(.)$ in $\mathcal{Q}$ is provided by that density in $\mathcal{Q}$ which minimizes the logarithmic divergence (2). We shall explore now some of the consequences of this result.

## 3. EXAMPLES

### 3.1. *Estimation*

Maximum likelihood estimation within a class of models can be viewed in a new light from this perspective. Indeed, if the true distribution $p(.)$ of a random vector $\mathbf{x}$ is to be approximated by a member $q(.|\theta)$ of a class of distributions indexed by $\theta \in \Theta$ using the information provided by a random sample $\{\mathbf{x}_1, \mathbf{x}_2, \ldots, \mathbf{x}_n\}$, one should minimize in $\Theta$ the value of

$$l(\theta) = \int p(\mathbf{x}) \log \frac{p(\mathbf{x})}{q(\mathbf{x}|\theta)} \, d\mathbf{x}$$

Since $p(.)$ is not known, this integral cannot be computed, but using a standard Monte Carlo technique, it may be approximated by

$$\frac{1}{n} \sum_{i=1}^{n} \log \frac{p(\mathbf{x}_i)}{q(\mathbf{x}_i|\theta)}$$

which is minimized by that value of $\theta$ which maximizes

$$\sum_{i=1}^{n} \log q(\mathbf{x}_i|\theta),$$

i.e. by the *maximum-likelihood* estimator.

This is the best available approximation if one insists on using a member of the family $q(.|\theta)$ in order to predict the value of $\mathbf{x}$. However, standard exchangeability arguments about the $\mathbf{x}_i$'s would typically imply that $p(\mathbf{x})$ must be of the form

$$p(\mathbf{x}) = \int q(\mathbf{x}|\theta) \, \pi(\theta) \, d\theta.$$

If the prior distribution $\pi(\theta)$ is then assumed to belong to a class $\pi(\theta|\omega)$ indexed by $\omega$, then the best approximation to $p(\mathbf{x})$ will be obtained as

$$p(\mathbf{x}\,|\,\hat{\omega}) = \int q(\mathbf{x}\,|\,\theta)\ \ \pi(\theta|\hat{\omega})\ \ d\theta$$

where $\omega$ should be replaced by the value which minimizes

$$\int p(\mathbf{x})\ \log \frac{p(\mathbf{x})}{\int q(\mathbf{x}\,|\,\theta)\,\pi(\theta|\omega)\,d\theta}\ d\mathbf{x}$$

or approximately, again using a Monte Carlo argument, by that value of $\omega$ which maximizes

$$\sum_{i=1}^{n} \log \int q(\mathbf{x}_i\,|\,\theta)\ \ \pi(\theta\,|\,\omega)\ \ d\theta$$

which is a form of *non-naîve empirical Bayes* estimator. Obviously, the argument may be extended to deeper hierarchies.

### 3.2 *Poisson Approximation of a Binomial Model*

The best Poisson approximation to a Binomial model $p(\mathbf{x})=\mathrm{Bi}(\mathbf{x}\,|\,n,\theta)$ is that which minimizes

$$l(\lambda\,|\,n,\theta) = \sum_{x=0}^{n} \mathrm{Bi}(x\,|\,n,\theta)\ \log \frac{\mathrm{Bi}(x\,|\,n,\theta)}{\mathrm{Po}(x\,|\,\lambda)}$$

where

$$\mathrm{Bi}(x\,|\,n,\theta) = \binom{n}{x}\ \theta^{x}\,(1-\theta)^{n-x}$$

$$\mathrm{Po}(x\,|\,\lambda) = \frac{\lambda^{x}}{x!}\ e^{-\lambda}.$$

This corresponds to that value of $\lambda$ which maximizes

$$\sum_{x=0}^{n} \mathrm{Bi}(x\,|\,n,\theta)\ \ \{x\ \log\lambda - \lambda - \log x!\} = n\theta\,\log\lambda - \lambda - E\{\log\ x!\}$$

which, as could be expected, is $\lambda=n\theta$. The resulting minimum expected loss is increasing in $\theta$ and decreasing in $n$; numerical computation shows, however, that the condition '$\theta$ small' is far more important than the condition '$n$ large' for the quality of the approximation.

### 3.3 *Normal Approximation*

The best normal approximation $\mathrm{N}(x\,|\,\mu,h)$ to a probability density $p(x)$ is obtained by minimizing

$$l(\mu,h) = \int p(x)\ \log \frac{p(x)}{\mathrm{N}(x\,|\,\mu,h)}\ dx.$$

It is easily seen that

$$\frac{\partial l}{\partial \mu}=0 \Rightarrow \mu = \int x p(x)\, dx = E[x]$$

$$\frac{\partial l}{\partial h}=0 \Rightarrow \frac{1}{h} = \int (x-\mu)^2 p(x)\, dx = V[x]$$

Thus, with this criterion, the best normal approximation to any distribution is obtained by fitting the first two moments. This may well be another characterization of the normal distribution since, typically, different solutions are obtained with other probability models. For instance, the best Beta approximation $Be(x|a,b)$ to a distribution $p(x)$ on $]0,1[$ is the solution to the system of equations

$$\int (\log x) p(x)\, dx = \varphi(a) - \varphi(a+b)$$

$$\int \{\log(1-x)\} p(x)\, dx = \varphi(b) - \varphi(a+b)$$

where $\varphi$ is the digamma function, i.e. that obtained by fitting $E[\log x]$ and $E[\log(1-x)]$.

### 3.4. *Transformations to Normality*

Suppose that given a random quantity x with density $p(x)$ it is desired to find a *tractable* monotone tranformation $y=f(x)$ whose distribution $p(y)= p(x)/|f'(x)|$ is as close to normality as possible.

Thus, a function f is desired which minimizes

$$l(f) = \int p(y) \log \frac{p(y)}{N\{y|E[y],V[y]\}}\, dy$$

since, from 3.3, the best normal approximation to $p(y)$ is that normal with the same first two moments as $p(y)$. The loss function $l(f)$ may be rewritten as

$$l(f) = \int p(y) \log p(y)\, dy + \frac{1}{2} \log\{2\pi e V[y]\}$$

where $p(y)= p(x)/|f'(x)|$. It follows that the result depends *both* on the entropy *and* the variance of the resulting distribution.

If, say, $p(x)=Be(x|a,b)$, $0<x<1$, and we consider the class of transformations

$$y = f(x); \quad f'(x) = x^{-\alpha}(1-x)^{-\beta}, \quad \alpha\geq 0, \beta\geq 0$$

which contains as particular cases the standard transformations

$y = x$ ($\alpha=\beta=0$, no transformation)

$y = 2 \sin^{-1}\sqrt{x}$ ($\alpha=\beta=0.5$, Fisher transformation)

$y = \log \frac{x}{1-x}$ ($\alpha=\beta=1$, logit transformation)

we find that, for $a$ and $b$ large compared to the transformation parameters $\alpha$ and $\beta$,

$$\frac{1}{V[y]} \cong -\frac{\partial^2}{\partial y^2} \log p(y) \Bigg|_{y = \text{Mode}[y]} \cong \frac{(a+b)^{3-2\alpha-2\beta}}{a^{1-2a} \; b^{1-2\beta}}$$

and

$$l(\alpha,\beta) = \frac{\Gamma(a+b)}{\Gamma(a)\Gamma(b)} + (a+\alpha-1)[\varphi(a)-\varphi(a+b)]$$

$$+ (b+\beta-1)[\varphi(b)-\varphi(a+b)] + \frac{1}{2}\log\{2\pi e V[y]\}$$

which is decreasing in both $\alpha$ and $\beta$. It follows that progressively better normalizing transformations are obtained for larger values of $\alpha$ and $\beta$, provided $a$ and $b$ are large enough for the first two moments of y to exist. For instance, for $\alpha=\beta=2$ one has

$$y = 2\log\frac{x}{1-x} + \frac{2x-1}{x(1-x)}$$

a correction to the logit transformation which should be better than any of the transformations listed above.

### 3.5. *Sensitivity Analysis*

In Bayesian inference, when either the prior information is rather vague or the sample size very large, the posterior distribution is typically insensitive to reasonably large changes in the prior. In these cases, it is possible to *approximate* the posterior distribution by a *reference* posterior (Bernardo, 1979b), thus bypassing the need for a more careful specification of the prior.

In terms of the model described in this paper, the loss which may be expected by performing such approximation is given by

$$\int p(\theta|D) \log\frac{p(\theta|D)}{\pi(\theta|D)}\, d\theta$$

where $\theta$ is the parameter of interest, $D$ the available data and $\pi(\theta|D)$ the corresponding reference posterior distribution. Thus, if $\boldsymbol{P}$ is the class of prior distributions which are compatible with elicited prior information,

$$\delta(\boldsymbol{P}) = \sup_{\boldsymbol{P}} \int p(D) \int p(\theta|D) \log\frac{p(\theta|D)}{\pi(\theta|D)}\, d\theta\, dD$$

is an appropriate measure of the *maximum expected loss* of the proposed approximation. The consequences of this view are explored in Bernardo (1986).

## 4. DISCUSSION

The basic ideas developed in this paper have long been part of Bayesian folklore. Thus, it has often been recognized that approximation problems should be treated as decision problems, that scoring rules provide interesting utility functions in inferential problems and that the ubiquitous logarithmic divergence is often a sensible measure of discrepancy. This paper is an attempt to organize this material from a strictly Bayesian decision-theoretical viewpoint, and to explore some of its most obvious implications.

We have argued that optimal approximation in statistics necessarily results from maximizing the expected value of a *proper* scoring rule, and we have characterized the conditions under which this reduces to *minimizing the logarithmic divergence*. Further work is necessary to characterize precisely those situations where other proper scoring rules are appropriate. We believe, however, that the systematic exploitation of the 'principle' of minimizing the logarithmic divergence in the myriad statistical problems where approximations are used will prove to be rewarding.

REFERENCES

Azcel, J. and Pfanzagl, J., 1966, Remarks on the measurement of subjective probability and information, *Metrika*, 11:91-105.

Bernardo, J.M., 1979a, Expected information as expected utility, *Ann. Statist.*, 7:686-690.

Bernardo, J.M., 1979b, Reference posterior distributions for Bayesian inference, *J. Roy. Statist. Soc. B*, 41:113-147 (with discussion).

Bernardo, J.M., 1986, Robustness and reference distributions. Invited paper at the *International Workshop on Bayesian Statistics*, Stresa.

Brier, G.W., 1950, Verification of forecasts expressed in terms of probability, *Month. Weather Rev.*, 78:1-3.

DeFinetti, B., 1962, Does it make sense to speak of good probability appraisers?, in: "*The Scientist Speculates: an Anthology of Partly Baked Ideas*", I.J. Good, ed., 357-364, Basic Books, New York.

Good, I.J., 1952, Rational decisions. *J. Roy. Statist. Soc. B*, 14:107-114.

Good, I.J., 1966, A derivation of the probabilistic explanation of information. *J. Roy. Statist. Soc. B*, 28: 578-581.

Good, I.J., 1971, Discussion of a paper by R.J. Buehler, in: "*Foundations of Statistical Inference*", V.P. Godambe and D.A. Sprott eds., 337-339, Holt, Rinehart and Winston, Toronto.

Kullback, S., 1959, "*Information Theory and Statistics*", Dover, New York (reprinted, 1978)

Lindley, D.V., 1956, On a measure of the information provided by an experiment, *Ann. Math. Statist.*, 27:986-1005.

McCarthy, J., 1956, Measurements of the value of information, *Proc. Nat. Acad. Sci. USA*, 42:654-655.

Savage, L.J., 1971, Elicitation of personal probabilities and expectations, *J. Amer. Statist. Assoc.*, 66: 783-801.

Shannon, C.E., 1948, A mathematical theory of communication, *Bell System Tech. J.*, 27:379-423, 623-656.

# RESTRICTED BAYES ESTIMATES FOR BINOMIAL PARAMETERS

James D. Broffitt*
University of Iowa
Iowa City, Iowa

## ABSTRACT

Let $\underset{\sim}{\theta} = (\theta_1,\cdots,\theta_k)$ be the parameters for k independent binomial random variables. We wish to estimate $\underset{\sim}{\theta}$ under the restriction $\underset{\sim}{\theta} \in R$ where R is a k-dimensional subset of the full parameter space $\{\underset{\sim}{\theta};\ 0 \le \theta_i \le 1,\ i = 1,\cdots,k\}$. Bayes estimators (means of posteriors) are developed for $\underset{\sim}{\theta}$ which correspond to prior distributions that assign probability one to the set R. Since the support of the resulting posterior is R, the posterior mean will be in R if R is a convex set. A bioassay example is given where the parameters are assumed to be increasing, or increasing and S-shaped.

## INTRODUCTION

In many estimation problems it may be *a priori* assumed that the parameters satisfy certain relationships. For example, in a bioassay experiment where $\theta_1,\cdots,\theta_k$ are probabilities of death at increasing dosage levels of a certain toxin, we may safely assume that $\theta_1 \le \cdots \le \theta_k$. Maximum likelihood estimation of $\underset{\sim}{\theta}$

---

*This research was partially supported by an NSERC grant from the Canadian government while the author was a visiting professor in the Department of Statistical and Actuarial Sciences, University of Western Ontario, London, Ontario.

under this restriction, commonly called isotonic estimation, has been well researched, and the estimates are often computed using the pool-adjacent-violators algorithm, see Barlow et al. (1972).

Restrictions of a more complex nature understandably lead to more difficult computational problems. Schmoyer (1984) computed the "sigmoid" mle of $\underset{\sim}{\theta}$ for a set of bioassay data. The sigmoid restriction assumes the parameters are increasing convex to the left of a certain point, and increasing concave to the right of that point. Such parameters will be called S-shaped.

A small amount of research has appeared on restricted Bayes estimation. Smith (1977) developed the Bayes estimator of $\theta_k$ under the isotonic assumption $\theta_1 \le \cdots \le \theta_k$. In his application, $\underset{\sim}{\theta}$ represented the reliability of a system measured at different stages of development. Broffitt (1984, 1986) found isotonic Bayes estimators for a fairly general model that were useful in estimating mortality rates. Sedransk et al. (1986) used importance sampling to compute the restricted Bayes estimate of the mean of a finite population. Their restriction specified that the population proportions be unimodal, i.e., $\theta_1 \le \cdots \le \theta_t \ge \theta_{t+1} \ge \cdots \ge \theta_k$.

In this paper attention is centered on the binomial data model, and Bayes estimators are developed under a general restriction. The main difficulty in applications is the numerical computation of the estimates. Direct calculation is used for the isotonic restriction while importance sampling is employed to compute the Bayes estimates under the S-shaped restriction. These techniques are applied to the bioassay data studied by Schmoyer.

## NOTATION

Throughout this paper $h(x|a,b)$ will denote the beta pdf with parameters $a$ and $b$, and $h(\underset{\sim}{x}|\underset{\sim}{a},\underset{\sim}{b})$ will denote $\prod_{i=1}^{k} h(x_i|a_i,b_i)$ where $\underset{\sim}{x} = (x_1,\cdots,x_k)$, $\underset{\sim}{a} = (a_1,\cdots,a_k)$, and $\underset{\sim}{b} = (b_1,\cdots,b_k)$. We will also use $\underset{\sim}{u}^{(i)}$ to represent a $k$-dimensional vector with a one in the $i^{th}$ position and zeros elsewhere.

Let $z_i$ be the observed value of a binomial random variable with parameters $n_i, \theta_i$, $i = 1, \cdots, k$. The likelihood function is

$$L(\underset{\sim}{\theta}) \propto \prod_{i=1}^{k} \theta_i^{z_i}(1-\theta_i)^{n_i-z_i}, \quad \underset{\sim}{\theta} \in R \tag{1}$$

where $R$ is a k-dimensional subset of $\Omega = \{\underset{\sim}{\theta};\ 0 \leq \theta_i \leq 1, i = 1, \cdots, k\}$.

Since we believe $\underset{\sim}{\theta} \in R$, we should select a prior distribution that assigns probability one to $R$. Let $Y_i$ have pdf $h(\cdot|\alpha_i, \beta_i)$ and let $Y_1, \cdots, Y_k$ be independent. The prior distribution is specified by $\underset{\sim}{\theta} \overset{D}{=} (\underset{\sim}{Y}|\underset{\sim}{Y} \in R)$, and accordingly the prior pdf is

$$\text{prior}(\underset{\sim}{\theta}) = h(\underset{\sim}{\theta}|\underset{\sim}{\alpha}, \underset{\sim}{\beta})/\rho(R), \quad \underset{\sim}{\theta} \in R \tag{2}$$

where $\rho(R) = P[\underset{\sim}{Y} \in R]$.

Combining (1) and (2) provides the posterior pdf,

$$\text{post}(\underset{\sim}{\theta}) = h(\underset{\sim}{\theta}|\underset{\sim}{a}, \underset{\sim}{b})/p(R), \quad \underset{\sim}{\theta} \in R, \tag{3}$$

where $a_i = \alpha_i + z_i$, $b_i = \beta_i + n_i - z_i$, and $p(R) = \int_R h(\underset{\sim}{x}|\underset{\sim}{a}, \underset{\sim}{b})d\underset{\sim}{x}$. From (3) it is clear that $(\underset{\sim}{\theta}|\underset{\sim}{z}) \overset{D}{=} (\underset{\sim}{X}|\underset{\sim}{X} \in R)$ where $X_i$ has pdf $h(\cdot|a_i, b_i)$ and $X_1, \cdots, X_k$ are independent. This demonstrates that the prior in (2) is conjugate.

Denote the posterior mean by $\underset{\sim}{\theta}^B(R) = (\theta_1^B(R), \cdots, \theta_k^B(R))$. Then;

$$\theta_i^B(R) = \int_R x_i h(\underset{\sim}{x}|\underset{\sim}{a}, \underset{\sim}{b})d\underset{\sim}{x}/p(R). \tag{4}$$

Since $xh(x|a,b) = [a/(a+b)]h(x|a+1, b)$, (4) reduces to

$$\theta_i^B(R) = \theta_i^B\, p^{(i)}(R)/p(R), \tag{5}$$

where $\theta_i^B = \theta_i^B(\Omega) = a_i/(a_i+b_i)$ is the unrestricted Bayes estimator of $\theta_i$, $\underset{\sim}{a}^{(i)} = \underset{\sim}{a} + \underset{\sim}{u}^{(i)}$ and $p^{(i)}(R) = \int_R h(\underset{\sim}{x}|\underset{\sim}{a}^{(i)}, \underset{\sim}{b})d\underset{\sim}{x}$.

The fundamental result given in (5) expresses the restricted Bayes estimator in a seemingly simple form. In applications, $p(R)$, and consequently $p^{(i)}(R)$, can be quite difficult to compute. An inspection of (5) provides no apparent indication that $\underset{\sim}{\theta}^B(R) \in R$; however, since the posterior distribution has support $R$, it follows that $\underset{\sim}{\theta}^B(R) \in R$ as long as $R$ is a convex set.

There may be a natural partition of $R$, i.e., $R = \bigcup_{t=1}^{m} R_t$ where the k-dimensional Lebesgue measure of $R_s \cap R_t$ is 0 if $s \neq t$. In this case it may be desirable to assign prior probabilities for each subset $R_t$, and to use different prior parameters over different subsets. Let $\pi_t$ be the prior probability of $R_t$, i.e., $\pi_t = P[\underset{\sim}{\theta} \in R_t]$, and let the pdf of $\underset{\sim}{\theta}$, conditioned on $\underset{\sim}{\theta} \in R_t$, be

$$\begin{aligned} \text{prior}(\underset{\sim}{\theta}|R_t) &\propto \prod_{i=1}^{k} h(\theta_i|\alpha_{it},\beta_{it}), \quad \underset{\sim}{\theta} \in R_t \\ &= I(\underset{\sim}{\theta} \in R_t)h(\underset{\sim}{\theta}|\underset{\sim}{\alpha}_t,\underset{\sim}{\beta}_t)/\rho_t(R_t), \end{aligned}$$

where $I(\cdot)$ is the indicator function, $\underset{\sim}{\alpha}_t = (\alpha_{1t},\cdots,\alpha_{kt})$, $\underset{\sim}{\beta}_t = (\beta_{1t},\cdots,\beta_{kt})$, and

$$\rho_t(R_t) = \int_{R_t} h(\underset{\sim}{x}|\underset{\sim}{\alpha}_t,\underset{\sim}{\beta}_t)d\underset{\sim}{x}.$$

The subscript on $\rho$ indicates that the prior parameters $\underset{\sim}{\alpha}_t$ and $\underset{\sim}{\beta}_t$ depend on t. Then

$$\begin{aligned} \text{prior}(\underset{\sim}{\theta}) &= \sum_{t=1}^{m} \pi_t \, \text{prior}(\underset{\sim}{\theta}|R_t) \\ &= \sum_{t=1}^{m} I(\underset{\sim}{\theta} \in R_t)\pi_t h(\underset{\sim}{\theta}|\underset{\sim}{\alpha}_t,\underset{\sim}{\beta}_t)/\rho_t(R_t). \end{aligned} \tag{6}$$

Combining (1) and (6) we have

$$\text{post}(\underset{\sim}{\theta}) \propto \sum_{t=1}^{m} I(\underset{\sim}{\theta} \in R_t)\pi_t c_t h(\underset{\sim}{\theta}|\underset{\sim}{a}_t,\underset{\sim}{b}_t)/\rho_t(R_t), \tag{7}$$

where $a_{it} = \alpha_{it}+z_i$, $b_{it} = \beta_{it}+n_i-z_i$, $\underset{\sim}{a}_t = (a_{1t},\cdots,a_{kt})$, $\underset{\sim}{b}_t = (b_{1t},\cdots,b_{kt})$, and

$$c_t = \prod_{i=1}^{k} \frac{\Gamma(\alpha_{it}+\beta_{it})\Gamma(a_{it})\Gamma(b_{it})}{\Gamma(\alpha_{it})\Gamma(\beta_{it})\Gamma(a_{it}+b_{it})}.$$

Therefore

$$\text{post}(\underset{\sim}{\theta}) = \frac{\sum_{t=1}^{m} I(\underset{\sim}{\theta} \in R_t)\pi_t c_t h(\underset{\sim}{\theta}|\underset{\sim}{a}_t,\underset{\sim}{b}_t)/\rho_t(R_t)}{\sum_{t=1}^{m} \pi_t c_t p_t(R_t)/\rho_t(R_t)}, \tag{8}$$

where $p_t(R_t) = \int_{R_t} h(\underset{\sim}{x}|\underset{\sim}{a}_t, \underset{\sim}{b}_t)d\underset{\sim}{x}$. The posterior mean of $\theta_i$ is

$$\theta_i^B(R) = \frac{\sum_{t=1}^{m} \pi_t c_t \theta_{it}^B \, p_t^{(i)}(R_t)/\rho_t(R_t)}{\sum_{t=1}^{m} \pi_t c_t p_t(R_t)/\rho_t(R_t)}, \tag{9}$$

where $\theta_{it}^B = a_{it}/(a_{it}+b_{it})$, $\underset{\sim}{a}_t^{(i)} = \underset{\sim}{a}_t + \underset{\sim}{u}^{(i)}$, and $p_t^{(i)}(R_t) = \int_{R_t} h(\underset{\sim}{x}|\underset{\sim}{a}_t^{(i)}, \underset{\sim}{b}_t)d\underset{\sim}{x}$. From (5), $\theta_{it}^B(R_t)p_t(R_t) = \theta_{it}^B \, p_t^{(i)}(R_t)$, which, when substituted into (9), yields

$$\theta_i^B(R) = \sum_{t=1}^{m} w_t \theta_{it}^B(R_t), \tag{10}$$

where $w_t = \dfrac{\pi_t c_t p_t(R_t)/\rho(R_t)}{\sum_{t=1}^{m} \pi_t c_t p_t(R_t)/\rho_t(R_t)}$. This demonstrates that the Bayes estimator $\theta_i^B(R)$ is a weighted average of the Bayes estimators $\theta_{it}^B(R_t)$, $t = 1,\cdots,m$, where the weight $w_t$ equals $P[\underset{\sim}{\theta} \in R_t|\underset{\sim}{z}]$, the posterior probabilty of $R_t$. Either (9) or (10) may be useful for computing $\theta_i^B(R)$.

If we assume $\alpha_{it} = \alpha_i$, $\beta_{it} = \beta_i$, $i = 1,\cdots,k$, $t = 1,\cdots,m$, so that the same set of prior parameters is used for each partition, then the subscript $t$ may be removed from $c_t$, $\rho_t$, $p_t$ and $\theta_{it}$. Then (9), e.g., becomes

$$\theta_i^B(R) = \theta_i^B \frac{\sum_{t=1}^{m} \pi_t p^{(i)}(R_t)/\rho(R_t)}{\sum_{t=1}^{m} \pi_t p(R_t)/\rho(R_t)}. \tag{11}$$

We close this section with the remark that posterior second moments are easily obtained in a similar manner. In particular, under the assumptions that led to (5),

$$E(\theta_i^2|\underset{\sim}{z}) = \left[\frac{a_i}{a_i+b_i}\right]\left[\frac{a_i+1}{a_i+b_i+1}\right]\frac{p^{(ii)}(R)}{p(R)},$$

where $\underset{\sim}{a}^{(ii)} = \underset{\sim}{a} + 2\underset{\sim}{u}^{(i)}$ and $p^{(ii)}(R) = \int_R h(\underset{\sim}{x}|\underset{\sim}{a}^{(ii)}, \underset{\sim}{b})d\underset{\sim}{x}$. Also, with $R$ in partition form and $\alpha_{it} = \alpha_i$, $\beta_{it} = \beta_i$,

$$E(\theta_i^2|\underset{\sim}{z}) = \left[\frac{a_i}{a_i+b_i}\right]\left[\frac{a_i+1}{a_i+b_i+1}\right]\frac{\sum_{t=1}^{m} \pi_t p^{(ii)}(R_t)/\rho(R_t)}{\sum_{t=1}^{m} \pi_t p(R_t)/\rho(R_t)}.$$

## ISOTONIC RESTRICTION

When $R = \{\underset{\sim}{\theta}: 0 \le \theta_1 \le \cdots \le \theta_k \le 1\}$, the Bayes estimate $\theta_i^B(R)$ may be computed using (5). This requires the computation of $p(R) = P[X_1 \le \cdots \le X_k]$ (and the similar probability $p^{(i)}(R)$) which may be done using the following result.

Let $X_i$ have pdf $h(\cdot|a_i, b_i)$, $X_1, \cdots, X_k$ be independent, $a_2, \cdots, a_k$ be integers, and $c_j = \sum_{n=j+1}^{k} (a_n + b_n - 1)$ for $j = 1, \cdots, k-1$. Then $P[X_1 \le \cdots \le X_k] =$

$$\sum_{i_{k-1}=0}^{a_k-1} f_{k-1}(i_{k-1}) \sum_{i_{k-2}=0}^{i_{k-1}+a_{k-1}-1} f_{k-2}(i_{k-2}) \cdots \sum_{i_1=0}^{i_2+a_2-1} f_1(i_1), \qquad (12)$$

where

$$f_j(i) = \frac{\Gamma(i+a_j)\Gamma(b_j+c_j-i)\Gamma(c_j+1)\Gamma(a_j+b_j)}{\Gamma(i+1)\Gamma(a_j)\Gamma(c_j+1-i)\Gamma(b_j)\Gamma(a_j+b_j+c_j)} .$$

The proof of (12) is similar to that of Corollary 2.1 in Broffitt (1984).

## S-SHAPED RESTRICTION

Suppose $\theta$ is a function of an independent variable $d$. Although numerous examples are possible, in bioassay $d$ refers to dosage level and $\theta(d)$ is the corresponding probability of death (or whatever event is being recorded). In practical examples it may or may not be appropriate to assume $\theta(0) = 0$. Schmoyer (1984) assumed $\theta(0) = 0$, and accordingly for the presentation in this section and the example to follow, we shall assume $\theta(0) = 0$. If this assumption is not desired, a slight modification is necessary, which is given in the appendix.

Without loss of generality let $0 = d_0 < d_1 < \cdots < d_k$, and let $\theta_i = \theta(d_i)$, $i = 0, \cdots, k$. Also let

$$s_i = (\theta_i - \theta_{i-1})/(d_i - d_{i-1}) , \qquad i = 1, \cdots, k,$$

and define

$$R_1 = \{\underset{\sim}{\theta}; s_1 \ge \cdots \ge s_k \ge 0\},$$

$$R_t = \{\underset{\sim}{\theta};\ 0 \le s_1 \le \cdots \le s_t \ge \cdots \ge s_k \ge 0\}, \quad t = 2,\cdots,k-1,$$

and

$$R_k = \{\underset{\sim}{\theta};\ 0 \le s_1 \le \cdots \le s_k\}.$$

Then $\underset{\sim}{\theta}$ is said to be S-shaped if $\underset{\sim}{\theta} \in \bigcup_{t=1}^{k} R_t$. If $\underset{\sim}{\theta}$ is S-shaped its elements must be nondecreasing and either convex ($R_k$), concave ($R_1$), or convex to the left and concave to the right ($R_2,\cdots,R_{k-1}$).

Unfortunately, $R$ is not a convex set (if $k \ge 3$). For example, let $k = 3$, $d_i - d_{i-1} = 1$, $i = 1,2,3$, $\underset{\sim}{\theta}_1 = (0.30, 0.44, 0.56)$, and $\underset{\sim}{\theta}_2 = (0.14, 0.30, 0.60)$. Then $\underset{\sim}{\theta} \in R_1$ and $\underset{\sim}{\theta}_2 \in R_3$, but $.5(\underset{\sim}{\theta}_1 + \underset{\sim}{\theta}_2) \notin \bigcup_1^3 R_t$. Since the posterior pdf has support $R$, which is not convex, the posterior mean, $\underset{\sim}{\theta}^B(R)$, need not be in $R$. This is a bit unsettling and should it happen, $\underset{\sim}{\theta}$ must not be estimated by the posterior mean. An alternative would be to subtract from $R$ that subset with the smallest posterior probability, and then recompute the posterior mean of $\underset{\sim}{\theta}$. This process could be repeated if necessary.

In the example to follow, $\theta_i^B(R)$ is computed using (11), which requires $\rho(R_t)$, $p(R_t)$, and $p^{(i)}(R_t)$. Since these three probabilities differ only in the parameters used in the beta densities, the process of computation is the same for each. Thus for simplicity our discussion will focus on $p(R_t)$.

Because of the complexity of $R_t$, some form of Monte Carlo is suggested as the computational method, but since $p(R_t)$ is extremely small, a rejection technique would be highly inefficient. These considerations suggest importance sampling as a viable solution.

The technique of importance sampling stems from the following observation: Let $\underset{\sim}{X}$ be a random vector with support $R_t$ and pdf $f(\cdot)$, and for simplicity let $h(\underset{\sim}{x}) = h(\underset{\sim}{x}|\underset{\sim}{a},\underset{\sim}{b})$. Then

$$\begin{aligned} p(R_t) &= \int_{R_t} h(\underset{\sim}{x})d\underset{\sim}{x} \\ &= \int_{R_t} [h(\underset{\sim}{x})/f(\underset{\sim}{x})]f(\underset{\sim}{x})d\underset{\sim}{x} \\ &= E_{\underset{\sim}{X}}[h(\underset{\sim}{X})/f(\underset{\sim}{X})]. \end{aligned}$$

The procedure is to generate $n$ independent observations on $\underset{\sim}{X}$,

$\underset{\sim}{x}_1, \cdots, \underset{\sim}{x}_n$ and approximate $p(R_t)$ by

$$\hat{p}(R_t) = n^{-1} \sum_{i=1}^{n} h(\underset{\sim}{x}_i)/f(\underset{\sim}{x}_i),$$

which is clearly an unbiased estimate. If possible, $f(\cdot)$ should be chosen so that $h(\underset{\sim}{X})/f(\underset{\sim}{X})$ has a minimal variance. Our algorithm for generating $\underset{\sim}{x}$ is based on the one given by Sedransk et al. (1986), and is detailed below:

1. Generate $U_1, \cdots, U_{k+1}$ iid with pdf $g(u) = e^{-u}$, $u > 0$.

2. Find $i^*$ $(1 \le i^* \le k)$ so that $U_{i^*} \ge U_i$ for $i = 1, \cdots, k$.

3. Exchange $U_{i^*}$ and $U_t$.

4. Sort $U_1, \cdots, U_t$ in increasing order.

5. Sort $U_t, \cdots, U_k$ in decreasing order.

6. Let $U_{(1)} \le \cdots \le U_{(t)} \ge U_{(t+1)} \ge \cdots \ge U_{(k)}, U_{k+1}$ be the result of steps (1) to (5).

7. Let $X_i = \dfrac{r_1 U_{(1)} + \cdots + r_i U_{(i)}}{r_1 U_{(1)} + \cdots + r_k U_{(k)} + \bar{r} U_{k+1}}$, $i = 1, \cdots, k$,

   where $r_i = d_i - d_{i-1}$ and $\bar{r} = \sum_{i=1}^{k} r_i/k$.

It can be shown that the resulting vector $\underset{\sim}{X} = (X_1, \cdots, X_k)$ has pdf

$$f(\underset{\sim}{x}) = \frac{(t-1)!(k-t)!k!k}{r_1 \cdots r_k \bar{r}} \left[\frac{x_1}{r_1} + \frac{x_2 - x_1}{r_2} + \cdots + \frac{x_k - x_{k-1}}{r_k} + \frac{1 - x_k}{\bar{r}}\right]^{-(k+1)}, \quad \underset{\sim}{x} \in R_t.$$

Notice that if $r_1 = \cdots = r_k$, i.e., the $d$'s are evenly spaced, then $\underset{\sim}{X}$ has the uniform density, $f(\underset{\sim}{x}) = (t-1)!(k-t)!k!k$. In any case $h(\underset{\sim}{x})/f(\underset{\sim}{x})$ is bounded, so the variance of $\hat{p}(R_t)$ is finite and may be made arbitrarily small by taking $n$ large enough.

## EXAMPLE

Table 1 lists the bioassay data and the resulting estimates. The superscripts M and B refer to maximum likelihood and Bayes respectively, while the arguments I and S denote the isotonic and S-shaped restrictions. Thus

$\theta^M = z/n$ is the unrestricted mle, $\theta^M(I)$ was obtained via the pool-adjacent-violaters algorithm, and $\theta^M(S)$ was taken from Schmoyer's paper. The Bayes estimates $\theta^B(I)$ were computed using (5) and (12) with the prior parameters $\alpha_i = \beta_i = 1$, $i = 1,\cdots,k$. These same values of $\alpha_i$ and $\beta_i$ together with $\pi_i = 1/k$, $i = 1,\cdots,k$ were used to compute $\theta^B(S)$. For this selection of prior parameters, $\underset{\sim}{\theta}^B(S) \in R_4$, and the approximation of the posterior probabilities $P[\underset{\sim}{\theta} \in R_t|\underset{\sim}{z}]$, $t = 1,\cdots,k$ are 0.00, 0.00, 0.00, 0.58, 0.29, 0.12, 0.00, and 0.00, respectively. Plots of these five estimates are displayed in figures 1 and 2.

Table 1. Data and estimates for the bioassay example.

| $d_i$ | $n_i$ | $z_i$ | $\theta_i^M$ | $\theta_i^M(I)$ | $\theta_i^M(S)$ | $\theta_i^B(I)$ | $\theta_i^B(S)$ |
|---|---|---|---|---|---|---|---|
| 8 | 30 | 0 | 0.000 | 0.000 | 0.000 | 0.016 | 0.009 |
| 16 | 40 | 1 | 0.025 | 0.025 | 0.025 | 0.043 | 0.036 |
| 24 | 40 | 2 | 0.050 | 0.050 | 0.050 | 0.088 | 0.105 |
| 28 | 10 | 5 | 0.500 | 0.425 | 0.390 | 0.357 | 0.299 |
| 32 | 30 | 12 | 0.400 | 0.425 | 0.448 | 0.456 | 0.439 |
| 48 | 20 | 16 | 0.800 | 0.733 | 0.677 | 0.680 | 0.715 |
| 64 | 10 | 6 | 0.600 | 0.733 | 0.892 | 0.752 | 0.838 |
| 72 | 10 | 10 | 1.000 | 1.000 | 1.000 | 0.930 | 0.871 |

To compute $\theta^B(S)$ we used (11) together with the importance sampling algorithm described in the preceding section. Ten sets of 1000 observations on $\underset{\sim}{X}$ were generated. The quantities $\rho(R_t)$, $p(R_t)$, and $p^{(i)}(R_t)$ were computed for each of these sets. Using (11) this provided 10 unbiased approximations of each estimate $\theta_i^B(S)$, $i = 1,\cdots,k$, from which means and variances were computed. The means were used as the final S-shaped Bayes estimates and are reported in table 1.

The variances, denoted by $SV_i$, provided a check on the accuracy of the importance sampling procedure. These are given in table 2 along with twice the corresponding standard errors, $2SE_i = 2(SV_i/10)^{1/2}$. Since the largest of these is 0.0096, we are reasonably sure that the differences between the computed estimates and the exact posterior means are less than 0.01.

Table 2 also contains the posterior variances of $\theta_i^B$,

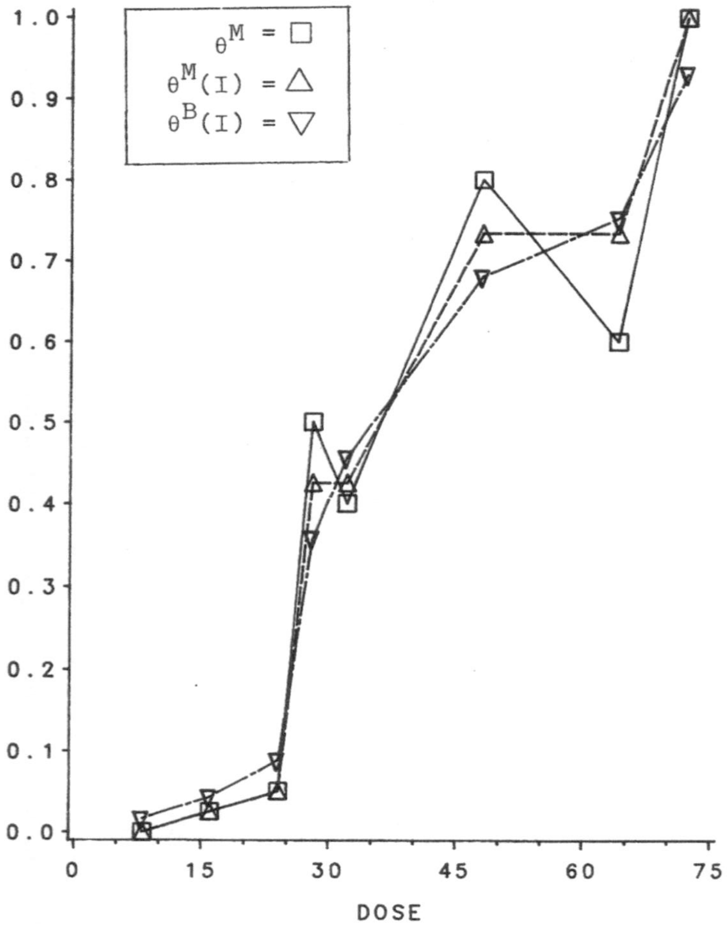


Figure 1. Plots of mle and isotonic estimates.

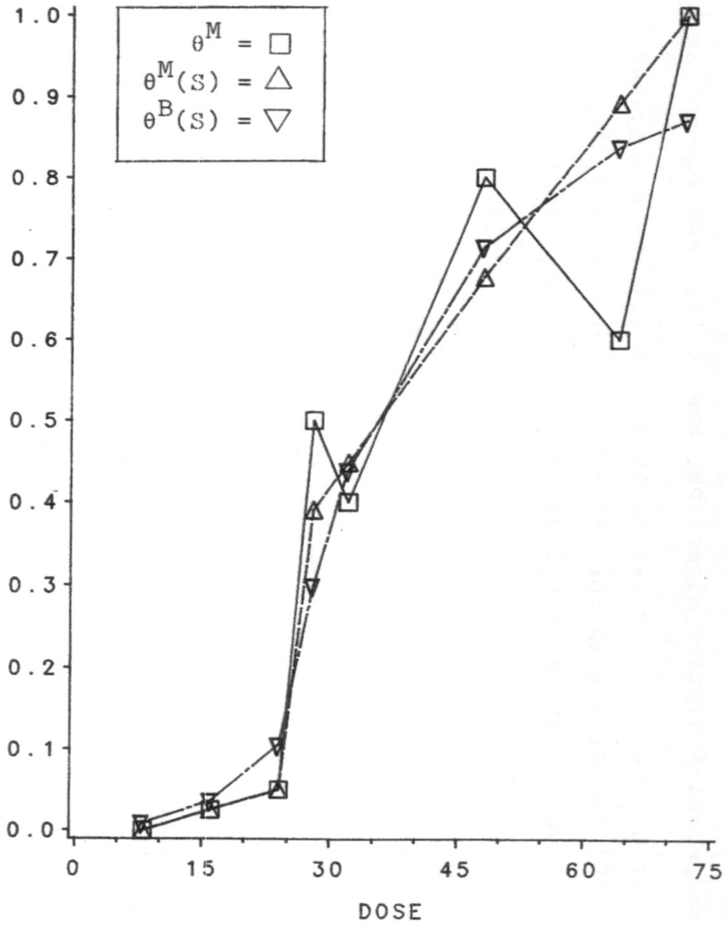


Figure 2. Plots of mle and S-shaped estimates.

$\theta_i^B(I)$, and $\theta_i^B(S)$, denoted by $V_i$, $V_i(I)$, and $V_i(S)$, respectively. Of course $V_i = a_i b_i/[(a_i+b_i)^2(a_i+b_i+1)]$. $V_i(I)$ was computed by exact formula, and $V_i(S)$ was approximated by importance sampling. Finally the ratios $V_i/V_i(I)$ and $V_i/V_i(S)$ furnish measures of the efficiency gained by imposing the isotonic or S-shaped restrictions.

Table 2. Variances and Efficiencies

| $SV\times10^4$ | 2SE | $V\times10^4$ | $V(I)\times10^4$ | $V(S)\times10^4$ | V/V(I) | V/V(S) |
|---|---|---|---|---|---|---|
| 0.009 | 0.0006 | 9.17 | 1.94 | 0.53 | 4.7 | 17.3 |
| 0.043 | 0.0013 | 10.55 | 5.15 | 3.25 | 2.0 | 3.2 |
| 0.434 | 0.0042 | 15.43 | 14.43 | 15.60 | 1.1 | 1.0 |
| 2.324 | 0.0096 | 192.31 | 77.17 | 75.27 | 2.5 | 2.6 |
| 1.995 | 0.0089 | 73.09 | 57.07 | 71.41 | 1.3 | 1.0 |
| 0.294 | 0.0034 | 76.36 | 62.08 | 38.47 | 1.2 | 2.0 |
| 0.333 | 0.0036 | 186.97 | 57.71 | 34.11 | 3.2 | 5.5 |
| 0.464 | 0.0043 | 58.76 | 34.80 | 46.18 | 1.7 | 1.3 |

## REFERENCES


Barlow, R.E., Bartholomew, D.J., Bremner, J.M., and Brunk, H.D., 1972, Statistical Inference Under Order Restrictions, Wiley, New York.

Broffitt, J.D., 1984, "A Bayes estimator for ordered parameters and isotonic Bayesian graduation," Scand. Actuarial J., 231-247.

Broffitt, J.D., 1986, "Isotonic Bayesian graduation with an additive prior," Actuarial Science-Festschrift in Honour of Professor V.M. Joshi's 70th Birthday, vol.6, I.B. MacNeill and G.J. Umphrey, eds., 19-40.

Schmoyer, R.L. 1984, "Sigmoidally constrained maximum likelihood estimation in quantal bioassay," J. Amer. Statist. Assoc. 79:448-453.

Sedransk, J., Monahan, J. and Chiu, H.Y., 1986, "Bayesian estimation of finite population parameters in categorical data models incorporating order restrictions," J. Royal Statist. Soc. Ser. B, 47:519-527.

Smith, A.M.F., 1977, "A Bayesian note on reliability growth during a development testing program," IEEE Trans. Reliability, R-26:346-347.

## APPENDIX

If the assumption $\theta(0) = 0$ is not made, then $\underset{\sim}{\theta}$ is said to be S-shaped if $\underset{\sim}{\theta} \in \bigcup_{t=2}^{k} R_t$, where

$$R_2 = \{\underset{\sim}{\theta}:\ s_2 \geq \cdots \geq s_k \geq 0\},$$

$$R_t = \{\underset{\sim}{\theta}:\ 0 \leq s_2 \leq \cdots \leq s_t \geq \cdots \geq s_k \geq 0\}, \quad t = 3,\cdots,k-1,$$

and

$$R_k = \{\underset{\sim}{\theta}:\ 0 \leq s_2 \leq \cdots \leq s_k\}.$$

The algorithm for generating an observation with support $R_t$ is as follows:

1. Generate $U_1,\cdots,U_{k+1}$ iid with pdf $g(u) = e^{-u}$, $u > 0$.
2. Find $i^*$ $(2 \leq i^* \leq k)$ so that $U_{i^*} \geq U_i$, $i = 2,\cdots,k$.
3. Exchange $U_{i^*}$ and $U_t$.
4. Sort $U_2,\cdots,U_t$ in increasing order.
5. Sort $U_t,\cdots,U_k$ in decreasing order.
6. Let $U_1,\ U_{(2)} \leq \cdots \leq U_{(t)} \geq \cdots \geq U_{(k)},\ U_{k+1}$ be the result of steps (1) to (5).
7. Let $X_i = \dfrac{\bar{r}U_1+r_2U_{(2)}+\cdots+r_iU_{(i)}}{\bar{r}U_1+r_2U_{(2)}+\cdots+r_kU_{(k)}+\bar{r}U_{k+1}}$, $i = 1,\cdots,k$,

   where $r_i = d_i-d_{i-1}$ and $\bar{r} = \sum_{i=2}^{k} r_i/(k-1)$.

The resulting vector $\underset{\sim}{X} = (X_1,\cdots,X_k)$ has pdf

$$f(\underset{\sim}{x}) = \frac{(t-2)!(k-t)!k!(k-1)}{\bar{r}r_2\cdots r_k\bar{r}}\left[\frac{x_1}{\bar{r}} + \frac{x_2-x_1}{r_2} + \cdots + \frac{x_k-x_{k-1}}{r_k} + \frac{1-x_k}{\bar{r}}\right]^{-(k+1)},$$

$$\underset{\sim}{x} \in R_t.$$

BAYESIAN PREVISION OF TIME SERIES

BY TRANSFER FUNCTION MODELS (°)

Francesco Carlucci

University of Rome
" La Sapienza "
Italy

Gino Zornitta

University of Venice
" Cà Foscari "
Italy

## 1. Introduction

Given the time series $\{y_t\}$, $t\in T\equiv\{1,2,\ldots,n\}$, which is assumed to be caused by the series $\{x_t\}$, $t\in T$, according to the transfer function model

(1.1) $y_t = \delta^{-1}(B)\omega(B)x_t + \phi^{-1}(B)\theta(B)u_t$

where B is the usual back-shift operator, and $\omega(B)=\omega_0-\omega_1B+\ldots-\omega_sB^s$, $\delta(B)=1-\delta_1B+\ldots-\delta_rB^r$, $\theta(B)=1-\theta_1B+\ldots-\theta_qB^q$, $\phi(B)=1-\phi_1B+\ldots-\phi_pB^p$ where $\{x_t\}$ is assumed to be generated by the ARMA process

(1.2) $\phi_x(B)x_t=\theta_x(B)a_t$

with $\phi_x(B)=1-\phi_{x1}B+\ldots-\phi_{xp_x}B^{p_x}$, $\theta_x(B)=1-\theta_{x1}B+\ldots-\theta_{xq_x}B^{q_x}$ and given a set H information regarding the unknown values of $\tilde{y}_{n+\tau}$ (°°) and $\tilde{x}_{n+\tau}$, $\tau\in T^*\equiv\{1,2,\ldots,m\}$, in this work previsions of these values are determined in de Finetti's sense, so that both the sample $\underline{y}_n=(y_1,y_2,\ldots,y_n)'$, $\underline{x}_n=(x_1,x_2,\ldots,x_n)'$, and the information H are taken into account. In the case of economic time series, this information may concern, for instance, either the causal relationship from the variable $x_t$ to $y_t$, or the period of the business cycle which influences the autoregressive schemes in submodels (1.1) and (1.2), or even the ARMA schemes on residuals. Furthermore, information H may concern the orders (s,r,p,q) and $(p_x,q_x)$ of submodels according to the opinion and experience of the model builder. Taking this information into account, "previsions" in de Finetti's (1974) subjective meaning rather

(°) This paper is due to common efforts of the two authors; nevertheless, Sections 3-4-6 may be attributed to G.Zornitta and Sections 5-7-8 to F.Carlucci. The remaining ones are common. A financial support by Consiglio Nazionale delle Ricerche (CTB N. 83.00075.10) has to be acknowledged.
(°°) A tilde (˜) over a variable denotes that its random character is taken into consideration.

than "projections" in the traditional sense are formulated.

Analytically, by use of a quadratic loss function, such previsions are defined as the mean values $E(\tilde{y}_{n+\tau} | \underline{y}_n, \underline{\underline{x}}_n, H)$, $\tau \in T^*$, of the conditional distributions of $\tilde{y}_{n+\tau}$, $\tau \in T^*$, given $\underline{y}_n$, $\underline{\underline{x}}_n$ and H.

Bayesian estimation in a transfer function model was developed by P. Newbold (1973), but he did not consider the prevision problem. On the other hand, Bayesian projection was studied by Monahan (1983) in the context of ARMA models. So, the present work's objectives could be reached by extending Monahan's results (which produce exact posterior distributions) to the case of transfer function models. But this way is not followed here for two reasons: firstly, posterior distributions are determined by means of non informative priors and therefore they loose de Finetti's subjective meaning; secondly, posterior distributions have an uncommon form and need a numerical cumbersome evaluation to be used. Therefore, in order to determine the previsive distributions, we prefer to utilize prior densities that are fully informative, as developed by F. Carlucci (1977) for a particular class of time series models, even if this procedure forces the likelihood function to be approximated.

The plan of the paper is as follows. Firstly, assumptions used in the work are specified, and then the likelihood function for parameters is determined following Newbold (1973) and supposing the orders of submodels as known. In Section 4 such orders are calculated by means of a Bayesian procedure grounded on probability odds. In order to make this procedure usable, it is necessary to reduce the likelihood function to a known form, and the transformation is performed in Section 5 by use of the usual Gauss-Newton method. In Section 6, posterior marginal distributions for the unknown parameters of the previsive model are determined on the ground of non-informative prior densities. Sections 7 and 8 concern previsions which are evaluated on the base of informative distributions, for the ARMA and the transfer function models, respectively.

## 2. Assumptions

For the model (1.1) - (1.2) let us make the following assumptions:

I - $\{\tilde{u}_t\}$ and $\{\tilde{a}_t\}$ are mutually uncorrelated normally distributed white noise processes, each with zero mean and variances $\sigma_u^2$ and $\sigma_a^2$, respectively.

II - All stationarity and invertibility conditions are satisfied.

III - All required initial values for $x_t$, $t \leq 0$, are known.

IV - Information H affects only the parameter prior density in the following way: If parameters, assumed independent, are referred to the representative model in the sample period, then their prior density is locally uniform. If parameters are referred to the previsive model, then their prior density is the Raiffa-Schlaifer natural conjugate one, in the normal-inverted gamma form; furthermore, parameters in submodels (1.1) and (1.2) are a priori independent.

## 3. Derivation of the likelihood function

In order to calculate the likelihood functions, it is firstly convenient to consider the model (1.1) - (1.2) written, by virtue of Assumption II, in the approximate form

(3.1) $y_t=\Omega(B)x_t+\theta(B)u_t$

(3.2) $x_t=\Theta_x(B)a_t$

where $\delta^{-1}(B)\cdot\omega(B)\simeq\Omega(B)=\Omega_0-\Omega_1B+\ldots-\Omega_kB^k$, $\phi^{-1}(B)\cdot\theta(B)\simeq\Theta(B)=$ $=1-\Theta_1B+\ldots-\Theta_QB^Q$, $\phi_x^{-1}(B)\cdot\theta_x(B)\simeq\Theta_x(B)=1-\Theta_{x1}B+\ldots-\Theta_{xQ_x}B^{Q_x}$ with k,Q, $Q_x$ sufficiently large so to get the desired approximation level.

By following Newbold (1973) and setting $\underline{\underline{u}}^*=(u_{1-Q},\ldots,u_{-1},u_0)'$ $\underline{\underline{u}}=(u^{*\prime},u_1,\ldots,u_n)'$, $\underline{\underline{x}}^*=(x_{1-k},\ldots,x_{-1},x_0)'$ we obtain

(3.3) $\underline{\underline{u}}=\underline{\underline{L}}_1\underline{\underline{y}}_n+\underline{\underline{L}}_2\underline{\underline{x}}_n+\underline{\underline{L}}_3\underline{\underline{x}}^*+\underline{\underline{L}}_4\underline{\underline{u}}^*$

where $\underline{\underline{L}}_1$, $\underline{\underline{L}}_2$, $\underline{\underline{L}}_3$ and $\underline{\underline{L}}_4$ are suitable matrices whose elements are functions of parameters $\underline{\underline{\Omega}}=(\Omega_0,\Omega_1,\ldots,\Omega_k)'$ and $\underline{\underline{\Theta}}=(\Theta_1,\Theta_2,\ldots,\Theta_Q)'$.

By virtue of Assumption I, the density for $\tilde{\underline{\underline{u}}}$ is $N(\underline{\underline{0}},\sigma_u^2\underline{\underline{I}}_{n+Q})$ and since the transformation from $\underline{\underline{y}}_n$ to $\underline{\underline{u}}$ has unitary Jacobian, the joint density for $\tilde{\underline{\underline{y}}}_n$ and $\tilde{\underline{\underline{u}}}^*$ is

(3.4) $p(\underline{\underline{y}}_n,\underline{\underline{u}}^*|\underline{\underline{\Omega}},\underline{\underline{\Theta}},\sigma_u,\underline{\underline{x}}_n,\underline{\underline{x}}^*)=(2\Pi\sigma^2)^{-(n+Q)/2}\exp\{-S_0(\underline{\underline{\Omega}},\underline{\underline{\Theta}})/2\sigma_u^2\}$

where

(3.5) $S_0(\underline{\underline{\Omega}},\underline{\underline{\Theta}})=\underline{\underline{u}}'\underline{\underline{u}}=S(\underline{\underline{\Omega}},\underline{\underline{\Theta}})+(\underline{\underline{u}}^*-\hat{\underline{\underline{u}}}^*)'\underline{\underline{L}}_4'\underline{\underline{L}}_4(\underline{\underline{u}}^*-\hat{\underline{\underline{u}}}^*)$

where $\hat{\underline{\underline{u}}}^*$ is the value for $\underline{\underline{u}}^*$that minimizes $S_0(\underline{\underline{\Omega}},\underline{\underline{\Theta}})$ and $S(\underline{\underline{\Omega}},\underline{\underline{\Theta}})=$ $=\sum_{t=1-Q}^{n}[E(\hat{\underline{\underline{u}}}_t|\underline{\underline{y}}_n,\underline{\underline{x}}_n,\underline{\underline{x}}^*,\underline{\underline{\Omega}},\underline{\underline{\Theta}})]^2$ is the minimum.

By taking (3.5) into account, we can factorize(3.4) into the two following densities

(3.6) $p(\underline{\underline{y}}_n|\underline{\underline{\Omega}},\underline{\underline{\Theta}},\sigma_u,\underline{\underline{x}}_n,\underline{\underline{x}}^*)=(2\Pi\sigma_u^2)^{-n/2}|\underline{\underline{L}}_4'\underline{\underline{L}}_4|^{-1/2}\exp\{-S(\underline{\underline{\Omega}},\underline{\underline{\Theta}})/2\sigma_u^2\}$

(3.7) $p(\underline{\underline{u}}^*|\underline{\underline{y}}_n,\underline{\underline{\Omega}},\underline{\underline{\Theta}},\sigma_u,\underline{\underline{x}}_n,\underline{\underline{x}}^*)=(2\Pi\sigma_u^2)^{-Q/2}|\underline{\underline{L}}_4'\underline{\underline{L}}_4|^{1/2}\exp\{-(\underline{\underline{u}}^*-\hat{\underline{\underline{u}}}^*)'\underline{\underline{L}}_4'\underline{\underline{L}}_4$ $(\underline{\underline{u}}^*-\hat{\underline{\underline{u}}}^*)/2\sigma_u^2\}$

By use of the same argument as above, we can determine the density for $\tilde{\underline{\underline{x}}}_n$. By writing $\underline{\underline{a}}^*=(a_{1-k-Q_x},\ldots,a_{-k})'$, $\underline{\underline{a}}=(\underline{\underline{a}}^*,a_{1-k},\ldots,a_0,\ldots,a_n)'$ in analogy to (3.3) we obtain

(3.8) $\underline{\underline{a}}=\underline{\underline{L}}_5\underline{\underline{x}}_n+\underline{\underline{L}}_6\underline{\underline{x}}^*+\underline{\underline{L}}_7\underline{\underline{a}}^*$

where $\underline{\underline{L}}_5$,$\underline{\underline{L}}_6$,and $\underline{\underline{L}}_7$, are suitable matrices whose elements are functions of parameters $\underline{\underline{\Theta}}_x=(\Theta_{x_1},\Theta_{x_2},\ldots,\Theta_{xQ_x})'$.

By virtue of Assumption I,the density for $\tilde{\underline{\underline{a}}}$ is $N(\underline{\underline{0}},\sigma_a^2\underline{\underline{I}}_{n+k+Q_x})$ and since the transformation from $\underline{\underline{x}}_n$ to $\underline{\underline{a}}$ has unitary Jacobian, the density for $\tilde{\underline{\underline{x}}}_n$ is , by use of the same reasoning as before and of Assumption III,

(3.9) $p(\underline{\underline{x}}_n|\underline{\underline{\Theta}}_x,\sigma_a,\underline{\underline{x}}^*)=(2\Pi\sigma_a^2)^{-n/2}|\underline{\underline{L}}_7'\underline{\underline{L}}_7|^{-1/2}\exp\{-S(\underline{\underline{\Theta}}_x)/2\sigma_a^2\}$

where $S(\underline{\underline{\Theta}}_x)=\sum_{t=1-k-Q_x}^{n}[E(\tilde{a}_t|\underline{\underline{x}}_n,\underline{\underline{x}}^*,\underline{\underline{\Theta}}_x)]^2$.

The same approach as above can be applied to submodels (1.1) and (1.2), obtaining

(3.10) $p(\underline{\underline{y}}_n|\underline{\underline{\beta}},\sigma_u,\underline{\underline{x}}_n,\underline{\underline{x}}^*)=(2\Pi\sigma_u^2)^{-n/2}|\underline{\underline{L}}_4'\underline{\underline{L}}_4|^{-1/2}\exp\{-S(\underline{\underline{\beta}})/2\sigma_u^2\}$

$$(3.11)\quad p(\underline{\underline{x}}_n|\underline{\underline{\gamma}},\underline{\underline{\sigma}}_a,\underline{\underline{x}}^*)=(2\Pi\sigma_a^2)^{-n/2}|\underline{\underline{L}}_7'\underline{\underline{L}}_7|^{-1/2}\exp\{-S(\underline{\underline{\gamma}})/2\sigma_a^2\}$$

where $\underline{\underline{\beta}}=(\beta_1,\beta_2,\ldots,\beta_{s+r+p+q})'=(\underline{\underline{\omega}}',\underline{\underline{\delta}}',\underline{\underline{\phi}}',\underline{\underline{\theta}}')'$ and $\underline{\underline{\gamma}}=(\gamma_1,\gamma_2,\ldots,\gamma_{p_x+q_x})'=(\underline{\underline{\phi}}_x',\underline{\underline{\theta}}_x')'$ are the parameters of submodels (1.1) and (1.2), respectively, and

$$(3.12)\quad S(\underline{\underline{\beta}})=\sum_{t=1-q}^{n}[E(\tilde{u}_t|\underline{\underline{y}}_n,\underline{\underline{x}}_n,\underline{\underline{x}}^*,\underline{\underline{\beta}})]^2$$

$$(3.13)\quad S(\underline{\underline{\gamma}})=\sum_{t=1-s-q_x}^{n}[E(\tilde{a}_t|\underline{\underline{x}}_n,\underline{\underline{x}}^*,\underline{\underline{\gamma}})]^2$$

with $\underline{\underline{x}}^*(x_{1-s},\ldots,x_{-1},x_0)'$ vector of known values ($^0$).

Now, by combining (3.10) and (3.11), we get

$$(3.14)\quad p(\underline{\underline{y}}_n,\underline{\underline{x}}_n|\underline{\underline{\beta}},\underline{\underline{\gamma}},\sigma_u,\sigma_a,\underline{\underline{x}}^*)=p(\underline{\underline{y}}_n|\underline{\underline{\beta}},\sigma_u,\underline{\underline{x}}_n,\underline{\underline{x}}^*)\cdot p(\underline{\underline{x}}_n|\underline{\underline{\gamma}},\sigma_a,\underline{\underline{x}}^*)$$

which is the likelihood function for the parameters of the complete model (1.1) - (1.2).

## 4. Model identification

The determination of the order for the model (1.1)-(1.2) that better fits time series $\{y_t\}$ and $\{x_t\}$ in the sample period, is performed separately for the submodels (1.1) and (1.2), by assuming that probability evaluations are independent. Furthermore, it may be observed that submodel (1.2) is a particular case of (1.1), with $\omega(B)=0$ identically: then, the procedure for the identification of submodel (1.1) can be used even for submodel (1.2).

Such a procedure starts from the consideration of a transfer function model of the least order $(\bar{s},\bar{r},\bar{p},\bar{q})$ and one of the largest $(\bar{\bar{s}},\bar{\bar{r}},\bar{\bar{p}},\bar{\bar{q}})$, with the orders subjectively evaluated so that one can reasonably believe that $\bar{s}<s<\bar{\bar{s}}$, $\bar{r}<r<\bar{\bar{r}}$, $\bar{p}<p<\bar{\bar{p}}$, and $\bar{q}<q<\bar{\bar{q}}$. Each model is denoted by $M_{hijk}$, where indexes correspond to s, r, p, and q, respectively.

The best model is chosen by using the principle of minimizing the prevision of the loss $\tilde{\ell}$

$$(4.1)\quad \min_{h'i'j'k'} E(\tilde{\ell}|\hat{M}_{h'i'j'k'})=\min_{h'i'j'k'}\sum_h\sum_i\sum_j\sum_k \ell(M_{hijk}|\hat{M}_{h'i'j'k'})\cdot p(M_{hijk}|\underline{\underline{y}}_n,\underline{\underline{x}}_n,\underline{\underline{x}}^*,H)$$

where $\bar{s}\leqslant h'\leqslant\bar{\bar{s}}$, $\bar{r}\leqslant i'\leqslant\bar{\bar{r}}$, $\bar{p}\leqslant j'\leqslant\bar{\bar{p}}$, $\bar{q}\leqslant k'\leqslant\bar{\bar{q}}$, and $\hat{M}_{h'i'j'k'}$ indicates the action of choosing the model $M_{h'i'j'k'}$ and $\ell(M_{hijk}|\hat{M}_{h'i'j'k'})$ is the loss associated with action $\hat{M}_{h'i'j'k'}$ when the true model is $M_{hijk}$; finally, $p(M_{hijk}|\underline{\underline{y}}_n,\underline{\underline{x}}_n,\underline{\underline{x}}^*,H)$ is the posterior probability of model $M_{hijk}$.

($^0$) In Newbold (1973) the argument of this point is developed with $\underline{\underline{x}}^*$ vector of unknown values.

Loss functions are subjectively evaluated and the posterior probability for $M_{hijk}$ is given by Bayes' theorem

$$(4.2)\quad p(M_{hijk}|\underline{\underline{y}}_n,\underline{\underline{x}}_n,\underline{\underline{x}}^*,H)\propto p(\underline{\underline{y}}_n|M_{hijk},\underline{\underline{x}}_n,\underline{\underline{x}}^*,H)\cdot p(M_{hijk}|H)$$

for every h, i, j, k, where $p(M_{hijk}|H)$ is the prior probability for model $M_{hijk}$ and $p(\underline{\underline{y}}_n|M_{hijk},\underline{\underline{x}}_n,\underline{\underline{x}}^*,H)$ is the likelihood. Since this function is formally equivalent to the density of $\tilde{\underline{\underline{y}}}_n$ under the hypothesis that $M_{hijk}$ is the true model, with parameters $\underline{\underline{\beta}}_{hijk}=(\beta_1,\beta_2,\ldots,\beta_{h+i+j+k})'$ and $\sigma_{hijk}$, it may be written in the form

$$(4.3)\quad p(\underline{\underline{y}}_n|M_{hijk},\underline{\underline{x}}_n,\underline{\underline{x}}^*,H) = \int p(\underline{\underline{y}}_n|\underline{\underline{\beta}}_{hijk},\sigma_{hijk},\underline{\underline{x}}_n,\underline{\underline{x}}^*)\cdot$$

$$p(\underline{\underline{\beta}}_{hijk},\sigma_{hijk}|\underline{\underline{x}}_n,\underline{\underline{x}}^*,H)\,d\underline{\underline{\beta}}_{hijk}d\sigma_{hijk}\propto\int\sigma^{-1}_{hijk}p(\underline{\underline{y}}_n\,\underline{\underline{\beta}}_{hijk},\sigma_{hijk},\underline{\underline{x}}_n,\underline{\underline{x}}^*)\cdot$$

$$d\underline{\underline{\beta}}_{hijk}\cdot d\sigma_{hijk} = J_{hijk}$$

where $p(\underline{\underline{y}}_n|\underline{\underline{\beta}}_{hijk},\sigma_{hijk},\underline{\underline{x}}_n,\underline{\underline{x}}^*)$ is not conditional on H because random sample does not depend on this information, and $p(\underline{\underline{\beta}}_{hijk},\sigma_{hijk}|\underline{\underline{x}}_n,\underline{\underline{x}}^*,H)$ is the locally uniform prior density for parameters of model $M_{hijk}$, proportional to $\sigma^{-1}_{hijk}$.

Then, the posterior probability of $M_{hijk}$, given by (4.2), is

$$(4.4)\quad p(M_{hijk}|\underline{\underline{y}}_n,\underline{\underline{x}}_n,\underline{\underline{x}}^*,H)\propto p(M_{hijk}|H)\cdot J_{hijk}$$

so that, if $\bar{\rho}_{hijk} = p(M_{hijk}|H)/p(M_{\overline{srpq}}|H)$, for every h,i,j,k,are the prior probability odds on $M_{hijk}$ against $M_{\overline{srpq}}$, subjectively evaluated, the posterior ones are

$$(4.5)\quad \rho_{hijk}= p(M_{hijk}|\underline{\underline{y}}_n,\underline{\underline{x}}_n,\underline{\underline{x}}^*,H)/p(M_{\overline{srpq}}|\underline{\underline{y}}_n,\underline{\underline{x}}_n,\underline{\underline{x}}^*,H)$$

$$\propto\bar{\rho}_{hijk}\cdot J_{hijk}\,/\,J_{\overline{srpq}}$$

for every h,i,j,k. By means of these posterior odds it is straightforward to compute the posterior probabilities $p(M_{hijk}|\underline{\underline{y}}_n,\underline{\underline{x}}_n,\underline{\underline{x}}^*,H)$ and the minimum (4.1).

Unfortunately, the computation of integrals in (4.5) is not an easy task because density (3.10) is not of a known type. Therefore, it has to be approximated, as it will be shown in the next Section.

## 5. Transformation of the sum of squares function and computation of the posterior odds

In order to reduce (3.10) and (3.11) to known density functions, it is convenient to transform the sums of squares (3.12) and (3.13) by using the classical Gauss-Newton procedure based on the expansion into a Taylor's series truncated at the first term around preliminary approximated values $\underline{\underline{\beta}}^0$ and $\underline{\underline{\gamma}}^0$ ($^0$).

($^0$) To be determined subjectively, for instance.

Then, by denoting by $[\underline{\underline{u}}]$ and $[\underline{\underline{a}}]$ the conditional expectations for $\tilde{\underline{\underline{u}}}=(\tilde{u}_{1-q},\ldots,\tilde{u}_{-1},\tilde{u}_0,\tilde{u}_1,\ldots,\tilde{u}_n)'$ and $\tilde{\underline{\underline{a}}}=(\tilde{a}_{1-s-q_x},\ldots,\tilde{a}_{-1},\tilde{a}_0,\tilde{a}_1,\ldots,\tilde{a}_n)'$, we have $[\underline{\underline{u}}]=[\underline{\underline{u}}^0]+\underline{\underline{D}}^0(\underline{\underline{\beta}}-\underline{\underline{\beta}}^0)$ and $[\underline{\underline{a}}]=[\underline{\underline{a}}^0]+\underline{\underline{G}}^0(\underline{\underline{\gamma}}-\underline{\underline{\gamma}}^0)$ where $[\underline{\underline{u}}^0]=$ $=E(\tilde{\underline{\underline{u}}}|\underline{\underline{y}}_n,\underline{\underline{x}}_n,\underline{\underline{x}}^*,\underline{\underline{\beta}}^0)$, $[\underline{\underline{a}}^0]=E(\tilde{\underline{\underline{a}}}|\underline{\underline{x}}_n,\underline{\underline{x}}^*,\underline{\underline{\gamma}}^0)$, and $\underline{\underline{D}}^0,\underline{\underline{G}}^0$, are $(n+q)\times(s+r+p+q)$ and $(n+s+q_x)\times(p_x+q_x)$ matrices, whose elements $d^0_{it}$ and $g^0_{it}$ are given by

$d^0_{it}=[-\partial[u_t]/\partial\beta_i]_{\beta_i=\beta_i^0}$, $i=1,2,\ldots,s+r+p+q$, $t=1,2,\ldots,n+q$,

$g^0_{it}=[-\partial[a_t]/\partial\gamma_i]_{\gamma_i=\gamma_i^0}$, $i=1,2,\ldots,p_x+q_x$, $t=1,2,\ldots,n+s+q_x$.

Going on with the usual iterative procedure, we get

(5.1) $S(\underline{\underline{\beta}})=[\underline{\underline{u}}]'[\underline{\underline{u}}]=\nu_y z_y^2+(\underline{\underline{\beta}}-\hat{\underline{\underline{\beta}}})'\underline{\underline{D}}'\underline{\underline{D}}(\underline{\underline{\beta}}-\hat{\underline{\underline{\beta}}})$

(5.2) $S(\underline{\underline{\gamma}})=[\underline{\underline{a}}]'[\underline{\underline{a}}]=\nu_x z_x^2+(\underline{\underline{\gamma}}-\hat{\underline{\underline{\gamma}}})'\underline{\underline{G}}'\underline{\underline{G}}(\underline{\underline{\gamma}}-\hat{\underline{\underline{\gamma}}})$

where $\nu_y=n-(s+r+p+q)$, $\nu_x=n-(p_x+q_x)$, $z_y^2=\nu_y^{-1}[\underline{\underline{u}}^0]'[\underline{\underline{u}}^0]$, $z_x^2=\nu_x^{-1}[\underline{\underline{a}}^0]'[\underline{\underline{a}}^0]$, and $\hat{\underline{\underline{\beta}}},\hat{\underline{\underline{\gamma}}},\underline{\underline{D}}$ and $\underline{\underline{G}}$ are the values for $\underline{\underline{\beta}},\underline{\underline{\gamma}},\underline{\underline{D}}^0$, and $\underline{\underline{G}}^0$ obtained at the last iteration.

The approximation (5.1) can be inserted in (3.10), so that we obtain

(5.3) $\int\sigma_u^{-(n+1)}\exp\{-[\nu_y z_y^2-(\underline{\underline{\beta}}-\hat{\underline{\underline{\beta}}})'\underline{\underline{D}}'\underline{\underline{D}}(\underline{\underline{\beta}}-\hat{\underline{\underline{\beta}}})]/2\sigma_u^2\}d\underline{\underline{\beta}}d\sigma\propto|z_y^2(D'D)^{-1}|^{-1/2}$

on integrating firstly with respect to $\sigma_u$ and then to $\underline{\underline{\beta}}$, by using the properties of multivariate Student $\underline{\underline{t}}$ distribution.

Since integrals $J_{hijk}$ in (4.3) are of the type (5.3), posterior odds (4.5) become

(5.4) $\rho_{hijk}\propto\bar{\rho}_{hijk}\, z_{\bar{\bar{s}}\bar{\bar{r}}\bar{\bar{p}}\bar{\bar{q}}}\, z_{hijk}^{-1}|\underline{\underline{D}}'_{hijk}\underline{\underline{D}}_{hijk}(\underline{\underline{D}}'_{\bar{\bar{s}}\bar{\bar{r}}\bar{\bar{p}}\bar{\bar{q}}}\underline{\underline{D}}_{\bar{\bar{s}}\bar{\bar{r}}\bar{\bar{p}}\bar{\bar{q}}})^{-1}|^{1/2}$

for every h,i,j,k, where $z_{\bar{\bar{s}}\bar{\bar{r}}\bar{\bar{p}}\bar{\bar{q}}},\underline{\underline{D}}_{\bar{\bar{s}}\bar{\bar{r}}\bar{\bar{p}}\bar{\bar{q}}}$, and $z_{hijk},\underline{\underline{D}}_{hijk}$ are associated with models $M_{\bar{\bar{s}}\bar{\bar{r}}\bar{\bar{p}}\bar{\bar{q}}}$ and $M_{hijk}$, respectively.

As noted before, the same procedure can be utilized for the determination of the order $(p_x,q_x)$ for the submodel (1.2). If it is believed that $\bar{p}_x<p_x<\bar{\bar{p}}_x$ and $\bar{q}_x<q_x<\bar{\bar{q}}_x$, and if $M_{jk}$ is the submodel with $p_x=j,q_x=k$, then posterior odds are

(5.5) $\rho_{jk}\propto\bar{\rho}_{jk}z_{\bar{\bar{p}}_x\bar{\bar{q}}_x}z_{jk}^{-1}|\underline{\underline{G}}'_{jk}\underline{\underline{G}}_{jk}(\underline{\underline{G}}'_{\bar{\bar{p}}_x\bar{\bar{q}}_x}\underline{\underline{G}}_{\bar{\bar{p}}_x\bar{\bar{q}}_x})^{-1}|^{1/2}$

for every j,k, where $z_{\bar{\bar{p}}_x\bar{\bar{q}}_x},\underline{\underline{G}}_{\bar{\bar{p}}_x\bar{\bar{q}}_x}$, and $z_{jk}$, $\underline{\underline{G}}_{jk}$ are associated with the models $M_{\bar{\bar{p}}_x\bar{\bar{q}}_x}$ and $M_{jk}$, respectively.

## 6. Posterior marginal densities for the parameters of the previsive model

Having identified the orders (s,r,p,q,) and $(p_x,q_x)$ of the best submodels (1.1) and (1.2), before evaluating the previsions it is necessary to determine the posterior densities for $\tilde{\underline{\underline{\beta}}},\tilde{\underline{\underline{\gamma}}},\tilde{\sigma}_u,\tilde{\sigma}_a$ of the previsive model, by use of Bayes'theorem and of prior densities. These are in the normal-inverted gamma form by virtue of the second part of Assumption IV

$$(6.1)\quad p(\underline{\underline{\beta}},\sigma_u|H) \propto \sigma^{-(\bar{\nu}_y+s+r+p+q+1)} \cdot \exp\{-[\bar{\nu}_y \bar{z}_y^2 + (\underline{\underline{\beta}}-\bar{\underline{\underline{\beta}}})' \underline{\underline{N}}_y (\underline{\underline{\beta}}-\bar{\underline{\underline{\beta}}})]/2\sigma_u^2\}$$

$$(6.2)\quad p(\underline{\underline{\gamma}},\sigma_a|H) \propto \sigma^{-(\bar{\nu}_x+p_x+q_x+1)} \cdot \exp\{-[\bar{\nu}_x \bar{z}_x^2 + (\underline{\underline{\gamma}}-\bar{\underline{\underline{\gamma}}})' \underline{\underline{N}}_x (\underline{\underline{\gamma}}-\bar{\underline{\underline{\gamma}}})]/2\sigma_a^2\}$$

for parameters $\tilde{\underline{\underline{\beta}}},\tilde{\sigma}_u$ and $\tilde{\underline{\underline{\gamma}}},\tilde{\sigma}_a$ respectively, where $\bar{\nu}_y,\bar{z}_y,\bar{\nu}_x,\bar{z}_x$ are values that, together with the vectors $\bar{\underline{\underline{\beta}}},\bar{\underline{\underline{\gamma}}}$, and the matrices $\underline{\underline{N}}_y$, $(s+r+p+q)\times(s+r+p+q)$, $\underline{\underline{N}}_x$, $(p_x+q_x)\times(p_x+q_x)$, are initially subjectively assigned in consequence of prior information H.

By virtue of the factorization (3.14) for the likelihood, Bayes'theorem can be applied separately to the densities of $\tilde{\underline{\underline{\beta}}},\tilde{\sigma}_u$ and $\tilde{\underline{\underline{\gamma}}},\tilde{\sigma}_a$. As for the former one, we get

$$(6.3)\quad p(\underline{\underline{\beta}},\sigma_u|\underline{y}_n,\underline{\underline{x}}_n,\underline{\underline{x}}^*,H) \propto p(\underline{\underline{\beta}},\sigma_u|H)\cdot p(y_n|\underline{\underline{\beta}},\sigma_u,\underline{\underline{x}}_n,\underline{\underline{x}}^*) \propto |\underline{\underline{L}}_4'\underline{\underline{L}}_4|^{-1/2}\cdot$$
$$\cdot \sigma_u^{-(\bar{\nu}_y+s+r+p+q+n+1)} \cdot \exp\{-[\bar{\nu}_y\bar{z}_y^2+(\underline{\underline{\beta}}-\bar{\underline{\underline{\beta}}})'\underline{\underline{N}}_y(\underline{\underline{\beta}}-\bar{\underline{\underline{\beta}}})+\nu_y z_y^2+(\underline{\underline{\beta}}-\hat{\underline{\underline{\beta}}})'\underline{\underline{D}}'\underline{\underline{D}}(\underline{\underline{\beta}}-$$
$$-\hat{\underline{\underline{\beta}}})]/2\sigma_u^2\} = |\underline{\underline{L}}_4'\underline{\underline{L}}_4|^{-1/2}\cdot\sigma_u^{-(\bar{\nu}_y+s+r+p+q+n+1)}\exp\{-[\bar{\nu}_y\bar{z}_y^2+\nu_y z_y^2+\varepsilon_y+(\underline{\underline{\beta}}-$$
$$-\underline{\underline{\beta}}^+)'\underline{\underline{R}}_y(\underline{\underline{\beta}}-\underline{\underline{\beta}}^+)]/2\sigma_u^2\}$$

where $\underline{\underline{R}}_y=\underline{\underline{D}}'\underline{\underline{D}}+\underline{\underline{N}}_y$, $\underline{\underline{\beta}}^+=\underline{\underline{R}}_y^{-1}(\underline{\underline{D}}'\underline{\underline{D}}\,\hat{\underline{\underline{\beta}}}+\underline{\underline{N}}_y\bar{\underline{\underline{\beta}}})$, $\varepsilon_y=\bar{\underline{\underline{\beta}}}'\underline{\underline{N}}_y\bar{\underline{\underline{\beta}}}+\hat{\underline{\underline{\beta}}}'\underline{\underline{D}}'\underline{\underline{D}}\hat{\underline{\underline{\beta}}}-\underline{\underline{\beta}}^{+'}\underline{\underline{R}}_y\underline{\underline{\beta}}^+$.

On integrating (6.3) with respect to $\sigma_u$, the posterior marginal density for $\tilde{\underline{\underline{\beta}}}$ is obtained

$$(6.4)\quad p(\underline{\underline{\beta}}|\underline{y}_n,\underline{\underline{x}}_n,\underline{\underline{x}}^*,H) \propto [\bar{\bar{\nu}}_y+(\underline{\underline{\beta}}-\underline{\underline{\beta}}^+)'\underline{\underline{R}}_y(\underline{\underline{\beta}}-\underline{\underline{\beta}}^+)/\bar{\bar{z}}_y^2]^{-(\bar{\bar{\nu}}_y+s+r+p+q)/2}$$

where $\bar{\bar{\nu}}_y=\bar{\nu}_y+n$ and $\bar{\bar{\nu}}_y\bar{\bar{z}}_y^2=\bar{\nu}_y\bar{z}_y^2+\nu_y z_y^2+\varepsilon_y$, in the multivariate Student $\underline{\underline{t}}$ form, with $\bar{\bar{\nu}}_y$ degrees of freedom.

As for the posterior densities for $\tilde{\underline{\underline{\gamma}}}$ and $\tilde{\sigma}_a$, the procedure is the same. The joint density is the following

$$(6.5)\quad p(\underline{\underline{\gamma}},\sigma_a|\underline{\underline{x}}_n,\underline{\underline{x}}^*,H) \propto |\underline{\underline{L}}_7'\underline{\underline{L}}_7|^{-1/2}\sigma_a^{-(\bar{\nu}_x+p_x+q_x+n+1)}\ .$$
$$\cdot\exp\{-[\bar{\nu}_x\bar{z}_x^2+\nu_x z_x^2+\varepsilon_x+(\underline{\underline{\gamma}}-\underline{\underline{\gamma}}^+)'\underline{\underline{R}}_x(\underline{\underline{\gamma}}-\underline{\underline{\gamma}}^+)]/2\sigma_a^2\}$$

whilst the marginal one for $\tilde{\underline{\underline{\gamma}}}$ is

$$(6.6)\quad p(\underline{\underline{\gamma}}|\underline{\underline{x}}_n,\underline{\underline{x}}^*,H) \propto [\bar{\bar{\nu}}_x+(\underline{\underline{\gamma}}-\underline{\underline{\gamma}}^+)'\underline{\underline{R}}_x(\underline{\underline{\gamma}}-\underline{\underline{\gamma}}^+)/\bar{\bar{z}}_x^2]^{-(\bar{\bar{\nu}}_x+p_x+q_x)/2}$$

where $\underline{\underline{R}}_x=\underline{\underline{G}}'\underline{\underline{G}}+\underline{\underline{N}}_x$, $\underline{\underline{\gamma}}^+=\underline{\underline{R}}_x^{-1}(\underline{\underline{G}}'\underline{\underline{G}}\,\hat{\underline{\underline{\gamma}}}+\underline{\underline{N}}_x\bar{\underline{\underline{\gamma}}})$, $\varepsilon_x=\bar{\underline{\underline{\gamma}}}'\underline{\underline{N}}_x\bar{\underline{\underline{\gamma}}}+\hat{\underline{\underline{\gamma}}}'\underline{\underline{G}}'\underline{\underline{G}}\,\hat{\underline{\underline{\gamma}}}-\underline{\underline{\gamma}}^{+'}\underline{\underline{R}}_x\underline{\underline{\gamma}}^+$,
$\bar{\bar{\nu}}_x=\bar{\nu}_x+n$, $\bar{\bar{\nu}}_x\bar{\bar{z}}_x^2=\bar{\nu}_x\bar{z}_x^2+\nu_x z_x^2+\varepsilon_x$

## 7. Previsions for the ARMA submodel

In order to evaluate the previsions $E(\tilde{y}_{n+\tau}|\underline{y}_n,\underline{\underline{x}}_n,\underline{\underline{x}}^*,H),\tau\in T^*$, it is firstly necessary to compute the previsions associated with the ARMA submodel (1.2), that is $E(\tilde{x}_{n+\tau}|\underline{x}_n,\underline{\underline{x}}^*,H)$.

To this end, we observe that variables $\tilde{x}_{n+\tau},\tau\in T^*$ can be expressed as functions of the sample and of residuals (current and past), according to the recursive relation

$$(7.1)\quad x_{n+\tau}=\eta_\tau+\underline{\psi}_\tau'\underline{\underline{a}}_\tau^f \qquad \tau\in T^*$$

where $\underline{\underline{a}}_\tau^f=(a_{n+1},a_{n+2},\ldots,a_{n+\tau})'$, $\underline{\underline{\psi}}_\tau=(\psi_{\tau-1},\psi_{\tau-2},\ldots,\psi_0)'$, $\psi_j=\sum_{i=1}^{j}\phi_{xi}\psi_{j-i}-\theta_j$, $j=0,1,\ldots,\tau-1$, with $\psi_0=1,\phi_{xi}=0$ for $i>p_x$, and

$\theta_{xi}=0$ for $i>q_x$. As for $\eta_\tau$, if we set $A=\sum_{j=\tau}^{p_x}\phi_{xj}x_{n+\tau-j}$, $B=\sum_{j=\tau}^{q_x}\theta_{xj}a_{n+\tau-j}$, $C=\sum_{j=1}^{\tau-1}\phi_{xj}\eta_{\tau-j}$, $F=\sum_{j=1}^{p_x}\phi_{xj}\eta_{\tau-j}$, it is

$$\eta_\tau = \begin{cases} A+B+C & \text{for} \quad \tau\leq q_x,\ \tau>p_x \\ -B+F & \text{"} \quad \tau\leq q_x,\ \tau>p_x \\ A\ \ +C & \text{"} \quad \tau>q_x,\ \tau\leq p_x \\ F & \text{"} \quad \tau>q_x,\ \tau>p_x \end{cases}$$

Now, if $\tilde{\underline{\underline{a}}}^f=(\tilde{a}_{n+1},\tilde{a}_{n+2},\ldots,\tilde{a}_{n+m})'$ is the vector of future residuals with distribution $N(\underline{\underline{0}},\sigma^2_a\underline{\underline{I}}_m)$ by Assumption I, by virtue of (6.6) the joint density for $\tilde{\underline{\underline{a}}}^f$ and parameters of the ARMA submodel (1.2) is

(7.2) $p(\underline{\underline{a}}^f,\underline{\underline{\gamma}},\sigma_a|\underline{\underline{x}}_n,\underline{\underline{x}}^*,H)\propto p(\underline{\underline{a}}^f|\underline{\underline{\gamma}},\sigma_a,\underline{\underline{x}}_n,\underline{\underline{x}}^*,H)\cdot p(\underline{\underline{\gamma}},\sigma_a|\underline{\underline{x}}_n,\underline{\underline{x}}^*,H)\propto$
$\propto\sigma_a^{-(\bar{\nu}_x+p_x+q_x+n+m+1)}\exp\{-[\underline{\underline{a}}^{f\prime}\underline{\underline{a}}^f+\bar{\nu}_x\bar{z}_x^2+\nu_x z_x^2+\varepsilon_x+(\underline{\underline{\gamma}}-\underline{\underline{\gamma}}^+)'\underline{\underline{R}}_x(\underline{\underline{\gamma}}-\underline{\underline{\gamma}}^+)]/2\sigma_a^2\}$

which, integrated firstly with respect to $\underline{\underline{\gamma}}$ and then with respect to $\sigma_a$, becomes

(7.3) $p(\underline{\underline{a}}^f|\underline{\underline{x}}_n,\underline{\underline{x}}^*,H)\propto[\bar{\bar{\nu}}_x+\underline{\underline{a}}^{f\prime}\underline{\underline{a}}^f/\bar{\bar{z}}_x^2]^{-(\bar{\bar{\nu}}_x+m)/2}$

in the multivariate Student $\underline{\underline{t}}$ form with $\bar{\bar{\nu}}_x$ d.o.f.

By virtue of (7.3) even the distributions of vectors $\tilde{\underline{\underline{a}}}^f_\tau$, $\tau\in T^*$, are in the same form, and owing to the properties of such distributions the linear combination $\underline{\underline{\psi}}'_\tau\tilde{\underline{\underline{a}}}^f_\tau$ is distributed as an univariate Student t with $\bar{\bar{\nu}}_x$ d.o.f., and since the Jacobian of the transformation (7.1) is unitary, the density for $\tilde{x}_{n+\tau}$, $\tau\in T^*$

(7.4) $p(x_{n+\tau}|\underline{\underline{\gamma}},\underline{\underline{x}}_n,\underline{\underline{x}}^*,H)\propto\left[\bar{\bar{\nu}}_x+(x_{n+\tau}-\eta_\tau)/\underline{\underline{\psi}}'_\tau\bar{\bar{z}}_x^2\,\underline{\underline{\psi}}_\tau\right]^{-(\bar{\bar{\nu}}_x+1)/2}$

is in the univariate Student t form with $\bar{\bar{\nu}}_x$ d.o.f., conditional on $\underline{\underline{\gamma}}$ also, since this parameter vector is a part of $\eta_\tau$.

Mean values of the distributions with density (7.4) are $E(\tilde{x}_{n+\tau}|\underline{\underline{\gamma}},\underline{\underline{x}}_n,\underline{\underline{x}}^*,H)=\eta_\tau$, $\tau\in T^*$, so that the previsions of $\tilde{x}_{n+\tau}$, $\tau\in T^*$, unconditionally on $\underline{\underline{\gamma}}$, are $E(\tilde{x}_{n+\tau}|\underline{\underline{x}}_n,\underline{\underline{x}}^*,H)=\int\eta_\tau\cdot p(\underline{\underline{\gamma}}|\underline{\underline{x}}_n,\underline{\underline{x}}^*,H)d\underline{\underline{\gamma}}$, $\tau\in T^*$, where $p(\underline{\underline{\gamma}}|\underline{\underline{x}}_n,\underline{\underline{x}}^*,H)$ is given by (6.7).

## 8. Previsions for the transfer function model

The evaluation of previsions $E(\tilde{y}_{n+\tau}|\underline{\underline{y}}_n,\underline{\underline{x}}_n,\underline{\underline{x}}^*,H)$ is similar to the one illustrated in the previous Section. Firstly, we write the transfer function submodel (1.1) in the form

(8.1) $\alpha(B)y_t= k(B)x_t + \mu(B)u_t$

where $\alpha(B)=\phi(B)\cdot\delta(B)=1-\alpha_1B+\ldots-\alpha_{p+r}B^{p+r}$, $k(B)=\phi(B)\cdot\omega(B)$ $=k_0-k_1B+\ldots-k_{p+s}B^{p+s}$, $\mu(B)=\delta(B)\cdot\theta(B)=1-\mu_1B+\ldots-\mu_{r+q}B^{r+q}$, are polynomials in the backshift operator B. Furthermore, we observe that variables $y_{n+\tau}$, $\tau\in T^*$, can be expressed as functions of the sample, of the $x_t$ and of residuals $u_t$, according to the recursive relation

$$(8.2)\qquad y_{n+\tau} = \Pi_\tau + \underline{\underline{\xi}}'_\tau \underline{\underline{x}}^f_\tau + \underline{\underline{\zeta}}'_\tau \underline{\underline{u}}^f_\tau \qquad \tau \in T^*$$

where $\underline{\underline{u}}^f_\tau = (u_{n+1}, u_{n+2}, \dots, u_{n+\tau})'$ , $\underline{\underline{x}}^f_\tau = (x_{n+1}, x_{n+2}, \dots, x_{n+\tau})'$
$\underline{\underline{\xi}}_\tau = (\xi_{\tau-1}, \xi_{\tau-2}, \dots, \xi_0)'$ , $\underline{\underline{\zeta}}_\tau = (\zeta_{\tau-1}, \zeta_{\tau-2}, \dots, \zeta_0)'$ and
$\xi_j = \sum_{i=1}^{j} \alpha_i \xi_{j-1} - \varepsilon_j$, $\zeta_j = \sum_{i=1}^{j} \alpha_i \zeta_{j-1} - \mu_j$, for $j=0,1,\dots,\tau-1$; with $\zeta_0=1$, $\xi_0 = k_0$, $\alpha_i=0$ for $i>p+r$; $\mu_i=0$ for $i>q+r$; and $k_i=0$ for $i>p+s$. As for $\Pi_\tau$ if we set $M = \sum_{j=\tau}^{p+r} \alpha_j y_{n+\tau-j}$, $U = \sum_{j=\tau}^{p+s} k_j x_{n+\tau-j}$, $V = \sum_{j=\tau}^{q+r} \mu_j u_{n+\tau-j}$, $W = \sum_{j=1}^{\tau-1} \alpha_j \Pi_{\tau-j}$, $Z = \sum_{j=1}^{p+r} \alpha_j \Pi_{\tau-j}$, it is

$$\Pi_\tau = \begin{cases} M-U-V+W & \tau\leq q+r,\ \tau\leq p+r,\ \tau\leq p+s \\ M\quad -V+W & \tau\leq q+r,\ \tau\leq p+r,\ \tau>p+s \\ -U-V+Z & \tau\leq q+r,\ \tau>p+r,\ \tau\leq p+s \\ -V+Z & \tau\leq q+r,\ \tau>p+r,\ \tau>p+s \\ M-U\quad +W & \tau>q+r,\ \tau\leq p+r,\ \tau\leq p+s \\ M\quad +W & \tau>q+r,\ \tau\leq p+r,\ \tau>p+s \\ -U\quad +Z & \tau>q+r,\ \tau>p+r,\ \tau\leq p+s \\ Z & \tau>q+r,\ \tau>p+r,\ \tau>p+s \end{cases}$$

At this point we cannot go on as in the ARMA case, because the term $\underline{\underline{\xi}}_\tau \underline{\underline{x}}^f_\tau$ is in (8.2) but not in (7.1): we have, therefore to transform it in function of residuals $\underline{\underline{a}}^f_\tau$ in the following manner

$$(8.3)\quad \underline{\underline{\xi}}'_\tau \underline{\underline{x}}^f_\tau = \underline{\underline{\xi}}'_\tau \underline{\underline{\eta}}_\tau + \underline{\underline{\xi}}'_\tau (\underline{\underline{\psi}}'_1 \underline{\underline{a}}^f_1, \underline{\underline{\psi}}'_2 \underline{\underline{a}}^f_2, \dots, \underline{\underline{\psi}}'_\tau \underline{\underline{a}}^f_\tau)' = \underline{\underline{\lambda}}'_\tau \underline{\underline{a}}^f_\tau + \underline{\underline{\xi}}'_\tau \underline{\underline{\eta}}_\tau$$

where $\underline{\underline{\lambda}}_\tau = (\lambda_1, \lambda_2, \dots, \lambda_\tau)'$ and $\lambda_j = \sum_{i=j}^{\tau} \xi_{j-1} \psi_{\tau-j}$ and (8.2) may be changed as follows

$$(8.4)\quad y_{n+\tau} = \Pi_\tau + \underline{\underline{\xi}}'_\tau \underline{\underline{\eta}}_\tau + \underline{\underline{\lambda}}'_\tau \underline{\underline{a}}^f_\tau + \underline{\underline{\zeta}}'_\tau \underline{\underline{\tilde{u}}}^f_\tau = \Pi_\tau + \underline{\underline{\xi}}'_\tau \underline{\underline{\eta}}_\tau + \underline{\underline{\chi}}'_\tau \underline{\underline{w}}^f_\tau$$

with $\underline{\underline{\chi}}'_\tau = (\underline{\underline{\lambda}}'_\tau, \underline{\underline{\xi}}_\tau)$ and $\underline{\underline{w}}^f_\tau = (\underline{\underline{a}}^{f'}_\tau, \underline{\underline{u}}^{f'}_\tau)'$. Now if $\underline{\underline{\tilde{u}}}^f = (\tilde{u}_{n+1}, \tilde{u}_{n+2}, \dots, \tilde{u}_{n+m})'$ is the vector of future residuals with distribution $N(\underline{\underline{0}}, \sigma^2_u \underline{\underline{I}}_m)$ by Assumption I, the joint density for $\underline{\underline{\tilde{u}}}^f$, $\underline{\underline{\tilde{a}}}^f$ and the parameters of the model (1.1)-(1.2) is, by virtue of parameters'independence.

$$(8.5)\quad p(\underline{\underline{u}}^f, \underline{\underline{a}}^f, \underline{\underline{\beta}}, \underline{\underline{\gamma}}, \sigma_u, \sigma_a | \underline{\underline{y}}_n, \underline{\underline{x}}_n, \underline{\underline{x}}^*, H) \propto p(\underline{\underline{u}}^f, \underline{\underline{a}}^f | \underline{\underline{\beta}}, \underline{\underline{\gamma}}, \sigma_u, \sigma_a, \underline{\underline{y}}_n, \underline{\underline{x}}_n, \underline{\underline{x}}^*, H) \cdot p(\underline{\underline{\beta}}, \sigma_u | \underline{\underline{y}}_n, \underline{\underline{x}}_n, \underline{\underline{x}}^*, H) \cdot p(\underline{\underline{\gamma}}, \sigma_a | \underline{\underline{x}}_n, \underline{\underline{x}}^*, H)$$

By integrating (8.5) with respect to $\underline{\underline{\gamma}}$ and $\underline{\underline{\beta}}$, and then to $\sigma_u$ and $\sigma_a$, we obtain

$$(8.6)\quad p(\underline{\underline{u}}^f, \underline{\underline{a}}^f | \underline{\underline{\beta}}, \underline{\underline{\gamma}}, \sigma_u, \sigma_a, \underline{\underline{y}}_n, \underline{\underline{x}}_n, \underline{\underline{x}}^*, H) \propto [\bar{\bar{\nu}}_y + \underline{\underline{u}}^{f'} \underline{\underline{u}}^f / \bar{\bar{z}}^2_y]^{-(\bar{\bar{\nu}}_y+m)/2} [\bar{\bar{\nu}}_x + \underline{\underline{a}}^{f'} \underline{\underline{a}}^f / \bar{\bar{z}}^2_x]^{-(\bar{\bar{\nu}}_x+m)/2}$$

and if we assume that $\bar{\bar{\nu}}_y = \bar{\bar{\nu}}_x = \bar{\bar{\nu}}$, the density for $\underline{\underline{\tilde{\omega}}}^f_\tau = (\underline{\underline{\tilde{a}}}^{f'}_\tau, \underline{\underline{\tilde{u}}}^{f'}_\tau)'$

$$(8.7)\quad p(\underline{\underline{w}}_\tau | \underline{\underline{\beta}}, \underline{\underline{\gamma}}, \sigma_u, \sigma_a, \underline{\underline{y}}_n, \underline{\underline{x}}_n, \underline{\underline{x}}^*, H) \propto [(\bar{\bar{\nu}} + \underline{\underline{u}}^{f'} \underline{\underline{u}}^f / \bar{\bar{z}}^2_y)(\bar{\bar{\nu}} + \underline{\underline{a}}^{f'} \underline{\underline{a}}^f / \bar{\bar{z}}^2_x)]^{-(\bar{\bar{\nu}}+m)/2}$$

is of the multivariate Student $\underline{\underline{t}}$ form with $\bar{\bar{\nu}}$ d.o.f.

Owing to the properties of such a distribution the linear combination $\underline{\underline{x}}'_{\tau}\tilde{\underline{\underline{w}}}^{f}_{\tau}$ is distributed as an univariate Student t with $\bar{\nu}$ d.o.f. and, as the Jacobian of the transformation (8.4) is unitary, the densities $p(y_{n+\tau}|\underline{\underline{\beta}},\underline{\underline{\gamma}},\sigma_u,\sigma_a,\underline{\underline{y}}_n,\underline{\underline{x}}_n,\underline{\underline{x}}^*,H)$, are of the same form, with mean values

$$(8.8)\quad E(\tilde{y}_{n+\tau}|\underline{\underline{\beta}},\underline{\underline{\gamma}},\underline{\underline{y}}_n,\underline{\underline{x}}_n,\underline{\underline{x}}^*,H)=\Pi_\tau+\underline{\underline{\xi}}'_\tau\underline{\underline{\eta}}_\tau \qquad \tau\in T^*$$

so that the previsions of $\tilde{y}_{n+\tau}$, $\tau\in T^*$, unconditional on $\underline{\underline{\beta}}$ and $\underline{\underline{\gamma}}$, are

$$(8.9)\quad E(\tilde{y}_{n+\tau}|\underline{\underline{y}}_n,\underline{\underline{x}}_n,\underline{\underline{x}}^*,H)=\int(\Pi_\tau+\underline{\underline{\xi}}'_\tau\underline{\underline{\eta}}_\tau)d\underline{\underline{\beta}}d\underline{\underline{\gamma}}$$

## 9. Concluding remarks

In the previous analysis, the identification and estimation of the transfer function model that better fits the time series $\{y_t\}$ and $\{x_t\}$ are performed by following a Bayesian procedure. By use of the same approach, previsions are evaluated in de Finetti's sense, so that both the sample and the prior information are taken into account. The results of the analysis are simple and easy utilizable because they are based on distributions of the multivariate Student $\underline{\underline{t}}$ type; therefore, even previsions are computed on the basis of Student's t. Subjective information can be introduced in a detailed manner, in such a way to modify, even substantially, the extrapolations based on the sample.

## REFERENCES

Carlucci, F., 1977, A Priori Information and Time Series Prevision, Metron, XXXV:337.

deFinetti, B., 1974, "Theory of Probability," Wiley, New York.

Monahan, J.F., 1983, Fully Bayesian Analysis of ARMA Time Series Models, J. of Econometrics, 21:307.

Newbold, P., 1973, Bayesian Estimation of Box-Jenkins Transfer Function-noise Models, Biometrika, 60:323.

PRIORS FOR EXPONENTIAL FAMILIES WHICH MAXIMIZE THE ASSOCIATION BETWEEN PAST AND FUTURE OBSERVATIONS

Donato M. Cifarelli
Istituto Metodi Quantitativi
Università L. Bocconi
Milano, Italy

Eugenio Regazzini
Dipartimento di Matematica
Università degli Studi
Milano, Italy

## 1. INTRODUCTION

Throughout the present paper, $\{X_n\}$ denotes a sequence of random quantities which are regarded as exchangeable, and which are assessed with a probability measure $P(\cdot)$ which is a member of the mixture-exponential family. To be precise, it will be presumed that the assessment $P(\cdot)$ for any finite subsequence $(X_1,\ldots,X_n)$ can be represented using the product of an identical non-degenerate parametric measure for each $X_i$, $P_\theta(\cdot)=P(\cdot|\tilde{\theta}=\theta)$, determined by

$$dP_\theta=\exp\{\theta x-M(\theta)\}d\mu \tag{1.1}$$

$\mu$ being a $\sigma$-finite measure on the class $\mathcal{B}$ of Borel sets of $\mathbb{R}$. It will always be assumed that the interior $\mathcal{X}^\circ$ of the convex hull $\mathcal{X}$ of the support of $\mu$ (in symbols: $\mathrm{supp}(\mu)$) is a nonempty open set (interval) in $\mathbb{R}$ and that $\{P_\theta;\theta\in\Theta\}$ is a regular exponential family (cf. Barndorff - Nielsen 1978, p.116). The latter condition implies that $\Theta=\{\theta:M(\theta)<\infty\}$ is an open interval in $\mathbb{R}$. Moreover, we will suppose that the set of the logically possible values of $\tilde{\theta}$ coincides with $\Theta$. Given such a particular frame, the present paper deals with the choice of a prior for (1.1); an excellent treatment of the same topic is included in Diaconis and Ylvisaker (1979, 1985). Our approach bases itself on the obvious remark that the choice of a prior establishes the strength of the dependence among the elements of the sequence $\{X_n\}$ and, consequently, the strength of the influence exercised by experience on our future predictions. This subjective standpoint is skilfully expounded in de Finetti (1937). More precisely, we will deal with the problem of singling out priors which maximize that influence when no attempt to quantify possible prior opinions about $\tilde{\theta}$ is made and one searches for priors for which there exists a function $\phi_n:\mathbb{R}^n\to\mathbb{R}$ satisfying

(1.2) $P(X_N \leq x, \phi_n(X_1,...,X_n) \leq x) = P(X_N \leq x) = P(\phi_n(X_1,...,X_n) \leq x)$

for all $N>n$, $n = 1,2,...,$ $x \in \mathbb{R}$.

Equalities (1.2) express a condition of perfect association between $X_N$ and $(X_1,...,X_n)$. In fact, if the probability distribution (p.d.) function of $X_N$ is proper [i.e. $\lim_{x\to-\infty} P(X_N \leq x) = 1 - \lim_{x\to+\infty} P(X_N \leq x) = 0$ ], then (1.2) holds iff:

$$P(|X_N - \phi_n(X_1,...,X_n)|>\varepsilon) = 0 \qquad \text{for all } \varepsilon > 0.$$

Whilst, if the same distribution is improper [i.e. $\lim_{x\to-\infty} P(X_N \leq x) > 0$, or/and $\lim_{x\to+\infty} P(X \leq x) < 1$] , (1.2) states that:

$$P(|X_N - \phi_n(X_1,...,X_n)|>\varepsilon \text{ and } |X_N|<A) = 0 \text{ for all } \varepsilon > 0 \text{ and } A \in X°;$$

consequently, the p.d. of $(X_N, \phi_n(X_1,...,X_n))$ will present masses concentrated on or adherent to (inf $X$, inf $X$) and (sup$X$, sup$X$); cf. de Finetti (1970).

It is worth stressing that our research does not aim at granting a particular logical status to the prior distributions characterized via (1.2). In fact, we deem it necessary to assess a prior distribution according to the actual information of an individual on $\tilde{\theta}$ and to the strength of the dependence that he wishes to establish between past and future observations (cf. de Finetti and Savage, 1962). Even more so, (1.2) cannot be seen as an "objective" principle in order to fix prior distributions. On the other hand, the adherence to the subjective Bayesian point of view does not lessen the interest in knowing priors which, by characterizing extremal attitudes such as that described by (1.2), may be considered as terms of a comparison in any actual assessment of a prior.

We conclude this section by summarizing the structure of the present paper. Section 2 describes a procedure in order to assign (finitely additive) priors, according to de Finetti's theory. Section 3 proves that the method is general enough to yield distributions satisfying (1.2). Section 4 includes some remarks about the application to a definition of the concept of noninformative prior and to a Bayesian justification of some classical inferential results. The Appendix contains the proofs of the theorems stated in the previous sections.

## 2. FINITELY ADDITIVE PRIORS FOR THE EXPONENTIAL FAMILY

The present section includes some preliminary remarks about the analytical representation of a class of priors that we will employ in the next sections. It is founded on a paper by Regazzini (1987) which assumes de Finetti's coherence condition (dF-coherence) as sole "axiom" for the theory of statistical inference.

Let $\rho$ be a $\sigma$-finite measure on $(\Theta, \mathcal{B}_\Theta)$, where $\mathcal{B}_\Theta = \mathcal{B} \cap \Theta$, and let $\{I_m\}$ be an increasing sequence in $\mathcal{B}_\Theta$ converging to $\Theta$ such that:

$0 < \rho(I_m) < \infty$ for all $m = 1,2,\ldots$ . The sequence $\{\rho_m\}$ defined by

$$(2.1) \quad \rho_m(B) = \rho(B \cap I_m)/\rho(I_m) \text{ , } B \in \mathcal{B}_\Theta \text{ , } m = 1,2,\ldots$$

is a sequence of probability measures on $(\Theta, \mathcal{B}_\Theta)$. Hence, in view of the hypotheses of the previous section, one can define an extension $P_m$ of $\rho_m$ to $\mathcal{B}^n \otimes \mathcal{B}_\Theta$ via the usual rule

$$(2.2) \quad P_m(\{(X_1 \ldots, X_n) \in A\} \cap \{\tilde{\theta} \in B\}) = \int_B \{\int_A \exp [n(\theta \bar{x}_n - M(\theta))] \cdot \prod_1^n \mu(dx_j)\} \cdot \rho_m(d\theta),$$

where A and B are elements of $\mathcal{B}^n$ and $\mathcal{B}_\Theta$ respectively, and $\bar{x}_n = \sum_1^n x_i/n$. Further extensions to

$$\mathcal{J} = \{B| \underline{x} : B \in \mathcal{B}_\Theta, \underline{x} \text{ is any realization of } (X_1, \ldots X_n)\} \cup$$

$$\cup \{\{X_N \in C\} | \underline{x} : C \in \mathcal{B}, \underline{x} \text{ is as above and } N > n\}$$

can be obtained according to the following rules of probability calculus:

$$(2.3) \quad P_m(\tilde{\theta} \in B| (X_1, \ldots, X_n) = \underline{x}) = \{ \int_{I_m} \exp [n(\theta \bar{x}_n - M(\theta))] \rho_m(d\theta)\}^{-1} \cdot \int_{B \cap I_m} \exp[ n(\theta \bar{x}_n - M(\theta))] \rho_m(d\theta) := q_{\underline{x}}^{(m)}(B),$$

$$(2.4) \quad P_m(X_N \in C| (X_1, \ldots, X_n) = \underline{x}) = \int_{I_m} \{\int_C \exp [\theta x - M(\theta)] \mu(dx)\} q_{\underline{x}}^{(m)}(d\theta)$$

By virtue of the results expounded in Regazzini (1985, 1987), (1.1) and (2.1)-(2.4) define a coherent probability $P_m$ (in the sense of de Finetti) on

$$\mathcal{K} = \mathcal{B}^n \otimes \mathcal{B}_\Theta \cup \mathcal{J} \cup \{C|\theta : C \in \mathcal{B}, \theta \in \Theta\}.$$

The employment of the probability law $P_m$ is appropriate to assign inferences when one supposes that $I_m$ constitutes the parameter space. In order to obtain assessments corresponding to $\Theta$ , one could consider $P = \lim P_m$, provided that such a limit exists. To make this idea precise, let us define the class $\mathcal{L} \subset \mathcal{K}$ on which the previous limit exists. In view of Theorem 5 in Regazzini (1985), P is a coherent probability on $\mathcal{L}$ and it can be extended to $\mathcal{K}$ by preserving coherence (cf. Theorem 4 in Regazzini 1985). Since $\sigma$-additivity can be destroyed in passing to the limit, P need not be $\sigma$-additive. This circumstance, in view of the developments of the next sections, induces us to revisit the concept of perfect association. For the sake of clearness, we will avail ourselves of an example from De Groot (1970, p.192):

$$dP_\theta(x)=(2\pi)^{-1/2}\exp\{-\frac{1}{2}(x-\theta)^2\}\cdot dx \qquad \theta\varepsilon\Theta = \mathbb{R},\ \mathcal{X} = \mathbb{R},$$

$$I_m = (-\alpha_m, \alpha_m) \qquad \alpha_m \uparrow \infty,$$

$\rho$ = Lebesgue measure.
Then, for $Z = \sqrt{n}(\bar{X}-\tilde{\theta})$, one obtains

$$
\begin{aligned}
P(Z \leq z|\bar{X} = x) &= \lim_{m\to\infty} P_m(Z \leq z|\bar{X} = x) = (2\pi)^{-1/2}\int_{-\infty}^{z} \exp[-\frac{x^2}{2}]\,dx \\
&= \lim_{m\to\infty} P_m(Z \leq z|\tilde{\theta} = \theta) = P(Z \leq z|\tilde{\theta} = \theta) \qquad (2.5)\\
&= \lim_{m\to\infty} P_m(Z \leq z) = P\ (Z \leq z)
\end{aligned}
$$

and it is easy to show that equalities of the same kind hold when $\{Z\leq z\}$ is replaced by $\{Z\varepsilon A\}$, A being any element of $\mathcal{B}$. Now, if one decides to consider Z independent of $\bar{X}(\tilde{\theta})$ whenever $P(Z\varepsilon A|\bar{X}=x)=P(Z\varepsilon A)$ $(=P(Z\varepsilon A|\tilde{\theta}=\theta))$ for all $A,B\varepsilon\mathcal{B}$ and x, $\theta\varepsilon\mathbb{R}$, then (2.5) states independence. De Groot reaches similar results by employing a "uniform improper prior" over IR and he maintains that they are inconsistent since "under any proper bivariate distribution of $\bar{X}$ and $\tilde{\theta}$ for which...Z and $\tilde{\theta}$ are independent, it would not be possible for the random variables Z and $\bar{X}$ also to be independent, unless the random variable Z is equal to a constant with probability 1".

This last circumstance can be described by saying that a constant c exists such that

$$(2.6)_1 \qquad P(\bar{X} \leq x \text{ and } \tilde{\theta}\leq x-c)= P(\bar{X}\leq x)=P(\tilde{\theta}\leq x-c) \text{ for all } x\,\varepsilon\,\mathbb{R}.$$

This description of the perfect linear correlation between $\bar{X}$ and $\tilde{\theta}$ is equivalent to the following condition

$$(2.6)_2 \qquad P(|\bar{X}-\tilde{\theta}-c|>\varepsilon)= 0 \text{ for all } \varepsilon>0$$

provided that the involved distributions are proper. Here the terms proper and improper (cf. Section 1 and de Finetti 1970 I, 6.4.11) are employed in a sense different from that commonly considered by most statisticians, according to which a prior distribution is improper if it turns out to be unbounded. According to the meaning employed by us, in the framework of finitely additive probabilities, improper distributions are real probability distribution functions which assign probabililies adherent to the extreme points of $\mathcal{X}$. Having said that, one notices that condition $(2.6)_1$ can be employed in order to define perfect linear correlation in the improper case also, whilst this does not occur for condition $(2.6)_2$. In our example, the distributions of $\bar{X}$ and $\tilde{\theta}$ are indeed improper $(P(\tilde{\theta}\leq x)=P(\bar{X}\leq x)= \frac{1}{2} = P(\tilde{\theta}>x) = P(\bar{X}>x)$ for all $x\,\varepsilon\mathbb{R})$ and it is easy to verify that they satisfy $(2.6)_1$. Hence, the inconsistency pointed out by De Groot does not arise w.r.t. definition $(2.6)_1$, the only one which, besides being equivalent to the traditional one in the proper case, is meaningful in the improper case also. Obviously, we succeeded in reaching such a conclusion because we introduced the "uniform prior on IR" through $\rho_m$ and a procedure which permits to evaluate the probability of a class of events larger than the domain of the posterior distribution. Such a procedure presents a few points of contact with Rényi's (1955) axiomatic theory of probability. On the contrary, the employment of improper priors following traditional statisticians' usage, does not generally enable one to obtain these evaluations, even if it produces coherent posteriors.

## 3. PRIORS MAXIMIZING THE ASSOCIATION BETWEEN $X_N$ AND $(X_1,\ldots,X_n)$

The present section shows that one can choose priors for (1.1) such that there exists a function $\phi_n: \mathbb{R}^n \to \mathbb{R}$ for which (1.2) holds. It is clear that, under the hypotheses expounded in the first part of Section 1, if the distribution of $\tilde{\theta}$ is proper, then no function $\phi_n$ exists satisfying (1.2). Hence, we look for a prior yielding (1.2) within the class of priors described in Section 2, and accept to compute the involved probabilities through the limit of $P_m$ under the condition:

(3.1) $\rho(\Theta) = \infty$ and for any compact interval $I \subset \Theta$, $m_0$ exists such that $I_{m_0} \supset I$

The main result of the present paper is represented by the following theorem, which provides a complete solution of the problem stated above.

Theorem 3.1: Let P be assessed on $L$ according to the previous section in such a way that (3.1) holds. Let $\psi_n$ be any real-valued increasing function defined on $X$ such that:

$$\lim_{x \downarrow \inf X} \psi_n(x) = \inf X , \qquad \lim_{x \uparrow \sup X} \psi_n(x) = \sup X ;$$

then (1.2) holds with $\phi_n(x_1,\ldots,x_n) = \psi_n(\bar{x}_n)$.

This theorem points out that there are very many priors satisfying (1.2). Among them it is interesting to analyse the case in which $\rho$ is determined by

(3.2) $\rho(A) = \int_A \exp\{n_o(x_o\theta - M(\theta))\}d\theta \qquad$ for all $A \in \mathcal{B}_\Theta$.

Diaconis and Ylvisaker (1979) have considered such a prior with $\rho(\Theta)<\infty$ (prior distribution conjugate to (1.1)) and they have shown that it is characterized through the property of linear predictive expectation: $E(X_N|X_1,\ldots,X_n) = a_n\bar{X}_n + b_n$. Cifarelli and Regazzini (1983) have proved that the same prior is characterized through the property of maximizing the dependence of $X_N$ on $\bar{X}_n$ (measured via Pearson's correlation ratio) among the priors which yield a fixed value in (0,1) for the ratio: $Var(E(X_n|\tilde{\theta}))/Var(X_n)$. The following theorems state that analogous results hold even if $\rho$ is determined by (3.2) with $\rho(\Theta)=\infty$. The first deals with the calculus of the regression function:

Theorem 3.2: If P is assessed on $L$ according to the previous section and $\rho$ is given by (3.2) with $x_o \in \bar{X}$ (the closure of $X$), then, for $(n+n_o)>0$ and $N>n$:

$$(3.3)\quad E(X_N|X_1 = x_1,\dots,X_n = x_n) = \begin{cases} \inf \mathcal{X}, \text{ if } \bar{x}_n \le n^{-1}\{(n+n_o)\cdot\inf\mathcal{X} - n_o x_o\} \\ \text{Sup } \mathcal{X}, \text{ if } \bar{x}_n \ge n^{-1}\{(n+n_o)\cdot\sup\mathcal{X} - n_o x_o\} \\ (n+n_o)^{-1}\cdot(n\bar{x}_n + n_o x_o), \text{ elsewhere.} \end{cases}$$

From this it follows that $E(X_N|X_1 = x_1,\dots,X_n = x_n)$ can be a candidate to represent the function $\psi_n$ of Theorem 3.1.

We will deal with the converse to Theorem 3.2 under special alternative assumptions about supp($\mu$) , i.e.:

$\underline{\underline{A}}$ : supp($\mu$) contains an open interval in $\mathbb{R}$;

$\underline{\underline{B}}$ : supp($\mu$) is a subset of $[0,\infty)$ or of $(-\infty,0]$ such that $\mu(\{0\})>0$.

Condition $\underline{\underline{A}}$ coincides with the one considered in Theorem 3 by Diaconis and Ylvisaker (1979). On the other hand, condition $\underline{\underline{B}}$ is both weaker than that of their Theorem 4, and weaker than condition $\underline{\underline{C}}$ of Theorem 1 in Cifarelli and Regazzini (1983), but it suffices to characterize (3.2) when (3.3) is demanded to hold for all n. In fact, in Theorem 1 by Cifarelli and Regazzini (1983), condition C is redundant; as a matter of fact, $\underline{\underline{B}}$ implies that $M(\theta)$ is monotonic and that $\exp\{-M(\theta)\}$ is bounded. Consequently, the next theorem is a generalization of Theorem 1 in our paper of 1983.

<u>Theorem 3.3:</u> Let $\mu$ satisfy $\underline{\underline{A}}$ or $\underline{\underline{B}}$; furthermore, let P be assigned according to Section 2 with a $\rho$ such that a positive integer $\nu$ and a sample $(x_1^*,\dots,x_\nu^*)$ exist for which

$$(3.4)\quad 0 < \int_\Theta \exp\{\nu(\theta\bar{x}_\nu^* - M(\theta))\}\rho(d\theta)<\infty.$$

Under these conditions, if (3.3) holds for all $n>\nu$, then:

$$\rho(d\theta)=c\cdot\exp(n_o\{x_o\theta - M(\theta)\})d\theta.$$

In particular, Theorems 3.2-3.3 enable us to characterize the classical "improper uniform" prior; in fact:

if $\mu$ satisfies $\underline{\underline{A}}$ or $\underline{\underline{B}}$; if P is assigned according to Section 2 in such a way that a positive integer $\nu$ and a sample $(x_1^*,\dots,x_\nu^*)$ exist for which (3.4) holds, then $E(X_N|X_1,\dots,X_n) = \bar{X}_n$ for all $n>\nu$, iff $\rho$ is defined by $\rho(d\theta)=cd\theta$.

## 4. CONCLUDING REMARKS

The attitude described in the previous sections, according to which one adopts priors maximizing the association between past and future observations, in our opinion, represents the kernel of almost all attempts made to define the concept of noninformative prior. In our view, any prior which emphasizes the role of observations by satisfying (1.2) can be considered as a candidate to represent vague prior information (cf. De Groot 1970, Ch. 10). Moreover it seems to us that condition (1.2) is more simple and natural than most requirements considered by the modern approaches to this problem (see Berger 1985, Ch. 3, and Dawid 1983, for recent reviews of them). In any case, since these formulations generally lead to consider priors represented through unbounded measures, the procedure expounded in Section 2 could be employed to frame them into de Finetti's theory of probability.

We conclude by stressing that the last statement of Section 3 provides a Bayesian justification of the "orthodox" estimator $\bar{X}_n$ of $M'(\theta)=E(X|\tilde{\theta}=\theta)$. In fact, since $E(X_N|X_1,\ldots,X_n)=E(M'(\tilde{\theta})|X_1,\ldots,X_n)$, from that statement we deduce that $\bar{X}_n$ is the Bayesian estimator of $M'(\tilde{\theta})$, for squared error loss, provided that P is assigned according to Section 2 with $\rho(d\theta)=cd\theta$. Results on the conditions for numerical equivalence between classical and Bayesian inference are generally founded on the use of priors with infinite mass. We think that the conclusions reached in the previous sections, suitably extended to general statistical models, will enable us to provide a substantial justification for this equivalence [via condition (1.2) which, in fact, could be interpreted as an attempt to provide a subjective probabilistic formulation of the paradigm of the sampling theory of inference, according to which only the observed data are taken into account] and to substitute priors with infinite mass by real p.d.'s according to Section 2. We will deal with these topics in a forthcoming paper.

## APPENDIX

### Proof of Theorem 3.1

After recalling that $\Theta=(\alpha,\beta)$, we will split the argument into several steps:

(1) for every $\varepsilon>0$ and any compact interval $I\subset X^\circ$, there exist $a, b \in \mathbb{R}$ such that $\alpha<a<b<\beta$, for which

$$\int_I \exp\{\theta x-M(\theta)\}\mu(dx)<\varepsilon, \qquad \text{for all } \theta\in(\alpha,a)\cup(b,\beta):=Q.$$

If $\beta=+\infty$, then from inequality (2.4) of Diaconis and Ylvisaker (1979):

$$\int_I \exp\{\theta x-M(\theta)\}\mu(dx) \leq \mu(A_y)^{-1}\cdot \int_I \exp\{\theta(x-y)\}\mu(dx)$$

where $y \in (\sup I, \sup X)$. Hence, from dominated convergence:

$$\lim_{\theta\to+\infty} \int_I \exp\{\theta x-M(\theta)\}\mu(dx)=0.$$

If $\beta<+\infty$, then $\lim_{\theta\to\beta^-} M(\theta)=+\infty$ (cf. Diaconis and Ylvisaker 1979, p. 273) and from $\exp\{\theta x\}\leq \exp\{|\theta|\max(|\inf I|, |\sup I|)\}$ , one deduces

$$0 \leq \limsup_{\theta\to\beta^-} \int_I \exp\{\theta x-M(\theta)\}\mu(dx)= 0.$$

Hence, in any case, given $\varepsilon>0$, b exists such that

$$0 \leq \int_I \exp\{\theta x - M(\theta)\}\mu(dx) < \varepsilon, \text{ for all } \theta \in (b,\beta).$$

A similar argument applies in order to determine a.

(2) $\quad 0 = \lim_{m\to\infty} P_m(X_N \varepsilon I) = P(X_N \varepsilon I) \quad$ for all N=1,2,...

In fact, if Q' denotes the complement of Q w.r.t.$\Theta$ :

$$P_m(X_N \varepsilon I) = \int_{I_m \cap Q'} P_\theta(I)\rho_m(d\theta) + \int_{I_m \cap Q} P_\theta(I)\rho_m(d\theta)$$

where, by virtue of (3.1), the first integral converges to 0 and, for the second one, (1) implies

$$0 \leq \int_{I_m \cap Q} P_\theta(I)\rho_m(d\theta) < \varepsilon \qquad \text{for all } m \geq 1.$$

(3) $\quad 0 = \lim_{m\to\infty} P_m(\bar{X}_n \varepsilon I) = P(\bar{X}_n \varepsilon I) \quad$ for all n=1,2...

The proof of (3) is analogous to that of (2) since

$$P_\theta(\bar{X}_n \varepsilon I) = \int_I \exp\{n[\theta x - M(\theta)]\}\mu_n(dx)$$

where $\mu_n$ is the image of $\mu$ induced by $\bar{X}_n$.

(4) Given $\varepsilon>0$ and $x_1, x_2 \in \mathbb{R}$, $Q=(\alpha,a) \cup (b,\beta)$ exists such that

$$P_\theta(X_N \leq x_1 \text{ and } \bar{X}_n \geq x_2),\ P(X_N \geq x_1 \text{ and } \bar{X}_n \leq x_2) < \varepsilon$$

for all $\theta \in Q$ and N>n. In fact:

$$P_\theta(X_N \varepsilon A \text{ and } \bar{X}_n \varepsilon B) = P_\theta(X_N \varepsilon A) P_\theta(\bar{X}_n \varepsilon B) \quad \text{for all } A,B \in \mathcal{B} \text{ and } \theta \in \Theta;$$

furthermore, arguing as in (1) and (3):

$$\lim_{\theta\to\beta^-} P_\theta(X_N \leq x) = \lim_{\theta\to\beta^-} P_\theta(\bar{X}_n \leq x) = \lim_{\theta\to\alpha^+} P_\theta(X_N \geq x) = \lim_{\theta\to\alpha^+} P_\theta(\bar{X}_n \geq x) = 0.$$

(5) $\quad 0 = P(X_N \leq x_1 \text{ and } \bar{X}_n \geq x_2) = P(X_N \geq x_1 \text{ and } \bar{X}_n \leq x_2)$;

in fact:

$$P_m(X_N \leq x_1 \text{ and } \bar{X}_n \geq x_2) = \int_{I_m \cap Q'} P_\theta(X_N \leq x_1) P_\theta(\bar{X}_n \geq x_2)\rho_m(d\theta) +$$

$$+ \int_{I_m \cap Q} P_\theta(X_N \leq x_1) P_\theta(\bar{X}_m \geq x_2)\rho_m(d\theta)$$

and the thesis follows from (4) by arguing as in (2).
In view of the previous results, we can state that the distribution of

$(X_N,\bar{X}_n)$ assigns the whole probability partly adherent and/or partly concentrated at the points $(\inf\mathcal{X}, \inf\mathcal{X})$, $(\sup\mathcal{X},\sup\mathcal{X})$; hence, given any monotonic function $\psi_n$ such that $\psi_n(\inf\mathcal{X})=\inf\mathcal{X}$ in the first case, $\psi_n(\sup\mathcal{X})=\sup\mathcal{X}$ in the second one, we see that (1.2) holds for $\phi_n(X_1,\dots X_n)=\psi_n(\bar{X}_n)$.

<u>Proof of Theorem 3.2</u>

Since $n+n_o > 0$ and

$$P_m(X_N \le x \mid X_1 = x_1,\dots,X_n=x_n) = \{\int_{I_m} \exp\{\theta(n\bar{x}_n+n_o x_o)-(n+n_o)M(\theta)\}\, d\theta\}^{-1}\cdot$$

$$\cdot \int_{I_m} P_\theta((-\infty,x]) \exp\{\theta(n\bar{x}_n+n_o x_o)-(n+n_o)M(\theta)\}\, d\theta,$$

if the denominator converges $(m\to\infty)$, then $(n+n_o)^{-1}(n\bar{x}_n+n_o x_o)\in\mathcal{X}°$ (cf. Theorem 1 of Diaconis and Ylvisaker 1979) and, in view of Theorem 2 of the same paper:

$$E(X_N|X_1=x_1,\dots,X_n=x_n)=(n+n_o)^{-1}(n\bar{x}_n+n_o x_o).$$

On the other hand, if the denominator diverges, then $(n+n_o)^{-1}(n\bar{x}_n+n_o x_o)\notin\mathcal{X}°$. Suppose that $(n+n_o)^{-1}.(n\bar{x}_n+n_o x_o) \ge \sup\mathcal{X}$; in such a case $x_1 = \sup\mathcal{X} <\infty$. Then, for $\theta\in(\alpha,\beta) = \Theta$ :

$$\exp\{M(\theta)\} = \int_{(-\infty,o)} \exp\{\theta x\}\mu(dx) + \int_{[o,x_1]} \exp\{\theta x\}\mu(dx)$$

and, if the first integral converges for $\alpha_o \in(\alpha,\beta)$, then it converges for all $\theta>\alpha_o$; as far as the second one is concerned, one obtains

$$\int_{[o,x_1]} \exp\{\theta x\}\mu(dx) \le \sup_{x\in [o,x_1]} e^{\theta x}\mu([o,x_1])<\infty \qquad \text{for all } \theta\in\mathbb{R}$$

since $\mu(A)< \infty$ for every compact subset of $\mathbb{R}$ (cf. Diaconis and Ylvisaker 1979, p. 272). Hence: $\beta = +\infty$. An analogous argument yields: $\inf\mathcal{X}>-\infty \Rightarrow\alpha=-\infty$. Now, in view of inequality (2.4) of Diaconis and Ylvisaker (1979):

$$\lim_{\theta\to\infty} \int_{(-\infty,x]} \exp\{\theta t-M(\theta)\}\mu(dt)=0, \text{ for all } x\in\mathcal{X}°.$$

Then, since for $m\to \infty$:

$$P_m(X_N\le x| X_1=x_1,.,X_n=X_n)\sim\{\int_\alpha^{\beta m} \exp\{\theta(n\bar{x}_n+n_o x_o)-(n+n_o)M(\theta)\}d\theta\}^{-1}.$$

$$.\int_\alpha^{\beta m} P_\theta((-\infty,x]) \exp\{\theta(n\bar{x}_n + n_o x_o)-(n+n_o)M(\theta)\}d\theta$$

one deduces: $P(X_N \le x|X_1=x_1,\dots X_n=x_n) = 0$ for all $x \in\mathcal{X}°$.
Hence, the whole probability is partly concentrated on and/or partly

adherent to $\sup X<+\infty$, consequently: $E(X_N|X_1=x_1,\dots X_n=x_n)=\sup X$ for all $(x_1,\dots,x_n)$ such that $\bar{x}_n \geq n^{-1}\{(n+n_o).\sup X - n_o x_o\}$.

If $(n+n_o)^{-1}(n\bar{x}_n + n_o x_o) \leq \inf X$, an analogous argument shows that

$$E(X_N|X_1=x_1,\dots X_n=x_n) = \inf X \text{ , if } \bar{x}_n \leq n^{-1}\{(n+n_o)\inf X - n_o x_o\}.$$

The second part of the proposition follows from the previous conclusions and from the obvious inequalities

$$n.\inf X \leq (n+n_o)\inf X - n_o \cdot x_o < (n+n_o)\cdot \sup X - n_o x_o \leq n \sup X.$$

Proof of Theorem 3.3

It consists of three steps.

(1) $\lim_{\theta\downarrow\alpha} M'(\theta)=\inf X$, $\lim_{\theta\uparrow\beta} M'(\theta)=\sup X$.

Firstly suppose $\beta < \infty$. In such a case: $\sup X = +\infty$ (cf. the proof of Theorem 3.2) and $\lim_{\theta\uparrow\beta} M(\theta) = +\infty$. Hence, if $M'(\beta^-)<\infty$ , then, given $a \in(\alpha,\beta)$:

$$\infty>\int_a^\beta M'(\theta)d\theta=M(\beta^-)-M(a) \text{ (a contradiction!).}$$

In other words: $\beta<+\infty \Longrightarrow \sup X = \lim_{\theta\uparrow\beta} M'(\theta) = +\infty$.

An analogous argument shows that

$\alpha>-\infty \Longrightarrow \inf X = \lim_{\theta\downarrow\alpha} M'(\theta) = -\infty$

Suppose now $\beta= +\infty$ and let $x_o$ be an element of $X^\circ$. Then:

$$M'(\theta) \geq x_o \{1- \int_{(-\infty,x_o)} \exp\{\theta x-M(\theta)\}\mu(dx)\} + \int_{(-\infty,x_o)} x \exp\{\theta x-M(\theta)\}\mu(dx)\}$$

and, in view of inequality (2.4) of Diaconis and Ylvisaker (1979), one can determine $x_A \in(x_o, \sup X)\cap A$ such that

$$0 \leq \int_{(-\infty,x_o)}\exp\{\theta x-M(\theta)\}\mu(dx) \leq \{\mu(A)\}^{-1}\cdot \int_{(-\infty,x_o)} \exp\{\theta(x-x_A)\}\mu(dx)$$

$$0 \leq \int_{(-\infty,x_o)}|x| \exp\{\theta x-M(\theta)\}\mu(dx) \leq\{\mu(A)\}^{-1} \int_{(-\infty,x_o)}|x|\exp\{\theta(x-x_A)\}\mu(dx)$$

Hence, by monotone convergence: $\sup X \geq M'(\beta^-) \geq x_o$ and the thesis follows from the arbitrariness of $x_o$. An analogous argument applies to $M'(-\infty)$ when $\alpha=-\infty$ , in order to state: $\inf X \leq M'(\alpha^+) \leq x_o$ for all $x_o \in X^\circ$.

(2) If $\underline{\underline{B}}$ holds, then $M(\theta)$ is strictly monotonic and $\exp\{-M(\theta)\}$ is bounded on $\Theta$. Indeed, (1) and $M''(\theta) > 0$, for all $\theta$, imply:

$M'(\alpha^+)=0< M'(\theta')<M'(\theta'')< M'(\beta^-)= \sup X$ for all $\theta' <\theta''$,

if $\underline{\underline{B}}$ holds with $\mathrm{supp}(\mu) \subset \mathbb{R}^+$;

$M'(\alpha^+)=\inf X < M'(\theta')< M'(\theta'')< M'(\beta^-)=0$ for all $\theta'< \theta''$,

if $\underline{\underline{B}}$ holds with $\mathrm{supp}(\mu)\subset \{\mathbb{R}^+\}\cup\{0\}$

In both cases $M(\theta)$ turns out to be strictly monotonic and, from the inequalities:

$$\exp\{M(\theta)\} = \int_X \exp\{\theta x\}\mu(dx)>\mu(\{0\})> 0,$$

one deduces that $\exp\{-M(\theta)\}$ is bounded on $\Theta$.

(3) In view of (3.4) and Theorem 1 by Diaconis and Ylvisaker (1979), we have for all samples $(x_1^*,\dots,x_\nu^*,x_{\nu+1},\dots,x_{\nu+k})$, $k\geq 1$, such that $\bar{x}_k=\sum_{i=1}^k x_{\nu+i}/k \in X°$:

$$\infty >\int_\Theta \exp\{\nu(\theta\bar{x}_\nu^* -M(\theta))\}\exp \{k(\theta\bar{x}_k -M(\theta))\}\rho(d\theta)=$$

$$= \int_\Theta \exp \{\theta ( \Sigma_1^\nu x_j^* + \sum_{j=1}^k x_{\nu+j} ) - (\nu+k)M(\theta)\}\rho(d\theta);$$

hence, without loss of generality, we can suppose:

$$(*) \quad n_o + \nu >0,\ \bar{x}_\nu^* \in X° ,\ \frac{n_o x_o + \nu\bar{x}_\nu^*}{\nu + n_o} \in X° .$$

Now, if (3.3) holds, we obtain for $N>\nu+k$ and $q_{\bar{x}_\nu^*}(d\theta) = e^{\nu(\theta\bar{x}_\nu^* - M(\theta))}\rho(d\theta)$:

$$E(X_N|X_1=x_1^*,\dots,X_\nu =x_\nu^* , X_{\nu+1}= x_{\nu+1},\dots, X_{\nu+k} = x_{\nu+k}) =$$

$$= \frac{\int_\Theta M'(\theta) \exp\{k(\theta\bar{x}_k - M(\theta))\}q_{\bar{x}_\nu^*}(d\theta)}{\int_\Theta \exp \{k(\theta\bar{x}_k - M(\theta))\} q_{\bar{x}_\nu^*}(d\theta)} = \frac{(\nu+k)\ \bar{x}_{\nu+k}+n_o x_o}{\nu + k + n_o},$$

where the latter equality holds if:

$$\frac{1}{\nu + k} \{(\nu+k+n_o) \inf X - n_o x_o\}< \bar{x}_{\nu+k}< \frac{1}{\nu + k} \{(\nu+k+n_o) \sup X - n_o x_o\} ,$$

i.e.

$$\inf X + \frac{\nu+n_o}{k} \{\inf X - \frac{n_o x_o +\nu\bar{x}_\nu^*}{\nu + n_o}\} < \bar{x}_k < \sup X + \frac{\nu+n_o}{k} \{\sup X - \frac{n_o x_o +\nu\bar{x}_\nu^*}{\nu + n_o}\} .$$

In view of (*), these inequalities are always satisfied and consequently:

$$\frac{\int_\Theta M'(\theta) \exp\{k(\theta\bar{x}_k - M(\theta))\}\, q_{\bar{x}^*_\nu}(d\theta)}{\int_\Theta \exp\{k(\theta\bar{x}_k - M(\theta))\}\, q_{\bar{x}^*_\nu}(d\theta)} = \frac{k}{\nu+k+n_o}\bar{x}_k + \frac{\nu\bar{x}^*_\nu + n_o x_o}{\nu+k+n_o}$$

for all $k \geq 1$ and $\bar{x}_k \in \bar{X}$.

Therefore, if supp($\mu$) satisfies $\underline{\underline{A}}$, from Theorem 3 by Diaconis and Ylvisaker (1979) we deduce

$$q_{\bar{x}^*_\nu}(d\theta) = c.\exp\{\theta(\nu\bar{x}^*_\nu + n\, x_o) - (\nu+n_o)M(\theta)\}d\theta;$$

on the other hand, if supp($\mu$) satisfies $\underline{\underline{B}}$, the same conclusion is reached through point C of Theorem 1 by Cifarelli and Regazzini (1983) and step (2) of the present proof.

Hence, in both cases:

$$q_{\bar{x}^*_\nu}(d\theta) = \exp\{\nu(\theta\bar{x}^*_\nu - M(\theta))\}\rho(d\theta)$$
$$= c\exp\{\theta(\nu\,\bar{x}^*_\nu + n_o x_o) - (\nu+n_o)M(\theta)\}d\theta,$$

which yields the thesis.

ACKNOWLEDGEMENTS

Research supported in part by Ministero della Pubblica Istruzione (40%, Progetto di Ricerca "Modelli probabilistici") and by CNR - GNAFA.

REFERENCES

Barndorff-Nielsen, O., 1978, "Information and exponential families in statistical theory", J. Wiley, New York.

Berger, James, O., 1980, "Statistical decision theory", Springer-Verlag, New York.

Cifarelli, Donato Michele and Regazzini, Eugenio, 1983, Qualche osservazione sull'uso di distribuzioni iniziali coniugate alla famiglia esponenziale, Statistica, 43:415.

Dawid, A.P., 1983, Invariant prior distributions in "Encyclopedia of statistical sciences" vol. 4 (S.Kotz and N.L. Johnson eds-in-chief) J. Wiley, New York.

de Finetti, Bruno, 1937, La prévision: ses lois logiques, ses sources subjectives, Annales de l'Inst. Henri Poincaré, 7:1.

de Finetti, Bruno, 1970, "Teoria della probabilità", Einaudi Editore, Torino.

de Finetti, Bruno and Savage, Leonard J., 1962, Sul modo di scegliere le probabilità iniziali, Biblioteca del "Metron", serie C: Note e Commenti, 1:81.

De Groot, Morris H., 1970, "Optimal statistical decisions", Mc Graw-Hill, New York.

Diaconis, Persi and Ylvisaker, Donald, 1979, Conjugate priors for exponential families, Ann. Statist., 7:269

Diaconis, Persi and Ylvisaker, Donald, 1985, Quantifying prior opinion in "Bayesian Statistics 2. Proceedings of the second Valencia international meeting", September 6/10, 1983 (Bernardo, J.M., De Groot, M.H. Lindley, D.V., Smith, A.F.M. eds) North-Holland, Amsterdam.

Regazzini, Eugenio, 1985, Finitely additive conditional probabilities, to appear in Rendiconti del Seminario Matematico e Fisico di Milano, 55.

Regazzini, Eugenio, 1987, de Finetti's coherence and statistical inference, to apper in The Annals of Statistics,15, No 2.

Rényi, Alfred, 1955, On a new axiomatic theory of probability, Acta Math. Acad. Sci. Hung., 6:285.

AMS (1980 subject classification): 62E10, 62A15, 60A05

Keywords and phrases: conjugate prior, de Finetti's coherent probabilities, (regular) exponential families, finitely additive probabilities, noninformative priors.

# CALIBRATING AND COMBINING PRECIPITATION PROBABILITY FORECASTS

Robert T. Clemen

College of Business
University of Oregon
Eugene, OR 97403
U. S. A.

Robert L. Winkler

Fuqua School of Business
Duke University
Durham, NC 27706
U. S. A.

## INTRODUCTION

Imagine a decision maker who has heard from one or more information sources regarding the probability of some future event and who desires to use this information to revise his personal beliefs concerning the event. One approach to this problem involves the decision maker treating the probabilities as data in a Bayesian inferential problem, the output of which is an updated probability regarding the event in question. The thorniest part of the Bayesian combination procedure is the assessment of a likelihood function by the decision maker to represent his beliefs regarding the quality of the information and, in the case of multiple sources, the nature of the dependence among the sources.

The Bayesian approach to the use of probabilities from various sources is now well established. Morris (1974, 1977) was the first to characterize the problem in Bayesian terms. Lindley, Tversky, and Brown (1979), French (1980), and Lindley (1982) developed models for the specific case of a single information source. Models for multiple information sources have been provided by French (1981), Winkler (1981), Lindley (1983, 1985), Agnew (1985), Chang (1985), Genest and Schervish (1985), and Clemen (1987), to name a few. Excellent reviews and critiques of this literature are available in French (1985) and Genest and Zidek (1986).

In this paper, we apply a Bayesian model for adjusting and combining discrete probabilities in the context of forecasting tomorrow's weather. Since 1966 the National Weather Service (NWS) of the United States has provided probability of precipitation (PoP) forecasts as the official precipitation forecasts. Meteorologists have studied these forecasts extensively (e.g., Murphy, 1985) and have shown consistently that the forecasts generally are well-calibrated (i.e., when the PoP forecast is x, the long-run frequency of measurable precipitation is approximately x). The NWS also uses a numerical-statistical model of the global atmospheric system to prepare "guidance" PoP forecasts about twelve hours prior to the issuance of the official forecasts. Thus, local forecasters may use guidance forecasts in the formulation of official PoP forecasts. Meteorologists have studied the relative performance of guidance and official forecasts; a review of this literature as well as a more complete overview of the forecasting process can be found in Murphy and Winkler (1984).

The fact that the guidance forecast is available for use by the official forecaster might lead us to suspect that the guidance and official forecasts are highly dependent, perhaps to the extent that the information provided by the guidance forecast is completely incorporated by the local forecaster in his formulation of the official PoP forecast. This issue was recently addressed by Clemen (1985) and Clemen and Murphy (1986a) with the conclusion that, while the official forecast apparently incorporates most of the information (in a statistical sense), it may be possible to improve the performance of the official forecast by combining the guidance and official forecasts through joint calibration (DeGroot and Fienberg, 1982, 1983). A follow-up study by Clemen and Murphy (1986b) showed that joint calibration indeed produced somewhat better forecasts, although the level of improvement was about the same as that obtained through individual frequency calibration of the official forecasts. In contrast, simply averaging the two forecasts, with or without calibration, performed somewhat better than joint calibration.

One problem with the joint calibration procedure is that it requires massive amounts of data to estimate the joint calibration function reliably. Thus, there is some motivation to turn to a modeling approach, using tractable probability models with known properties to represent the stochastic nature of the official and guidance forecasts. Lindley (1982) provided a model for the Bayesian calibration of discrete probabilities, using as likelihood functions normal distributions for the log odds of the stated probabilities, conditioned on the occurrence or non-occurrence of the event. Lindley (1985) and Chang (1985) extended this model to the case of probabilities from multiple sources.

Our objective in this paper is to apply the normal log-odds model to the calibration and combination of official and guidance PoP forecasts. Thus, we require two normal distributions for the forecast log odds of rain, one given that it rains tomorrow and one given that it does not rain. The prior probability is simply the climatological probability (long-run frequency) of precipitation. The analysis includes some measures of the degree to which the calibrated and combined forecasts improve on the official forecast and how well they perform relative to the uncalibrated forecasts and simple combinations thereof.

The paper is organized as follows. First, we discuss the calibration of individual probabilities via the normal log-odds model, describe the data, and present the analysis of the uncalibrated and calibrated probabilities. Next, we deal with models and empirical results regarding the combination of probabilities. We conclude with a discussion of the results and their implications.

## CALIBRATING PROBABILITIES

### The Normal Log-Odds Model

Suppose that an information source provides probability $p$, his probability that measurable precipitation (rain) will occur tomorrow. We will be interested in the corresponding log odds $q = \log[p/(1-p)]$. If the prior probability of rain is the climatological probability of rain, denoted by $\gamma$, the posterior log-odds $q^*$ of rain can be found using Bayes' theorem in log-odds form:

$$(1) \qquad \begin{aligned} q^* &= \log[\mathrm{P}(\text{rain}|q) / \mathrm{P}(\text{no rain}|q)] \\ &= \log[L(q|\text{rain}) / L(q|\text{no rain})] + \log[\gamma/(1-\gamma)], \end{aligned}$$

where $L(q|\text{rain})$ and $L(q|\text{no rain})$ are the likelihood functions given rain and no rain, respectively.

Following Lindley (1982), we model the likelihood functions $L(q|\text{rain})$ and $L(q|\text{no rain})$ with normal distributions having means $\mu_1$ and $\mu_0$ and variances $\sigma_1^2$ and $\sigma_0^2$, respectively. Substituting the expressions for the two normal distributions into (1) and manipulating the expression algebraically, we obtain an expression that is quadratic in $q$:

$$(2) \qquad \begin{aligned} q^* = \{&\log(\sigma_0^2/\sigma_1^2) - (\sigma_1^{-2} - \sigma_0^{-2})q^2 + 2(\sigma_1^{-2}\mu_1 - \sigma_0^{-2}\mu_0)q \\ &- \sigma_1^{-2}\mu_1^2 + \sigma_0^{-2}\mu_0^2\}/2 + \log[\gamma/(1-\gamma)]. \end{aligned}$$

If $\sigma_1 = \sigma_0 = \sigma$, the expression simplifies to a linear form in $q$ :

(3) $$q^* = \{2(\mu_1 - \mu_0)q - (\mu_1^2 - \mu_0^2)\} / 2\sigma^2 + \log[\gamma/(1-\gamma)].$$

The parameters $\mu_1$, $\mu_0$, $\sigma_1$, and $\sigma_0$ (or $\sigma$) can be estimated from historical data and substituted into the expression. When these estimates are supplied, (2) or (3) provides a way to find $q^*$, which represents the calibrated log odds. [We use the term "calibrated" in the same subjective sense as Lindley (1982).] This approach yields essentially the same results as a full Bayesian analysis with a diffuse prior distribution on the normal parameters.

Table 1. NWS offices for which guidance and official forecasts were analyzed in this study. Also shown are the overall sample climatological probabilities of measurable precipitation in the cool and warm seasons in the respective locations.

| | Climatological probabilities of precipitation | |
|---|---|---|
| Office | Cool season | Warm season |
| Albany, NY | .24 | .24 |
| Atlanta, GA | .20 | .19 |
| Boston, MA | .23 | .22 |
| Dallas, TX | .13 | .13 |
| Denver, CO | .12 | .16 |
| Des Moines, IA | .18 | .23 |
| Phoenix, AZ | .08 | .04 |
| Portland, OR | .40 | .21 |

The Data

The data analyzed in this paper consist of guidance and official PoP forecasts for eight NWS offices in various parts of the United States. These data, covering the period from April 1972 through September 1983, were provided by the NWS Techniques Development Laboratory. The offices are shown in Table 1, along with the climatological probabilities of measurable precipitation in the respective areas for the warm (April-September) and cool (October-March) seasons. Guidance and official forecasts are made twice each day, in the morning and evening (in conjunction with the so-called 0000 and 1200 GMT cycle times). On each occasion, forecasts are formulated for three consecutive 12-hour periods, or lead times. These lead times are 12-24 hours, 24-36 hours, and 36-48 hours after the guidance forecast is issued.

Meteorologists traditionally analyze the warm and cool seasons separately because of differences in weather patterns. Moreover, the characteristics of the forecasts vary considerably with lead time (e.g., forecasts are less accurate as the lead time increases). While there may be some circumstances under which the characteristics of forecasts vary with cycle time, Clemen and Murphy (1986a) found virtually no such differences. Thus, forecasts for both cycle times were aggregated for our analysis. For each station, then, we analyzed six different kinds of forecasts, corresponding to six combinations of season and lead time.

For the analysis, the data set was divided into two subsets. The first seven years of data (April 1972 - March 1979) were used to fit the log-odds model, and this fitting was done separately for each combination of station, season, and lead time. The sample means and

Table 2. Average quadratic scores (MSEs) for climatology, uncalibrated, and calibrated forecasts.

| Forecast type: | Climatology | Official | Official Calibrated Variances Equal | Official Calibrated Variances Unequal | Guidance | Guidance Calibrated Variances Equal | Guidance Calibrated Variances Unequal | Sample Size |
|---|---|---|---|---|---|---|---|---|
| COOL SEASON | | | | | | | | |
| 12-24 hours | | | | | | | | |
| Albany | 0.1720 | 0.1022 | 0.1016 | 0.0928 | 0.0976 | 0.1004 | 0.0931 | 894 |
| Atlanta | 0.1631 | 0.0740 | 0.0794 | 0.0748 | 0.0811 | 0.0904 | 0.0808 | 1058 |
| Boston | 0.1641 | 0.0822 | 0.0871 | 0.0753 | 0.0721 | 0.0825 | 0.0705 | 911 |
| Dallas | 0.1005 | 0.0632 | 0.0635 | 0.0654 | 0.0658 | 0.0676 | 0.0654 | 1036 |
| Denver | 0.1045 | 0.0593 | 0.0617 | 0.0588 | 0.0596 | 0.0643 | 0.0594 | 1061 |
| Des Moines | 0.1341 | 0.0805 | 0.0827 | 0.0804 | 0.0778 | 0.0803 | 0.0778 | 1030 |
| Phoenix | 0.0656 | 0.0329 | 0.0341 | 0.0308 | 0.0399 | 0.0430 | 0.0378 | 1002 |
| Portland | 0.2442 | 0.1172 | 0.1352 | 0.1185 | 0.1301 | 0.1559 | 0.1352 | 1036 |
| 24-36 hours | | | | | | | | |
| Albany | 0.1690 | 0.1104 | 0.1100 | 0.1039 | 0.1181 | 0.1173 | 0.1137 | 894 |
| Atlanta | 0.1605 | 0.0932 | 0.0959 | 0.0935 | 0.1011 | 0.1062 | 0.1002 | 1058 |
| Boston | 0.1682 | 0.0945 | 0.1001 | 0.0912 | 0.0937 | 0.1013 | 0.0924 | 911 |
| Dallas | 0.1000 | 0.0782 | 0.0770 | 0.0791 | 0.0719 | 0.0754 | 0.0738 | 1036 |
| Denver | 0.0939 | 0.0672 | 0.0674 | 0.0684 | 0.0692 | 0.0706 | 0.0690 | 1061 |
| Des Moines | 0.1336 | 0.0878 | 0.0922 | 0.0881 | 0.0878 | 0.0943 | 0.0904 | 1030 |
| Phoenix | 0.0624 | 0.0405 | 0.0415 | 0.0407 | 0.0466 | 0.0491 | 0.0460 | 1002 |
| Portland | 0.2434 | 0.1512 | 0.1629 | 0.1512 | 0.1582 | 0.1765 | 0.1604 | 1036 |
| 36-48 hours | | | | | | | | |
| Albany | 0.1688 | 0.1218 | 0.1187 | 0.1157 | 0.1250 | 0.1224 | 0.1211 | 894 |
| Atlanta | 0.1505 | 0.0964 | 0.1006 | 0.0967 | 0.1074 | 0.1093 | 0.1078 | 1058 |
| Boston | 0.1608 | 0.1036 | 0.1089 | 0.1037 | 0.1107 | 0.1144 | 0.1117 | 911 |
| Dallas | 0.1039 | 0.0817 | 0.0811 | 0.0807 | 0.0833 | 0.0863 | 0.0865 | 1036 |
| Denver | 0.1001 | 0.0789 | 0.0795 | 0.0782 | 0.0820 | 0.0840 | 0.0820 | 1061 |
| Des Moines | 0.1274 | 0.1019 | 0.1032 | 0.1021 | 0.0995 | 0.1038 | 0.1048 | 1030 |
| Phoenix | 0.0622 | 0.0460 | 0.0472 | 0.0460 | 0.0520 | 0.0533 | 0.0516 | 1002 |
| Portland | 0.2433 | 0.1639 | 0.1739 | 0.1633 | 0.1696 | 0.1881 | 0.1754 | 1036 |
| WARM SEASON | | | | | | | | |
| 12-24 hours | | | | | | | | |
| Albany | 0.1867 | 0.1094 | 0.1165 | 0.1089 | 0.1112 | 0.1194 | 0.1138 | 794 |
| Atlanta | 0.1507 | 0.0995 | 0.1056 | 0.1033 | 0.1028 | 0.1096 | 0.1066 | 1156 |
| Boston | 0.1749 | 0.1107 | 0.1165 | 0.1101 | 0.1125 | 0.1200 | 0.1144 | 927 |
| Dallas | 0.1053 | 0.0790 | 0.0814 | 0.0821 | 0.0805 | 0.0846 | 0.0842 | 1171 |
| Denver | 0.1329 | 0.0959 | 0.1006 | 0.1005 | 0.0971 | 0.1051 | 0.1038 | 1282 |
| Des Moines | 0.1676 | 0.1206 | 0.1288 | 0.1237 | 0.1225 | 0.1305 | 0.1275 | 1205 |
| Phoenix | 0.0420 | 0.0366 | 0.0355 | 0.0354 | 0.0368 | 0.0358 | 0.0359 | 1320 |
| Portland | 0.1563 | 0.0932 | 0.1013 | 0.0941 | 0.1030 | 0.1164 | 0.1055 | 1335 |
| 24-36 hours | | | | | | | | |
| Albany | 0.1755 | 0.1239 | 0.1261 | 0.1207 | 0.1309 | 0.1307 | 0.1253 | 794 |
| Atlanta | 0.1462 | 0.1057 | 0.1088 | 0.1111 | 0.1112 | 0.1131 | 0.1092 | 1156 |
| Boston | 0.1678 | 0.1244 | 0.1234 | 0.1198 | 0.1242 | 0.1263 | 0.1218 | 927 |
| Dallas | 0.1072 | 0.0906 | 0.0911 | 0.0934 | 0.0913 | 0.0928 | 0.0917 | 1171 |
| Denver | 0.1323 | 0.1074 | 0.1088 | 0.1100 | 0.1060 | 0.1098 | 0.1097 | 1282 |
| Des Moines | 0.1618 | 0.1282 | 0.1312 | 0.1282 | 0.1268 | 0.1321 | 0.1287 | 1205 |
| Phoenix | 0.0360 | 0.0356 | 0.0347 | 0.0381 | 0.0351 | 0.0342 | 0.0345 | 1320 |
| Portland | 0.1538 | 0.1122 | 0.1200 | 0.1123 | 0.1160 | 0.1226 | 0.1188 | 1335 |
| 36-48 hours | | | | | | | | |
| Albany | 0.1731 | 0.1400 | 0.1361 | 0.1340 | 0.1390 | 0.1383 | 0.1351 | 794 |
| Atlanta | 0.1519 | 0.1098 | 0.1158 | 0.1161 | 0.1170 | 0.1208 | 0.1202 | 1156 |
| Boston | 0.1622 | 0.1284 | 0.1267 | 0.1257 | 0.1266 | 0.1288 | 0.1263 | 927 |
| Dallas | 0.1036 | 0.0932 | 0.0928 | 0.0936 | 0.0898 | 0.0924 | 0.0912 | 1171 |
| Denver | 0.1336 | 0.1118 | 0.1140 | 0.1170 | 0.1117 | 0.1165 | 0.1148 | 1282 |
| Des Moines | 0.1554 | 0.1302 | 0.1325 | 0.1309 | 0.1285 | 0.1336 | 0.1329 | 1205 |
| Phoenix | 0.0394 | 0.0377 | 0.0368 | 0.0397 | 0.0375 | 0.0356 | 0.0359 | 1320 |
| Portland | 0.1513 | 0.1178 | 0.1248 | 0.1187 | 0.1178 | 0.1237 | 0.1201 | 1335 |

Table 3. Average percentage improvements in MSE for the uncalibrated and calibrated forecasts. The upper (lower) figure in each cell gives improvement relative to the official forecast (climatology).

| Forecast type: | Climatology | Official | Official Calibrated Variances Equal | Official Calibrated Variances Unequal | Guidance | Guidance Calibrated Variances Equal | Guidance Calibrated Variances Unequal |
|---|---|---|---|---|---|---|---|
| COOL SEASON | | | | | | | |
| 12-24 hours | -87.34 | 0.00 | -4.89 | 2.37 | -3.32 | -12.49 | -2.13 |
| | 0.00 | 45.94 | 43.51 | 47.20 | 44.39 | 39.75 | 45.06 |
| 24-36 hours | -54.76 | 0.00 | -2.81 | 0.67 | -3.68 | -9.29 | -3.50 |
| | 0.00 | 34.74 | 33.05 | 35.11 | 32.54 | 28.97 | 32.59 |
| 36-48 hours | -39.07 | 0.00 | -2.11 | 0.85 | -5.09 | -8.62 | -6.37 |
| | 0.00 | 27.57 | 26.16 | 28.20 | 24.04 | 21.52 | 23.07 |
| WARM SEASON | | | | | | | |
| 12-24 hours | -46.67 | 0.00 | -4.77 | -1.48 | -2.80 | -9.43 | -5.77 |
| | 0.00 | 30.77 | 27.73 | 29.78 | 28.97 | 24.70 | 26.99 |
| 24-36 hours | -27.57 | 0.00 | -1.59 | -1.42 | -1.41 | -3.38 | -1.11 |
| | 0.00 | 20.76 | 19.65 | 19.43 | 19.80 | 18.35 | 20.02 |
| 36-48 hours | -21.38 | 0.00 | -1.02 | -1.36 | 0.16 | -1.81 | -0.50 |
| | 0.00 | 17.06 | 16.35 | 15.90 | 17.32 | 15.81 | 16.86 |
| Overall Average Improvement | -46.13 | 0.00 | -2.86 | -0.06 | -2.69 | -7.50 | -3.23 |
| | 0.00 | 29.47 | 27.74 | 29.27 | 27.84 | 24.85 | 27.43 |

variances of the relative frequency distributions of log odds corresponding to official and guidance probabilities conditional on rain and no rain were used as estimates of the model parameters. For each lead time/season combination at each station, forecasts and observations for over 2000 occasions were available for the April 1972-March 1979 period. The remaining four and one-half years of data (April 1979-September 1983) were used to evaluate the calibrated probabilities generated from the model as well as the official and guidance forecasts. Even though no fitting was necessary for the official and guidance forecasts, their evaluation was based only on the April 1979-September 1983 period to facilitate comparison with the performance of the calibrated probabilities.

## Calibrating Probabilities: Empirical Results

In our analysis of individual forecasts, we investigated the following probabilities:

1) Official and guidance forecasts.
2) Calibrated official and guidance forecasts using the normal log-odds model with equal variances.
3) Calibrated official and guidance forecasts using the normal log-odds model with unequal variances.

For each type of forecast, we computed average scores for each of the six combinations of season and lead time using a quadratic scoring rule. The average quadratic score is equivalent to a mean square error (MSE); a lower score therefore indicates better performance.

The MSEs are presented in Table 2, and average percentage improvements for the different types of forecasts over the official forecast and climatology are given in Table 3. First, note from Table 3 that every type of forecast easily outperformed climatology. As anticipated, the improvements over climatology were greater as the lead time decreased and for the cool season as opposed to the warm season.

Next, in looking at the raw, uncalibrated probabilities, we see from Table 3 that the guidance forecasts generally performed worse than the official forecasts. Overall, the guidance forecasts performed 2.69% worse than the official forecasts. The differences in the table may not seem large, but they are equivalent to changes that have occurred over a period of a few years as forecasts have improved (see Murphy and Sabin, 1986). Recall that the local forecasters have access to the guidance forecasts before they formulate the official forecasts; hence the better performance of the official forecasts is not surprising.

Tables 2 and 3 also show that for both official and guidance forecasts, the raw forecasts outperformed the calibrated forecasts. The calibrated forecasts using the normal log-odds model with equal variances were particularly weak, resulting in a 2.86% overall increase in MSE for the calibrated official forecasts as compared with the raw official forecasts and a 4.68% increase for the calibrated guidance forecasts as compared with the raw guidance forecasts. Relaxing the assumption of equal variances led to improvements, resulting in forecasts only slightly worse than the raw forecasts.

In comparing the models with equal and unequal variances, it is helpful to look at the means and standard deviations of log odds conditional on rain and no rain. The differences between the means and the ratios of the standard deviations for the two conditional distributions are given in Table 4. The standard deviations given no rain were, for the most part, larger than those given rain. For example, with a 12-24 hour lead time in the cool season at Portland, these standard deviations for the official forecast were 2.361 and 1.750. The two normal distributions of log odds in this case are shown in Figure 1.

Of course, the unequal-variances model offers more flexibility than the equal-variances model. From (3), the equal-variances model gives calibrated log odds linear in the uncalibrated log odds. This implies calibration curves shaped like the solid curve in Figure 2, which shows calibration curves (in probabilities, not log odds) for the case of the official forecast at Portland during the cool season and for the 12-24 hour lead time. The unequal-variances model given by (2) adds a quadratic term and is less restrictive in terms of the shape of the resulting calibration curve. In Figure 2, the dashed curve (the unequal-variances model) seems much more consistent with typical frequency calibration curves for PoP forecasts (e.g., Murphy, 1985) than does the solid curve. The corresponding frequency calibration data are included in Figure 2 for comparative purposes.

In summary, among the calibrated and uncalibrated forecasts, the official forecasts performed best. Overall, the guidance forecasts were not quite as good as the local forecasts. Calibration via the equal-variances model produced the worst results. The unequal-variances model did better, producing forecasts that were roughly comparable to the raw probabilities.

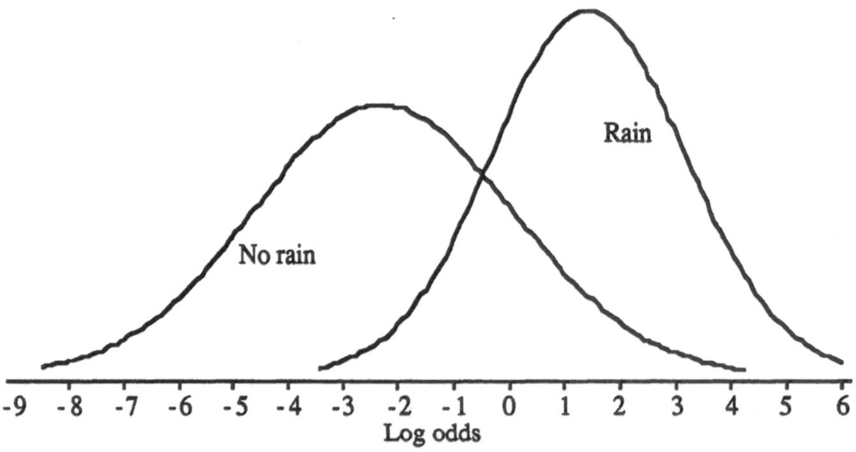


Figure 1. Distributions of official forecast log odds conditional on rain and no rain at Portland (cool season, 12-24 hour lead time).

Table 4. Differences between estimated means and ratios of estimated standard deviations for the distributions of log odds conditional on rain and no rain. The upper figure in each cell gives the difference between the means ($\mu_1 - \mu_0$), and the figure in parentheses gives the ratio of the standard deviations ($\sigma_1/\sigma_0$).

| | OFFICIAL FORECAST | | | | | |
|---|---|---|---|---|---|---|
| | COOL SEASON | | | WARM SEASON | | |
| | 12-24 hours | 24-36 hours | 36-48 hours | 12-24 hours | 24-36 hours | 36-48 hours |
| Albany | 3.69 (0.92) | 2.85 (0.90) | 2.09 (0.81) | 3.18 (0.83) | 2.37 (0.81) | 1.61 (0.79) |
| Atlanta | 4.82 (0.94) | 3.74 (0.76) | 2.85 (0.73) | 2.60 (0.76) | 2.06 (0.63) | 1.50 (0.67) |
| Boston | 4.57 (0.94) | 3.60 (0.88) | 2.66 (0.83) | 3.26 (0.85) | 2.46 (0.84) | 1.78 (0.76) |
| Dallas | 3.74 (1.02) | 2.94 (1.00) | 2.13 (0.90) | 2.32 (0.70) | 1.71 (0.72) | 1.16 (0.75) |
| Denver | 3.60 (0.97) | 2.60 (0.80) | 1.97 (0.73) | 1.85 (0.86) | 1.31 (0.85) | 0.96 (0.79) |
| Des Moines | 3.51 (1.14) | 2.26 (1.02) | 1.37 (0.97) | 2.44 (0.96) | 1.58 (0.78) | 0.97 (0.90) |
| Phoenix | 4.16 (0.85) | 3.31 (0.71) | 2.83 (0.76) | 2.62 (0.57) | 2.10 (0.57) | 1.76 (0.63) |
| Portland | 3.73 (0.74) | 2.78 (0.73) | 2.00 (0.74) | 3.71 (0.83) | 2.85 (0.75) | 2.21 (0.77) |

| | GUIDANCE FORECAST | | | | | |
|---|---|---|---|---|---|---|
| | COOL SEASON | | | WARM SEASON | | |
| | 12-24 hours | 24-36 hours | 36-48 hours | 12-24 hours | 24-36 hours | 36-48 hours |
| Albany | 3.40 (0.86) | 2.82 (0.92) | 2.08 (0.79) | 2.60 (0.77) | 1.94 (0.80) | 1.34 (0.77) |
| Atlanta | 3.95 (0.89) | 3.25 (0.70) | 2.04 (0.63) | 1.93 (0.77) | 1.68 (0.71) | 1.16 (0.74) |
| Boston | 4.02 (0.95) | 3.39 (0.95) | 2.38 (0.80) | 2.51 (0.92) | 2.03 (0.93) | 1.39 (0.80) |
| Dallas | 2.90 (0.96) | 2.27 (0.80) | 1.83 (0.71) | 1.63 (0.79) | 1.23 (0.86) | 0.85 (0.83) |
| Denver | 2.72 (0.77) | 2.20 (0.75) | 1.70 (0.74) | 1.64 (0.97) | 1.31 (0.92) | 1.14 (0.78) |
| Des Moines | 3.14 (0.98) | 2.21 (0.88) | 1.50 (0.77) | 1.78 (0.94) | 1.36 (0.86) | 0.96 (0.86) |
| Phoenix | 3.18 (0.65) | 2.65 (0.75) | 2.37 (0.74) | 1.80 (0.92) | 1.67 (0.74) | 1.14 (0.95) |
| Portland | 2.62 (0.66) | 2.16 (0.71) | 1.59 (0.64) | 3.03 (0.77) | 2.36 (0.77) | 1.85 (0.79) |

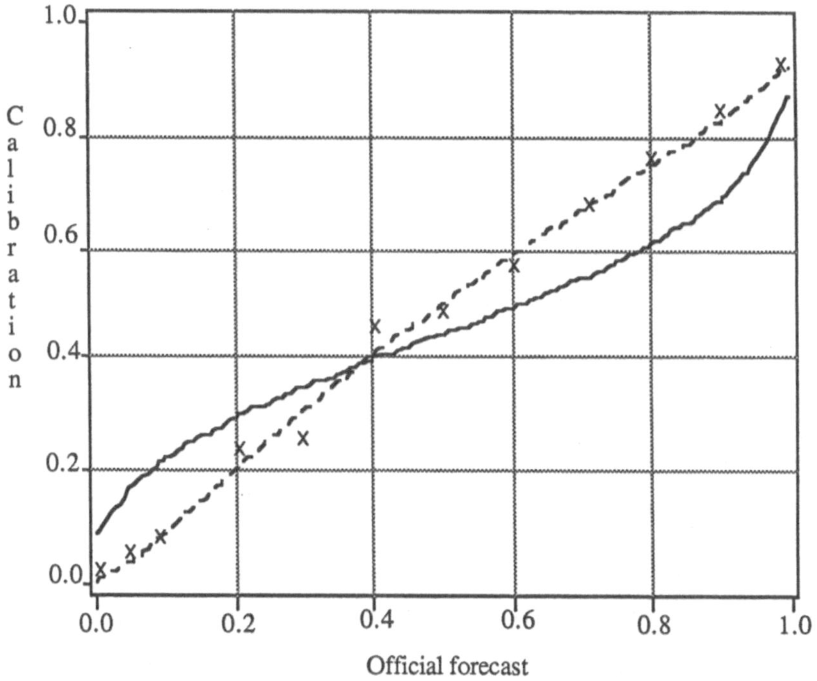


Figure 2. Calibration curves for the official forecast at Portland (cool season, 12-24 hour lead time). The solid line is the calibration curve from the equal-variances model, and the dashed line is the calibration curve from the unequal-variances model. For comparison, the direct frequency calibration data (represented by x's) are included.

## COMBINING PROBABILITIES

### The Multivariate Log-Odds Model

The multivariate log-odds model is a straightforward generalization of the univariate log-odds model discussed above. Let $p_i$ denote information source $i$'s probability of rain, $q_i$ the corresponding log odds, and $\boldsymbol{q} = (q_1, \ldots, q_k)'$ the vector of log odds from $k$ experts, where a prime indicates transposition. The likelihood functions $L(q|\text{rain})$ and $L(q|\text{no rain})$ are modeled with normal distributions having mean vectors $M_1$ and $M_0$ and covariance matrices $\Sigma_1$ and $\Sigma_0$, respectively. Now the multivariate counterparts of (2) and (3) are

(4) $$q^* = \{\log(|\Sigma_0| / |\Sigma_1|) - \boldsymbol{q}'(\Sigma_1^{-1} - \Sigma_0^{-1})\boldsymbol{q} + 2\boldsymbol{q}'(\Sigma_1^{-1}M_1 - \Sigma_0^{-1}M_0) - M_1'\Sigma_1^{-1}M_1 + M_0'\Sigma_0^{-1}M_0\} / 2 + \log[\gamma/(1-\gamma)]$$

and

(5) $$q^* = \boldsymbol{q}'\Sigma^{-1}(M_1 - M_0) - (M_1 + M_0)'\Sigma^{-1}(M_1 - M_0) / 2 + \log[\gamma/(1-\gamma)].$$

These models assume unequal and equal ($\Sigma_1 = \Sigma_0 = \Sigma$) covariance matrices, respectively. For brevity, we will refer to the models as having unequal or equal variances.

### The Data

The data set is as described above, as is the separation into subsets for fitting and for evaluation of the different approaches. For the combination of forecasts through the multivariate log-odds model, the correlations between the log odds from the official forecasts and the log odds from the guidance forecasts, conditional on rain and no rain, were estimated in addition to the means and variances. These estimates are shown in Table 5. Table 6 gives sample sizes used in estimating the parameters of both likelihood functions for each combination of station, season, and lead time

### Combining Probabilities: Empirical Results

In the analysis of combined forecasts, we considered the following combination techniques:

1) Simple averages of probabilities and simple averages of log odds (subsequently transformed back into probabilities).
2) Simple averages of calibrated probabilities and simple averages of calibrated log odds.
3) The combined forecast using the multivariate normal log-odds model with equal variances.
4) The combined forecast using the multivariate normal log-odds model with unequal variances.

Again, we computed the average quadratic scores (MSEs) for each combining technique for the six season/lead time combinations. These MSEs are presented in Table 7, and the average percentage improvements for the different techniques over the official forecast and over climatology are shown in Table 8.

The simple averages of the raw, uncalibrated probabilities and log odds performed well relative to the other techniques. The average of log odds consistently performed slightly better than the average of probabilities, with average percentage improvement over the official forecast performance ranging from 1.42% to 4.14%, depending on the season and lead time. Overall, the MSE for the average of log odds was 2.79% lower than that of the official forecast.

Table 5. Estimated correlations for the multivariate distributions of log odds conditional on rain and no rain. The upper (lower) figure gives the estimated correlation of the official and guidance log odds conditional on rain (no rain).

| | COOL SEASON | | | WARM SEASON | | |
|---|---|---|---|---|---|---|
| | 12-24 hours | 24-36 hours | 36-48 hours | 12-24 hours | 24-36 hours | 36-48 hours |
| Albany | 0.75<br>0.60 | 0.70<br>0.62 | 0.70<br>0.59 | 0.65<br>0.68 | 0.66<br>0.68 | 0.69<br>0.64 |
| Atlanta | 0.66<br>0.69 | 0.72<br>0.70 | 0.69<br>0.65 | 0.68<br>0.68 | 0.69<br>0.72 | 0.64<br>0.69 |
| Boston | 0.76<br>0.65 | 0.77<br>0.68 | 0.74<br>0.61 | 0.72<br>0.70 | 0.66<br>0.66 | 0.67<br>0.65 |
| Dallas | 0.64<br>0.65 | 0.57<br>0.60 | 0.57<br>0.55 | 0.54<br>0.53 | 0.54<br>0.55 | 0.45<br>0.47 |
| Denver | 0.48<br>0.69 | 0.62<br>0.62 | 0.59<br>0.59 | 0.70<br>0.61 | 0.61<br>0.62 | 0.61<br>0.56 |
| Des Moines | 0.69<br>0.66 | 0.75<br>0.63 | 0.70<br>0.62 | 0.70<br>0.71 | 0.77<br>0.72 | 0.72<br>0.69 |
| Phoenix | 0.62<br>0.73 | 0.64<br>0.69 | 0.68<br>0.66 | 0.52<br>0.58 | 0.40<br>0.55 | 0.23<br>0.48 |
| Portland | 0.61<br>0.78 | 0.64<br>0.78 | 0.63<br>0.75 | 0.68<br>0.72 | 0.74<br>0.67 | 0.69<br>0.61 |

Table 6. Sample sizes used to estimate the parameters of the normal distributions conditional on rain and no rain. The upper (lower) figure gives the number of occurrences when rain (no rain) occurred.

| | COOL SEASON | | | WARM SEASON | | |
|---|---|---|---|---|---|---|
| | 12-24 hours | 24-36 hours | 36-48 hours | 12-24 hours | 24-36 hours | 36-48 hours |
| Albany | 458<br>1487 | 465<br>1480 | 468<br>1477 | 498<br>1556 | 497<br>1557 | 494<br>1560 |
| Atlanta | 416<br>1714 | 422<br>1708 | 416<br>1714 | 423<br>1809 | 425<br>1807 | 424<br>1808 |
| Boston | 487<br>1639 | 491<br>1635 | 508<br>1618 | 471<br>1698 | 469<br>1700 | 456<br>1713 |
| Dallas | 266<br>1831 | 271<br>1826 | 268<br>1829 | 303<br>1893 | 293<br>1903 | 294<br>1902 |
| Denver | 263<br>1885 | 263<br>1885 | 266<br>1882 | 383<br>1879 | 372<br>1890 | 370<br>1892 |
| Des Moines | 386<br>1773 | 391<br>1768 | 397<br>1762 | 514<br>1733 | 507<br>1740 | 505<br>1742 |
| Phoenix | 179<br>1915 | 178<br>1916 | 177<br>1917 | 81<br>2074 | 80<br>2075 | 74<br>2081 |
| Portland | 865<br>1296 | 868<br>1293 | 871<br>1290 | 471<br>1781 | 465<br>1787 | 463<br>1789 |

Table 7. Average quadratic scores (MSEs) for the combined forecasts.

| Forecast type: | Average Prob. | Average Calibrated Prob. | Average Log Odds | Average Calibrated Log Odds | Combined Variances Equal | Combined Variances Unequal | Sample Size |
|---|---|---|---|---|---|---|---|
| COOL SEASON | | | | | | | |
| 12-24 hours | | | | | | | |
| Albany | 0.0934 | 0.0882 | 0.0934 | 0.0884 | 0.0939 | 0.0900 | 894 |
| Atlanta | 0.0737 | 0.0735 | 0.0729 | 0.0727 | 0.0780 | 0.0736 | 1058 |
| Boston | 0.0716 | 0.0681 | 0.0710 | 0.0676 | 0.0780 | 0.0681 | 911 |
| Dallas | 0.0607 | 0.0612 | 0.0606 | 0.0610 | 0.0616 | 0.0643 | 1036 |
| Denver | 0.0554 | 0.0540 | 0.0554 | 0.0538 | 0.0584 | 0.0551 | 1061 |
| Des Moines | 0.0760 | 0.0759 | 0.0754 | 0.0754 | 0.0773 | 0.0769 | 1030 |
| Phoenix | 0.0354 | 0.0323 | 0.0353 | 0.0321 | 0.0341 | 0.0314 | 1002 |
| Portland | 0.1177 | 0.1423 | 0.1171 | 0.1208 | 0.1346 | 0.1165 | 1036 |
| 24-36 hours | | | | | | | |
| Albany | 0.1079 | 0.1048 | 0.1067 | 0.1042 | 0.1180 | 0.1068 | 894 |
| Atlanta | 0.0933 | 0.0928 | 0.0933 | 0.0929 | 0.0953 | 0.0968 | 1058 |
| Boston | 0.0882 | 0.0870 | 0.0870 | 0.0861 | 0.0939 | 0.0873 | 911 |
| Dallas | 0.0713 | 0.0725 | 0.0719 | 0.0724 | 0.0732 | 0.0770 | 1036 |
| Denver | 0.0646 | 0.0642 | 0.0638 | 0.0632 | 0.0651 | 0.0658 | 1061 |
| Des Moines | 0.0851 | 0.0868 | 0.0846 | 0.0862 | 0.0882 | 0.0852 | 1030 |
| Phoenix | 0.0422 | 0.0411 | 0.0421 | 0.0410 | 0.0414 | 0.0408 | 1002 |
| Portland | 0.1482 | 0.1499 | 0.1483 | 0.1501 | 0.1621 | 0.1486 | 1036 |
| 36-48 hours | | | | | | | |
| Albany | 0.1170 | 0.1145 | 0.1166 | 0.1171 | 0.1142 | 0.1184 | 894 |
| Atlanta | 0.0975 | 0.0981 | 0.0968 | 0.1023 | 0.0987 | 0.1039 | 1058 |
| Boston | 0.1004 | 0.1028 | 0.1001 | 0.1078 | 0.1040 | 0.1074 | 911 |
| Dallas | 0.0785 | 0.0809 | 0.0781 | 0.0816 | 0.0790 | 0.0800 | 1036 |
| Denver | 0.0771 | 0.0771 | 0.0775 | 0.0803 | 0.0783 | 0.0778 | 1061 |
| Des Moines | 0.0983 | 0.1011 | 0.0982 | 0.1018 | 0.0999 | 0.1023 | 1030 |
| Phoenix | 0.0476 | 0.0472 | 0.0475 | 0.0497 | 0.0475 | 0.0476 | 1002 |
| Portland | 0.1611 | 0.1645 | 0.1610 | 0.1647 | 0.1734 | 0.1609 | 1036 |
| WARM SEASON | | | | | | | |
| 12-24 hours | | | | | | | |
| Albany | 0.1062 | 0.1083 | 0.1055 | 0.1073 | 0.1113 | 0.1074 | 794 |
| Atlanta | 0.0980 | 0.1029 | 0.0969 | 0.1019 | 0.1027 | 0.1023 | 1156 |
| Boston | 0.1070 | 0.1091 | 0.1061 | 0.1084 | 0.1126 | 0.1081 | 927 |
| Dallas | 0.0762 | 0.0810 | 0.0757 | 0.0807 | 0.0783 | 0.0810 | 1171 |
| Denver | 0.0933 | 0.0999 | 0.0925 | 0.0991 | 0.0974 | 0.0981 | 1282 |
| Des Moines | 0.1178 | 0.1228 | 0.1182 | 0.1225 | 0.1259 | 0.1212 | 1205 |
| Phoenix | 0.0358 | 0.0351 | 0.0355 | 0.0350 | 0.0347 | 0.0359 | 1320 |
| Portland | 0.0930 | 0.0936 | 0.0927 | 0.0932 | 0.1004 | 0.0908 | 1335 |
| 24-36 hours | | | | | | | |
| Albany | 0.1232 | 0.1202 | 0.1229 | 0.1199 | 0.1236 | 0.1202 | 794 |
| Atlanta | 0.1057 | 0.1062 | 0.1053 | 0.1058 | 0.1073 | 0.1151 | 1156 |
| Boston | 0.1192 | 0.1168 | 0.1187 | 0.1163 | 0.1203 | 0.1198 | 927 |
| Dallas | 0.0878 | 0.0889 | 0.0977 | 0.0883 | 0.0887 | 0.0921 | 1171 |
| Denver | 0.1039 | 0.1071 | 0.1038 | 0.1070 | 0.1054 | 0.1124 | 1282 |
| Des Moines | 0.1251 | 0.1263 | 0.1250 | 0.1261 | 0.1279 | 0.1277 | 1205 |
| Phoenix | 0.0345 | 0.0346 | 0.0341 | 0.0341 | 0.0341 | 0.0379 | 1320 |
| Portland | 0.1088 | 0.1097 | 0.1084 | 0.1096 | 0.1159 | 0.1080 | 1335 |
| 36-48 hours | | | | | | | |
| Albany | 0.1352 | 0.1320 | 0.1353 | 0.1321 | 0.1330 | 0.1356 | 794 |
| Atlanta | 0.1102 | 0.1143 | 0.1097 | 0.1139 | 0.1136 | 0.1249 | 1156 |
| Boston | 0.1230 | 0.1228 | 0.1230 | 0.1228 | 0.1234 | 0.1327 | 927 |
| Dallas | 0.0885 | 0.0899 | 0.0885 | 0.0894 | 0.0896 | 0.0919 | 1171 |
| Denver | 0.1091 | 0.1134 | 0.1091 | 0.1135 | 0.1109 | 0.1158 | 1282 |
| Des Moines | 0.1269 | 0.1298 | 0.1269 | 0.1297 | 0.1294 | 0.1302 | 1205 |
| Phoenix | 0.0367 | 0.0364 | 0.0362 | 0.0359 | 0.0358 | 0.0398 | 1320 |
| Portland | 0.1139 | 0.1142 | 0.1141 | 0.1152 | 0.1194 | 0.1162 | 1335 |

Table 8. Average percentage improvements in MSE for the combined forecasts. The upper (lower) figure in each cell gives improvement relative to the official forecast (climatology).

| Forecast type: | Average Prob. | Average Calibrated Prob. | Average Log Odds | Average Calibrated Log Odds | Combined Variances Equal | Combined Variances Unequal |
|---|---|---|---|---|---|---|
| COOL SEASON | | | | | | |
| 12-24 hours | 3.70<br>48.08 | 5.87<br>49.25 | 4.14<br>48.31 | 6.42<br>49.52 | -0.36<br>46.03 | 5.55<br>48.95 |
| 24-36 hours | 2.79<br>36.66 | 3.22<br>36.90 | 3.22<br>36.92 | 3.68<br>37.20 | -1.15<br>34.16 | 1.80<br>35.90 |
| 36-48 hours | 1.71<br>28.85 | 0.74<br>28.14 | 1.95<br>29.01 | 0.94<br>28.28 | 0.07<br>27.72 | -0.91<br>27.03 |
| WARM SEASON | | | | | | |
| 12-24 hours | 2.34<br>32.41 | -0.71<br>30.33 | 2.93<br>32.82 | -0.13<br>30.72 | -1.79<br>29.78 | 0.08<br>30.78 |
| 24-36 hours | 2.45<br>22.75 | 2.16<br>22.47 | 1.42<br>21.88 | 2.60<br>22.85 | 0.88<br>21.59 | -1.35<br>19.50 |
| 36-48 hours | 2.91<br>19.53 | 1.87<br>18.70 | 3.07<br>19.67 | 2.04<br>18.87 | 1.81<br>18.70 | -2.52<br>15.07 |
| Overall Average Improvement | 2.65<br>31.38 | 2.19<br>30.96 | 2.79<br>31.44 | 2.59<br>31.24 | -0.09<br>29.66 | 0.44<br>29.54 |

A somewhat more sophisticated approach to combining the probabilities involves first calibrating them separately and then averaging. We first calibrated both official and guidance log odds using the fitted calibration functions from the unequal-variances model given by (2). Then we calculated the average of the log odds and the average of the probabilities. From Tables 7 and 8, these combinations performed slightly worse than the averages of the uncalibrated forecasts. The average of calibrated log odds was the better performer of the two, with average percentage improvement over the official forecasts ranging from -0.13% to 6.42% and an overall average improvement of 2.59%

Finally, we combined the official and guidance forecasts using the multivariate log-odds model. As with calibration of the individual forecasts, we used the model with and without the assumption of equal variances. The results in Tables 7 and 8 indicate that the unequal-variances model was the better of the two models, with overall performance about the same as that of the official forecasts.

The explanation for the difference in performance between the two models is similar to that given above for the calibration models. We have already argued that the variances do not appear to be equal, and those arguments apply here as well. Examination of the estimated correlations in Table 5 reveals no discernable patterns due to station, season, lead time, or the occurrence of rain.

The equal-variances model results in a linear combination of the forecast log odds. On the other hand, the unequal-variances model includes quadratic terms, resulting in more flexibility in the shape of the combining function. As an example, the estimated combining functions for Portland in the cool season and with a 12-24 hour lead time are shown in Figure 3. The contours for the equal-variances model demonstrate a two-dimensional version of the curve seen in the case of the equal-variances calibration model (Figure 2). For the unequal-variances model, the slope of the combining function is quite steep for large values of the guidance forecast. In fact, suppose the guidance forecast is large and the official forecast takes a value around 0.5. In this region, an increase in the guidance probability could lead to a substantial decrease in the combined probability. While this behavior appears to be counterintuitive, it occurs primarily in areas of the grid where observations are unlikely; most often the official and guidance forecasts are not too dissimilar. Indeed, the behavior of the unequal-variances model near the 45° line appears to be quite reasonable, considerably more so than the behavior of the equal-variances combining function in this area.

To summarize, the simple average of the uncalibrated log odds performed the best of the combining techniques, followed by the average of uncalibrated probabilities. Calibrating the log odds or probabilities and then averaging performed slightly worse. Finally, using the full multivariate log-odds model gave the poorest results of the combining techniques, roughly equivalent to the official forecasts.

## DISCUSSION

In our analysis of individual probability of precipitation forecasts, the official forecasts were better than the guidance forecasts. Moreover, calibration of either type of forecast via a log-odds model failed to yield performance improvements. The model with unequal variances was preferable to that with equal variances, but the raw forecasts were still just as good or better. This contrasts with Clemen and Murphy (1986b), who found that direct frequency calibration, as opposed to the modeling approach, led to very slight performance improvements (roughly on the order of 1%).

The results from the analysis of the combined forecasts indicated that modeling failed to improve forecast performance. In this case, simple averages (no modeling) performed best. Averaging the calibrated forecasts represents a moderate amount of modeling, and this approach performed slightly worse than no modeling. Finally, the full multivariate log-odds model gave the poorest results. In contrast, Clemen and Murphy (1986b) found that a simple combining method, averaging not the raw forecasts or the model-calibrated forecasts, but instead the frequency-calibrated forecasts, performed slightly better than any other combining technique (with the average of raw forecasts being next best).

What went wrong with the normal model? It is tempting to suggest that the multivariate normal model for log odds does not provide an adequate fit to the data, and this is probably a

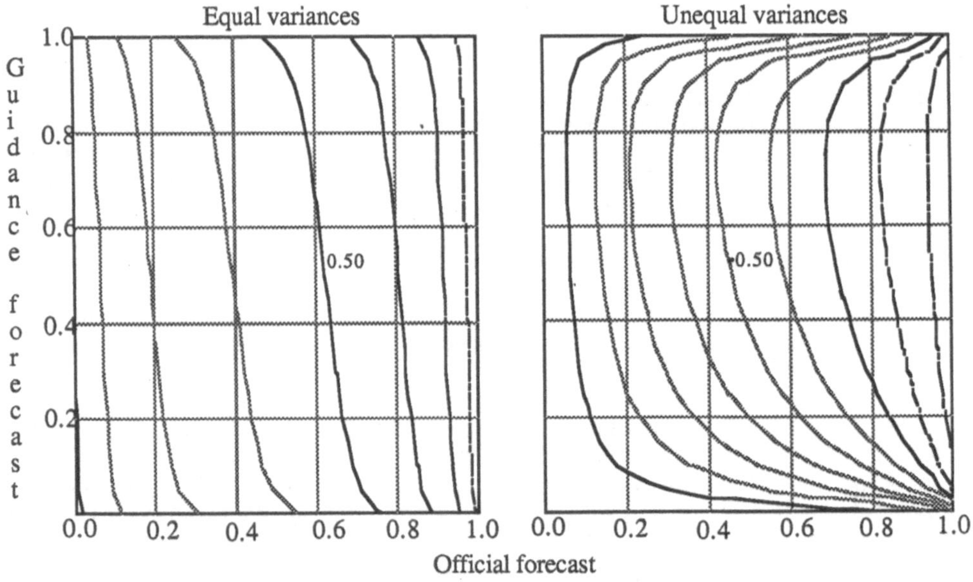


Figure 3. Combined PoP forecasts for Portland (cool season, 12-24 hour lead time). The contours show the combined PoP forecast given official and guidance probabilities. The contour interval is 0.10, with the 0.50 contour marked.

relevant consideration. Figure 4 shows the actual and modeled distributions of official forecast probabilities given rain and no rain for Portland in the cool season with a 12-24 hour lead time. Note that the actual empirical distributions are skewed, but the normal log-odds model yields distributions that are even more skewed.

Despite the apparent non-normality of the empirical distributions in this case, however, the normal log-odds calibration technique with unequal variances resulted in calibrated probabilities close for the most part to the frequency-calibrated probabilities (see Figure 2). Nevertheless, a different family of distributions might yield more promising results.

Another possible factor contributing to the relatively poor performance of the models is that weather forecasters are already well calibrated (Murphy, 1985). Thus, any attempt to improve on their performance via calibration could be expected to yield small improvements at best. In a similar vein, if the official forecasts were able to incorporate fully the information contained in the guidance forecasts, then combinations of the two forecasts should not be expected to improve on the official forecasts. Clemen and Murphy (1986a) found that official forecasts incorporated most of the information contained in the guidance forecasts; hence, combining techniques might result in only slight performance improvements (the averaging techniques), no improvement (the multivariate log-odds model), or possibly performance deterioration.

Our finding that simple forecasting methods do better empirically than more complex methods is one that has been reported elsewhere and in other contexts. For example, Armstrong (1984) surveyed empirical results regarding the performance of various forecasting models and concluded that simpler methods tend to do better. Results by Makridakis and Winkler (1983), Clemen and Winkler (1986), and others indicate that simply averaging forecasts is a robust combination technique; averaging seems to perform well compared to more complex approaches in a large variety of forecasting situations.

In the weather forecasting situation, a more detailed investigation, possibly with models other than the normal log-odds model, might provide more insight into the calibration and combination of probabilities. The question of whether our results with precipitation probabilities would generalize to other situations, possibly with forecasts that are less well-calibrated and less similar, is difficult to answer. Unfortunately, large sets of probability forecasts are not readily available for analysis. Perhaps, in the spirit of de Finetti, the use of personal probabilities to quantify beliefs regarding observable events and variables will become more widespread and we will eventually be able to learn more about the relative merits of modeling vis-a-vis non-modeling approaches under various circumstances.

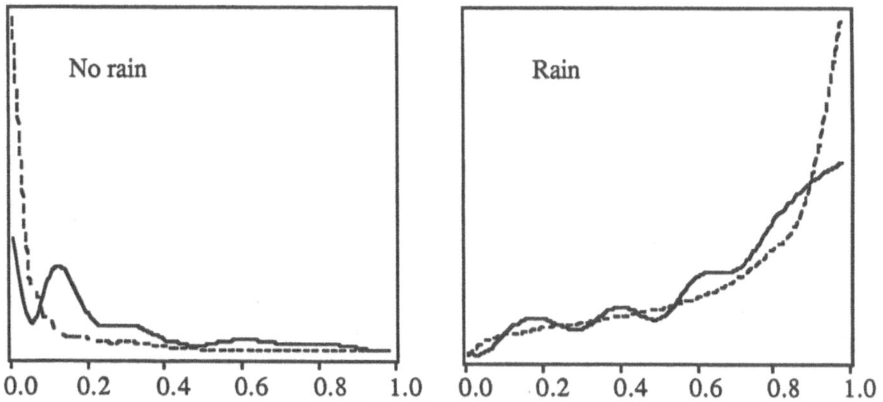


Figure 4. Empirical (solid line) and modeled (dashed line) relative frequency distributions for official PoP forecasts, conditional on rain and no rain, at Portland (cool season, 12-24 hour lead time).

ACKNOWLEDGMENTS

Gary M. Carter of the NWS Techniques Development Laboratory provided the PoP forecasts and verification data analyzed in this study. This work was supported in part by the National Science Foundation under Grants IST-8600788 and ATM-8507495.

REFERENCES

Agnew, C. E., 1985, Multiple probability assessments by dependent experts. *J. Amer. Statist. Assoc. 80*, 343-347.

Armstrong, J. S., 1984, Forecasting by extrapolation: conclusions from 25 years of research, *Interfaces 14 (6)*, 52-66.

Chang, K., 1985, "Combination of Opinions: The Expert Problem and the Group Consensus Problem," Ph.D. Dissertation, University of California, Berkeley, CA.

Clemen, R. T., 1985, Extraneous expert information, *J. Forecast. 4*, 329-348.

Clemen, R. T., 1987, Combining overlapping information, *Mgmt. Sci.*, in press.

Clemen, R. T., and Murphy, A. H., 1986a, Objective and subjective precipitation probability forecasts: statistical analysis of some interrelationships, *Weather and Forecasting 1*, 56-65.

Clemen, R. T., and Murphy, A. H., 1986b, Objective and subjective precipitation probability forecasts: improvements via calibration and combination, *Weather and Forecasting 1*, in press.

Clemen, R. T., and Winkler, R. L., 1986, Combining economic forecasts, *J. Bus. Econ. Stat. 4*, 39-46.

DeGroot, M. H., and Fienberg, S. E., 1982, Assessing probability assessors: calibration and refinement, in: "Statistical Decision Theory and Related Topics III," S. S. Gupta and J. O. Berger, eds., Academic Press, New York, 291-314.

DeGroot, M. H., and Fienberg, S. E., 1983, The comparison and evaluation of forecasters, *The Statistician 32*, 12-22.

French, S., 1980, Updating of belief in the light of someone else's opinion, *J. R. Statist. Soc. Ser. A 143*, 43-48.

French, S., 1981, Consensus of opinion, *Eur. J. Oper. Res. 7*, 332-340.

French, S., 1985, Group consensus probability distributions: a critical survey, in: "Bayesian Statistics 2," J. M. Bernardo, M. H. DeGroot, D. V. Lindley, and A. F. M. Smith, eds., North Holland, Amsterdam, 183-201.

Genest, C., and Schervish, M. J., 1985, Modeling expert judgments for Bayesian updating, *Ann. Stat. 13*, 1198-1212.

Genest, C., and Zidek, J. V., 1986, Combining probability distributions: a critique and an annotated bibliography, *Stat. Sci. 1*, 114-148.

Lindley, D. V., 1982, The improvement of probability judgments, *J. R. Statist. Soc. Ser. A 145*, 117-126.

Lindley, D. V., 1983, Reconciliation of probability distributions, *Oper. Res. 31*, 866-880.

Lindley, D. V., 1985, Reconciliation pf discrete probability distributions, in: "Bayesian Statistics 2," J. M. Bernardo, M. H. DeGroot, D. V. Lindley, and A. F. M. Smith, eds., North Holland, Amsterdam, 375-390.

Lindley, D. V., Tversky, A., and Brown, R. V., 1979, On the reconciliation of probability assessments, *J. R. Statist. Soc. Ser. A 142*, 146-180.

Makridakis, S., and Winkler, R. L., 1983, Averages of forecasts: some empirical results, *Mgmt. Sci. 29*, 987-996.

Morris, P. A., 1974, Decision analysis expert use, *Mgmt. Sci. 20*, 1233-1241.

Morris, P. A., 1977, Combining expert judgments: a Bayesian approach, *Mgmt. Sci. 23*, 679-693.

Murphy, A. H., 1985, Probabilistic weather forecasting, in: "Probability, Statistics, and Decision Making in the Atmospheric Sciences," A. H. Murphy and R. W. Katz, eds., Westview Press, Boulder, CO, 337-377.

Murphy, A. H., and Sabin, T. E., 1986, Trends in the quality of National Weather Service forecasts, *Weather and Forecasting 1*, 42-55.

Murphy, A. H., and Winkler, R. L., 1984, Probability forecasting in meteorology, *J. Amer. Statist. Assoc. 79*, 489-500.

Winkler, R. L., 1981, Combining probability distributions from dependent information sources, *Mgmt. Sci. 27*, 479-488.

COHERENT DISTRIBUTIONS AND LINDLEY'S PARADOX (*)

Guido Consonni
I.M.Q.
L. Bocconi University
Milano, Italy

Piero Veronese
Dottorato Ricerca Trento University
L. Bocconi University
Milano, Italy

SUMMARY

A Bayesian test of the simple null hypothesis $H_0:\theta=\theta_0$ versus the composite alternative $H_1:\theta\neq\theta_0$ is performed using finitely additive prior distributions in order to investigate the so-called Lindley's paradox. In particular two priors for $\theta$ under $H_1$ are considered. The first represents a coherently non-informative distribution which is shown to correctly yield the "paradox" because of the overall induced distribution of $\theta$. The second, through the use of adherent masses to $\theta_0$, does instead avoid Lindley's paradox.

## 1. INTRODUCTION

1.1 Let $X_1, X_2 \dots, X_n$ be, given $\theta \in \mathcal{R}$, independently and identically distributed (i.i.d.) normal random variables with mean $\theta$ and known variance $\sigma^2$. We consider a Bayesian test of the simple null hypothesis $H_0:\theta=\theta_0$ versus the composite alternative $H_1:\theta\neq\theta_0$. Henceforth we shall take, without loss of generality, $\theta_0=0$. To avoid a trivial solution it is necessary to assign a positive mass, $\epsilon$ say, to $\theta=0$. Furthermore it is common practice to distribute the remaining mass $(1-\epsilon)$ on an interval $I\in\mathcal{R}$ according to a continuous cumulative distribution function (cdf) G, see DeGroot (1970, pp. 238-239). Letting $\bar{x}= \Sigma x_i/n$, the result of this Bayesian analysis is typically summarized by the posterior odds $Q(H_0|\bar{x})=$ $=P(\theta=0|\bar{x})/P(\theta\neq0|\bar{x})$ which can also be usefully written as $\epsilon L_0/((1-\epsilon)L_1)$ where $\epsilon/(1-\epsilon)$ are the prior odds and $L_0/L_1$ is the so called likelihood ratio, where $L_0$ is the likelihood of $H_0$ and $L_1$ is the overall likelihood of $H_1$, i.e.

$$\int_{-\infty}^{\infty} n^{1/2}/\sigma \ \varphi(n^{1/2}(\bar{x}-\theta)/\sigma)dG(\theta)$$

with $\varphi(x)$ denoting the density of the standard normal evaluated at x. To decide in favour of either hypothesis one needs a loss function. If a>0 is the loss of rejecting $H_0$ when $H_0$ is true and b>0 that of accepting $H_0$ when $H_0$ is false, the corresponding loss function will be denoted by $l_{ab}$, and a Bayesian test will reject $H_0$ if $Q(H_0|x)<b/a$. (When

* Research supported by Ministero Pubblica Istruzione (60% grants) and Consiglio Nazionale delle Ricerche.

a=b=1 we have a "0-1 loss function" which implies rejection of $H_0$ when the posterior odds are less than one). In a sampling theory context $H_0$ will be rejected if $|x| \geq z_\alpha \sigma/n^{1/2}$ where $z_\alpha$ is the suitable quantile of the standard normal corresponding to significance level $\alpha$.
Suppose now, that we observe a value of $\bar{x}$ just significant at level $\alpha$, i.e. $\bar{x}=k_\alpha\sigma/n^{1/2}$ where $k_\alpha$ is either $z_\alpha$ or $-z_\alpha$, then Lindley (1957), see also Jeffreys (1948), showed that, for $n \to \infty$ a Bayesian test would firmly accept $H_0$, since $P(\theta=0|\bar{x}=k_\alpha \sigma/n^{1/2})$ tends to one, and this occurs for any $\alpha$ and any $\epsilon$. This is referred to as Lindley's paradox. Actually Lindley's paradox arises whenever the prior distribution of $\theta$ under $H_1$ is fairly flat relative to the likelihood independently of the value taken on by n, see Shafer (1982) and Hill (1982). Nevertheless when n is sufficiently large this condition is typically satisfied since in this case only a small set of parameter values is strongly suggested by the data.

1.2 As it is known no paradox appears when the null hypothesis is modified into $H_0^*:-d\leq\theta\leq d$ for some positive small d. Indeed such a hypothesis seems the most natural in many applications, where the real issue is not whether $\theta$ is actually zero but, rather, whether it is very small. This implies that values which are negligibly different from zero are conceptually indistinguishable from it. Nevertheless there seems to be instances where the null hypothesis must be simple since the specific value zero of $\theta$ arises naturally and "is fundamentally different from any value $\theta\neq0$, however near to zero it might be", see Lindley (1957, p. 189). Examples of this situation may be found in parapsychology and genetics (Lindley, 1957).
This paper will examine the problem of testing $H_0:\theta=0$ versus $H_1:\theta\neq0$ from a Bayesian viewpoint using coherent finitely-additive prior distributions which will be shown to be particularly suitable to handle such a type of problem. Furthermore the analysis will be generally performed under the assumption that the observable random variables $X_1,\ldots X_n$ are, given $\theta$, i.i.d. according to the exponential family.
Basically, we first discuss the purported non-informativity of a traditional prior on $\theta$, next we suggest a coherent prior which seems especially relevant in this case and which does not necessarily lead to the paradox.

More specifically section 2 reviews basic aspects of finitely additive distributions; section 3 discusses the use of an "improper" prior under the alternative hypothesis and reveals its inadequacy for testing purposes, section 4 suggests a suitable coherent prior, derives the corresponding posterior probability of $H_0$ (which does not yield Lindley's paradox) and finally illustrates in detail the special case in which the statistical model is assumed to be normal.

## 2. FINITELY ADDITIVE DISTRIBUTIONS ON THE PARAMETER SPACE

Two basic aspects of de Finetti's (1974) approach to probability theory are represented by a betting scheme and a coherence principle, which only requires finite-additivity, so that the usual assumption of $\sigma$-additivity, though acceptable, is not necessary.
A typical feature of finitely-additive distributions is that they may present so-called adherent masses. In order to clarify their nature, let X be a random variable and let $F(x)=P(X\leq x)$ be its cdf. It is worth noticing at this stage that F need not be right-continuous, contrary to what happens in the traditional $\sigma$-additive framework. Similarly $F_l(x)=P(X<x)$ need not be left-continuous. Thus in order to characterize the probability distribution of X both F and $F_l$ are necessary.

If $\quad F(x^-)= \lim_{t\to x^-} F(t), \quad$ and $\quad F(x^+)= \lim_{t\to x^+} F(t)$

then one can show that for all $x\in\mathcal{R}$: $F(x^-)\le F_l(x)\le F(x^+)$. From this set of inequalites it is possible to define the concept of adherent mass. More precisely if $F_l(x)-F(x^-)=p_l(x)>0$, then $p_l(x)$ is said to be the probability adherent to the left of x (briefly: left-adherent to x); if $F(x^+)-F(x)=p_r(x)>0$, then $p_r(x)$ is said to be the probability right-adherent to x; finally if $F(x)-F_l(x)=p_c(x)>0$, then $p_c(x)$ is said to be the probability concentrated on x. Futher if $F(x)=p_0$ for all $x\le x_0$ (say), then F presents (right) adherent mass $p_0$ to $-\infty$; if $P(X>x)=1-F(x)=p_1$ for all $x\ge x_1$ (say), then F presents (left) adherent mass $p_1$ to $\infty$.
In order to perform a Bayesian test of hypothesis we shall need to assign a prior distribution on the parameter space $\Theta\subseteq\mathcal{R}$. Since we shall not restrict our attention to $\sigma$-additive priors, we briefly describe below a way to assign coherent prior distributions on $\Theta$. The basic idea of this method, originally proposed by de Finetti (1974), is to employ a gradual procedure consisting in assigning a prior to proper subsets $\Theta_k$ of $\Theta$, and then obtaining the prior on $\Theta$ by passing to the limit. This procedure is justified since coherence is always preserved when passing to the limit. If the limit does not exist, the distribution of $\theta$ will have to be assigned directly making sure that it is consistent with previous assignments on $\Theta_k$.
Following the approach described in Regazzini and Cifarelli (1986), let $\mathcal{B}_\Theta$ be the Borel class of $\Theta$ and let $\{\Theta_k\}$ k=1,2,... be a sequence of elements of $\mathcal{B}_\Theta$ converging from below to $\Theta$, i.e. $\Theta_k\uparrow\Theta$. Further let f be a non-negative $\mathcal{B}_\Theta$-measurable function such that, for a given $\sigma$-finite measure $\tau$ on $(\Theta, \mathcal{B}_\Theta)$,

$$0<I_k=\int_{\Theta_k} f(\theta)d\tau(\theta)<\infty \quad \text{for each} \quad k\ge 1,$$

and with the understanding that

$$P_{\Theta_k}(\theta\le\theta^*)=F_{\Theta_k}(\theta^*)=(I_k)^{-1}\int_{\Theta_k\cap(-\infty,\theta^*]} f(\theta)d\tau(\theta) \tag{2.1}$$

represents, for each real $\theta^*$, the probability of $(\theta\le\theta^*)$ conditional on the hypothesis that $\theta\in\Theta_k$. It can be seen that, for each fixed $\Theta_k$, $F_{\Theta_k}$ is a distribution function that can generate on $(\Theta,\mathcal{B}_\Theta)$ a probability measure. This probability measure will be taken as the distribution on $\mathcal{B}_\Theta$ conditional on $(\theta\in\Theta_k)$. If now $\lim_{k\to\infty} F_{\Theta_k}(\theta)$ exists, then we shall take this limit as the distribution function of $\theta$.
Notice that the function f which appears in (2.1) is not in general the density of $\theta$, although it is the density of $\theta$ conditional on $(\theta\in\Theta_k)$.
In the sequel we shall have $\Theta=(A,B)$ $-\infty\le A<B\le\infty$ and shall take $\Theta_k=(a_k,b_k)$ with $a_k\to A$ and $b_k\to B$ for $k\to\infty$. For simplicity we shall omit the subscript k and consider intervals of type (a,b) with $a\to A$ and $b\to B$. A device which will prove useful consists in linking a and b by setting, say, b=b(a) for a suitable function b. In this case we shall write $\Theta_k=\Theta_a=(a,b(a))$ with $a\to A$ and $b(a)\to B$ for $a\to A$. For an application to $\Theta=(-\infty,\infty)$ which gives rise to a finitely-additive analog of the so-called "uniform prior over the real line" see Consonni and Veronese (1986).
Let now $\mathcal{X}\subseteq\mathcal{R}^n$ be the sample space and $\{P_\theta:\theta\in\Theta\}$ a family of $\sigma$-additive probability distributions on the class of Borel sets of $\mathcal{X}$,

$\mathcal{B}_{\mathcal{X}}$, dominated by a $\sigma$-finite measure $\mu$ defined on $(\mathcal{X}, \mathcal{B}_{\mathcal{X}})$. The density of $P_\theta$ with respect to (w.r.t.) $\mu$ will be denoted by $p_\theta$.

The distribution function of $\theta$ given $x \in \mathcal{X}$ conditional on the hypothesis $(\theta \in \Theta_k)$ is as usual represented by

$$P_{\Theta_k}(\theta \leq \theta^* \mid x) = \frac{\int_{(-\infty,\theta^*] \cap \Theta_k} f(\theta) p_\theta(x) d\tau(\theta)}{\int_{\Theta_k} f(\theta) p_\theta(x) d\tau(\theta)}, \quad \theta^* \in \Theta \text{ and } x \in \mathcal{X} \quad (2.2)$$

As for the prior on $\theta$, the posterior distribution function of $\theta$, i.e. conditional on the sure event $\theta \in \Theta$, will be obtained by a passage to the limit (if it exists) for $k \to \infty$ in (2.2).

## 3. BAYESIAN TEST USING AN "IMPROPER" PRIOR UNDER THE ALTERNATIVE HYPOTHESIS

3.1 It is a well-known fact that using a diffuse improper prior over under the alternative hypothesis $H_1$ leads to unsatisfactory results. This fact was discovered, under a normal sampling distribution, by Jeffreys (1948) who argued that "even if $H_0$ were true, it would not ordinarily be the case that $\theta = \theta_0$ exactly, and any discrepancy between $\theta$ and $\theta_0$ would for large n (...) lead to rejection of $H_0$", see Hill (1982, p. 346).

On the other hand he went on remarking that the choice of a proper uniform distribution on any finite interval (-K,K), with K sufficiently large, was also unsatisfactory although for an opposite reason. Indeed, in this case, any data would lead to acceptance of $H_0$, see again Hill (1982, p. 346).

The discrepancy between the conclusions obtained under the two cases mentioned above is suspicious, for it would seem sensible that, at least for large K, the uniform over (-K,K) should offer results similar to the ones which hold under a "uniform" prior over $\mathcal{R}$. The impropriety of the prior on $\theta$ under $H_1$ is particularly relevant in this case since, when considering the posterior odds

$$Q(H_0|x) = \frac{P(\theta=0)}{P(\theta\neq0)} \cdot \frac{L_0}{L_1},$$

the term $L_1$, which should represent the density of x given $H_1$, can be taken to be any positive constant.

This remark was made by DeGroot (1982) who went on arguing that improper priors "are never appropriate for tests of significance. Under no circumstances should they be regarded as representing ignorance".

While agreeing on the fact that improper priors are not appropriate, we wish to remark that, if employed in a suitable context, diffuse priors can indeed be said to represent ignorance see Veronese and Consonni (1986).

The point that must be made absolutely clear, however, is the distinction between the distribution of $\theta$ under $H_1$, which can represent ignorance, and that of $\theta$ overall, which, when $\theta|H_0$ is degenerate, will be shown to become strongly informative as indeed it ought to be. This distinction was overlooked also by Shafer (1982, p. 326) who apparently did not realize that the more non-informative the distribution of $\theta$ under $H_1$ is, the more informative the overall distribution of $\theta$ becomes: this is indeed perfectly sensible and not paradoxical. By using only finitely additive priors, we shall reexamine the whole issue and i) study the nature of the prior on $\theta$ when $\theta$ under $H_1$ is assumed to be "uniform over $\mathcal{R}$" and ii) see how this prior naturally implies strong acceptance of $H_0$

contrary to Jeffreys's claims thus providing a reconciliation between the conclusions under the "uniform" over $\mathcal{R}$-case and the uniform over (-K,K)-one.

3.2 As usual let $\epsilon$ be the mass concentrated on $\theta=0$, so that it remains to distribute the remaining mass $1-\epsilon$ over $\mathcal{R}-\{0\}$. Clearly the cdf of $\theta|H_0$ is 0 for $\theta^*<0$ and 1 for $\theta^*\geq 0$. In order to assign the cdf $\theta|H_1$ we shall follow the method described in section 2, so that on each finite interval (-a,a) we assume a uniform distribution. We thus have

$$P_a(\theta\leq\theta^*|\theta\neq 0)=\begin{cases} 0 & \theta^*<-a \\ (\theta^*+a)/2a & -a\leq\theta^*<a \\ 1 & \theta^*\geq a \end{cases}$$

whence, since $\quad P_a(\theta\leq\theta^*|\theta=0)=P(\theta\leq\theta^*|\theta=0),\quad$ we have

$$P_a(\theta\leq\theta^*)=P_a(\theta\leq\theta^*|\theta=0)\epsilon + P_a(\theta\leq\theta^*|\theta\neq 0)(1-\epsilon)=$$

$$=\begin{cases} 0 & \theta^*<-a \\ (1-\epsilon)(\theta^*+a)/2a & -a\leq\theta^*<0 \\ \epsilon+(1-\epsilon)(\theta^*+a)/2a & 0\leq\theta^*<a \\ 1 & \theta^*\geq a \end{cases}$$

Consequently the prior cdf on $\theta$ is given by

$$P(\theta\leq\theta^*)=\lim_{a\to\infty} P_a(\theta\leq\theta^*)=\begin{cases} (1-\epsilon)/2 & \theta^*<0 \\ (1+\epsilon)/2 & \theta^*\geq 0 \end{cases} \tag{3.1}$$

Notice that (3.1) is a finitely additive cdf which presents a mass concentrated on $\theta=0$ and equal adherent mass $(1-\epsilon)/2$ to $-\infty$ and $\infty$. Having written down explicity the prior of $\theta$ it is immediate to realize that this prior is highly informative since, while concentrating a mass on the origin, it assigns probability zero to any finite interval not including the origin. We thus have a prior distribution which is markedly different from the traditional non-informative priors employed for inferential purposes in order to emphasize the role of the observations. Indeed with a prior of type (3.1), all the data can do is to change the value of the three masses, but obviously the posterior distribution of $\theta$ will still assign probability zero to any finite interval not including the origin.

If in particular the model is assumed to belong to the exponential family written in the natural parametrization whose density w.r.t. a $\sigma$-finite measure $\mu$ is expressed by

$$p_\theta(x)=\exp(\theta x-M(\theta)) \qquad \theta\in\Theta\subseteq\mathcal{R},\ x\in\mathcal{X}\subseteq\mathcal{R} \tag{3.2}$$

then we can further specify the structure of the posterior distribution of $\theta$ when the prior is of type (3.1). Because of a well known result, if

$X_1,\ldots,X_n$ are, given $\theta$, i.i.d. according to (3.2), then $T=\Sigma X_i$ is sufficient and T has density of type (3.2) (with $M(\theta)$ replaced by $nM(\theta)$) w.r.t. the convolution measure of order n, $\mu_n$.

First of all consider the posterior cdf of $\theta$ conditional on $\theta\in(-a,a)$ and on T=t, and let $p_\theta(t)$ be the sampling distribution of T. Then because of (2.2)

$$P_a(\theta\leq\theta^*|t)=\begin{cases} 0 & \theta^*<-a \\ \dfrac{\int_{-a}^{\theta^*} p_\theta(t)d\theta}{\dfrac{2a\,\epsilon\, p_0(t)}{1-\epsilon}+\int_{-a}^{a} p_\theta(t)d\theta} & -a\leq\theta^*<0 \\ \dfrac{\dfrac{2a\,\epsilon\, p_0(t)}{1-\epsilon}+\int_{-a}^{\theta} p_\theta(t)d\theta}{\dfrac{2a\,\epsilon\, p_0(t)}{1-\epsilon}+\int_{-a}^{a} p_\theta(t)d\theta} & 0\leq\theta^*<a \\ 1 & \theta^*\geq a \end{cases} \tag{3.3}$$

To obtain the posterior distribution of $\theta$ one must compute the limit for $a\to\infty$ of (3.3). If $I(a)=\int_{-a}^{a} p_\theta(t)d\theta\to C<\infty$, then one immediately concludes that the posterior distribution is degenerate on $\theta=0$. If however I(a) diverges for $a\to\infty$, then it becomes essential to check whether also the numerator of (3.3) diverges. When this is the case the limit operation can be easily carried out invoking de l'Hospital's rule and using the following results valid whenever $p_\theta$ is as in (3.2):

$$p_{|\theta|}(x)\to 0 \quad \text{for} \quad |\theta|\to\infty \tag{3.4}$$

for all $x\in\mathfrak{X}$ except when $x=c=\min\mathfrak{X}$ or $x=d=\max\mathfrak{X}$, c and d finite, with $\mu\{c\}$ and $\mu\{d\}$ positive. In this case we have

$$p_\theta(c)\to 1/\mu\{c\} \qquad \text{for } \theta\to-\infty \qquad \text{and}$$
$$p_\theta(d)\to 1/\mu\{d\} \qquad \text{for } \theta\to\infty.$$

For a proof see Veronese and Consonni (1986).
One can thus conclude that the posterior distribution of $\theta$ is always degenerate on zero except when t=nc or t=nd. (Notice that this case may arise if and only if $X_i$ is discrete and the $x_i$'s are respectively either all c or all d).
So, going back to our hypothesis test, if the $X_i$'s are continuous, then $H_0$ is always accepted whatever the data and the sample size. Similarly $H_0$ is always accepted if the $X_i$'s are discrete except when all observations are equal to either of the (finite) boundary values. In this case indeed the posterior distribution of $\theta$ will generally exhibit, beyond a concentrated mass on 0, an adherent mass to either $-\infty$ or $\infty$. For an illustration of these points see Consonni and Veronese (1986).

## 4. A FINITELY ADDITIVE PRIOR ON $\theta$ (UNDER THE ALTERNATIVE HYPOTHESIS) WHICH AVOIDS LINDLEY'S PARADOX

4.1 As we mentioned in the introduction to this paper, Lindley's paradox arises whenever the prior distribution of $\theta$ under $H_1$ is fairly flat relatively to the likelihood.

Suggestions to overcome Lindley's paradox have been proposed, for

example, by Bernardo (1980) and Shafer (1982). Both however are not immune from criticism, see Dempster (1980), Jaynes (1980) and Hill (1982), Lindley (1982). Surely, as already recalled in subsect. 1.2, it is possible to avoid Lindley's paradox by turning the null simple hypothesis into one composite. When this is not possible, however, it means that the value $\theta=0$ has a special status with respect to all other points. As a consequence one must ensure that numerical proximity be not mixed up with logical proximity which, because of the very nature of the problem, is nonsensical.
We can therefore conclude that data generated by values of $\theta$ close to, but distinct from, zero should not provide evidence in favour of $H_0$, but rather be interpreted in favour of $H_1$. In order to achieve this we need to reconsider the prior distribution of $\theta$ and, as it will appear, the notion of adherent probability will play a significant role.

4.2 From the remarks of the previous subsection it follows that the prior distribution on $\theta$ must take into special consideration the point zero and points very close to zero; specifically let $c_2\epsilon$ be the mass concentrated on zero, and $c_1\epsilon$, $c_3\epsilon$ ($c_i\geq 0$, $c_1+c_2+c_3=1$) the mass adherent, respectively, to the left and to the right of zero.
As usual, let $H_0:\theta=0$ and $H_1:\theta\neq 0$. We thus have $P(\theta=0)=c_2\epsilon$, while the distribution of $\theta|H_1$ will present adherent mass $(c_1+c_3)\epsilon/(1-c_2\epsilon)$ to zero and will distribute the remaining mass according to a continuous cdf G. Notice that if $c_2=1$ the standard Bayesian setting is recovered.
If $c_2<1$, then our set-up amounts to a weakening of $H_0$ because of the presence of adherent masses to zero which favour $H_1$.
The prior cdf of $\theta$ can thus be described as:

$$P(\theta\leq\theta^*)=\begin{cases}(1-\epsilon)G(\theta^*) & \theta^*<0\\ (1-\epsilon)G(0)+(c_1+c_2)\epsilon & \theta^*=0\\ (1-\epsilon)G(\theta^*)+\epsilon & \theta^*>0\end{cases}\tag{4.1A}$$

and

$$P(\theta<\theta^*)=\begin{cases}(1-\epsilon)G(\theta^*) & \theta^*<0\\ (1-\epsilon)G(0)+c_1\epsilon & \theta^*=0\\ (1-\epsilon)G(\theta^*)+\epsilon & \theta^*>0\end{cases}\tag{4.1B}$$

Furthermore the conditional cdf of $\theta$ given $H_0$, $G^*(\cdot\,|\theta=0)$, is degenerate on zero, while the cdf of $\theta$ given $H_1$ is

$$G^*(\theta^*|\theta\neq 0)=P(\theta\leq\theta^*|\theta\neq 0)=\begin{cases}\dfrac{1-\epsilon}{1-c_2\epsilon}\,G(\theta^*) & \theta^*<0\\ \dfrac{1}{1-c_2\epsilon}\,(1-\epsilon)G(0)+c_1\epsilon & \theta^*=0\\ \dfrac{1-\epsilon}{1-c_2\epsilon}\,G(\theta^*)+\dfrac{\epsilon(1-c_2)}{1-c_2\epsilon} & \theta^*>0\end{cases}$$

To obtain the posterior odds

$$Q(H_0|x)=\frac{P(\theta=0)}{P(\theta\neq 0)}\cdot\frac{L_0}{L_1}=\frac{c_2}{1-c_2\epsilon}\cdot\frac{L_0}{L_1}$$

we have to compute $L_0$ and $L_1$. If the density w.r.t. to a $\sigma$-finite measure of the sample X given $\theta$ is $p_\theta(x)$, then

$$L_0 = \int_{-\infty}^{\infty} p_\theta(x)\,dG^*(\theta|\theta=0) = p_0(x) \tag{4.2}$$

and

$$L_1 = \int_{-\infty}^{\infty} p_\theta(x)\,dG^*(\theta|\theta\neq 0) \tag{4.3}$$

$$= \frac{\epsilon(1-c_2)}{1-c_2\epsilon}\, p_0(x) + \frac{1-\epsilon}{1-c_2\epsilon} \int_{-\infty}^{\infty} p_\theta(x)\,dG(\theta)$$

Setting the last integral equal to p(x) we thus have

$$Q(H_0|x) = \frac{c_2\epsilon}{1-c_2\epsilon} \quad \frac{(1-c_2\epsilon)p_0(x)}{\epsilon(1-c_2)p_0(x)+(1-\epsilon)p(x)} \tag{4.4}$$

It can be easily checked that if $c_2 \leq 1/2$, then $Q(H_0|x)<1$, i.e. $H_0$ is rejected under a 0-1 loss function. Notice that this result holds for any model $p_\theta$, for any sample realization x and for any value of $\epsilon$.
It follows that a prior for which the proportion $c_2$ of mass concentrated on $\theta=0$ is less than that adherent to the same point $(c_1+c_3)$ leads always to posterior odds which are less than unity and so $H_0$ is always rejected under a 0-1 loss function.
More generally under an $l_{ab}$ loss function (see subsect. 1.1) if $c_2 \leq b/(a+b)$, $Q(H_0|x)$ is always less than b/a, which leads to rejection of $H_0$. Since data play no role when $c_2 \leq b/(a+b)$ it follows in particular that Lindley's paradox does not arise, so that given a just significant observation both a sampling theory statistician and a Bayesian will reject $H_0$. Actually one does reject $H_0$ independently of the data not only when $c_2 \leq b/(a+b)$ but also for greater values of this threshold depending on the model, the prior and the sample size. For an illustration relative to the normal case see Consonni and Veronese (1986).

4.3 In this subsection we shall pursue the analysis assuming that a just significant $\bar{x}$ has been observed and implicitly accepting that $c_2 > b/(a+b)$, so that $\bar{x}$ is offered a chance to play a role.
Suppose that observations $X_1, X_2, \ldots, X_n$ are conditionally on $\theta$ i.i.d. according to the natural exponential family (3.2), so that it is meaningful to speak of a just significant sample statistic, in particular $\bar{x}$.
Remembering that $E_\theta(X_i)=M'(\theta)$ and $Var_\theta(X_i)=M''(\theta)$, for n large $\bar{X}$ will be approximately normally distributed with mean $M'(\theta)$ and variance $M''(\theta)/n$, we shall now derive the approximate expression for $L_0$ and $L_1$ and compute the approximate posterior odds for large n.
Recalling that $\varphi(x)$ is the density of the standard normal evaluated at x and using a prior like (4.1) we have

$$L_0 \simeq \frac{n^{\frac{1}{2}}}{(M''(0))^{\frac{1}{2}}}\ \varphi\left(\frac{n^{\frac{1}{2}}}{(M''(0))^{\frac{1}{2}}}\ (\bar{x}-M'(0))\right)$$

and

$$L_1 \simeq \frac{1}{1-c_2\epsilon}\left\{(1-c_2)\epsilon\ \frac{n^{\frac{1}{2}}}{(M''(0))^{\frac{1}{2}}}\ \varphi\left(\frac{n^{\frac{1}{2}}}{(M''(0))^{\frac{1}{2}}}\ (\bar{x}-M'(0))\right)+ \right. \qquad (4.5)$$

$$\left. +(1-\epsilon)\int_{-\infty}^{\infty}\frac{n^{\frac{1}{2}}}{(M''(\theta))^{\frac{1}{2}}}\ \varphi\left(\frac{n^{\frac{1}{2}}}{(M''(\theta))^{\frac{1}{2}}}\ (\bar{x}-M'(\theta))\right)dG(\theta)\right\}$$

Resorting to arguments similar to those of Jeffreys's and Hill's (1982), we shall provide an approximation for the integral appearing in (4.5). Since n is large and $M''(\theta)/n$ is small for each $\theta$, the integral can be regarded as a normal distribution for $M'(\theta)$ centred on $\bar{x}$ and with negligible variance. As a consequence $M'(\theta)$ will be approximately degenerate on $\bar{x}$ and so $\theta$ will be approximately degenerate on $M'^{-1}(\bar{x})$ (notice that since $M''(\theta)=Var_\theta(X_1)>0$ for each $\theta$, $M'$ is strictly increasing and so $M'^{-1}$ is well defined). Thus, letting g be the density of G the integral in (4.5) can be seen to become $g(M'^{-1}(\bar{x}))$.

We can finally write the approximate expression for the posterior odds

$$Q(H_0\,|\,\bar{x}) \simeq \frac{c_2\epsilon\ \dfrac{n^{\frac{1}{2}}}{(M''(0))^{\frac{1}{2}}}\ \varphi\left(\dfrac{n^{\frac{1}{2}}}{(M''(0))^{\frac{1}{2}}}\ (\bar{x}-M'(0))\right)}{\epsilon(1-c_2)\ \dfrac{n^{\frac{1}{2}}}{(M''(0))^{\frac{1}{2}}}\ \varphi\left(\dfrac{n^{\frac{1}{2}}}{(M''(0))^{\frac{1}{2}}}\ (\bar{x}-M'(0))\right)+(1-\epsilon)\ \ g(M'^{-1}(\bar{x}))}$$

If now $\bar{x}$ is just significant at level $\alpha$ , i.e. $\bar{x}=M'(0)+k_\alpha\ (M''(0)/n)^{\frac{1}{2}}$, then for $n\to\infty$, $Q(H_0|x)$ tends to $c_2\epsilon\ /\ \epsilon(1-c_2)\ =c_2/(1-c_2)$, i.e. $P(\theta=0|\bar{x})$ tends to $c_2$. We thus see that the only presence of an adherent mass to zero in the prior is sufficient to avoid the paradox which implied $P(\theta=0|\bar{x})\to 1$.

Indeed if $c_2\leq b/(a+b)$ then $H_0$ is rejected under an $l_{ab}$ loss function, consistently with the result of subsection 4.2 which held true independently of the observations. If, however, $c_2>b/(a+b)$, then $H_0$ is accepted under the previous loss function but not necessarily in general, contrary to Lindley's result.

Remarks

i) An appreciation of the sensitivity to n and $c_2$ of the rejection region for the normal case may be found in Consonni and Veronese (1986). In that paper, moreover, assuming a just significant observation at level $\alpha$ , the highest value of $c_2$ which leads to rejection of $H_0$ is derived for selected sample sizes. Typically agreement between sampling and Bayesian theory is easier (i.e. no restriction is imposed on $c_2$) when $\alpha$ is small (e.g. 0.01 or 0.001) and n is not very large.

ii) As we have seen the value of $c_2$, which implicitly gives $c_1+c_3$, is particularly relevant throughout the analysis. As we know $c_2$ represents the proportion of the mass concentrated on, whereas $c_1+c_3$ is that of the mass adherent to, $\theta=0$. By assigning directly the probability that $\theta=0$ and the probability that $\theta$ is adherent to zero according to the definitions of sect. 2, one can recover $c_2$ and $c_1+c_3=1-c_2$.

On the other hand it is sometimes more natural to start from the mass $\epsilon$ and then to subdivide it into two components, i.e. the concentrated and the adherent one. This usually occurs when $\theta$ is a measurable quantity, so that its prior distribution can be assigned on the basis of an empirical distribution function of available data before any further statistical processing (e.g. hystogram smoothing) has occurred.
Because of several considerations (e.g. provenance, quality, reliability of data) one is typically led not to give full credit to the distribution as such, although it remains the only empirical source which can be usefully exploited. So if prior knowledge does not allow one to modify the basic structure of the data (e.g. to transfer portions of frequencies from one datum to another), the only option which is able to incorporate this natural skepticism on data quality is to lower the amount of concentrated mass on each single point by transferring a part of it into adherent mass. These considerations have implications on the prior distribution on $\theta$ with special reference to the point $\theta=0$, because it is only there that we have to distinguish carefully between zero and values which are only approximately zero.
In practice if $\epsilon$ is the frequency of $\theta=0$, then $c_2$ can be regarded as the discounting factor which summarizes our opinion on the quality of the data.
This type of reasoning can, in our opinion, be usefully applied to the forensic case discussed in Lindley (1977) and Shafer (1982), since the role of $c_2$ is easily understood and its value can be reasonably supplemented by an expert witness.
Finally notice that, under an $l_{ab}$ loss function, it will be usually sensible to assign a value of $c_2$ higher than $b/(a+b)$, in order to allow the data to influence our decision. Typically this condition should be satisfied if enough credence is given to the available data.

Acknowledgments: We wish to thank E. Regazzini for constant advice and useful discussions.

## REFERENCES

Bernardo, J.M., 1980, A bayesian analysis of classical hypothesis testing, (with discussion), in: "Bayesian Statistics", J.M. Bernardo, M.H. DeGroot, D.V. Lindley, A.F.M. Smith, eds., Univ. Press, Valencia.

Consonni, G., and Veronese, P., 1986, Coherent distributions and Lindley's paradox, Studi Statistici, n. 15, Istituto di Metodi Quantitativi, Università L. Bocconi, Milano.

de Finetti, B., 1974, "Theory of Probability", vol.1, J. Wiley, New York.

DeGroot, M.H., 1970, "Optimal statistical decisions", McGraw-Hill, New York.

DeGroot, M.H., 1982, in discussion of G. Shafer, 1982.

Dempster, A.P., 1980, in discussion of J.M. Bernardo, 1980.

Hill, B.M., 1982, in discussion of G. Shafer, 1982.

Jaynes, E.T., 1980, in discussion of J.M. Bernardo, 1980.

Jeffreys, H., 1948, "Theory of probability", second edition, Oxford Univ. Press, London.

Lindley, D.V., 1957, A statistical paradox, Biometrika, 44:187.

Lindley, D.V., 1977, A problem in forensic science, Biometrika, 64:207.

Lindley, D.V., 1982, in discussion of G. Shafer, 1982.

Regazzini, E. and Cifarelli, D.M., 1986, Priors for exponential families which maximize the association between past and future observations, in this volume.

Shafer, G., 1982, Lindley's paradox, (with discussion), J.A.S.A., 77:325.

Veronese, P. and Consonni, G., 1986, Non informative priors for exponential families, Studi Statistici n. 14, Istituto di Metodi Quantitativi, Università L. Bocconi, Milano.

MEAN VARIANCE, EXPECTED UTILITY AND RUIN PROBABILITY IN REINSURANCE DECISIONS:

SUGGESTIONS AND COMMENTS ON THE LINE OF DE FINETTI'S SEMINAL WORK

Luciano Daboni and Flavio Pressacco

Department of Applied Mathematics B. de Finetti
University of Trieste
Italy

## 1. INTRODUCTION

Roughly speaking risk theory in insurance concerns the survival of (a branch of) an insurance company over some specified time horizon.* The key goal variable is usually the ruin probability of the company along that time horizon. While practical everyday problems involve mid term (e.g. five or ten years) horizon, theoretical models are mainly concerned with single period problems, or at the other extreme with (asymptotic) evaluations over an infinite time horizon. The usually relevant control variables are the initial reserve fund and the safety loading coefficient placed to obtain tariff insurance premiums. A third prominent control variable, sometimes implicitly considered, is the reinsurance strategy of the firm.

In some sense reinsurance turns out to be at least in the short period the only one really manageable, as is not easy to adequate the reserve fund and on the other side the market conditions dictate loading charges at least under more or less perfect competition. Undoubtedly a milestone in the analysis of the role of proportional reinsurance in controlling ruin probability of an insurance company is B. de Finetti's paper (de Finetti, 1940). The author's treatment concerns both the single period and the infinite horizon problem. As regards the first point de Finetti's solution is based on a two stage model, where efficiency and optimality goals are clearly defined and separatedly pursued, each on its respective stage (see chapter 2).

As we shall see in chapter 3 de Finetti's paper is to be seen as an early anticipation of H. Markovitz (Markovitz, 1952) well known two stage mean variance approach to the portfolio selection problem. The relevance of the theoretical reinsurance model to practical companies behaviour is

* An exhaustive treatment of the really involved subject matter of risk theory is given e.g. in Buhlmann (1970), Gerber (1979), Seal (1969). An historical very interesting resume with claryfying comments is offered by chapter 18 in Borch (1974).

shortly discussed in para 4. To solve the infinite horizon problem de Finetti exploits old classical results of probability theory going back to De Moivre and concerning the so called gambler's ruin problem. The key idea, recalled in chapter 6, is to properly transform the original unfair sequence (due to loading charges) of random gains of the insurance company so as to obtain a fair process, or with modern terminology to derive a martingale from a submartingale process.

As we shall see later while accomplishing this goal a constant $\beta$ representing the risk level of a portfolio (as well as of single contracts) is rather naturally derived, such that an evaluation of the asymptotic ruin probability is $\exp(-G_0/\beta)$, with $G_0$ as the initial reserve of the company. Then de Finetti's risk level turns out to be a simple and meaningful counterpart of the "adjustment coefficient" well known in Lundberg's collective risk theory.* It is well known that besides ruin probability another prominent decision criterion to solve economic problems under uncertainty is the expected utility one.**

Another goal of this paper is to show and discuss elegant and interesting connections between the expected utility approach (especially in case of exponential utility) and de Finetti's solutions both for the single period (chapter 5) and for the asymptotic approach (para 7). There is another interesting connection between the risk level $\beta$ and the risk tolerance B, characterizing the utility function of the company will be derived and discussed .

## 2. DE FINETTI'S TREATMENT OF THE SINGLE PERIOD PROBLEM

Let us shortly recall de Finetti's approach to the solution of the single period retention problem (in case of proportional reinsurance) of an insurance company. The company has a portfolio with n insured risks, whose respective claims are described by random variables $X_h$, h=1,.....,n. Suppose that insurance and proportional reinsurance markets follow working rules (conveniently simplifying reality) such that the expected value and the mean square deviation of the single risk retained are homogeneous linear functions of the retained quotas $q_h$ ***, so that formally denoting by $G_h(q_h)$ the above random profit and by: $m_h = E(G_h(1))$ and $\sigma_h = \sigma(G_h(1))$, it is: $E(G_h(q_h)) = q_h m_h$; $\sigma(G_h(q_h)) = q_h \sigma_h$ .

---

* See the previously referred resume in Borch (1974)

** Of course it is impossible to give account of the applications of utility theory to insurance problems. Several chapter e.g. in Buhlmann (1970) and Gerber(1979)are devoted to the point, while almost the entire treatment in Borch (1974) is based on modern utility theory.

***Alternatively a non homogeneous linear function such that $E(G_h(q_h)) = q_h \mu_h + \pi_h$ with $\mu_h$ as the reinsurer expected gain in case of complete reinsurance and $\pi_h$ as the difference (possibly negative) $m_h - \mu_h$, could be used. Results later derived are not altered, except for the obvious formal variations.

Writing $G(\underline{q}) = \sum_h G_h(q_h))$, if the usual simple no correlation assumption among the $X_h$ is accepted we have:

$$E(G(\underline{q}) = \sum_h q_h m_h \; ; \qquad \sigma(G(\underline{q})) = (\sum_h q_h^2 V_h)^{1/2}.$$

An efficient proportional retention strategy is defined by de Finetti as one that minimizes the ruin probability of the company for any given value of the expected profit retained. Now the single period ruin probability is, leaving aside financial factors and keeping account only of technical ones, simply the probability that the losses (negative profits) are greater than the initial reserve $G_0$ of the company or formally that: $G(\underline{q}) \leqslant - G_0$. The above probability is that of the event: $(G_0 - E(G)) / \sigma(G) \leqslant - (G_0 + E(G)) / \sigma(G)$ that the standardized overall gain is lesser than the opposite of the so called stability index of the company.

As stressed by de Finetti if the distribution of the overall profit is of the "same type" for any choice of $\underline{q}$ the above probability is for any given value of E(G) an increasing function of $\sigma(G)$. After that an efficient retention strategy is found as the one that minimizes the variance of the company's single period profit for any given value of the expected profit retained, and the whole set of efficient strategies is obtained by solving the following set of constrained minimization problems:

$$\min_{0 \leqslant \underline{q} \leqslant 1} \sum q_h^2 V_h \qquad \text{sub} \quad \sum q_h m_h = \bar{E} \tag{1}$$

de Finetti offers a (rather involved) proof that the optimal solution is given by:

$$q_h^+ = \begin{cases} l(\bar{E})(m_h/V_h) & \text{if } < 1 \\ 1 & \text{otherwise} \end{cases} \tag{2}$$

with $l(\bar{E})$ a, common for any h, piecewise linearly increasing function of $\bar{E}$.

Today, the same results are obtained as a simple exercise on Kuhn Tucker conditions in quadratic programming problems. Once the efficient set has been found, there is still to select a single point among the efficient ones. To this purpose de Finetti suggests to fix the maximum value of ruin probability judged as acceptable by the company and choose the efficient solution corresponding to that level. He offers an exhaustive discussion of the connections between ruin probability and the behaviour of the l function. The point is bypassed here.

## 3. CONNECTIONS AMONG DE FINETTI'S SOLUTION AND MARKOVITZ MEAN VARIANCE APPROACH

We claim that the approach so far quickly resumed is with some minor differences the same applied, more than a dozen years later, by H. Markovitz to solve the portfolio selection problem, now universally known as mean variance criterion. Indeed even if forced by the need to keep as the key goal variable the ruin probability, de Finetti operationally follows a two stage approach, where in the first stage a mean variance (formally a mean-ruin probability) efficient set is defined on purely objective basis, while the second stage is devoted to select on the basis of a subjective tradeoff between the two parameters involved the optimal subjective solution as a specific point of the efficient set.

Going back now to the solution (2) , it is interesting to look at what happens if we consider the restriction of the efficient set to its interior, that is where $\underline{q}$ belongs to the open (0,1) hypercube. It is immediate to check that solutions belonging to this restriction are characterized by constant ratios among the optimal retention quotas (relative retentions) , independent from $\overline{E}$ , and given by:

$$q_i^+/q_j^+ = (m_i/V_i)/(m_j/V_j) \qquad \forall\, i,j \qquad i \neq j \qquad (3)$$

In turn this is a partial (provided its validity is restricted as above punctualized) reinsurance counterpart of the existence of an unique optimal risky portfolio in financial theory. Moreover something partially corresponding to the well known separation theorem is reached. * The optimal retention quotas are chosen according to (3) and then absolute retention limits are chosen. The first choice does not involve (is separated from) any preference evaluation about the risk return tradeoff. In our opinion these facts open the way to gain some new insights about the working of markets for proportional reinsurance and their links (allready stressed elsewhere) with the C.A.P.M. model of equilibria on asset markets. We do not enter here into details concerning this point. The interested reader is referred to Pressacco (1986).

## 4. THEORY AND PRACTICE IN PROPORTIONAL REINSURANCE

It is convenient to introduce here another simple assumption, that direct collected premiums are computed on the basis of the mean value principle with a common loading coefficient $\lambda$ , so that $m_h = \lambda E(X_h)$, and moreover that the random claims of the portfolio are distributionally obtained by homogeneous linear transforms of a base variable or formally that $X_h \overset{d}{=} t_h X_1$, $t_h > 1$, $h = 2,\ldots\ldots, n$. Denoting by $S_h$ the maximum possible loss for contract h, and keeping account that $m_h = \lambda\, t_h E(X_1)$for any h, expression (3) of the optimal ratios between retention quotas becomes (as

* Fundamental papers concerning mutual fund separation in financial theory are those of Cass and Stiglitz (1970) and Ross (1978). Resuming roughly, financial separability means that the choice of the risky portfolio is the same for any investor belonging to some family of decision makers. Thus this choice is independent or separated from the particular preference system (parameter of the utility function) characterizing an investor within the family. Any efficient portfolio is then obtained as a proper combination of the sure prospect and the risky portfolio.

easily checked) :

$$q_i^+/q_j^+ = t_j/t_i = S_j/S_i \qquad (4)$$

But for an efficient restricted solution this means nothing but the constance for any h of the product $S_h q_h^+$ at some level $R^+$. In reinsurance practice $R^+$ is known as retention level and the above reasoning makes clear under what conditions a practical proportional reinsurance strategy based on the choice of a single retention level enyois efficiency properties in a mean variance environment.

## 5. CONNECTIONS BETWEEN EXPECTED UTILITY APPROACH AN DE FINETTI SOLUTIONS FOR THE SINGLE PERIOD PROBLEM

Of course the expected utility criterion could be generally applied without restrictions as a different tool to solve the optimal retention problem. Alternatively, if we want it to be coherent with a mean variance approach, so that to be specifically useful in the second stage of the procedure determining indirectly a risk return tradeoff schedule, some restrictions ought to be placed either on utility functions or (and) on claims distributions. Concerning this point it is well known that coherence is surely granted either by assuming that utility functions are quadratic or that claims distribution is multivariate normal.* In the last case simple formulas are derived provided moreover that the utility function of the company's gain belongs to the exponential family.

For a detailed treatment of the quadratic case the reader is referred to e.g. Daboni (1986), we shall treat here in some detail the combined normal exponential case, that is with utility function: $u(G) = B(1-\exp(-G/B))$, $B>0$, where as well known B is the constant value of the risk tolerance function, that is the risk tolerance parameter of the insurance company. Being:

$$E(u(G)) = B(1-E(\exp(-(\Sigma G_h(q_h)/B))) \qquad (5)$$

the constrained maximization for any fixed B is found minimizing the expectation:

$$E(\exp-(\Sigma G_h(q_h)/B))=E(\Pi \exp-(G_h(q_h)/B)) \qquad (6)$$

Owing to the no correlation assumption, this gives rise to a set of n univariate constrained problems:

$$\min_{0\leq q_h\leq 1} \Psi_{G_h(q_h)}(-1/B) \qquad (7)$$

with $\Psi_G(.)$ as the moment generating function of the retained random gain on the h-th contract. Keeping account of the normality of $G_h$** and after

---

* See Tobin (1958) and Markovitz (1952)

** Alternatively relaxing the normality assumption, and keeping expansion of the cumulant function of $G_h(q_h)$ truncated at the third degree, we obtain the following approximation holding for the general case:

$$q_h^+ = \begin{cases} Bm_h(1-m_h\gamma/2V_h)V_h^{-1} & \text{if } < 1 \\ 1 & \text{otherwise} \end{cases}$$

with $\gamma$ as the asymmetry coefficient of $G_h(q_h)$.

some simple computation the solution is given by:

$$q_h^+ = \begin{cases} B\, m_h/V_h & \text{if} < 1 \\ 1 & \text{otherwise} \end{cases} \tag{8}$$

It is interesting to note that a two side reading is offered by this solution. For a given value of B it provides an answer to the question of selecting the optimal solution (absolute retention) in the efficient set, while leaving B undermined it gives immediately a compact expression for the whole efficient set that looks like the one obtained by de Finetti (compare with (2) ) . Of course this way relative retentions are immediately derived.

## 6. DE FINETTI'S TREATMENT OF THE INFINITE HORIZON PROBLEM

As previously said de Finetti's idea is to make recourse to old results of probability theory, concerning the gambler's ruin problem. Indeed de Finetti suggests to look at an insurance company as a gambler playing an infinite sequence $G_t$, t=1,..... of independent unfair (due to safety loading charges) bets, so that the story of its fortune is described by a submartingale process. To manage computationally the problem de Finetti needs fair transforms $Y_t$ of $G_t$ and looks for that purpose to $Y_t = \exp(-\alpha_t G_t) - 1$ . Given $G_t$, there is (under not binding restrictions) an unique value $\alpha_t^+ (\neq 0)$ providing the desired fairness for $Y_t$; it satisfies $\alpha_t^+ \cdot E(G_t) < 0$. Moreover a sum of unfair independent random numbers with a common "fairness coefficient" $\alpha$ , has in turn exactly $\alpha$ as its "fairness coefficient". After that an evaluation of the asymptotic ruin probability of an insurance company, seen as playing a sequence of games with the same fairness coefficient $\alpha_t^+ = \alpha^+$ for any t, is given by:

$$p^+ = \begin{cases} \exp(-\alpha^+ G_0) & \alpha^+ > 0 \\ 1 & \text{otherwise} \end{cases} \tag{9}$$

Then the surprisingly simple end of the story is that to reach a given goal $p^+$ concerning the asymptotic ruin probability, the company may apply to any future contract a retention strategy such that the retained random gain $G_h(q_h)$ has a fairness coefficient:

$$\alpha^+ = -(1/G_0)\ln p^+ . \tag{10}$$

In his paper de Finetti prefers to work with another index $\beta^+ = 1/\alpha^+$ . The reason is simple: given $G_0$, the ruin probability is an increasing normalized function of $\beta^+$ , going from 0 to 1 as $\beta^+$ goes from 0 to ∞, so that $\beta^+$ deserves the name of "livello di rischiosità" (hence risk level) of a random variable or of a portfolio whose contracts all have the same risk level. As said in the introduction, $\beta^+$ appears a meaningful counterpart of the "adjustment coefficient" well known in scandinavian collective risk theory approach to ruin probability evaluations.

Finally and before passing to next para 7 it is convenient to remark that under normality of the $G_h(q_h)$ the optimal single absolute retention quotas, indirectly derived through the request that the retained random gain have risk level not greater then the one associated to the fairness coefficient $\alpha^+$ appearing in (10) are given by:

$$q_h^+ = \begin{cases} 2\beta^+ m_h/V_h & \text{if} < 1 \\ 1 & \text{otherwise} \end{cases} \qquad (11)$$

## 7. EXPONENTIAL UTILITY AND ASYMPTOTIC RUIN PROBABILITY

The somewhat astonishing similarity of (8) and (11) reveals another interesting link between two key parameters of the reinsurance problem, that is the final risk level $\beta^+$ of a group of retained contracts granting that a given asymptotic ruin probability goal is reached, and the constant risk tolerance B of that company. Indeed comparison of (8) and (11) makes clear a one to one correspondence between the B coefficient guiding company's decision based on a (myopic) expected utility criterion, and the goal riskness level $\beta^+$ of the retained quotas of the single contracts, and thus finally with the ruin probability goal. Couples of $(B, \beta^+)$ obtained on this basis induce exactly the same absolute retention strategy either if decisions are aimed to utility maximization or to ruin probability control. That happens because a choice based on exponential utility grants (as revealed by (8) , at least under normality and independence assumptions) that any contract is reduced to a common risk level ( see (11) ), and thus indirectly but unambiguously determines an asymptotic evaluation of a ruin probability.

Conversely a goal expressed in ruin probability terms, to be traduced through the derived common risk level $\beta^+$ , is coherent with an exponential utility approach and determines a unique value of the risk tolerance coefficient, such that an optimal utility based decision strategy attains the ruin probability chosen as starting point. It is interesting to remark that even if generally referred to an infinite horizon problem, the above results are formally still holding for whatever temporal horizon provided that:
.) the risk tolerance coefficient remains invariant throughout the whole period considered, and:
..) the number of the contracts on the horizon is high enough, so that the asymptotic evaluation of the ruin probability is a good proxy of the true value for a finite number of contracts. In principle then the horizon could be even a single period one. But be cautious that, this being the case, the ruin probability obtained is not simply the probability to be insolvent at the end of the year with all contracts expired, as in classical risk theory; on the contrary ruin may appear after any number of contracts, irrespective of the results of the next ones.

Obviously at the time de Finetti was writing his fundamental paper the modern utility theory was still in mind of J. Von Neumann and O. Morgenstern. To complete our picture we want to signal that even without spending some room for any formal or verbal proof, in another paper some years later (see de Finetti (1952) ), he shows to be well aware of the crucial connections previously discussed. Indeed after a short discussion of the properties of the exponential utility and undoubtedly with this idea in mind, he says explicitly that: "the risk level criterion derived by the ruin probability criterion induces the same behaviour coming from the expected utility in the exponential case".

REFERENCES

Borch, K., 1974, "The Mathematical Theory of Insurance", Lexington Books, London .

Bühlmann, H.,1970, "Mathematical Methods in Risk Theory", Springer Verlag, New York.

Cass, D., Stiglitz, J., 1970, The structure of investor preferences and asset returns and separability in portfolio selection: a contribution to the pure theory of mutual funds, Journal of Economic Theory 2, 122-160.

Daboni, L., 1986, Il problema dei pieni di B. de Finetti rivisitato con il criterio dell'utilità attesa, Rendiconti del Comitato per gli Studi e la Programmazione Economica XXIV. 41-57.

De Finetti, B.,1940, Il problema dei pieni, Giornale Istituto Italiano Attuari 11, 1 - 88.

De Finetti, B., 1952, Sulla preferibilità, Giornale degli Economisti e Annali di Economia 6, 3-27.

Gerber, H., 1979, "An Introduction to Mathematical Risk Theory" Huebner Foundation 8, Philadelphia.

Markovitz, H., 1952, Portfolio Selection, Journal of Finance 6, 77-91.

Pressacco, F., 1986, Separation theorems in proportional reinsurance, Insurance and Risk Theory, D. Reidel, 209-215.

Ross, S., 1978, Mutual Fund Separation in Financial Theory-The separating Distributions, Journal of Economic Theory 17, 254-286.

Seal, H., 1969, "Stochastic Theory of Risk Business", J.Wiley, N.York.

Tobin, J., 1958, Liquidity preference as behaviour toward risk, Review of Economic Studies 25, 65-85.

# A PRODUCT OF MULTIVARIATE T DENSITIES AS UPPER BOUND FOR THE POSTERIOR KERNEL OF SIMULTANEOUS EQUATION MODEL PARAMETERS*

Herman K. van Dijk

Erasmus University, Rotterdam and
Center for Operations Research and Econometrics (CORE),
Université Catholique de Louvain

## INTRODUCTION

The linear simultaneous equation model (SEM) is one of the best known models in econometrics. It is used in several areas, for instance, in micro-economic modelling for the description of the operation of a market for a particular economic commodity and in macro-economic modelling for the description of the interrelations between a large number of macro-economic variables. [See, e.g., Hausman (1983) for a recent survey of the linear SEM.]

A linear simultaneous equation model contains, usually, some exactly known structural coefficients and some unrestricted structural coefficients. Suppose that the prior information on the unrestricted coefficients is taken from a noninformative approach. Then one can derive that the kernel of the posterior density of the unrestricted coefficients has the same functional form as the so-called concentrated likelihood function [see, e.g., Drèze and Richard (1983) or Zellner (1971, p. 272)]. This kernel is, however, not proportional to a density with known properties. In an earlier paper [Van Dijk (1985), hereafter referred to as HVD] I studied the global properties of this posterior kernel (or likelihood function) in the structural parameter space. One of the results of the HVD paper is that, given certain conditions, the posterior kernel of the unrestricted structural coefficients of a linear simultaneous equation model is dominated by a matricvariate Student t density,

* I am indebted to Luc Bauwens and Teun Kloek for helpful discussions. Any errors are my own responsibility.

multiplied by a constant. [For details on properties of this density, see Dickey (1967) or Zellner (1971, Appendix B5).] However, in the derivation of this result I did not make full use of the zero restrictions that appear in many simultaneous equation models. Consider, for instance, the case of a market model for an agricultural commodity. Weather conditions will, probably, appear in a supply equation for this commodity, but, in most cases, not in a demand equation. Personal income may appear in a demand equation but, in most cases, not in a supply equation.[1] So, zero restrictions reflect a priori considerations on the variables that are excluded from the different equations.

In the present paper I make explicit use of the zero restrictions mentioned above. As a result one can derive that under certain conditions the matricvariate Student t bound can be replaced by an other upper bound function that is a product of multivariate Student t densities that are defined, in a certain sense, in a recursive way. This bound has as an advantage that the multivariate Student t density possesses known properties [see, e.g., Zellner (1971, Appendix B2)] and it is comparable with the bound derived by Drèze and Richard (1983, p. 596).

PRELIMINARIES

The linear simultaneous equation model (SEM) can be written as

$$XA = U \tag{2.1}$$

where X is a T×(G+K) matrix of T observations on G+K variables and A is a (G+K)×G matrix of parameters some of which are known a priori [see below]; U is a T×G matrix of disturbances. The matrices X and A are partitioned as follows.

$$X = (Y \;\; Z), \qquad A = \binom{B}{\Gamma} \tag{2.2}$$

where Y is a T×G submatrix of X that refers to G endogenous variables and Z is a T×K submatrix of X that refers to the K predetermined variables. The matrix A has been partitioned in a similar way as the matrix X so that the left hand side of (2.1) can be written as

$$XA := YB + Z\Gamma \tag{2.3}$$

1. For a viewpoint that (almost) all variables should appear in all equations see, e.g., Sims (1980).

I shall make use of the set of standard assumptions with respect to the linear SEM [see, e.g., Hausman (1983), and the references cited there]. These assumptions may be summarized as follows. (i) The determinant of B does not vanish. (ii) The T rows of U are independently distributed with a common normal distribution that has mean zero and positive definite covariance matrix $\Sigma$. (iii) Current values of the disturbances are independently distributed from current and lagged values of predetermined variables. (iv) The data matrix X has full column rank. So, $T \geq G+K$.

The prior information on the structural parameters is summarized as follows. The elements of A are partitioned into two subsets. The first subset contains the a priori restricted elements, which are denoted by the vector $\phi$. The second subset contains the unrestricted elements of A, denoted by the vector $\theta$. So, one has

$$A = A(\theta, \phi = \phi_0) \tag{2.4}$$

where $\phi = \phi_0$ indicates that the elements $\phi$ take a particular value $\phi_0$. The stochastic prior information on $(\theta, \Sigma)$ is given as

$$p(\theta, \Sigma) \propto |\Sigma|^{-\frac{1}{2}h} \qquad (h \geq 1) \tag{2.5}$$

where h is usually taken as an integer. Well known values for h are $h = G+1$, $h = 2G$ or $h = G+K+1$. In HVD I derived a bound on h given as $h \geq 2G + K$ as a condition for existence of the zero-th order moment of the vector $\theta$. The <u>marginal</u> posterior density of the unrestricted coefficients $\theta$, given the data X and the exactly known elements $\phi_0$, can be written as

$$p(\theta|X, \phi_0) \propto \|B\|^T |U'U|^{-\frac{1}{2}(T+h-G-1)} \tag{2.6}$$

For details, see HVD, Zellner (1971, Chapter 9) or Drèze and Richard (1983, p.562).

Before we start with the main result of this paper we need the following theorem. For a proof, see HVD, Theorem 4.

<u>THEOREM 1</u>. Given $R(X) = G+K$, it follows that U'U is a positive definite symmetric matrix and

$$\|B\|^T |U'U|^{-\frac{1}{2}(T+h-G-1)} \leq c|U'U|^{-\frac{1}{2}(h-G-1)} \qquad (c > 0) \tag{2.7}$$

if and only if $R(A) = G$.

Some comments on the role of the positive constant c are given in th next section. One can use (2.6) and (2.7) and write

$$p(\theta|X, \phi_0) \leq c|U'U|^{-\frac{1}{2}(h-G-1)} \tag{2.8}$$

The upper bound function given at the right hand side of (2.8) is the starting point for the analysis of this paper.

## A DOMINATING FUNCTION FOR THE POSTERIOR DENSITY OF $\theta$

Let $a_j$, $j = 1,\ldots,G$, be the j-th column of the matrix of structural coefficients [see equation (2.1)]. The typical j-the equation of the mode (2.1) can be written as

$$Xa_j = u_j \qquad (j = 1,\ldots,G) \tag{3.1}$$

Suppose that the exact restrictions with respect to the elements of $a_j$ ar either zero restrictions or the normalization restriction. One can make use of the following notation.

$$y_j - W_j\theta_j = u_j \qquad (j = 1,\ldots,G) \tag{3.2}$$

The T-vector $y_j$ is the j-the column of the matrix Y of (2.2) and it consists of T observations on the j-the endogenous variable. I assume tha the diagonal elements of the matrix B of (2.2) are equal to unity due to the normalization restrictions. The matrix $W_j$ is defined as

$$W_j := (Y_j \;\; Z_j) \tag{3.3}$$

where the $T \times g_j$ matrix $Y_j$ contains observations on $g_j$ current endogenous variables that are present as explanatory variables in equation (3.2). So $G - g_j - 1$ endogenous variables are excluded from this equation. The $T \times k_j$ matrix $Z_j$ contains the T observations on the $k_j$ predetermined variable present in the j-equation. So, $K - k_j$ predetermined variables are exclude from this equation. The matrix $(y_j \;\; W_j)$ has full column rank. The parameter vector $\theta_j$ contains the $\ell_j = g_j + k_j$ parameters of interest. Not that the $\ell$-vector $\theta$ is given as

$$\theta' := (\theta_1', \ldots, \theta_j', \ldots, \theta_G'), \qquad \ell = \sum_{j=1}^{G} \ell_j \tag{3.4}$$

Next, I construct a proof of the proposition that under certain conditions the posterior kernel $p(\theta|X, \phi_0)$ [equation (2.6)] is dominated by a constant times a product of multivariate Student t densities that are defined in a recursive way.

The first step of the proof is as follows. Start with the right hand side of (2.8). The matrix U is restricted to have full column rank so that U'U is a positive definite symmetric matrix [see Theorem 1]. Partition the matrix U as

$$U = (u_j \quad U_j) \tag{3.5}$$

where $u_j$ is the T-vector of disturbances of the j-th equation [see (3.1)] and $U_j$ is the remaining submatrix of U after $u_j$ has been deleted. Note that one may reorder the columns of U in such a way that the j-th column is moved to the position of the first column without affecting the value of $|U'U|$. Make use of

$$\begin{aligned} |U'U| &= |(u_j \; U_j)'(u_j \; U_j)| \qquad (j = 1,\ldots,G) \\ &= \begin{vmatrix} u_j'u_j & u_j'U_j \\ U_j'u_j & U_j'U_j \end{vmatrix} \\ &= |U_j'U_j|(u_j'M_ju_j) \end{aligned} \tag{3.6}$$

with

$$M_j = I - U_j(U_j'U_j)^{-1}U_j' \tag{3.7}$$

Given that U'U is PDS, it follows that $U_j'U_j$, $u_j'u_j$ and $u_j'M_ju_j$ are PDS.

As a second step, substitute $u_j = y_j - W_j\theta_j$ [equation (3.2)] in $u_j'M_ju_j$. A well known decomposition yields

$$\begin{aligned} u_j'M_ju_j &= (y_j - W_j\theta_j)'M_j(y_j - W_j\theta_j) \\ &= \hat{\sigma}_j^2 + (\theta_j - \hat{\theta}_j)'W_j'M_jW_j(\theta_j - \hat{\theta}_j) \end{aligned} \tag{3.8}$$

with

$$\hat{\theta}_j = (W_j'M_jW_j)^{-1}W_j'M_jy_j \tag{3.9}$$

$$\hat{\sigma}_j^2 = (y_j - W_j\hat{\theta}_j)'M_j(y_j - W_j\hat{\theta}_j) \tag{3.10}$$

under the condition that $W_j'M_jW_j$ is a positive definite matrix. In the next section I discuss conditions under which $W_j'M_jW_j$ is a PDS matrix and $\hat{\sigma}_j^2$ is positive. By making use of these properties and by using (3.5)-(3.10), one can derive in a straightforward way that

$$|U'U|^{-\frac{1}{2}(h-G-1)} = |U_j'U_j|^{-\frac{1}{2}(h-G-1)}p(\theta_j|\hat{\theta}_j, \hat{V}_j, \lambda_j) \tag{3.11}$$

$$\times|W_j'M_jW_j|^{-\frac{1}{2}}(\hat{\sigma}_j^2)^{-\frac{1}{2}\lambda_j}C_j^{-1}$$

with

$$p(\theta_j|\hat{\theta}_j, \hat{V}_j, \lambda_j) = C_j\lambda_j^{\frac{1}{2}\lambda_j}\left|\frac{\lambda_jW_j'M_jW_j}{\hat{\sigma}_j^2}\right|^{\frac{1}{2}} \tag{3.12}$$

$$\times[\lambda_j + (\theta_j - \hat{\theta}_j)'[\frac{\lambda_jW_j'M_jW_j}{\hat{\sigma}_j^2}](\theta_j - \hat{\theta}_j)]^{-\frac{1}{2}(\lambda_j+\ell_j)}$$

and

$$\lambda_j = h - G - 1 - \ell_j \tag{3.13}$$

and

$$C_j = \frac{\Gamma[\frac{1}{2}(\lambda_j + \ell_j)]}{\pi^{\frac{1}{2}\ell_j}\Gamma(\frac{1}{2}\lambda_j)} \tag{3.14}$$

Equation (3.12) is equal to a multivariate Student t density of the $\ell_j$-vector $\theta_j$ under the following conditions

$$W_j'M_jW_j \text{ is PDS, } \hat{\sigma}_j^2 > 0, \ \lambda_j > 0 \tag{3.15}$$

[see, e.g., Zellner (1971, Appendix B2)]. The location parameters are denoted by $\hat{\theta}_j$, which is defined in (3.9), and the scale parameters are denoted by the matrix $\hat{V}_j$, which is given as

$$\hat{V}_j = \hat{\sigma}_j^2(\lambda_jW_j'M_jW_j)^{-1} \tag{3.16}$$

By making use of the definition of $M_j$, equation (3.7), and by redefining $U_j$ as $U_j = (u_{j+1}, \ldots, u_G)$ [compare the text below (3.5)], it is seen that $\hat{\theta}_j$ and $\hat{V}_j$ depend on the unrestricted parameters $\theta_{j+1}, \ldots \theta_G$, of the simultaneous equation system (2.1). Therefore, $p(\theta_j|\hat{\theta}_j, \hat{V}_j, \lambda_j)$ is a <u>conditional</u> multivariate Student t density of $\theta_j$ given values of $\theta_{j+1}, \ldots, \theta_G$.

As a third step, the second line in (3.11) is analyzed. The determinant of $W_j'M_jW_j$ can be rewritten by making use of a well known rule for the evaluation of the determinant of a partitioned matrix. That is, given $W_j'W_j$ and $U_j'U_j$ are PDS matrices, it follows that

$$\begin{vmatrix} W_j'W_j & W_j'U_j \\ U_j'W_j & U_j'U_j \end{vmatrix} = |U_j'U_j||W_j'M_jW_j| = |W_j'W_j||U_j'\bar{M}_jU_j| \tag{3.17}$$

where

$$\bar{M}_j = I - W_j(W_j'W_j)^{-1}W_j' \tag{3.18}$$

One can use (3.17) and derive that

$$|W_j'M_jW_j|^{-\frac{1}{2}} = |W_j'W_j|^{-\frac{1}{2}}|U_j'U_j|^{\frac{1}{2}}|U_j'\bar{M}_jU_j|^{-\frac{1}{2}} \tag{3.19}$$

where $|W_j'W_j| > 0$. Substitution of (3.19) in (3.11) yields

$$\begin{aligned} |U'U|^{-\frac{1}{2}(h-G-1)} &= |U_j'U_j|^{-\frac{1}{2}(h-G-2)}p(\theta_j|\hat{\theta}_j, \hat{V}_j, \lambda_j) \\ &\times |U_j'\bar{M}_jU_j|^{-\frac{1}{2}}(\hat{\sigma}_j^2)^{-\frac{1}{2}\lambda_j}C_j^{-1}|W_j'W_j|^{-\frac{1}{2}}. \end{aligned} \tag{3.20}$$

In the next section I discuss the conditions under which

$$|U_j'\bar{M}_jU_j|^{-\frac{1}{2}}(\hat{\sigma}_j^2)^{-\frac{1}{2}\lambda_j}C_j^{-1} \leq K_j \tag{3.21}$$

where $K_j$ is a positive number. Then one can write

$$|U'U|^{-\frac{1}{2}(h-G-1)} \leq K_j|U_j'U_j|^{-\frac{1}{2}(h-G-2)}p(\theta_j|\hat{\theta}_j, \hat{V}_j, \lambda_j) . \tag{3.22}$$

One can use this inequality for $j = 1, \ldots, G-1$. This yields

$$|U'U|^{-\frac{1}{2}(h-G-1)} \leq K \prod_{j=1}^{G-1} p(\theta_j|\hat{\theta}_j, \hat{V}_j, \lambda_j)(u_G'u_G)^{-\frac{1}{2}(h-2G)} \tag{3.23}$$

where $K = \Pi_{j=1}^{G-1} K_j$ is a positive number and $\lambda_j = h - G - j - \ell_j$. Note the difference with (3.13). The sum of squared posterior residuals of the G-th equation can be decomposed in a similar way as done in (3.8). Then one can obtain

$$(u_G'u_G)^{-\frac{1}{2}(h-2G)} = C^{-1}p(\theta_G|\hat{\theta}_G, \hat{V}_G, \lambda_G) \tag{3.24}$$

where $p(\theta_G|\hat{\theta}_G, \hat{V}_G, \lambda_G)$ is a marginal Student t density with parameters

$$\hat{\theta}_G = (W_G'W_G)^{-1}W_G'y_G, \; \hat{V}_G = \hat{\sigma}_G^2(\lambda_G W_G'W_G)^{-1}, \; \lambda_G = h - 2G - \ell_G \tag{3.25}$$

and

$$\hat{\sigma}_G^2 = (y_G - W_G\hat{\theta}_G)'(y_G - W_G\hat{\theta}_G) \tag{3.26}$$

The numerical constant C is the same as given in (3.14) with index G instead of index j. Note that the location and scale parameters of $p(\theta_G|\hat{\theta}_G, \hat{V}_G, \lambda_G)$ depend only on the given data $y_G$, $W_G$ and the degrees of freedom parameter $\lambda_G$.

As a final step, one makes use of (2.8), (3.23) and (3.24). Then one can obtain that

$$p(\theta|X, \phi_0) \leq K^* \prod_{j=1}^{G-1} p(\theta_j|\hat{\theta}_j(\theta_{j+1},\ldots,\theta_G), \hat{V}_j(\theta_{j+1},\ldots,\theta_G), \lambda_j) \tag{3.27}$$
$$\times \, p(\theta_G|\hat{\theta}_G, \hat{V}_G, \lambda_G)$$

where $K^*$ is a positive number. In (3.27) it is explicitly indicated that the location and scale parameters of the conditional multivariate Student t density of the parameters $\theta_j$ of the j-th equation depend on the values of the unrestricted parameters $\theta_{j+1}$, ..., $\theta_G$ of the equations j+1, ..., G.

## INTEGRABILITY CONDITIONS

Apart from the standard set of assumptions for the linear SEM, I have made use of the following conditions [compare Theorem 1, (3.15), (3.21) and the line below (3.23)].

(i) $R(A) = G$, (ii) $\hat{\sigma}_j^2 > 0$, $(j = 1,\ldots,G-1)$

(iii) $W_j'M_jW_j$ is PDS, (iv) $U_j'\bar{M}_jU_j$ is PDS, $(j = 1,\ldots,G-1)$ (4.1)

(v) $\lambda_j = h-G-j-\ell_j > 0$ , $(j = 1,\ldots,G)$.

The conditions (i) and (ii) are not independent. That is, if (i) holds then it follows that $u_j'M_ju_j > 0$ and hence, in particular, $\hat{\sigma}_j^2 = \hat{u}_j' M \hat{u}_j > 0$, with $\hat{u}_j = y_j - W_j\hat{\theta}_j$. [See (3.8) and (3.10)]. Conditions (iii) and (iv) are related in the sense that if (iii) holds, then it follows that (iv) holds. This can be derived using (3.17). Given $|U_j'U_j| > 0$ and $|W_j'W_j|$ is a positive constant, it follows that $|U_j'\bar{M}_jU_j| > 0$ if $|W_j' M_jW_j| > 0$.
Condition (iii) may be interpreted as follows. Let

$$\hat{V}_j = W_j - U_j\hat{\Delta}_j, \quad \hat{\Delta}_j = (U_j'U_j)^{-1}U_j'W_j \tag{4.2}$$

The restriction $\hat{V}_j'\hat{V}_j = W_j'M_jW_j$ is PDS implies that the explanatory variables in the j-th equation cannot linearly depend on the posterior residuals of the disturbances of equations j+1,...,G. This condition has to be verified in practice for particular models.

One may distinguish between two classes of simultaneous equation models. In the first class of models one has that R(A) = G everywhere in the prior region of $\theta$ and $|W_j'M_jW_j| \geq \varepsilon > 0$ everywhere in the prior region of $\theta$ for j=1,..., G-1. One may verify this for, e.g., Johnston's model [see Johnston (1963)] and for Klein's Model I [see Klein (1950)]. In the second class of models it may occur that R(A) becomes less than G and $|W_j'M_jW_j|$ tends towards zero in the prior region of $\theta$. [See Drèze and Richard (1983,p.533) for an example of a market model where this may occur.] Then one can make use of the following solution. Truncate the uniform prior of $\theta$ in such a way that it is zero on an open subset of the prior region where $|A'X'XA| < \varepsilon_0$, $\varepsilon_0 > 0$ and $|W_j'M_jW_j| < \varepsilon_j$, $\varepsilon_j > 0$, j=1, ..., G-1. This implies that the positive constant $K^*$ [see (3.27)] depends on $\varepsilon_0$ and $\varepsilon_j$, j=1,..., G-1. One may investigate the sensitivity of $K^*$ by varying the value of $\varepsilon_0$ and $\varepsilon_1,\ldots,\varepsilon_{G-1}$, which may be an unattractive approach in practice. Therefore, the first class of models, where the constant $K^*$ is a given positive number everywhere in the prior region of $\theta$ is the more relevant case.

Another condition is the degrees of freedom restriction $\lambda_j > 0$, j=1,...,G. This implies a bound on the prior parameter h given as

$$h > \sup \{(G+j+\ell_j),\ j=1,\ldots,G\} \tag{4.3}$$

This bound is essentially the same as a degrees of freedom bound derived by Drèze and Richard (1983,p.566). In the HVD paper I derived the bound h > 2G+K. A sufficient condition for (4.3) is $h > 2G + \ell_{max}$. It follows that the bound in the HVD paper is larger then the present bound if $K > \ell_{max}$, which can be restated as $K - k_{max} > g_{max}$ in the equation with $\ell_{max}$. This condition is equal to the classical order condition for identification for the equation with the largest number of explanatory variables.

I conclude this paper with two remarks. First, the results of this paper may be extended to a linear SEM with identities. This analysis has been deleted from the present paper due to space limitations. Second, the results of the paper are part of a larger project on existence conditions for posterior moments of simultaneous equation model parameters. In a forthcoming revision and extension of the HVD paper I shall discuss the use of the results of this paper for the existence of the moments mentioned above.

## REFERENCES


Dickey, J.M., 1967, Matricvariate generalizations of the multivariate t distribution and the inverted multivariate t distribution, The Annals of Mathematical Statistics 38, 511-518.

Drèze, J.H. and J.F. Richard, 1983, Bayesian analysis of simultaneous equation systems, Chapter 9, in: Z. Griliches and M. Intriligator, eds., Handbook of econometrics, vol. 1, North Holland, Amsterdam.

Hausman, J.A., 1983 Specification and estimation of simultaneous equation models, Chapter 7 in: Z. Griliches and M. Intriligator, eds., Handbook of econometrics, vol. 1 (North Holland, Amsterdam).

Johnston, J.J., 1963, "Econometric methods," first edition (McGraw-Hill, New York

Klein, L.R., 1950, "Economic fluctuations in the United States, 1921 - 1941," Wiley, New York .

Sims C., 1980, Macroeconomics and reality, Econometrica 48, 1-48.

Van Dijk, H.K., 1985, Existence conditions for posterior moments of simultaneous equation model parameters, Report 8551 of the Econometric Institute, Erasmus University Rotterdam.

Zellner A., 1971, "An introduction to Bayesian inference in econometrics," Wiley, New York .

A COMMON MODEL SELECTION CRITERION

N.R. Draper and
I. Guttman

University of Wisconsin
University of Toronto

## 1. INTRODUCTION

We consider the linear model situation

$$\underset{\sim}{y} = \underset{\sim}{X}_t \underset{\sim}{\beta}_t + \underset{\sim}{\varepsilon}_t \qquad (1.1)$$

where $\underset{\sim}{y} = (y_1, y_2, \ldots, y_n)'$ is an $n \times 1$ vector of response observations, $\underset{\sim}{X}_t$ is an $n \times p_t$ matrix of predictor variable values, $n > p_t$, $\underset{\sim}{\beta}_t$ is a $p_t \times 1$ vector of regression parameters to be estimated and $\underset{\sim}{\varepsilon}_t$ is distributed $N(\underset{\sim}{\delta}_t, \sigma^2 I_n)$. We shall distinguish between problems in which (a) $\underset{\sim}{\delta}_t = \underset{\sim}{0}$ for all $t$, and (b) $\underset{\sim}{\delta}_t = (\underset{\sim}{0}', \underset{\sim}{a}_t')'$, for all $t$, where the elements of the $\underset{\sim}{a}_t$ are non-zero and each $\underset{\sim}{a}_t$ vector is size $k \times 1$ where, typically, $k << n/2$. The generic notation $t$ denotes a general indexing which will be made specific for particular problems to be discussed below. Each choice of $t$ will provide a model $M_t$, say, defined by (1.1). The general problem, given a specific indexing system for $t$, is to decide, from data made available on $\underset{\sim}{y}$ and the $\underset{\sim}{X}_t$'s, which $M_t$ "best represents" the data.

Among the multitude of problems covered by the above, we distinguish four specific areas.

a. Outlier problems with spuriousity caused by shift of mean. Suppose we fear the presence of $k$ spurious observations, $k$ fixed and pre-selected. Then the indexing $t$ runs over all possible choices of $k$ from $n$ observations, and the $\underset{\sim}{X}_t$'s are permutations of $n$ specified rows of a matrix $\underset{\sim}{X}$ used to generate the data. We may write

$$\underset{\sim}{y} = \begin{bmatrix} \underset{\sim}{y}_1 \\ \underset{\sim}{y}_2 \end{bmatrix} = \begin{bmatrix} \underset{\sim}{X}_{1t} \\ \underset{\sim}{X}_{2t} \end{bmatrix} \underset{\sim}{\beta} + \begin{bmatrix} \underset{\sim}{\varepsilon}_{1t} \\ \underset{\sim}{\varepsilon}_{2t} \end{bmatrix} \qquad (1.2)$$

where the $k \times 1$ vector $\underset{\sim}{y}_2 = (y_{i_1}, y_{i_2}, \ldots, y_{i_k})'$, $i_1 < i_2 < \ldots < i_k$, is associated with the $k$ spurious observations, and the $(n-k)\times 1$ vector $\underset{\sim}{y}_1$ is $\underset{\sim}{y}$, but with the elements of $\underset{\sim}{y}_2$ deleted. Assuming $E(\underset{\sim}{\varepsilon}_{1t}) = \underset{\sim}{0}$, $E(\underset{\sim}{\varepsilon}_{2t}) = \underset{\sim}{a}_t$ now brings us into case (b), with $p_t = p$.

b. Change point problems. We partition items of length $n$ into two parts of lengths $m$ and $(n-m)$ to express (1.1) as

$$\begin{bmatrix} \underset{\sim}{y}_1 \\ \underset{\sim}{y}_2 \end{bmatrix} = \begin{bmatrix} \underset{\sim}{X}_{mt} & \underset{\sim}{0} \\ \underset{\sim}{0} & \underset{\sim}{X}_{n-m,t} \end{bmatrix} \begin{bmatrix} \underset{\sim}{\beta}_1 \\ \underset{\sim}{\beta}_2 \end{bmatrix} + \begin{bmatrix} \underset{\sim}{\varepsilon}_{1t} \\ \underset{\sim}{\varepsilon}_{2t} \end{bmatrix}, \tag{1.3}$$

where $\underset{\sim}{X}_{mt}$ is $m \times p_1$, $\underset{\sim}{X}_{n-m,t}$ is $(n-m) \times p_2$, and $\underset{\sim}{\beta}_i$ is $p_i \times 1$, i=1,2. The index $t$ runs sequentially over $m$, $p_1 < m < n-p_2$. We shall define $\underset{\sim}{X}_{1t} = (\underset{\sim}{X}_{mt}, \underset{\sim}{0})$, $\underset{\sim}{X}_{2t} = (\underset{\sim}{0}, \underset{\sim}{X}_{n-m,t})$ for notational purposes. Assuming $E(\underset{\sim}{\varepsilon}_{1t}) = \underset{\sim}{0}$, i=1,2, now brings us into case (a), with $p_t = p_1 + p_2 = p$.

c. Join Problem. Consider the following motivating example. Suppose, in one dimension $x$, we wish to fit a quadratic model in $x$ for $x \le \alpha$ and a straight line model in $x$ for $x \ge \alpha$, with continuity $x = \alpha$. (An alternative description is the fitting of a quadratic and straight line spline continuous at the join $\alpha$.) There are three parameters in the quadratic, two parameters in the straight line, and one parameter $\alpha$, making a total of six, in addition to $\sigma^2$. However, continuity at $\alpha$ imposes one restriction.

Suppose, given an $\alpha$, $m$ observations $(x_1,y_1), (x_2,y_2), \ldots, (x_m,y_m)$ are such that $x_1 \le x_2 \le \ldots \le x_m \le \alpha$, while $(x_{m+1},y_{m+1}), \ldots, (x_n,y_n)$ are such that $\alpha < x_{m+1} \le \ldots \le x_n$. Then by applying the continuity restriction

$$\gamma_o + \gamma_1\alpha + \gamma_2\alpha^2 = \delta_o + \delta_1\alpha \tag{1.4}$$

and eliminating $\delta_o = \gamma_o + (\gamma_1-\delta_1)\alpha + \gamma_2\alpha^2$, we can rewrite the model, conditional on $\alpha$, in the form (1.2) with $\underset{\sim}{\beta} = (\gamma_o,\gamma_1,\gamma_2,\delta_1)'$,

$$\underset{\sim}{X}_{1t} = \begin{bmatrix} 1 & x_1 & x_1^2 & 0 \\ \cdots & & & \\ 1 & x_m & x_m^2 & 0 \end{bmatrix} \tag{1.5}$$

$$\underset{\sim}{X}_{2t} = \begin{bmatrix} 1 & \alpha & \alpha^2 & x_{m+1}-\alpha \\ \cdots & & & \\ 1 & \alpha & \alpha & x_n-\alpha \end{bmatrix} \tag{1.6}$$

Assuming $E(\underset{\sim}{\varepsilon}_{it}) = \underset{\sim}{0}$, i=1,2, produces case (a) with $p_t = p = 4$. The indexing $t$ is related to the possible choices for values of $\alpha$. For example, if the $x_i$ are equally spaced, one value of $\alpha$, for example, the midpoint, may be chosen in each gap between successive x's, so that the choices are equally spaced. For unequally spaced $x_i$'s, the number of values of $\alpha$ in each gap can be chosen proportional to the gap width and equally spaced throughout. Or, the values of $\alpha$ can be very densely spaced, approximating a continuous functional choice of $\alpha$.

Note that, in problems (1)-(3) listed above, $p_t = p$ for all $M_t$. For our fourth problem below, this will not be true.

d. Variable selection problem. Suppose we are given a set $\underset{\sim}{X} = (\underset{\sim}{1}, \underset{\sim}{x}_1, \ldots, \underset{\sim}{x}_q)$ of $(q+1)$ predictor variable vectors and an associated response vector $\underset{\sim}{y}$, all vectors being of dimension $n \times 1$ and with $\underset{\sim}{1} = (1,1,\ldots,1)'$. Let model $M_t$ be defined by a selected submatrix $\underset{\sim}{X}_t = (\underset{\sim}{1}, \underset{\sim}{x}_{i_1}, \ldots, \underset{\sim}{x}_{i_r})$, $0 \leq r \leq q$. Thus $t$ indexes the $2^q$ possible choices of models, which can all be described in form (1.1), with $\underset{\sim}{\beta}_t = (\beta_0, \beta_{i_1}, \ldots, \beta_{i_r})'$ and $p_t = r+1$. Assuming $E(\underset{\sim}{\varepsilon}_{it}) = \underset{\sim}{0}$, i=1,2, produces case (a).

## 2. THE CASE $p_t = p$

For cases (1), (2) and (3) of Section 1, we may proceed as follows. Because it includes case (a), we treat case (b). We obtain a posterior probability for model $M_t$ in the general situation. Maximization of this probability over the indexing set of $t$ will determine our choice of "best" model, and/or the entire set of probabilities can be reviewed.

From the case (b) assumptions in section (1.1), we obtain the likelihood function as proportional to

$$\sigma^{-n} \exp\{-[S_t + (\underset{\sim}{\beta}_t - \hat{\underset{\sim}{\beta}}_{1t})'\underset{\sim}{X}'_{1t}\underset{\sim}{X}_{1t}(\underset{\sim}{\beta}_t - \hat{\underset{\sim}{\beta}}_{1t}) + (\underset{\sim}{a}_t - \underset{\sim}{y}_2 + \underset{\sim}{X}_{2t}\underset{\sim}{\beta}_t)'(\underset{\sim}{a}_t - \underset{\sim}{y}_2 + \underset{\sim}{X}_{2t}\underset{\sim}{\beta}_t)]/(2\sigma^2) , \tag{2.1}$$

where $\underset{\sim}{X}_{1t}$ is $(n-k) \times p_t$, $\underset{\sim}{X}_{2t}$ is $k \times p_t$, and $\underset{\sim}{X}'_t = (\underset{\sim}{X}'_{1t}, \underset{\sim}{X}'_{2t})'$. Also we partition $\underset{\sim}{y}' = (\underset{\sim}{y}'_1, \underset{\sim}{y}'_2)'$ correspondingly. Furthermore,

$$\hat{\underset{\sim}{\beta}}_{1t} = (\underset{\sim}{X}'_{1t}\underset{\sim}{X}_{1t})^{-1}\underset{\sim}{X}'_{1t}\underset{\sim}{y}_1, \qquad S_t = (\underset{\sim}{y}_1 - \underset{\sim}{X}_{1t}\hat{\underset{\sim}{\beta}}_1)'(\underset{\sim}{y}_1 - \underset{\sim}{X}_{1t}\hat{\underset{\sim}{\beta}}_1). \tag{2.2}$$

Assuming the prior information to be non-informative and of the form

$$p(M_t, \underset{\sim}{\beta}_t, \sigma^2, \underset{\sim}{a}_t) \propto \sigma^{-2} \tag{2.3}$$

we obtain the posterior $p(M_t, \underset{\sim}{\beta}_t, \sigma^2, \underset{\sim}{a}_t | \underset{\sim}{y})$ by combining (2.1) and (2.3). Integrating out successively $\underset{\sim}{a}_t$, $\underset{\sim}{\beta}_t$, $\sigma^2$ yields the marginal posterior

$$p(M_t|\underset{\sim}{y}) = C|(\underset{\sim}{X}'_{1t}\underset{\sim}{X}_{1t})^{-1}|^{\frac{1}{2}} S_t^{-(n-k-p)/2} , \tag{2.4}$$

where the constant C is such that the sum of terms on the right hand side, summed over index t, is one. Note that, if $k = 0$, $\underset{\sim}{X}_{1t} = \underset{\sim}{X}_t$. Also, if desired, a conjugate prior could replace (2.3), with appropriate changes throughout. The result (2.4) has wide applicability. In Sections 3-5 we turn to its use in the specific applications (1)-(3) already mentioned. When $p_t \neq p$ as in case (4), care must be taken in specifying the prior distribution $p(M_t,\underset{\sim}{\beta}_t,\sigma^2,\underset{\sim}{a}_t)$. This is discussed in Section 6.

## 3. OUTLIER PROBLEMS

As we see from subparagraph (1) of Section 1, the notation for this case corresponds exactly with the general notation in Section 2, given $p_t = p$. A special case of the probability (2.4) was previously given by Guttman (1973) for the no-predictor-variables case and by Guttman, Dutter and Freeman (1978) for the regression case, and used by them (a) for given k, to determine which k observations were most likely to be spurious, (b) for given k, as weights in determining estimates for $\underset{\sim}{\beta}_t$ and $\sigma^2$ and also their posterior distributions, and (c) to develop a procedure for estimating k. For related comments, see Beckman and Cook (1983, pp. 138-139).

## 4. CHANGE POINT PROBLEMS

For these problems, $k = 0$, $p_t = p$, and the $\underset{\sim}{X}_{1t}$ are as defined in subparagraph (2) of Section 1. Eq. (2.4) simplifies to

$$p(M_t|\underset{\sim}{y}) = C'\{|\underset{\sim}{X}'_{mt}\underset{\sim}{X}_{mt}||\underset{\sim}{X}'_{n-m,t}\underset{\sim}{X}_{n-m,t}|\}^{-\frac{1}{2}} S_t^{-(n-p)/2} , \tag{4.1}$$

where C' is the appropriate normalizing constant. For the special case of change of mean value only, $\underset{\sim}{X}_{mt}$ and $\underset{\sim}{X}_{n-m,t}$ are vectors of 1's of length m and (n-m) respectively. This problem has been discussed by Guttman and Menzefricke (1982). Formula (4.1) now enables more general change point problems to be tackled, but we do not discuss them here.

## 5. A JOIN PROBLEM

Because of the difficulty in stating the join problem in its full generality (but see below) we first give a numerical example for the specific join problem outlined in subparagraph (3) of Section 1 which involves one predictor variable x, one join, and quadratic and straight line functions continuous at their join.

Example 1.

Our data consist of 32 observations on $y$ = boy's height/weight ratio taken at equally spaced values of the predictor variable $x$ = age in months, for $x = 0.50(1)31.5$. These are part of a larger set of Eppright et al. (1972), which were used by Gallant and Fuller (1973) and by Draper and Smith (1981, p. 286). The actual $y$ values, multiplied by 100 and corresponding to ascending $x$, are 46, 47, 56, 61, 61, 67, 68, 78, 69, 74, 77, 78, 75, 80, 78, 82, 77, 80, 81, 78, 87, 80, 83, 81, 88, 81, 83, 82, 82, 86, 82, and 85.

In view of the equal spacing of the x-values, we choose to evaluate (2.4) at the intergers $\alpha = 2,\ldots,30$. Twenty-nine values of (2.4) can thus be found, of which nine are essentially zero and eight more lie below 0.01. The remaining 12 values $p_\alpha$, for $\alpha = 6(1)17$ are as follows:

| | | | | | | | |
|---|---|---|---|---|---|---|---|
| $\alpha$ = | 6 | 7 | 8 | 9 | 10 | 11 | (continued) |
| $p_\alpha$ = | 0.047 | 0.187 | 0.157 | 0.106 | 0.114 | 0.104 | (continued) |
| $\alpha$ = | 12 | 13 | 14 | 15 | 16 | 17 | |
| $p_\alpha$ = | 0.084 | 0.063 | 0.046 | 0.032 | 0.023 | 0.015 | |

We see from these numbers that the modal estimate of the join appears to lie to the right of the seventh observation. This estimate of $\alpha$ could be further refined by using a grid finer than the integer values we have chosen. It must be remembered that, in this problem, unlike the change point problem, the tabled values are simply an approximation to a continuous posterior distribution. The latter can be evaluated to any accuracy desired. The present accuracy appears adequate for the problem at hand. The solution is compatible with those of Gallant and Fuller (1973) in which an additional 40 observations for $x = 32.5(1)71.5$ are used and continuity of the spline function and of the slope of the spline function are assumed at the join, and of Draper and Smith (1981, pp. 582-583) for both 72 and 32 observations which apply to straight line and straight line functions continuous at their join.

Generalizations

1. If only one predictor $x$ is involved, generalization consists of a spline function with $r$ joins, $r \geq 2$, $\alpha_1, \alpha_2, \ldots, \alpha_r$, say. Eq. (2.4) then defines an r-dimensional posterior probability function whose maximum and/or characteristics may be obtained. The polynomial functional forms between the joins , and the restrictions at the join points, affect the form of the $\underset{\sim}{X}$-matrices in (2.4), but not the dimensionality of the posterior (2.4).

2. For an $\ell$-dimensional predictor space $x_1, x_2, \ldots, x_\ell$ with $r_j$ joins in the $x_j$ space, Eq. (2.4) represents a posterior $\Sigma_{j=1}^{\ell} r_j$ dimensions, whatever the polynomial functional forms may be between the joins.

## 6. VARIABLE SELECTION PROBLEM

In this application, the $M_t$ are the $2^q$ regression models alluded to in (4) of Section 1. An alternative description of $M_t$ in which the specification of the prior plays an important role is as follows: Suppose $\underset{\sim}{y}$ is generated as usual by

$$\underset{\sim}{y} = \underset{\sim}{X}\underset{\sim}{\beta} + \underset{\sim}{\varepsilon} = [\underset{\sim}{1}, \underset{\sim}{x}_1, \ldots, \underset{\sim}{x}_q](\beta_o, \beta_1, \ldots, \beta_q)' + \underset{\sim}{\varepsilon} \tag{6.1}$$

where $\underset{\sim}{\varepsilon} \sim N(\underset{\sim}{0}, \sigma^2 I_n)$, but that we may write

$$\begin{aligned} \underset{\sim}{y} &= [\beta_o \underset{\sim}{1} + \beta_{i_1} \underset{\sim}{x}_{i_1} + \ldots + \beta_{i_r} \underset{\sim}{x}_{i_r}] + [\beta_{j_1} \underset{\sim}{x}_{j_1} + \ldots + \beta_{j_{q-r}} \underset{\sim}{x}_{j_{q-r}}] + \underset{\sim}{\varepsilon} \\ &= \underset{\sim}{X}_t \underset{\sim}{\beta}_t + \underset{\sim}{X}_{\bar{t}} \underset{\sim}{\beta}_{\bar{t}} + \underset{\sim}{\varepsilon}, \end{aligned} \tag{6.2}$$

where our prior information is such that

$$p(M_t, \underset{\sim}{\beta}_t, \underset{\sim}{\beta}_{\bar{t}}, \sigma^2) = p(\sigma^2) p(\underset{\sim}{\beta}_t, \underset{\sim}{\beta}_{\bar{t}} | M_t) p(M_t) \tag{6.3}$$

with

$$p(\sigma^2) \propto (\sigma^2)^{-(\frac{\nu_o}{2}+1)} \exp\{-\nu_o s_o^2/(2\sigma^2)\}, \tag{6.3a}$$

$$\begin{aligned} p(\underset{\sim}{\beta}_t, \underset{\sim}{\beta}_{\bar{t}} | M_t) &\propto |\underset{\sim}{C}_1^{(t)}|^{\frac{1}{2}} \exp\{-\tfrac{1}{2}(\underset{\sim}{\beta}_t - \underset{\sim}{\beta}_{to})' \underset{\sim}{C}_1^{(t)} (\underset{\sim}{\beta}_t - \underset{\sim}{\beta}_{to})\} \\ &\quad \times |\underset{\sim}{C}_2^{(t)}|^{\frac{1}{2}} \exp\{-\tfrac{1}{2} \underset{\sim}{\beta}_{\bar{t}}' \underset{\sim}{C}_2^{(t)} \underset{\sim}{\beta}_{\bar{t}}\} \end{aligned} \tag{6.3b}$$

and

$$p(M_t) \propto |\underset{\sim}{C}_2^{(t)}|^{\frac{1}{2}}, \tag{6.3c}$$

where $\nu_o$, $s_o^2$, $\underset{\sim}{\beta}_{to}$ and $\underset{\sim}{C}_i^{(t)}$, i=1,2, are all given, and where $\underset{\sim}{C}_1^{(t)}$ and $\underset{\sim}{C}_2^{(t)}$ are such that $\underset{\sim}{C}_1^{(t)} \simeq \underset{\sim}{0}$, $(\underset{\sim}{C}_2^{(t)})^{-1} \simeq \underset{\sim}{0}$, i.e., their elements are all small. Thus $M_t$ implies through (6.2) together with (6.3 a,b,c), that $x_{i_1}, \ldots, x_{i_r}$ are "important" variables, and that with high probability, the effects $\beta_{j_s}$ of the $x_{j_s}$, $s = 1, \ldots, q-r$, are expected to be zero. [Of course $(i_1, \ldots, i_r) U (j_1, \ldots, j_{q-r}) = (1, \ldots, q)$]. The prior distribution of $\underset{\sim}{\beta}_{\bar{t}} = (\beta_{j_1}, \ldots, \beta_{j_{q-r}})'$ is concentrated about zero with high precision given by $\underset{\sim}{C}_t^{(2)}$ and the degree of belief of the experimenter that $M_t$ holds, given the parameters $\underset{\sim}{\beta}_t$, $\underset{\sim}{\beta}_{\bar{t}}$, is proportional to the square root of the generalized precision matrix of $\underset{\sim}{\beta}_{\bar{t}}$, as stated in (6.3c).

We further assume that $\underset{\sim}{C}_1^{(t)}, \underset{\sim}{C}_2^{(t)}$ are such that $|\underset{\sim}{C}_1^{(t)}| \times |\underset{\sim}{C}_2^{(t)}|$ is constant, independent of $t$, which is to say that the prior generalized precision for $\underset{\sim}{\beta}_t$ and $\underset{\sim}{\beta}_{\bar{t}}$, and hence the generalized variance, is the same for all $M_t$. We believe that this is a sensible requirement in view of the fact that the experimenter does not know which model $M_t$ holds, so that for any $M_t$, his knowledge about the $\beta$'s, as measured by the generalized precision, should be the same as for any other model, say $M_{t'}$.

We now combine the prior (6.3 a,b,c) with the likelihood specified by $M_t$ in (6.2), to find the posterior of $M_t$, $\underset{\sim}{\beta}_t$, $\underset{\sim}{\beta}_{\bar{t}}$, $\sigma^2$. We then integrate over the same $(q+1)$ dimensional estimation space. Of course, as we range over the different $M_t$'s, the order of integration of the $(q+1)$ $\beta$'s varies, but nevertheless, we do integrate over all of the estimation space of dimension $q+1$, and this integration is done with respect to proper priors.

After the integrating out of $\underset{\sim}{\beta}_t$, $\underset{\sim}{\beta}_{\bar{t}}$, the resulting posterior of $(M_t,\sigma^2)$ depends on $\underset{\sim}{C}_1^{(t)}$ and $\underset{\sim}{C}_2^{(t)}$. We then approximate $p(M_t,\sigma^2|\text{data})$ by taking limits as $\underset{\sim}{C}_1^{(t)} \to \underset{\sim}{0}$, $\underset{\sim}{C}_2^{(t)} \to \infty$, $|\underset{\sim}{C}_1^{(t)}||\underset{\sim}{C}_2^{(t)}|$ constant, as assumed earlier. For the non-informative case for $\sigma^2$ (i.e., letting $\nu_o \to 0$, $s_o^2 \to \infty$ such that $\nu_o s_o^2 \to 0$) this provides (all details are given in Draper and Guttman (1986))

$$p(M_t|\text{data}) = K\gamma(n;p_t)|\underset{\sim}{X}_t'\underset{\sim}{X}_t|^{-\frac{1}{2}} S_t^{-\frac{n-p_t}{2}} \tag{6.4}$$

with

$$S_t = \underset{\sim}{y}'[I-\underset{\sim}{X}_t(\underset{\sim}{X}_t'\underset{\sim}{X}_t)^{-1}\underset{\sim}{X}_t']\underset{\sim}{y}, \tag{6.4a}$$

$$\gamma(n;p_t) = 2^{-p_t/2}\,\Gamma\left(\frac{n-p_t}{2}\right), \tag{6.4b}$$

$$K^{-1} = \sum_t \gamma(n;p_t)|\underset{\sim}{X}_t'\underset{\sim}{X}_t|^{-\frac{1}{2}} S_t^{-\frac{n-p_t}{2}}, \tag{6.4c}$$

and where $p_t$ is the dimension of the estimation space defined by the span of the columns of $\underset{\sim}{X}_t$ (e.g., for (6.2), $p_t = r+1$, etc.).

Equation (6.4) provides us with a model selection procedure. For other related recent work on comparing two linear models, one of which includes the other, see Smith and Spiegelhalter (1980) and Spiegelhalter and Smith (1982), Mitchell and Beauchamp (1986), and Trader (1983).

## 7. DISCUSSION

We have received many comments on an earlier draft of this paper, including the comment that "the primary conclusion (6.4)... is not invariant to scale changes in either dependent or independent variables." This is true. The underlying reason is to be found in the prior assumptions on the model parameters which are conditions on the $\beta$'s in the metrics in which they are defined. We argue that (6.4) is a perfectly reasonable outcome of the prior assumptions, while understanding that some readers will object to an answer that varies if the metrics are changed. It would of course be theoretically possible to present prior information in an invariant way (e.g., in terms of the $\beta_i x_i$ rather then the $\beta_i$) but that approach, which is currently under study, also presents difficulties.

In general, if we change scales by factors $f_i$ so that $x_i^* = x_i/f_i$, $\beta_i^* = \beta_i f_i$, the effect is eventually transmitted to (6.4) as a factor $\Pi f_i$, the product being taken over those $f_i$ whose subscripts are included in the notation "subscript t"; see under (6.3c). This non-constant $\Pi f_i$ also affects (6.4c). (The value of $S_t$ is unchanged if $y$ is not transformed from the original metric, but the prior conditions on the $C_j^{(t)}$ matrices are affected by making scale changes in the $x$'s.)

Note that exactly the same difficulty appeared in Spiegelhalter and Smith (1982, p. 378), in which only two models $M_0$ and $M_1$ were considered. In that paper, a "thought experiment" led to the fixing of a ratio $c_0/c_1$ which corresponds, in our notation, to the choice of the factor ratio $f_1/f_2$ when only two models are considered.

The effects of rescaling can be large as we see in an example using the Hald data.

Example 2. (Hald data)

This well-known set of 13 observations on a response and four predictors has proved to be exceptionally popular for illustrative examples, perhaps because it is small and yet awkward. See, for example, Hald (1952, p. 647), Daniel and Wood (1980, p. 89), and Draper and Smith (1981, p. 630).

Evaluating (6.4) for all 16 possible models using the metrics of the original data, we obtain nine values which are essentially zero and seven others as follows:

$$p_{12} = 0.335, \qquad p_{14} = 0.086,$$
$$p_{123} = 0.161,\ p_{124} = 0.153,\ p_{134} = 0.123,\ p_{234} = 0.022, \tag{7.1}$$
$$p_{1234} = 0.120,$$

where $p_{123}$ denotes $p(M_{123}|\underset{\sim}{y})$ and $M_{123}$ is the model $E(y) = \beta_o + \beta_1 X_1 + \beta_2 + \beta_3 X_3$, and so on. The probabilities add to 1.000. We see that, in the original metrics, our method favors, in order, the sets 12, 123, 124, 134, 14, 234. Most other selection procedures also favor 12 first.

Suppose, however, we decide (for example) to code all the x's to $x^*$'s so that

$$\sum_{j=1}^{n} (x_{ij}^* - \bar{x}_i^*)^2/n = 1. \tag{7.2}$$

Then

$$\beta_i^* = \beta_i \{ \sum_{j=1}^{n} (x_{ij} - \bar{x}_i)^2/n \}^{\frac{1}{2}} , \tag{7.3}$$

and $x_i^* = x_i \beta_i / \beta_i^*$, so that $f_i = \beta_i^* / \beta_i$ as expressed in (7.3). For the Hald data, $f_1 = 5.6516$, $f_2 = 14.9504$, $f_3 = 6.1538$, and $f_4 = 16.0815$. The prior information formulas (6.3b) and (6.3c) are now re-phrased in terms of the $\beta_i^*$ not the $\beta_i$. In place of (7.1) we now find that (the others are zero to three decimal places):

$$\begin{aligned} &p_{12} = 0.020, \qquad\qquad\qquad p_{14} = 0.006, \\ &p_{123} = 0.058,\ p_{124} = 0.145,\ p_{134} = 0.048,\ p_{234} = 0.023, \\ &\qquad\qquad p_{1234} = 0.701. \end{aligned} \tag{7.4}$$

The change in emphasis is interesting, and makes it clear that, <u>in the new metric</u>, it is unreasonable to regard any of the $\beta_i^*$ as being substantively smaller than the others, because the model with highest posterior probability involves all four predictor variables.

There is, however, for the Hald data, substantial reason to regard the original predictor variable metrics as eminently sensible ones. The original x's are four cement ingredients expressed as percentages of a mixture and, in fact, $\Sigma x_i = 100\%$, approximately. Changing the metrics to satisfy (7.2) would not make much practical sense.

We now look at another set of data where the predictor variable metrics appear to be natural ones.

<u>Example 3</u>. (Rutting Data)

Thirty-one observations were taken on six predictor variables and a response. The data are given by Daniel and Wood (1980, p. 109) and are used as an exercise by Draper and Smith (1981, p. 375). Sixty-four values of

(6.4) can thus be evaluated; 39 are essentially zero, and 17 more lie below 0.01. The remaining 8 are as follows:

$$p_{12} = 0.140,$$

$$p_{123} = 0.021, \quad p_{124} = 0.041, \quad p_{126} = 0.490,$$

$$p_{1236} = 0.070, \quad p_{1246} = 0.165, \quad p_{1256} = 0.018,$$

$$p_{12346} = 0.035.$$

We see that our method favors, in order, the sets 126, 1246, 12, 1236, 124, 12346, 123, 1256, the emphasis being on the first three. The superiority of sets 126 and 1246 also emerges from the other selection procedures used in the references quoted. The ambiguity of whether or not to include the dummy variable 4 with variables, 1, 2, and 6 has been discussed by Daniel and Wood (1980, pp. 96-100).

Again, our method works well compared with other proceedures and, as in the Hald case, the predictor variables seem to be in sensible and natural units: $x_1$ = log (viscosity of asphalt), $x_2$, $x_3$, $x_5$, and $x_6$ are percentages of material or voids, while $x_4$ is a dummy.

In summary, our Bayesian selection procedure has both virtues and drawbacks. On the one hand, we have avoided the problems that arose in some previous Bayesian work because of the different dimensionalities of the β-spaces as different models are considered, and we have developed a procedure valid for any given specification of the metrics of the $\beta_i$. On the other hand our procedure is not invariant to these choices of metrics. One could argue that it need not be; such a viewpoint would perhaps not appeal to those used to thinking in terms of the standard types of selection procedures, because these are based on quantities (such as extra sums of squares) that are invariant to x-metric choice. A Bayesian procedure parallel to these would thus need to have prior information specified and incorporated in a similarly invariant manner. However, we can argue that our method accurately reflects the prior information in the chosen metrics. We can also question whether prior information should be invariant in the metrics used. For a related discussion of Bayesian difficulties see Atkinson (1978).

Two other selection procedures (discussed, for example, by Stone, 1979) use criteria of the form

$$C_q = \ell n(\text{maximum likelihood}) - qp \tag{7.5}$$

where $p$ is the number of parameters in the model being considered. When $q = \frac{1}{2}\ell n\ n$, we have Schwarz's (1978) criterion; when $q = 1$, Akaike's (1973). $C_q$ simply "penalizes" the likelihood for the number of parameters. When the errors $\underset{\sim}{\varepsilon} \sim N(\underset{\sim}{0}, \underset{\sim}{I}\sigma^2)$, these criteria reduce to

$$\text{constant} - \tfrac{1}{2}[n\ell n\{\text{Residual SS}\} + 2qp]. \tag{7.6}$$

For the Hald data, and for models in the subscript order [see (7.1)], -; 1, 2, 3, 4; 12, 13, 14, 23, 24, 34; 123, 124, 134; 234; 1234, the values of the square bracket quantity in (7.6) are as below. (Smaller is better here. All figures have been rounded to integers except when necessary to establish relative sizes.)

Schwarz' Criterion

105; 98, 94, 104, 93; 60.46, 100, 64, 86, 96, 75; 60.62, 60.58, 72, 66; 63.

Akaike's Criterion

105; 96, 92.5, 102, 92.2; 59, 98, 62, 84, 94, 73; 58.4, 58.3, 69, 64; 60.

We see that Schwarz's criterion favors models in order 12, 124, 123, 1234, 14, 234,..., while Akaike's order is 124, 123, 12, 1234, 14, 234, .... In both cases there is not much to choose between 12 and 1234. Models 1, 2, 3, 4, 13, and 24 are decisively excluded by both criteria.

## 8. ACKNOWLEDGEMENTS

Conversations with Dennis Lindley and T.J. Mitchell were very helpful. This work was partially sponsored by the United States Army under Contract No. DAAG29-80-C-0041, by the Wisconsin Alumni Research Fund through the University of Wisconsin Graduate School, and by NSERC (Canada) under Grant No. A8743.

## REFERENCES

Akaike, H., 1973, Information theory and an extension of the maximum likelihood principle. 2nd Int. Symp. Inf. Theory (B.N. Petrov and F. Czaki, eds.), pp. 267-281. Budapest: Akad. Kiadó.

Atkinson, A.C., 1978, Posterior probabilities for choosing a regression model. Biometrika, 65, 39-48.

Beckman, R.J. and Cook, R.D., 1983, Outlier ..........s. Technometrics, 25, 119-149.

Daniel, C. and Wood, F.S., 1980, Fitting Equations to Data, 2nd ed. New York: Wiley.

Draper, N.R. and Guttman, I., 1986, Model Selection Problems. Technical Report, Department of Statistics, University of Toronto.

Draper, N.R. and Smith, H. (1981). Applied Regression Analysis, 2nd ed. New York: Wiley.

Eppright, E.S., Fox, H.M., Fryer, B.A., Lamkin, G.H., Vivian V.M. and Fuller, E.S., 1972, Nutrition of infants and preschool children in the north central region of the United States of America. World Review of Nutrition and Dietetics, 14, 269-332.

Gallant, A.R. and Fuller, W.A., 1973, Fitting segmented polynomial regression models whose join points have to be estimated. J. Amer. Statist. 68, 144-147.

Guttman, I., 1973, Care and handling of univariate or multivariate outliers in detecting spuriosity - a Bayesian approach. Technometrics, 15, 723-738.

Guttman, I., Dutter, R. and Freeman, P.R., 1978, Care and handling of univariate outliers in the general linear model to detect spuriosity - a Bayesian approach. Technometrics, 20, 187-193.

Guttman, I. and Menzefricke, U., 1982, On the use of loss functions in the changepoint problem, Ann. Instit. Statist. Math., 34, A, 319-326.

Hald, A., 1952, Statistical Theory with Engineering Applications, New York: Wiley.

Mitchell, T.J., and Beauchamp, J.J., 1986, Algorithms for Bayesian Variable Selection in Regression. Proceedings of the Eighteenth Symposium on the Interface, held March 19-21, 1986; T.J. Boardman, Editor, American Statistical Association.

Schwarz, G., 1978, Estimating the dimension of a model. Ann. Statist., 6, 461-464.

Smith, A.F.M. and Spiegelhalter, D.J., 1980, Bayes factors and choice criteria for linear models. J.R. Statist. Soc., B-42, 213-220.

Spiegelhalter, D.J. and Smith, A.F.M., 1982, Bayes factors for linear and log-linear models with vague prior information. J.R. Statist. Soc., B-44, 377-387.

Stone, M., 1979, Comments on model selection criteria of Akaike and Schwarz. J.R. Statist. Soc., B-41, 276-278.

Trader, R.L., 1983, A Bayesian predictive approach to the selection of variables in multiple regression. Communications in Statistics, A12, 1553-1567.

# PREDICTIVE SCREENING METHODS IN BINARY RESPONSE MODELS

Ian R. Dunsmore
University of Sheffield
U.K.

Richard J. Boys
University of Newcastle upon Tyne
U.K.

## SUMMARY

Screening procedures are used in order to improve the 'quality' of individuals retained in some way. In this paper we present a Bayesian predictive approach to screening for binary response data. We discuss its relationship with diagnosis and classification problems. The methods are illustrated by an example of medical screening for Conn's syndrome.

## INTRODUCTION

Discrimination and classification problems form an important area of both statistical theory and practice. Since the linear discriminant method was introduced by Fisher (1936), much has been written and developed. Extensive accounts are given, for example, by Anderson (1958) and Hand (1981). One field of application is the medical situation, where the problem of diagnosis of the form of disease from which a patient suffers is often of paramount importance. A discussion of such problems with consideration of criteria for evaluation of the discriminant rules is provided in a series of papers by Habbema, Hilden and Bjerregaard in 1978 and 1981, whilst a comparison of different techniques in a particular application is given in Titterington et al. (1981).

In diagnosis problems the aim is to assess the category t of a future individual who has provided us with a set $\underset{\sim}{x}$ of feature variables. A data set $(\underset{\sim}{x}_1,t_1),(\underset{\sim}{x}_2,t_2),\ldots,(\underset{\sim}{x}_n,t_n)$ is available for n individuals whose categories are known with certainty. Since we are concerned with statements or forecasts about an observable quantity, a predictive approach seems to be the most sensible avenue of attack. We extend here the predictive methods developed by Geisser (1964) and Aitchison and Dunsmore (1975, Ch.11), where the basic aim is to derive a diagnostic probability function $p(t|\underset{\sim}{x},\text{data})$.

We illustrate the methodology within the framework of a data set described in Aitchison and Dunsmore (1975,Ex.1.7). Conn's syndrome is a rare form of hypertension. Two forms of the syndrome exist, namely:

A: benign tumour in the adrenal cortex, (adenoma),
B: a more diffuse condition of the adrenal glands, (bilateral hyperplasia).

The treatment for A is a surgical operation to remove the adrenal gland. For B drug therapy is the recognised treatment, and surgery is inadvisable. For the purpose of illustration we wish to diagnose the form of disease ($t=1$ for A, $t=0$ for B) on the basis of the three concentrations (meq/ℓ) in blood plasma:sodium (Na), potassium (K) and carbon dioxide ($CO_2$). The data are given in table 1, and we use log (concentrations) for the basic variables $\underset{\sim}{x}$, as this transformation removes much of the skewness apparent in the data. For the undiagnosed patient we wish to assess which form of the disease is appropriate. It is clearly important that we are fairly sure that patients for whom surgery is decided do in fact have form A of the syndrome.

Table 1 : Conn's Syndrome Data

log(concentrations,meq/ℓ) in blood plasma

| | | Na | K | $CO_2$ |
|---|---|---|---|---|
| | Patient | $x_1$ | $x_2$ | $x_3$ |
| Type A | 1 | 4.9459 | 0.8329 | 3.4112 |
| | 2 | 4.9628 | 1.1314 | 3.2995 |
| | 3 | 4.9416 | 1.0986 | 3.2958 |
| | 4 | 4.9836 | 1.0296 | 3.4965 |
| | 5 | 4.9323 | 1.2809 | 3.1822 |
| | 6 | 4.9677 | 1.1314 | 3.3322 |
| | 7 | 4.9222 | 0.9163 | 3.3878 |
| | 8 | 4.9488 | 0.9163 | 3.4012 |
| | 9 | 4.9684 | 0.8755 | 3.4720 |
| | 10 | 4.9740 | 1.0647 | 3.3844 |
| | 11 | 4.9381 | 0.8329 | 3.2581 |
| | 12 | 4.9698 | 0.7885 | 3.5175 |
| | 13 | 4.9767 | 0.9933 | 3.4965 |
| | 14 | 4.9431 | 1.1314 | 3.3707 |
| | 15 | 4.9747 | 1.0647 | 3.3105 |
| | 16 | 4.9345 | 1.1314 | 3.4468 |
| | 17 | 4.9754 | 0.6419 | 3.5116 |
| | 18 | 4.9816 | 1.3083 | 3.3105 |
| | 19 | 4.9698 | 0.7885 | 3.4965 |
| | 20 | 4.9663 | 0.9933 | 3.3142 |
| Type B | 21 | 4.9438 | 1.4586 | 3.1527 |
| | 22 | 4.9488 | 1.1632 | 3.2189 |
| | 23 | 4.9502 | 1.2809 | 3.2504 |
| | 24 | 4.9558 | 1.0986 | 3.0910 |
| | 25 | 4.9663 | 1.4351 | 3.3250 |
| | 26 | 4.9395 | 1.2238 | 3.3322 |
| | 27 | 4.9495 | 1.2809 | 3.2189 |
| | 28 | 4.9488 | 1.3350 | 3.2581 |
| | 29 | 4.9452 | 1.1939 | 3.2958 |
| | 30 | 4.9416 | 1.2809 | 3.2581 |
| | 31 | 4.9416 | 1.4816 | 3.2426 |

SCREENING

The predictive approach developed in Aitchison and Dunsmore (1975, Ch.11) evaluates the diagnostic probabilities $P(A|\underset{\sim}{x},\text{data})$ and $P(B|\underset{\sim}{x},\text{data})$.

An assignment is then made on the basis of the predictive odds - perhaps diagnosing the form of disease corresponding to the larger probability. Misclassification costs could be incorporated in a decision theoretic approach, and within the Conn's syndrome context these could be large because of the radically different nature of the relevant treatments. An equivalent procedure, and the criterion which we develop here, is to formulate a decision rule which ensures that the predictive probability that a patient for whom we decide on surgery is in fact of type A takes some prespecified value $\delta_p$, which will be close to 1; i.e. we make

$$P(A|\text{decide on surgery}) = \delta_p. \tag{1}$$

Clearly we would also be interested in P(A|decide on drug therapy).

Such a decision rule lies within the context of predictive screening as derived in Boys and Dunsmore (1986,1987). We wish to use the feature vector $\underset{\sim}{x}$ to attempt to screen out the B cases (t=0) and to retain the A cases (t=1). Retention corresponds to deciding on surgery in (1), and so we need to determine a specification region $C_{\underset{\sim}{x}}$ such that

$$p(t=1|\underset{\sim}{x}\in C_{\underset{\sim}{x}},\text{data}) = \delta_p. \tag{2}$$

We frame the problem in terms of linear models by restricting attention to specification regions of the form

$$C_{\underset{\sim}{x}} = \{\underset{\sim}{x} : \underset{\sim}{a}'\underset{\sim}{x} \geq w\} \quad ,$$

where constants $\underset{\sim}{a}$ and $w$ are to be determined to satisfy (2). In the derivation of $C_{\underset{\sim}{x}}$ we use standardized variables in order to eliminate problems of dimensionality or scale. Without loss of generality we impose the restriction $\underset{\sim}{a}'\underset{\sim}{a}=1$, since clearly an indeterminacy would result otherwise. The problem of multiple solutions does however remain, and so we seek the values of a and w which satisfy (2) and which minimize the (predictive) error probability

$$\varepsilon_p = p(t=1|\underset{\sim}{x}\notin C_{\underset{\sim}{x}},\text{data}). \tag{3}$$

Two modelling approaches for the joint distribution of t and $\underset{\sim}{x}$ have been discussed; see Dawid (1976) and Aitchison and Begg (1976). These are the sampling paradigm and the diagnostic paradigm. In the former models of the form $p(\underset{\sim}{x}|t,\underset{\sim}{\eta})p(t|\underset{\sim}{\psi})$ are used, whilst in the latter attention is concentrated on $p(t|\underset{\sim}{x},\underset{\sim}{\xi})\ p(\underset{\sim}{x}|\underset{\sim}{\phi})$. The sampling framework is more useful for situations in which polynomial or interaction effects are required, whilst the diagnostic model is more robust against selection biases.

SAMPLING MODEL

Within the sampling framework the conditional predictive probability of t required is given by

$$p(t=1|\underset{\sim}{x}\in C_{\underset{\sim}{x}},\text{data}) = \frac{p(t=1|\text{data})\int_{C_{\underset{\sim}{x}}} p(\underset{\sim}{x}|t=1,\text{data})d\underset{\sim}{x}}{\sum_{i=o}^{1} p(t=i|\text{data})\int_{C_{\underset{\sim}{x}}} p(\underset{\sim}{x}|t=i,\text{data})d\underset{\sim}{x}} \, ,$$

so that for the evaluation of $C_{\underset{\sim}{x}}$ from (2) and (3) we require both the predictive forms $p(\underset{\sim}{x}|t,\text{data})$ and $p(t|\text{data})$.

For the Conn's syndrome data an underlying normality assumption within the two groups for the log(concentration) vector seems reasonable, i.e.

$$p(\underset{\sim}{x}|t=i,\underset{\sim}{\eta}) \sim N(\underset{\sim}{\mu}_i,\underset{\sim}{\Sigma}_i) \qquad (i=0,1)$$

Summary statistics from the data set are

$$n_1 = 20 \qquad\qquad n_0 = 11$$

$$\overline{\underset{\sim}{x}}_1 = \begin{pmatrix} 4.96 \\ 1.00 \\ 3.38 \end{pmatrix}, \qquad \overline{\underset{\sim}{x}}_0 = \begin{pmatrix} 4.95 \\ 1.29 \\ 3.24 \end{pmatrix},$$

and

$$\underset{\sim}{S}_1 = \begin{pmatrix} 0.035 & -0.028 & 0.073 \\ -0.028 & 2.974 & -1.026 \\ 0.073 & -1.026 & 0.920 \end{pmatrix} \text{x}10^{-2}, \qquad \underset{\sim}{S}_0 = \begin{pmatrix} 0.006 & 0.000 & 0.003 \\ 0.000 & 1.546 & 0.186 \\ 0.003 & 0.186 & 0.503 \end{pmatrix} \text{x}10^{-2}$$

There appears to be no strong reason to assume that the covariance matrices $\Sigma_1$ and $\Sigma_0$ are equal. With vague prior assumptions on $\underset{\sim}{\eta}$ the predictive densities are of Student form (see Boys and Dunsmore, 1987), namely

$$p(\underset{\sim}{x}|t=i,\text{data}) \propto \left\{1 + \frac{n_i}{n_i^2-1}(\underset{\sim}{x}-\overline{\underset{\sim}{x}}_i)'\underset{\sim}{S}_i^{-1}(\underset{\sim}{x}-\overline{\underset{\sim}{x}}_i)\right\}^{-\frac{1}{2}n_i} \quad (i=0,1).$$

Similarly with a vague prior on the arrival parameter $\psi$ the predictive function for t is given by

$$p(t=1|\text{data}) = \frac{20}{31}, \qquad p(t=0|\text{data}) = \frac{11}{31}.$$

The solution of (2) and (3) for $\delta_p=0.95$ then provides the optimal specification region

$$\{\underset{\sim}{x} : 0.91x_1 - 0.24x_2 + 0.34x_3 \geq 5.34\}.$$

Although we have achieved a value of $\delta_p=0.95$ we note that $\varepsilon_p$ is rather large at 0.25. The predictive probability $\beta_p=P(\underset{\sim}{X}\in C_{\underset{\sim}{x}}|\text{data})$ that an individual is assigned to surgery is 0.56. Without screening the predictive probability $\gamma_p=p(t=1|\text{data})$ that an individual is of type A is 0.65.

The dimensionality of the problem and therefore the scale of the computational effort can be reduced significantly if we summarize the feature vector initially through some linear score function $D(\underset{\sim}{x})$, such as Fisher's linear discriminant, a principal component or the first crimcoord (Gnanadesikan, 1977,p.86). The specification region $C_{\underset{\sim}{x}}$ is then of the form

$$C_{\underset{\sim}{x}}=\{\underset{\sim}{x} : D(\underset{\sim}{x}) \geq \text{constant}\} \tag{4}$$

where only the constant is now unknown.

For example suppose we use Fisher's linear discriminant

$$D(\underset{\sim}{x}) = (\overline{\underset{\sim}{x}}_1-\overline{\underset{\sim}{x}}_0)'\underset{\sim}{S}^{-1}\underset{\sim}{x},$$

where $\underset{\sim}{S}$ is the pooled covariance matrix based on the data. This reduces the problem to 1-dimension, and the assumption of normality of D within the two groups, albeit with different variances $\sigma_1^2$ and $\sigma_0^2$, seems reasonable.

In table 2 we compare the specification regions and summary statistics for the two methods - I: full multivariate, II: univariate with Fisher's linear discriminant. We notice that although the form of $C_{\underset{\sim}{x}}$ seems rather different, the values of $\varepsilon_p$, $\beta_p$ and $\gamma_p$ are remarkably consistent. It is not surprising that the form is different since it is well known that D is not as good a discriminator when $\Sigma_1 \neq \Sigma_0$ Out of interest we show in table 2 the forms of regions if we use III: full multivariate with $\Sigma_1 = \Sigma_0$, and IV: univariate with Fisher's linear discriminant with $\sigma_1 = \sigma_0$. Whilst III provides an almost identical $C_{\underset{\sim}{x}}$ to II the summary statistics $\varepsilon_p$, $\beta_p$ and $\gamma_p$ vary considerably.

## DIAGNOSTIC MODEL

Within the diagnostic framework the conditional predictive probability of t required is given by

$$p(t=1|\underset{\sim}{x}\in C_{\underset{\sim}{x}},\text{data}) = \frac{\int_{C_{\underset{\sim}{x}}} p(t=1|\underset{\sim}{x},\text{data})\; p(\underset{\sim}{x}|\text{data})d\underset{\sim}{x}}{\int_{C_{\underset{\sim}{x}}} p(\underset{\sim}{x}|\text{data})d\underset{\sim}{x}},$$

so that for the evaluation of $C_{\underset{\sim}{x}}$ from (2) and (3) we require both the oredictive forms $p(t|\underset{\sim}{x},\text{data})$ and $p(\underset{\sim}{x}|\text{data})$.

The linear logistic model with

$$p(t=1|\underset{\sim}{x},\underset{\sim}{\xi}) = \frac{e^{\xi_0+\underset{\sim}{\xi}_1'\underset{\sim}{x}}}{1+e^{\xi_0+\underset{\sim}{\xi}_1'\underset{\sim}{x}}}$$

is the popular candidate for the condidtional probability. Copas' (1983) plots suggest that linearity in the $x_i$'s is not too unreasonable an assumption to make for the Conn's syndrome data, although we return to this point later.

No simple analytically tractable prior for $\underset{\sim}{\xi}$ presents itself in this logistic model. We must resort to numerical integration for a specificed $p(\underset{\sim}{\xi})$ - in four dimensions for the Conn's syndrome data - or consider some approximations. Here we follow the second approach, and use the approximate (asymptotic) normality of the posterior $p(\underset{\sim}{\xi}|\underset{\sim}{x},\text{data})$. Further discussion is provided in Boys and Dunsmore (1987), where a third approach, suggested by Bernardo (1983), in which $p(t=1\ \underset{\sim}{x},\text{data})$ is forced to logistic form, is also mentioned.

The assumption of normality for $p(\underset{\sim}{x}|\phi)$ appears to be reasonable here - although of course strictly this is at odds with the normality assumptions in the sampling approach. A vague prior on $\underset{\sim}{\phi}$ leads to a Student predictive density.

$$p(\underset{\sim}{x}|\text{data}) \propto \left\{1 + \frac{n}{n^2-1}(\underset{\sim}{x}-\bar{\underset{\sim}{x}}_T)'\underset{\sim}{S}_T^{-1}(\underset{\sim}{x}-\underset{\sim}{x}_T)\right\}^{-\frac{1}{2}n}$$

where $\bar{x}_T$ and $S_T$ are the sample mean vector and sample covariance matrix of the complete data set of n=31 individuals, namely

$$\bar{\underset{\sim}{x}}_T = \begin{pmatrix} 4.96 \\ 1.10 \\ 3.33 \end{pmatrix} \quad , \quad \underset{\sim}{S}_T = \begin{pmatrix} 0.027 & -0.092 & 0.081 \\ -0.092 & 4.476 & -1.600 \\ 0.081 & -1.600 & 1.244 \end{pmatrix} x10^{-2} \quad .$$

The solution of (2) and (3) for $\delta_p$=0.95 then provides the optimal specification region shown in table 2 as method V. Whilst the form of $C_{\underset{\sim}{x}}$ is similar to the multivariate sampling method I, the performance, especially of $\varepsilon_p$, is disappointing.

Table 2 Optimal specification regions of form $\{\underset{\sim}{x} : a_1x_1+a_2x_2+a_3x_3 \geq w\}$ together with summary measures $\varepsilon_p, \beta_p, \gamma_p$.

| | $a_1$ | $a_2$ | $a_3$ | $w$ | $\varepsilon_p$ | $\beta_p$ | $\gamma_p$ |
|---|---|---|---|---|---|---|---|
| Sampling | | | | | | | |
| I : multivariate | 0.91 | -0.24 | 0.34 | 5.34 | 0.25 | 0.56 | 0.65 |
| II : linear discriminant | 0.76 | -0.43 | 0.48 | 4.87 | 0.26 | 0.56 | 0.65 |
| III : multivariate, $\Sigma_1=\Sigma_0$ | 0.77 | -0.43 | 0.47 | 4.95 | 0.39 | 0.46 | 0.65 |
| IV : linear discriminant $\sigma_1=\sigma_0$ | 0.76 | -0.43 | 0.48 | 4.94 | 0.39 | 0.46 | 0.65 |
| Diagnostic | | | | | | | |
| V : multivariate | 0.96 | -0.18 | 0.22 | 5.30 | 0.38 | 0.49 | 0.66 |
| VI : $\hat{\xi}_0+\hat{\underset{\sim}{\xi}}_1'\underset{\sim}{x}$ | 0.93 | -0.26 | 0.26 | 5.23 | 0.38 | 0.49 | 0.66 |

We again consider reducing the dimensionality of the analysis by summarizing the data $\underset{\sim}{x}$ to produce a specification region as in (4). One obvious candidate here is

$$D(\underset{\sim}{x}) = \hat{\xi}_0 + \hat{\underset{\sim}{\xi}}_1'\underset{\sim}{x} \quad ,$$

where $\hat{\xi}_0, \hat{\underset{\sim}{\xi}}_1$ are the maximum likelihood estimates of $\underset{\sim}{\xi}_0$ $\underset{\sim}{\xi}_1$. The results are shown in VI in table 2. It is very gratifying in this example to find that the results for V and VI are comparatively close, since the reduction in computing time achieved through VI is quite considerable.

It is perhaps not too surprising that the diagnostic model used has not performed as well as the multivariate sampling model I. There was evidence there that $\Sigma_1 \neq \Sigma_0$. For such situations a quadratic logistic model would be more appropriate (Anderson, 1975). The Copas plots do in fact suggest that there may be a quadratic effect in variable $x_1$, so that the logistic model could be improved by incorporating terms $x_1^2, x_1x_2$ and $x_1x_3$. The computational aspects of the analysis then become most unwieldy.

## EXTENSIONS

The predictive screening models used here can be adapted for use in other situations. Within the diagnostic setting it may be that the categorization t=1 or t=0 is based on some underlying (perhaps latent) variable y such that there exists a specification region $C_y$ with

$$t=1 \leftrightarrow y \in C_y ,$$
$$t=0 \leftrightarrow y \notin C_y .$$

Then we observe $(y,\underset{\sim}{x})$ and need to choose $C_{\underset{\sim}{x}}$ such that

$$P(y \in C_y | \underset{\sim}{x} \in C_{\underset{\sim}{x}}, \text{data}) = \delta_p$$

perhaps so as to minimize

$$\varepsilon_p = P(y \in C_y | \underset{\sim}{x} \notin C_{\underset{\sim}{x}}, \text{data}).$$

The analysis within a normal model framework for $(y,\underset{\sim}{x})$ is given in Boys and Dunsmore (1986).

Other extensions which are at present under investigation deal with the sequential selection of screening variables and decision theoretic models with criteria involving expected utilities.

An important point to emphasize from the paper is that we are using a predictive approach. We plead guilty however to using a global approach in that we condition over the region $\underset{\sim}{x} \in C_{\underset{\sim}{x}}$. In effect we average $p(t=1|\underset{\sim}{x}, \text{data})$ over a conditonal distribution on $\underset{\sim}{x}$. The pure predictive approach should be local, i.e.

select $C_{\underset{\sim}{x}}$ such that

$$p(t=1|\underset{\sim}{x},\text{data}) \begin{cases} \geq \delta_p & \text{for } \underset{\sim}{x} \in C_{\underset{\sim}{x}}, \\ < \delta_p & \text{for } \underset{\sim}{x} \notin C_{\underset{\sim}{x}}. \end{cases}$$

We are at present investigating such models, and in defence of using the global approach we appeal to the mind of Sherlock Holmes, who said

'While the individual man is an insoluble puzzle, in the aggregate he becomes a mathematical certainty. You can, for example, never foretell what one man will do, but you can say with precision what an average number will be up to.'

(A. Conan Doyle. 'The Sign of Four')

## REFERENCES


Aitchison, J. & Begg, C.B. (1976), Statistical diagnosis when basic cases are not classified with certainty, Biometrika 63, 1-12.

Aitchison, J. & Dunsmore, I.R. (1975), Statistical Prediction Analysis, Cambridge : Cambridge University Press.

Anderson, J.A. (1975), Quadratic logistic discrimination, Biometrika 62, 149-154.

Anderson, T.W. (1958), An Introduction to Multivariate Statistical Analysis, New York : Wiley.

Bernardo, J.M. (1983), Bayesian logistic diagnostic distributions, Technical report, Universidad de Valencia.

Boys, R.J. & Dunsmore, I.R. (1986), Screening in a normal model, J.R. Statist.Soc. B 48, 60-69

Boys, R.J. & Dunsmore, I.R. (1987), Diagnostic and sampling models in screening, Biometrika (to appear)

Copas, J.B. (1983). Plotting p against x. Appl.Statist. 32, 25-31.
Dawid, A.P. (1976). Properties of diagnostic data distributions. Biometrics 32, 647-658.
Fisher, R.A. (1936). The use of multiple measurements in taxonomic problems. Ann.Eugen. 7, 179-188.
Geisser, S. (1964). Posterior odds for multivariate normal classifications. J.R. Statist.Soc.B 26, 69-76.
Gnanadesikan, R. (1977). Methods for Statistical Data Analysis of Multivariate Observations. New York : Wiley.
Habbema, J.D.F. & Hilden, J. (1981). The measurement of performance in probabilistic diagnosis IV. Meth.Inform.Med. 20, 80-96.
Habbema, J.D.F., Hilden, J. & Bjerregaard, B. (1978) The measurement of performance in probabilistic diagnosis I. Meth.Inform.Med. 17, 217-226.
Habbema, J.D.F., Hilden, J. & Bjerregaard, B. (1981). The measurement of performance in probabilistic diagnosis V. Meth.Inform.Med. 21, 97-100.
Hand, D.J. (1981) Discrimination and Classification. Chichester : Wiley.
Hilden, J., Habbema, J.D.F. & Bjerregaard, B. (1978a). The measurement of performance in probabilistic diagnosis II. Meth.Inform.Med. 17, 227-237.
Hilden, J., Habbema, J.D.F. & Bjerregaard, B. (1978b). The measurement of performance in probabilistic diagnosis III. Meth.Inform.Med. 17, 238-246.
Titterington, D.M., Murray, G.D., Murray, L.S., Spiegelhalter, D.J., Skene, A.M., Habbema, J.D.F. & Gelpke, G.J. (1981). Comparison of discrimination techniques applied to a complex data set of head injured patients(with discussion). J.R.Statist Soc. A 144, 145-175.

# DE FINETTI'S PROBABILISTIC APPROACH AND THE THEORY OF EXPECTATIONS IN ECONOMICS

Massimo de Felice
Facolta di Economia e Commercio
University of Bari
Bari, Italy

Gianluigi Pelloni
Department of Economics & Commerce
University of Hull
Hull, HU6 7RX, U.K.

The rational expectations revolution has re-proposed the necessity of a deeper analysis of the role expectations play in economic model building. It would be an "intellectual fraud" to claim that the expectations controversy" was triggered by the debate generated by the rational expectations hypothesis. The beginning of such a controversy can be located in the 20's and 30's with the issue of the works of Keynes, Knight and of the Austrian and Swedish schools.

In this paper we shall show Bruno de Finetti's contribution to this controversy and put it in an updated perspective.

His contribution was timely, constant and extremely original. de Finetti supplied a definitive account of the neo-Bayesian approach in "Probabilismo" (1931b) where the notion of exchangeability was re-presented and clarified within a thorough introduction to the philosophical underpinnings of the subjectivist paradigm. His survey of the other "points of view" (Richard von Mises, Keynes, Jeffreys, Borel, Reichenbach, Kolmogorov, Wald) offered critical hints for subsequent research (as well as providing refined polemical strategies), whereas his sharp treatment of utility analysis enhanced the operational and pragmatic content of his approach.

Unfortunately, de Finetti's contribution was largely neglected and, at the time, almost passed unnoticed in the economic profession, because most of his works were published in Italian and those which were published in French were not easily accessible in that they were written for mathematicians.

Many economists refer to de Finetti's neo-Bayesianism as only an historical and cultural curiosity and reveal a preference for those "adhockeries for mathematical convenience" harshly criticized by de Finetti.

In this paper it will be shown how de Finetti's approach, in addition to the occasional citations, can be a powerful tool for interpreting the methodological debate triggered by rational expectations and also offers a broader perspective for solving the crucial difficulties characterizing the research agenda on the theory of expectations.

1. Arrow (1951, ch.1, pp.2-4) identifies three "dramatic breaks"

characterizing the modern formal approach to the theory of decision making under uncertainty: Von Neumann and Morgenstern's view which, resuming Ramsey's pioneering works, leads to a new understanding of the role of expected utility maximization; the modern theory of statistical inference as developed by Neymann, Pearson and Wald; the "new formulation" of the whole problem of uncertainty suggested by Shackle.

These fundamental insights came at about the same time as the technical developments of the mathematical theory of probability (in particular Kolmogorov's axiomatic treatment of probability as a branch of measure theory) which caused a paradoxical departure from the results relevant to behaviour in the face of uncertainty.

In this very perspective, de Finetti's approach to the calculus of probability and decision making under uncertainty takes on a crucial role and fulfils Arrow's wish for "a much clearer understanding of the problem".

2. In "Probabilismo," de Finetti (1931b) gives a definitive account of the subjectivist approach to probability theory(1). In that paper, de Finetti puts his approach in a broad epistemological perspective by referring to the influence of Italian pragmatism(2), Mach's positivism(3) and certain insights of Poincaré on his thinking(4), while carrying out an in depth analysis of concepts which are still essential to current research programmes on the theory of expectations.

In "Probabilismo," probability theory emerges as an unique and general method for dealing with decision making under uncertainty. Probability is no longer "a thing in itself", but a purely epistemological concept with a relative and subjective value, relevant also for those who are only concerned with practical applications (de Finetti, 1931b, p.26).

Within this framework de Finetti removes "the fetish" of true or false probability, as a meaningless statement claiming that the observation of

---

(1) This long essay provides an exhaustive discussion of the foundations of probability theory and springs from a shorter paper (completed by April 1928) which was set aside by de Finetti because of the many difficulties encountered in getting his point of view understood let alone accepted (c.f. de Finetti, 1931b, p.5, note 2). "Probabilismo", although written without formulae and mathematical expressions, utilizes and discusses analytical results previously obtained by de Finetti(1929, 1930b, 1931a).

(2) As suggested by Papini the cultural position of Italian pragmatists, such as Caleroni and Vailati, can be summarized by their concern "to teach the prudence and tricks by means of which it is possible to succeed in formulating propositions that have a meaning". According to the pragmatist approach the meaning of each statement is given by the set of predictions and expectations, embodied in it (c.f. Calderoni and Vailati, 1909). For an interpretation of de Finetti's pragmatism c.f. De Felice, 1981.

(3) "My point of view is ... the analogue of Mach's positivism, where by 'positive fact' we mean that we can use our own subjective opinions" (de Finetti, 1931, p.3; authors' translation).

(4) With respect to some subjects that we shall discuss in due course, it is important to point out that de Finetti (1931b, p.6) quotes Poincaré so as to strengthen his claims that probability calculus, and not logic is the key to an understanding of scientific method.

a frequency can prove a probability assessment(5). Rather it is correct to speak in terms of the probability of a single event, since the concept of proofs of the same phenomenon is arbitrary, as is that of elements of the same class in logic (de Finetti 1931b, p.16). Thus probability, as the psychological perception of an individual, can be measured and subjected to mathematical formalization (de Finetti, 1931b, pp.39-41). de Finetti's frame work is completed by the introduction of the notion of exchangeability, relating the concept of subjective probability to the problems of classical statistical inference while removing all the metaphysical apparatus of constant but unknown probabilities, of independent proofs and of hypothetical values of probability (de Finetti 1931b, p.36).

3. The approach of "Probabilismo" is corroborated and enhanced by de Finetti's reviews of other probability theorist's writings. In these papers he often reaffirms the "total" (i.e. general and universal) nature of probability calculus as opposed to the "incomplete" (i.e. partial and specific) points of view which aim to limit its applicability to specific fields and problems (de Finetti, 1938a, p.15; 1941,(p.2). Subjectivism is defended as a natural and irremovable concept(6) and the absurdity of the verification of a probability assessment through the observation of a frequency is again discussed with respect to Borel's "Traité"(de Finetti, 1939, pp.8-12), von Mises's notion of 'kollectiv' (de Finetti 1937a; in particular the section "La frequenza limite e il teorema di Bernoulli) and Wald's critiques (de Finetti 1938a, pp.8-12). The axiomatic approaches are analysed in great detail (de Finetti 1949, 1951) and criticized for their lack of practical sense(7).

There is instead a substantial area of agreement with Cambridge Probability Theorists (de Finetti, 1938b). For de Finetti, as for Keynes and Jeffreys, probability is a conditional concept so that "... the probability of an event makes sense only if it is relative to a specific body of knowledge" (p.351) and "opinion is not generated by experience, but experience tells us the circumstances in which we find ourselves, from which we may select an opinion related to experience" (p.355).

However, in spite of this similarity of views essentially concerning the problem of induction, de Finetti disagrees with Keynes and Jeffreys about the subjective meaning of the notion of probability. In fact both Cantabrigian authors, as members of the logical school, claim that probability expresses a degree of implication between a proposition and a specific body of knowledge (or between two propositions) and that this degree of implication is unique. Moreover Keynes, in contrast with Jeffreys (1931, pp.222-4), holds the opinion that not all probabilities are quantifiable and that they cannot always be ordered. For Keynes given two probabilities we may face three distinct kinds of situations: one in which we can assign a numerical measure to our degrees of belief, one in which, although we cannot measure them, we can still assert that one is bigger than

(5) In de Finetti's perspective as well, no concept pertaining to probability can be introduced a priori but must always be defined with respect to the probability assessment, e.g. you cannot attach an a priori meaning, as an hypothesis, to the notion of independent events.

(6) "I do not care whether an individual is normal and thinks equally probable the ninety numbers in a lottery or whether he is superstitious and assignes a higher probability to the numbers he dreamed about; what is essential are the mathematical laws with which these evaluations are combined in order to obtain other evaluations" (de Finetti 1937a,p.14; authors' translation).

(7) This critique is resumed in de Finetti's treatise (1970, p.728)

the other and finally one in which no comparison of magnitude is possible (Keynes, 1929,p.36). Thus probabilities are only partially ordered. When numerical values can be assigned Keynes subscribes a frequentist view and appeals to the principle of insufficient reason (Keynes, 1921,p.44). Keynes's approach, vis-a-vis de Finetti's subjectivistic framework, is not capable of fully translating probability theory in probability calculus and does not envisage probability theory as a unique and general method for dealing with decision making under uncertainty (de Finetti, 1938b,p.359).

4. In 1931 de Finetti also tackles the problem of the mean from Chisini's point of view and among other results obtains the general expression for associative means (de Finetti-Nagumo-Kolmogerov theorem)(8). Exploiting this result, he suggests an approach to utility theory alternative and symmetric to the Von Neumann and Morgenstern's axiomatic one (9), so completing the subjectivist approach to decision making in the face of uncertainty (10). In fact the introduction of the utility function "helps to reconcile the more general coherent behaviour in a probabilistic sense with classical behaviour based on mathematical expectation (i.e. on the notion of fair bet)" (de Finetti, 1952,p.18, authors' translation).

5. de Finetti's critique of Neyman and Pearson's theory, discussed with reference to Abraham Wald's work, is consistent with the basic tenets of "Probabilismo".

The aim of the Neyman-Pearson school is that of developing a theory of statistical induction based on purely objectivistic foundations so that probability has no other meaning - "not even for distraction or convenience sake" - than that of a long run frequency.

de Finetti's critical analysis (1951, 1959) focuses on the foundations of the programme and emphasises the lack of practical meaning of the method. From a frequentist point of view "to accept a hypothesis ... is not to attribute to it any kind of probability or plausibility; such acceptance is a mechanical act, based not on a judgement of its actual validity, but on the frequent validity of the method from which it was derived". In other words it is "the criterion followed by the man who buys a suit of brand A that he considers defective instead of buying a suit of brand B that he considers non-defective, because he knows from statistics that A has a smaller percentage of defective suits than B. For him, the direct comparison of the two suits has no value at all, since it concerns only individual cases (de Finetti, 1959, p.33).

Wald's 'involuntary' revolutionary work goes beyond the objectivist approach.(11) It identifies the fundamental characteristic of a decision in its economic effect, makes effective Neyman's concept of inductive behaviour and discloses that the choice is between decisions rather than hypotheses (de Finetti, 1951,p.190). But from the point of view of

(8) de Finetti: (1931b) pp.381-383, c.f. in particular notes 5 and 6.

(9) This approach is developed and enhanced by Daboni (1984).

(10) In Von Neumann and Morgenstern's approach the notion of probability is accepted as something already known. They use "the perfectly well founded interpretation of probability as frequency in long runs which gives directly the necessary numerical foothold." For criticism of this approach c.f. de Finetti (1952, p.15).

(11) In a sense Wald, in order to find a conceptual framework for objectivist statistics wound up by destroying it rather than justifying it (c.f. de Finetti, 1959, p.37).

subjectivism another step is required; the choice of an initial distribution(12). To choose an admissible decision rule simply means to choose a Bayesian rule, that is, at least implicitly, an initial opinion (13): "Wald should have asked himself whether ... such a rule should be chosen not arbitrarily but as reflecting a real initial opinion", (de Finetti, 1959, p.48).

6. At this stage it is clear that the "dramatic breaks" identified by Arrow were already present and well-posed (devoid of inconsistencies and misunderstandings) in de Finetti's neo-Bayesian and neo-Bernoullian view. As many years before(14) the difficulty of accepting and correctly understanding was still there but on a larger scale. It is exactly in this sense that the "new formulations", such as that suggested by Shackle, are justifiable only in an historical perspective, since they appeared during a period of total bewilderment in the field of probability. A bewilderment that, as de Finetti pointed out (1952, p.14), discouraged those who should have envisaged in probability theory the foundations of their doctrines

Although during the last twenty years econometricians and economists have been very interested in Bayesian ideas and methods (c.f. Zellner, 1985), that same bewilderment expressed by de Finetti in the early fifties is still creating the crucial difficulties characterizing many research programmes in the theory of expectations in economics.

7. It is paradoxical that the Austrian school upholds the role of subjectivism in economic method yet at the same time considers probability calculus inapplicable because economic events are single and non-repetitive.

Lachman's itinerary (1976) is sufficient to clarify this paradox(15); the cultural climate and references of the Austrian research programme are the same as those of "Probabilismo" (the references to Mach and Poincaré, the recognition that Schumpeter applied Mach's positivistic methodology to economics, the relevance assigned to pragmatically based methodological instruments)(16). The role of uncertainty, as a basic feature of the economic world, is crucial(17); events are singular and the standard of subjectivism is extended from tastes to expectations so that "the formation of expectations is an act of our mind by means of which we try to catch a glimpse of the unknown; each one of us catches a different glimpse". Nevertheless this attitude towards uncertainty, instead of leading to a spontaneous application of the subjectivistic approach to probability in its "total" view, leads to a flat refusal of the calculus of probability. The frequentist mistake creates an unsurmountable obstacle. Shackle (1972) summarises his position in the title of section 34.40, "Probability concerns groups of events, not single critical choice". Ludwig von Mises shows a

---

(12) We conform to de Finetti's terminology and use the word initial instead of a priori.

(13) For a superb discussion of the technical and epistemological issues involved in the problem of intial probabilities c.f. de Finetti and Savage (1962).

(14) c.f. note 1.

(15) Further insights are to be found in Lachman (1977, pp.20-34) and in Kirzner (1982).

(16) c.f. Lachman (1976, p.56).

(17) By paraphrazing Shackle and von Mises, we may say that we live in a kaleidic world and there is no stability in the course of human events and consequently no safety, c.f. Lachman (1976), p.67).

similar preoccupation when he distinguishes between class and case probability: "Case probability is a peculiar feature of our dealing with problems of human action. Here any reference to frequency is inappropriate, as our statements always deal with unique events which as such ... are not members of any class ... Case probability is not open to any kind of numerical evaluation."(18)

This lack of understanding about foundations neutralizes the potentialities of the Austrian method and leads to either operational impotence in the face of uncertainty or to those falsifications de Finetti warned us about. Shackle's solution(19) is founded on fictitous arguments (such as the distinctions between "unique trial", "isolated trial" and "crucial trial" or the hypothesis that economic agents make decisions by focusing their attention on the consequences of the most favourable and the most unfavourable of the possible cases) and is empty of any empirical content:" rather than a criterion it is more an attempt at describing certain specific behaviours; as such I do not know how much it conforms to reality" (de Finetti, 1969, p.120; authors' translation)

8. As we have pointed out above, closely related to the rejection of the "total" view of probability theory is the distinction between risk and uncertainty. This distinction, originally suggested by Keynes (1921), part 1; 1937) and Knight (1921, ch.7) is once again of some interest in framing the contemporary research agenda in the theory of expectations(20), in spite of Stigler's specification(21) and Friedman's critique (which explicitly refer to de Finetti's subjectivist and "total" approach(22).

---

(18) The same problem is put forth by Hicks (1984, p.14): "An experimental science can make use of the probability calculus, which makes it possible to enunciate useful laws in terms of ...... numerical probabilities. There is no clear meaning of a numerical probability except in relation to a sequence of experiments, experiments which are willing to treat as cases of the same phenomenon." Moreover, elsewhere Hicks (1979, p.105) agrees with the position of Cambridge probability theorists and in particular "on those points where Keynes and Jeffreys appear to differ, I generally find myself on the side of Keynes."

(19) This "solution" has found some support in the economic profession: e.g. Turvey believes that Shackle defines a new treatment of uncertainty and van de Graaf & Baumol claim that he "develops a quite devastating criticism of the orthodox probability approach to expectations to be found in most theoretical discussions by professional economists". (c.f. Turvey et.al. 1949)

(20) An illuminating example is Frydman & Phelps's treatment (1983) of the distinction between measurable uncertainty and true uncertainty.

(21) Stigler in the "Introduction" to the 1971 edition of Knight's work claims "... tradition has assigned a distinction between risks (capable of actual treatment) and uncertainty (stochastic events not capable of such treatment) as Knight's contribution. Fortunately this is an extreme caricature of his work, because modern analysis no longer views the classes as different in kind". (Stigler 1971, p.XIV). Also LeRoy and Singell (1986) have rejected the distinction between measurable and unmeasurable probabilities as correctly representing Knight's approach.

(22) c.f. Pelloni (1986a)

The modern role of the Knight-Keynes tradition is clearly expressed by Lucas's view (1976, 1977)(23) which supports risk as an explanatory and operational tenet of his equilibrium theory of business cycle and of his critique of standard economic policy evaluation(24). According to Lucas the rational expectations hypothesis(25) makes sense only when it refers to recurrent events, thus only in a frequentist context. The neo-Bayesian approach is rejected a priori because of its lack of "empirical content" in the sense that "without some way of inferring what an agent's subjective view of the future is, this hypothesis is of no help in understanding his behaviour." (Lucas, 1977, p.15). In these terms subjective Bayesianism is reduced to an empirical psychological theory of degrees of belief, which can accommodate for psychotic behaviour given a sufficiently abnormal view of relevant probabilities.

Actually Lucas in interpreting neo-Bayesianism in this way, rather than as a theory of decision making under uncertainty, makes a common interpretative mistake.(26) The identification and the rebuttal of this fallacy (and its implications) are present in de Finetti's writings, though referring to different contexts.(27) The psychotic behaviour argument is in reality a fictitious problem entailing a non "total" view of probability calculus.(28) Possible references to insurability are logically and operationally irrelevant since in theory any random prospect can be covered by transferring it to another individual or institute willing to take it on; in a concrete practical sense references to insurability might be appropriate and well-posed but empty of any general conceptual meaning, since they would reflect contingent or institutional situations (c.f. de Finetti & Emanuelli, 1967).

The essence of the question is that we are dealing with a problem of decision theory and from a subjectivist viewpoint a decision making criterion must have general validity since it is deduced not from specific empirical hypothesis but from general logical conditions of coherence (c.f. de Finetti, 1969, p.35). It is nonsense to restrict the criterion validity (particularly on the basis of uncessary and ill-defined hypthesis) to the situations of risk, since for the remaining situations decisions would be left to intuition or to arbitrary choices from among more or less fictitious criteria

(23) Meltzer (1982) as well sees the Knight-Keynes tradition as an alternative model that can be combined with the method of rational expectations, in this way subscribing as conceptually discriminant the existence of insurable and uninsurable risks.

(24) LeRoy and Singell (1986) label Lucas as Keynesian from a methodological point of view. Ironically, in a different perspective, Lucas can be viewed as a neo-Austrian, c.f. Laidler (1982).

(25) Muth's (1961, p.316) original definition states that "expectations of firms (or, more generally, the subjective probability distribution of outcomes) tend to be distributed, for the same information set about the prediction of the theory (or, the "objective" probability distribution of outcomes)".

(26) A similar misinterpretation is due to Solow (1984).

(27) C.f. Pelloni (1986b) for a discussion of the REH in a neo-Bayesian perspective.

(28) de Finetti discussed this point several times, e.g. the aforementioned (note 6) remark about superstition and de Finetti (1937b, p.71, note (e)). For further details c.f. Furst (1978, pp.114-120).

expressing partial reasons of possible preference. If the situations of risk are these which show more "regularity" and less "dissimilarity" of evaluation between individuals then the risk-uncertainty distinction has a purely descriptive (and accessory) meaning and is incapable of distinguishing a priori the applicability or not of an operational scheme.

9. Muth's reference to an "objective" probability as a standard of rationality has a methodological content which goes beyond the frequentist interpretation. It can be seen as a situation of "consensus" of expectations, guaranteed, for instance, by economic theory or by a public prediction structure such as that of Grunberg and Modigliani.(29) In other words, it might refer to a situation of consensus with respect to a forecasting system and imply a judgement about the goodness (in the sense of fitting the facts) of forecasts generated by this forecasting system.(30) In this perspective the issue of rational expectations is linked with the problem of empirical evaluation of models. In particular, research on calibration as the natural criterion of empirical validity can suggest a way to define in a less ambiguous manner "the correct objective probability forecasts".(31)

It is well known that, from the point of view of subjectivism, probability, as a measure of degrees of belief, cannot be corroborated or falsified by facts(32) as "a scientific theory, in the sense of law, is not a statement whose truth or falsity is objectively decidable" (de Finetti, 1971, p.88). All this does not mean that subjectivists do not recognize the important role of the problem of giving a clear and unambiguous meaning to the concept of measure of success, as a measure of the goodness of evaluation of a prediction (either an individual or a forecasting system). The method is implicit in the definition of probability as betting odds. The method of employing scoring rules (to which also David (1984) refers, pp.21-24) "gives, in fact, a direct behavioural meaning to the familiar expression of a belief in terms of a numerical probability, leads automatically to an overall comparison between the outcomes of different personal evaluations" and so "the accumulated loss ... is indeed a thoroughly concrete measure of success" (de Finetti, 1962, p.360). Consistent with the approach foundations, the operational meaning of this measure is guaranteed: "I find no difficulty in admitting that any form of comparison between probability evaluations (of myself or of other people) and actual events may be an element influencing my further judgement, of the same status as any other kind of information" (de Finetti, 1962, p.360).

The critical and recurrent remark that the subjectivist paradigm, when facing situations of interpersonal evaluations and collective choice, is

(29) A suggestion for a non-frequentist interpretation of Muth's hypothesis can be found in De Felice & Pelloni (1982, pp.68 - 71).

(30) About these issues Box (1980) is of extreme interest. Illuminating remarks can be found in Zellner (1985).

(31) Dawid's research programme on calibration (1982, 1984a, 1984b) shows similarities with the REH which might be interesting for further research. For discussion and criticism of Dawid's work c.f. Lindley (1980, pp. 31-32; 1982) and Oakes (1985).

(32) This problem is thoroughly discussed in de Finetti's treatise (ch.5, section 9). However he had already dealt with it in his critique of Borel and von Mises (de Finetti, 1937, 1939). Thus if Lad (1984) is right in locating the origins of the calibration question in Frechet then paradoxically some of the answers and objections are antecedent to the origins.

impotent (since from this extreme viewpoint, it would seem that any set of forecasts is as good as any other) was already rebutted by de Finetti, without adding or modifying anything of his original approach as presented in "Probabilismo". Thus even for the problem of goodness of evaluation, which is still unsettled and represents the heart of the methodological debate triggered by rational expectations, de Finetti supplied if not the answer at least a broader perspective useful in avoiding dangerous pitfalls.

## REFERENCES


Arrow, K.J., 1951, Alternative Approaches to the Theory of Choice in Risk-Taking Situations, as reprinted in "Essays in the Theory of Risk-Bearing", North Holland, Amsterdam, 1971.

Box, G.E.P., 1980, Sampling and Bayes' Inference in Scientific Modelling and Robustness, JRSS, A143, Part 4.

Calderoni, M., Vailati, G., 1909, Le origini e l'idea fondamentale del pragmatismo, Psicologia Applicata, n.1.

Daboni, L., On the Axiomatic Treatment of the Utility Theory, Metroeconomica, XXXVI, n.2-3.

Dawid, A.P., 1982, "The Well-Calibrated Bayesian", JASA, 77, n.379.

Dawid, A.P., 1984, Statistical Theory. The Frequential Approach, JRSS,A147.

Dawid, A.P., 1985, Calibration-based empirical probability, Ann. Statist. 13, n.4.

De Felice, M. 1981, Prefazione, in de Finetti, B., "Scritti (1926-1930)". CEDAM, Padova.

De Felice, M., Pelloni, G., 1982, "Aspettative Razionali, Teoria Economica e Politiche di Stabilizzazione". Isedi-Mondadori, Milano.

de Finetti, B., 1929, Funzione caratteristica di un fenemeno aleatorio, Atti, Congresso Internaz. Matematici, Zanichelli, Bologna.

de Finetti, B., 1930a, Fondamenti logici del ragionamento probabilistico, Azzognidi, Bologna.

de Finetti, B., 1930b, Problemi determinati e indeterminati nel calcolo della probabilità, Rendiconti, Acc. Naz. Lincei, XII, 9.

de Finetti, B., 1931a, Sul Significato soggettivo della probabilità, Fundamenta Mathematicae, 17.

de Finetti, B., 1931b, "Probabilismo", Perrella, Napoli.

de Finetti, B., 1931c, Sul concetto di media, GIIA, 2,3.

de Finetti, B., 1937a, Statistica e probabilità nella concezione di R. von Mises, Supplemento Statistico, fasc. 2-3.

de Finetti, B., 1937b, La prévision: ses lois logiques, ses sources subjectives, Ann. Inst. H. Poincaré, Paris, 7 as translated in Kyburg, H.E., Smokler, H.E., (eds),"Studies in Subjective Probability".

de Finetti, B., 1938a, Resoconto critico del colloquio di Ginevra intorno alla teoria della probabilità, GIIA, IX, n.1.

de Finetti, B.,1938b, Probabilisti di Cambridge, Supplemento Statistico, 3, as translated in The Manchester School, LIII, n.4, 1985.

de Finetti, B., 1939, Punti divista: Emile Borel, Supplemento Statistico, 4.

de Finetti, B., 1941, Punti di vista: Hans Reichenbach, Statistica, n.1.

de Finetti, B., 1949, Sull' impostazione assiomatica del calcolo della probabilità , Ann. Triestini, XIX

de Finetti, B., 1951a, Aggiunta alla Nota sull'Assionatica della Probabilità, Ann. Triestini, XX

de Finetti, B., 1951b, L'Opera di Abraham Wald e l'assestamento concettuale della statistica matematica, Statistica, fasc. II.

de Finetti, B., 1952, Sulla Preferibilità, Giorn. Econ. e Ann. Economia, Nov. - Dec.

de Finetti, B., 1959, "La probabilità e la statistica nei rapporti con l'induzione", Ist. Matematico, Univ. of Rome, as translated in de Finetti, B.,"Probability Induction and Statistics", Wiley, 1972.

de Finetti, B.,1962, Does it Make Sense to Speak of "Good Probability

Appraisers", in Good, I.J., (ed), "The Scienties Speculates", Heineman, London.
de Finetti, B., 1967, L'incertezza nella economia, in "Trattato Italiano di Economica", XVI, UTET, Torino.
de Finetti, B., 1970,"Teoria della Probabilità", Einaudi, Torino.
de Finetti, B., 1971, Probabilità di una teoria e probabilità dei fatti, in "Studi in onore di Guiseppe Pompily", Oderisi, Gubbio.
de Finetti, B., Savage, L.J., 1962, Sul modo di scegliere le probabilità iniziali, Bibliotecoa del Metron, Roma.
Friedman, M., 1976, "Price Theory", Aldine, Chicago.
Frydman, R., Phelps, E.S., 1983, Introduction, in "Individual forecasting and aggregate outcomes", CUP, Cambridge.
Fürst, D., 1978, Tre uomini e un'urna, Proceedings of the Conference: "Induzione, Probabilita, Statistica", Venezia, Sept.
Hicks, J., 1979, "Causality in Economics", Blackwell, Oxford.
Hicks, J., 1984, The "New Causality": An Explanation, OEP, 36
Jeffreys, H., 1931, "Scientific Inference", CUP, London
Keynes, J.M., 1921, "A Treatise on Probability", as reprinted in Keynes's Collected Writings, VIII.
Keynes, J.M., 1937, "The General Theory of Employment", as reprinted in Keynes's Collected Writings, XIV.
Kirzner, I.M., (ed), 1982, "Method, Process and Austrian Economics. Essays in Honor of L. Von Mises", Lexington Books, Massachussets.
Knight, F.H., 1921, "Risk, Uncertainty and Profit", as reprinted by UCP, 1971.
Lachman, L.M., 1976, From Mises to Shackle: An Essay, JEL, XIV
Lachman, L.M., 1977, "Capital, Expectations and Market Process", Andrews and McMeel, Kansas City.
Lad, F., 1984, The Calibration Question, BJPSc, 35.
Laidler, D.E.W., 1982,"Monetarist Perspectives", Philip Allan, Oxford.
LeRoy, S.F., Singell, L.D., jr. 1986, Knight on Risk and Uncertainty, mimeo, Univ. of California, S. Barbara, Feb.
Lindley, D.V., 1980, The Bayesian Approach to Statistics, Univ. of California, Berkeley.
Lindley, D.V., 1982, Comment, JASA, 77, n.379.
Lucas, R.E., 1976, Econometric Policy Evaluation: A Critique, Carnegie-Rochester Conference Series, 1, North Holland, Amsterdam.
Lucas, R.E., 1977, Understanding Business Cycles, Carnegie-Rochester Conference Series, 5, North Holland, Amsterdam.
Meltzer, A.H., 1982, Rational Expectations, Risk Uncertainty and Market Responses, mimeo.
Muth, J.F., 1961, Rational Expectations and the Theory of Price Movements, Econometrica, 29, n.6.
Stokes, D., 1985, Self-Calibrating Priors Do Not Exist, JASA, 80, n.390.
Pelloni, G., 1986a, A Note on Friedman and the Neo-Bayesian Approach, Hull Economic Research Paper, n.129, forthcoming in The Manchester School.
Pelloni, G., 1986b, The Neo-Bayesian Approach and Rational Expectations, mimeo, forthcoming in Economica Politica.
Shackle, G.L.S., Epistemics and Economics, CUP, Cambridge, 1972
Solow, R.M., 1984, Review of Chick, V., "Macroeconomics after Keynes, A Reconstruction of the General Theory, JPE, 52, n.4.
Stigler, G.J., Introduction, in Knight (1921).
Turvey, R., de V.Graaf, J., Baumol, W., Shackle, G.L.S., 1949, Three Notes on "Expectations in Economics", Economica, Nov.
Zellner, A., 1985, Bayesian Econometrics, Econometrica, 53, n.2.

SOME CHARACTERISTICS OF BAYESIAN DESIGNS

Klaus Felsenstein

Technische Universität Wien
Wien, Austria

## 1. INTRODUCTION

A considerable number of stochastic models comprise the potentiality of selecting the experimental conditions. A control-variable influences the observations and likewise the gained information about some parameter or in a more Bayesian mode of expression the 'state of nature'. Reaching our goal of increasing the information demands a model-formulation with independence between the parameter and the chosen control variable or with a concrete functional connection that seems defendable. The choice of an appropriate likelihood is aggraviated by specifying how the distribution of the observations is altered by different levels of the control variable. An even more difficult problem is the valuation of information and precision. Each measure of information has to stand many discussions about its shortcomings and hardly any can be employed generally.

It's again the Bayesian approach that offers a reasonable conception of experimental design. Especially decision theory covers rational methods of solving design problems. We give a brief survey of posing the problem and use entropy as a measure of information.

## 2. THE MODEL

The distribution of some observable (multivariate) random quantity $X_\upsilon$ depends upon a number of controlled factors summarized as an element out of a set of designs $\upsilon \in \mathfrak{V}$ and upon a parameter $\theta$. $f(x|\theta,\upsilon)$ denotes the density of $X_\upsilon$.

A function $\delta(x,\upsilon)$ into a space of strategies $\mathfrak{A}$ is called a decision rule and a nonnegative function $L(\theta,a), a \in \mathfrak{A}$, represents a loss-function. A strategy consists of the choice of a design $\upsilon$ and a decision rule $\delta$, therefore the loss could be written as $L(\theta,\upsilon,\delta(.,\upsilon))$.

The goal of experimental design is to minimize the resulting Bayes-risk with respect to a prior distribution of $\theta$ with density $\pi(\theta)$.

$$r(\pi,\mathfrak{a}) = r(\pi,\mathfrak{v},\delta(.,\mathfrak{v})) := \mathbb{E}_{X_\mathfrak{v},\theta} L(\theta,\mathfrak{v},\delta(X_\mathfrak{v},\mathfrak{v}))$$

( $\mathbb{E}_{X_\mathfrak{v},\theta}$ denotes the expectation with respect to the joint distribution of $(X_\mathfrak{v},\theta)$. )

A Bayes-strategy $\mathfrak{a}^*$ fulfills the equation

$$r(\pi,\mathfrak{a}^*) = \inf_{\mathfrak{A}} r(\pi,\mathfrak{a}) \quad .$$

The task of finding a Bayes-strategy can be decomposed into two optimization problems. If $\delta^*(.,\mathfrak{v})$ is the Bayes-decision for a concrete design $\mathfrak{v}$ then the design $\mathfrak{v}^*$ that minimizes the posterior Bayes-risk

$$\hat{r}(\pi,\mathfrak{v}) = \mathbb{E}_{X_\mathfrak{v},\theta} L(\theta,\mathfrak{v},\delta^*(X,\mathfrak{v}))$$

leads to the Bayes-strategy $\mathfrak{a}^* = (\mathfrak{v}^*,\delta^*(.,\mathfrak{v}^*))$ .

Subjectivity enters this fairly general concept through the prior distribution and the loss-function. The latter measures the error in the decision on the parameter as well as the drawback of a specified design. Overall high costs and effort due to $\mathfrak{v}$ are punished with the loss. Often L is decomposed into the decision loss and costs

$$L(\theta,\mathfrak{v},\delta(.,\mathfrak{v})) = L(\theta,\delta(.,\mathfrak{v})) + C(\mathfrak{v}) \quad .$$

Example. Most of the research concerning design problems has been done for linear regression models. In this case the design determines the moments of the random quantity X. Assume

$$\mathbb{E}(X_v|\theta) = \psi(v)'\theta \qquad \text{and}$$

$$\mathrm{Var}(X_v|\theta) = (\alpha.\lambda(v))^{-1} \quad ,$$

where $v \in V \subset \mathbb{R}^m$, $\psi: \mathbb{R}^m \to \mathbb{R}^r$, $\theta \in \mathbb{R}^r$ .$\alpha > 0$ is some unknown constant called the precision of the regression model. The known function $\lambda: \mathbb{R}^m \to \mathbb{R}^+$ measures the efficiency of the design v. A vector of observations $\underline{X}_\mathfrak{v} = (X_{v_1},\ldots,X_{v_n})$ leads to

$$\underline{X}_\mathfrak{v} = F\,\theta + \underline{e}_\mathfrak{v} \tag{2.1}$$

with $\mathfrak{v} = (v_1,\ldots,v_n)$, the design matrix $F = F_\mathfrak{v} = (\psi(v_1),\ldots,\psi(v_n))'$. The vector of errors $e_\mathfrak{v}$ is supposed to have a multivariate normal distribution $N(\underline{0},\alpha^{-1}\Sigma)$ with $\Sigma = \mathrm{diag}(\lambda^{-1}(v_1),\ldots,\lambda^{-1}(v_n))$. The actual parameter is $(\theta,\alpha)$. Estimation of $\theta$ represents a decision $\delta$ and the loss L is a quadratic form

$$L(\theta,\alpha,\delta) = h(\alpha)\,(\theta-\delta)'\,\Lambda\,(\theta-\delta)$$

with $h > 0$ and $\Lambda \in \mathbb{R}^{r\times r}$ and positive definite.

Both of the following assumptions I) or II) concerning the prior distribution of $(\theta,\alpha)$

I) The conditional prior $\theta|\alpha$ is normal $N(\mu,\alpha^{-1}\Phi)$.
II) Only linear estimators are examined and the prior distribution satisfies $\mathbb{E}(\theta|\alpha) = \mu$ and $\mathrm{Cov}(\theta|\alpha) = \alpha^{-1}\Phi$ .

yield the Bayes-estimation $\theta_B$ of the Parameter $\theta$.

$$\theta_B = \tilde{\Phi}\ (F'\Sigma^{-1}\underline{X}_\upsilon + \Phi^{-1}\mu) \quad \text{where}$$

$$\tilde{\Phi} = (F'\Sigma^{-1}F + \Phi^{-1})^{-1}.$$

The inverse of the posterior covariance $\tilde{\Phi}$ is called the Bayes-information matrix.

Now the interest centers around the characteristics of a Bayes-design which has smallest risk.

$$\hat{r}(\pi,\upsilon) = r(\pi,\upsilon,\theta_B) = \mathbb{E}_{X_\upsilon,\theta,\alpha} h(\alpha)(\theta-\theta_B)'\Lambda(\theta-\theta_B) =$$

$$\mathbb{E}_{\theta,\alpha} h(\alpha)\ \mathbb{E}_{X_\upsilon|\theta,\alpha}(\theta_B-\theta^* - \tilde{\Phi}\Phi^{-1}(\mu-\theta))'\Lambda(\theta_B-\theta^* - \tilde{\Phi}\Phi^{-1}(\mu-\theta)) =$$

$$\mathbb{E}_{\theta,\alpha} h(\alpha)\ \mathbb{E}_{X_\upsilon|\theta,\alpha} tr[\Lambda(\theta_B-\theta^*)(\theta_B-\theta^*)' + \Lambda\tilde{\Phi}\Phi^{-1}(\mu-\theta)(\mu-\theta)'\Phi^{-1}\tilde{\Phi}]$$

with $\theta^* = \mathbb{E}_{X_\upsilon|\theta,\alpha}(\theta_B)$.

Since $\mathbb{E}_{X_\upsilon|\theta,\alpha}(\theta_B-\theta^*)(\theta_B-\theta^*)' = \frac{1}{\alpha}\ \tilde{\Phi}F'\Sigma^{-1}F\tilde{\Phi}$ it follows that

$$\hat{r}(\pi,\upsilon) = \mathbb{E}_\alpha\ h(\alpha)\alpha^{-1}\ tr[\Lambda(\tilde{\Phi}(F'\Sigma^{-1}F + \Phi^{-1})\tilde{\Phi})]$$

$$= tr(\Lambda\tilde{\Phi})\ \mathbb{E}_\alpha\ h(\alpha)\ \alpha^{-1}.$$

Thus the prior of $\alpha$ and the function h(.) should satisfy

$$\mathbb{E}_\alpha\ h(\alpha)\ \alpha^{-1} < \infty .$$

Under these conditions the Bayes-design corresponds to an A-optimal design in the usual sense applied to the posterior covariance matrix multiplied by the loss matrix $\Lambda$.

$\mathfrak{M}^+$ and $\mathfrak{M}$ denote the set of positive and non-negative definite r×r matrices respectively. The function $A(M)=tr(\Lambda M^{-1})$ is convex for $M\in\mathfrak{M}^+$. A is bounded on the open subset of matrices $\{M+1/2\Phi^{-1}|M\in\mathfrak{M}^+\}$ and therefore continuous. Hence the function $B(M) = tr[\Lambda(M+\Phi^{-1})^{-1}]$ is continuous on $\mathfrak{M}$. If now $\psi$ and $\lambda$ are continuous and V is a compact subset of $\mathbb{R}^m$ then a Bayes-design and a corresponding information matrix exist since a continuous function attains its infimum over a compact set.

## 3. ENTROPY

In the Bayesian point of view the choice of a distribution $p(\theta)$ for the parameter describes a decision procedure. The Bayes rule p* minimizes the posterior Bayes-risk, indicated by

$$\mathbb{E}_{\theta|X}\ L(\theta,p(.)) .$$

If we try to advance coherently the posterior distribution $\pi(\theta|x)$ should be the Bayes-decision p*. Hence a loss-function with

$$\inf_p \mathbb{E}_\theta L(\theta,p) = \mathbb{E}_\theta\ L(\theta,\pi)$$

is deemed appropriate for the prior $\pi$. Assuming differentability of L it is well known that there are a constant c and a real function $\tau$ such that $L(\theta,p(\theta)) = c \log p(\theta) + \tau(\theta)$.

All suitable loss-functions lead to a Bayes-risk which is related to Shannon's measure of information for a density p. The entropy of a density p is defined by

$$H(p) := \int -\log p(\theta)\ \ p(\theta)\ \ d\theta .$$

In that context L can take negative values too. We should better use utility functions and keep L in conformity with the notation.

The concept of entropy loses some of its shortcomings if it is not used as an absolute measure of information but as a distance measure of distributions.

We return to the design problem and choose the loss

$$L(\theta,\nu,\pi(\theta|X_\nu)) = \log \pi(\theta|X_\nu) - \log \pi(\theta) .$$

Then the posterior Bayes-risk can be written as

$$\hat{r}(\pi,\nu) = \mathbb{E}_{X_\nu}\mathbb{E}_{\theta|X_\nu} \log \pi(\theta|X_\nu) - \log \pi(\theta)$$

$$= \mathbb{E}_{X_\nu} -H(\pi(.|X_\nu)) + H(\pi) .$$

The Bayes-risk coincides with the expectation with respect to the marginal distribution of $X_\nu$ of the reduction of entropy comparing the prior and the posterior distribution. Since the inequality

$$\int \log f(t) \; g(t)\, dt \leq \int \log g(t) \; g(t)\, dt$$

holds for densities f and g it follows that the Bayes-risk is non-negative.

Naturally the Bayes-design $\nu^*$ has to maximize the expected increase of information,

$$\hat{r}(\pi,\nu^*) = \inf_\nu \hat{r}(\pi,\nu) .$$

$\nu^*$ is called entropy-optimal design for the prior $\pi$. An immediate consequence of that definition is

<u>Theorem 3.1</u>. If $\mathbb{E}_{X|\theta} \log \pi(\theta|X_{\nu^*}) \geq \mathbb{E}_{X|\theta} \log \pi(\theta|X_\nu)$ holds for all designs $\nu$ then $\nu^*$ is entropy-optimal.

Of course replacing the observation $X_\nu$ by a sufficient statistic for $\theta$ leaves the entropy information unchanged. Define two designs $\nu_1$ and $\nu_2$ by $X_{\nu_1} = (X,t)$ and $X_{\nu_2} = t$ where $t(X)$ is a sufficient statistic then

$$\hat{r}(\pi,\nu_1) = \iint \log \frac{f(x,t|\theta)}{g(x,t)} \; f(x,t|\theta) \; \pi(\theta) \; d\theta \; d(x,t)$$

$$= \iint \log \frac{f(x|t,\theta) f(t,\theta)}{g(x|t) \; g(t)} \; f(x,t|\theta) \; \pi(\theta) \; d\theta \; d(x,t)$$

$$= \iint \log \frac{f(x|t,\theta)}{g(x|t)} \; f(x,t|\theta) \; \pi(\theta) \; d\theta \; d(x,t)$$

$$+ \iint \log \frac{f(t|\theta)}{g(t)} \; f(t|\theta) \; \pi(\theta) \; d\theta \; dt .$$

Since t is sufficient the densities fulfill $f(x|t,\theta) = g(x|t)$ and only the second integral remains which is the Bayes-risk of the design $\nu_2$.

Similarly the entropy-information of a design $\nu$ for $\theta$ and another parameter $\theta'$ which is a bijective transformation of $\theta$, $\theta' = T(\theta)$, is the same. This is an obvious consequence of the transformation of densities.

<u>Example</u>. We consider the classical regression model (2.1) and want to characterize an entropy-optimal design $\nu^*$. It turns out that in the Bayesian sense D-optimal designs are entropy-optimal. The parameter of interest is $\theta$ and no prior distribution of the precision $\alpha$ is specified now.

Theorem 3.2. If the parameter $\theta$ in the regression model (2.1) has a normal prior distribution $N(\mu,\Phi)$ then $\upsilon^*$ is entropy-optimal iff det $\tilde{\Phi}_{\upsilon*}$ is minimal. $\tilde{\Phi}_{\upsilon*}$ is the posterior covariance matrix.

Proof: The posterior is a normal distribution with the mean

$$\theta_B = \tilde{\Phi}(F'\Sigma^{-1}\underline{X}_\upsilon + \Phi^{-1}\mu) \quad \text{and the covariance}$$

$$\tilde{\Phi} = (F'\Sigma^{-1}F + \Phi^{-1})^{-1} .$$

Hence $\log \pi(\theta|X_\upsilon) \propto -1/2 \log\det\tilde{\Phi} - 1/2\ (\theta-\theta_B)\tilde{\Phi}^{-1}(\theta-\theta_B)$
and

$$\mathbb{E}_{\theta|x} \log \pi(\theta|x_\upsilon) = C - 1/2 \log\det\tilde{\Phi} - 1/2\,\mathbb{E}_{\theta|x}\mathrm{tr}(\tilde{\Phi}^{-1}\tilde{\Phi}).$$

Interchange of expectation and trace yields

$$\mathbb{E}_{\theta|x} \log \pi(\theta|x_\upsilon) = C - 1/2 \log\det\tilde{\Phi} - r/2$$

where C is independent of $\upsilon$. The right term is independent of x. Thus $\upsilon^*$ is entropy-optimal iff det $\tilde{\Phi}_{\upsilon*}$ is a minimum.

## 4. NON-INFORMATIVE PRIORS

Many techniques have been proposed for specifying a prior even when hardly any usable information is available. In this case the determination of the prior should not insert unintentional restrictions for the parameter. The entropy concept is adapted for the construction of such prior densities.

We are looking for a distribution that maximizes the information of the data

$$I(\pi) := -\mathbb{E}_\theta H(f(.|\theta)) + H(\pi) .$$

The prior $\hat{\pi}$ is said to be non-informative if $I(\pi)$ is maximal for $\hat{\pi}$. This solution $\hat{\pi}$ can be described by

$$\hat{\pi}(\theta) \propto \exp(-H(f(.|\theta)) . \quad \text{(See Zellner(1977).)}$$

If the observations are derived from a location family then the non-informative prior density is constant. The data density $f(x|\theta)$ depends on $\theta$-x only and therefore $H(f(.|\theta))$ is constant. Location-scale families with density

$$f(x|\theta_1,\theta_2) = \frac{1}{\theta_2}\, g(\frac{x-\theta}{\theta_2}1) \quad , \quad g>0$$

have the non-informative prior

$$\hat{\pi}(\theta_1,\theta_2) \propto \frac{1}{\theta_2} .$$

Now interest centers on the optimal design under non-informative prior. In case of location families $\upsilon^*$ turns out to be the design that maximizes the entropy of the marginal density of the data. The observations shouldn't contain any other systematical structure but the location parameter in order to avoid confounding of different effects.

Theorem 4.1. Suppose $X_\upsilon$ has a location distribution and the prior is non-informative then $\upsilon^*$ is entropy-optimal iff it maximizes $H(m_\upsilon(x))$ where $m_\upsilon(x)$ represents the marginal density of $X_\upsilon$.

Proof: $\mathbb{E}_{X_\upsilon} H(\pi(.|X_\upsilon)) = -\int\int \log\pi(\theta|x)\ \pi(\theta|x)\ d\theta\ m_\upsilon(x)\ dx$

$= -\int\pi(\theta)\ [\ \int f(x|\theta,\upsilon)\ \log\pi(\theta) + f(x|\theta,\upsilon)\ \log f(x|\theta,\upsilon)$

$- f(x|\theta,\upsilon)\ \log m_\upsilon(x)\ dx\ ]\ d\theta$ .

$\pi(\theta)$ and $I(\pi)$ are constant and the entropy $H(f(.|\theta))$ is independent of $\theta$. Hence the integral equals

$$c_1 + \int\int f(x|\theta,\upsilon)\ \log m_\upsilon(x)\ dx\ \ c_2\ d\theta =$$

$$c_1 + c_3 \int m_\upsilon(x)\ \log m_\upsilon(x)\ \ dx\ .$$

The constants $c_1, c_2$ and $c_3$ don't depend on $\upsilon$ and the integral is maximal if $H(m_\upsilon(x))$ is maximal.

Example. Since $\underline{Y}_\upsilon := (F'F)^{-1}F'\underline{X}_\upsilon$ has a normal distribution with mean $\theta$ and is a sufficient statistic the regression model described in (2.1) belongs to a location family. The constant (improper) prior leads to a normal posterior with

mean $\theta^* = (F'F)^{-1}F'\underline{X}_\upsilon$

and covariance $\tilde{\Phi} = \alpha^{-1}(F'F)^{-1}$ if $\Sigma = I_r$ .

Following the ideas in theorem 3.2 we obtain a similar comparison. In this case the entropy-optimal design coincides with a D-optimal design in the classical approach.

## REFERENCES


Bandemer H., Pilz J., 1978, Optimum experimental design for a Bayes estimator in linear regression, Transactions of the Eighth Prague Conference, A:93.

Bernardo J., 1979, Expected information as expected utility, Ann. Statist., 7:686.

Lindley D. V., 1956, On a measure of information provided by an experiment, Ann. Math. Statist., 27:986.

Pilz J., 1983, "Bayesian Estimation and Experimental Design in Linear Regression Models", Teubner, Leipzig.

Zellner A., 1977, Maximal Data Information Prior Distributions, in: "New developments in the applications of Bayesian Methods", Aykac A., Brumat C., ed., North-Holland, Amsterdam.

# THE ANALYSIS OF MULTIPLE CHOICE TESTS IN EDUCATIONAL ASSESSMENT

Simon French

Statistical Laboratory
Department of Mathematics
University of Manchester
Manchester M13 9PL, U.K.

## ABSTRACT

Multiple choice tests are much, but not exclusively, used in the British public examinations system. The analysis of results from such tests has been subject to much debate, particularly concerning the appropriateness of latent trait models.

In this paper I adopt an entirely subjectivist approach. I believe the purpose of a public examination is not to measure in some objective sense the performances of candidates, but rather to report the judgements of examiners as to those performances. It is the examiners' judgements that are modelled by marks and grades, not something directly about the candidates themselves. Adopting this viewpoint, I make two groups of comments pertinent to multiple choice tests.

First, if one is to use latent trait models to analyse candidate responses, then one must be clear as to the meaning of parameters within the models. I argue that latent trait variables are technical devices which encode certain expectations about the data, but other than that they have no physical meaning. Because of this view, I shall argue that latent trait models are appropriate for critically evaluating assumptions about examination data, but are inappropriate for the purpose of ranking candidates' work to report and grade individual performances.

Second, one should consider in what form to elicit responses from the candidates. De Finetti suggested that candidates should respond with their probability of the correctness of each possible answer to an item and that these responses should be assessed by means of a scoring rule. However, such schemes have many problems: the difficulty of getting candidates, still at school, to accept the inevitability of uncertainty in their lives; the problem of calibration, because they are unlikely to be equally good probability assessors. Perhaps more serious is the difficulty that a scoring rule which encourages a candidate to honestly reveal his beliefs may not reflect the manner in which the examiners wish to judge the candidate.

## INTRODUCTION

As Bayesians we pride ourselves that our approach to analyses is coherent - both in the technical sense of coherence and its everyday sense. Yet, sometimes we confine our attention to the analysis of a model without considering the implications of our philosophy for the generation of the model itself: indeed, for our conception of the context in which the model is developed and analysed. The essence of the Bayesian or subjectivist approach, it seems to me, is the recognition that, as individuals, we continually have to express beliefs, preferences, etc. and behave according to these judgements. We wish these judgements to be as rational, as consistent, and, perhaps, as fair as possible and we seek ways of thinking which help us achieve this. A Bayesian is not simply someone who updates a prior by a likelihood and then maximises an expected utility. He is someone who thinks carefully about how to encourage - he hopes, ensure - consistency and coherence in his judgements and actions.

In the following, I wish to explore the implications of this view for the manner in which we should conceive of certain aspects of the British public examination system. The British system of GCE, CSE and, from 1988, GCSE examinations is subject-based. There is no requirement to pass in groups of subjects; each candidate is awarded a separate grade for each subject taken. The form of examination, naturally, varies from subject to subject. Apart from sitting formal examination papers, candidates may be required to submit coursework or projects, or to be assessed practically or orally. Within the formal papers, they may be required to write essays; answer short, structured questions; or - this will be our concern - answer multiple choice items. Given my prejudices expressed above, I shall not refer to multiple choice testing by its other name: objective testing. The interested reader may find descriptions of the British examination system in Christie and Forrest (1981) and Mathews (1985).

Theories of educational assessment and the concepts they use seem to have been developed, by and large, from the psychometric theories used to analyse psychological tests, i.e. IQ tests, etc. They have at their base a belief that inside every candidate lies something that might be called his 'ability or level of achievement in the subject being examined', that this entity can be quantified on an objective unidimensional scale, and that the purpose of the examination is to gather evidence from which it may be estimated. The overall mark or grade awarded is an estimate of the candidate's ability or achievement. I and others do not accept this: we have been arguing for another, entirely different view of the examination process. Our arguments may be found in French (1981, 1985), French et al (1986a, b) and Vassiloglou (1982, 1986): here I only summarise our conclusion.

Firstly, the purpose of a public examination is not to measure in some objective sense something directly about the candidate; but rather it is to report the judgements of the examiners. Secondly, the examiners do not make judgements about something that they postulate to exist within the candidate, his ability or achievement or what-have-you; but rather they are concerned with the quality of performance within the candidate's script and, in the case of practicals, coursework and orals, within the processes observed during the assessment.

That examinations are meant to report examiners' judgements is a view that is entirely consistent with the subjectivist approach, and I shall support that no further here. It does, of course, only make some sort of operational sense if there is a general consensus among examiners, but that I believe to exist and to be engendered by various procedures within the examination system and, indeed, the wider educational context. However, the view that it is performances that are judged and not some intrinsic quality of the candidate does require brief comment.

Behind many other conceptions of the examination process is the belief that one should allow for the day-to-day variability of candidates. Some days a candidate will be 'on form' and others he will not. I find it conceptually impossible, however, to attribute some of the qualities and flaws within a candidate's script to his general achievement or lack of it, and others to his day to day variability. Hence I believe it is impossible for the examiners to do other than assess the performance that they have observed. Whether one says it is the candidate's achievement on the day or the quality of performance in his script which is assessed is perhaps a finer point of language. I do have a distinct preference for the latter terminology, however: it reduces the temptation to construct a model of the mental processes of the candidate.

When judging the quality of a performance, the examiners express something very akin to a value judgement. To do this fairly and consistently they need normative techniques akin to those used to assess multi-attribute value functions to guide decision makers' preferences (French, 1981, 1985; and Vassiloglou, 1984). Before they can make such judgements, though, they need to understand how their examination components performed. Were there any unforeseen biases; e.g. were any optional questions distinctly 'harder' than the alternatives? To foster such understanding the examiners may investigate their data statistically. How should any parameters within such an analysis be interpreted? It is to answering this question that the next section is directed, but considering only the case of a multiple choice components.

## LATENT TRAIT MODELS

Consider then the case of a multiple choice component. On this there are a number of questions or items, for each of which several alternative answers are offered. Candidates must select the one that they believe to be the correct answer. Usually there is only one truly correct answer among those offered: however, in variants some incorrect answers are designed to be more 'sensible' than others: in yet other variants candidates have to select several answers each of which satisfies some given condition. Given that candidates' scripts take such simple forms, essentially sequences of ticks and crosses, it is not surprising that many statistical models have been developed to describe, analyse and summarise them: see Lord and Novick (1968) and Weiss and Davison (1981). Moreover, such models invariably contain candidate parameters which are highly correlated with the number of correct answers that a candidate is expected to give. These parameters have naturally been called abilities, and there is a strong temptation for examiners to fit the models to their data, thus estimating candidates' abilities, and then to grade candidates according to these. Whether they should do so has been a matter of some controversy (see e.g. Goldstein, 1979; Wood, 1978; and Wright, 1977). I believe that a Bayesian approach can explicate matters greatly.

Before the examiners see any candidates' scripts, they have certain expectations. Precisely what expectations will depend on many circumstances and will certainly vary from examination to examination. Since these expectations will influence the judgements that the examiners eventually make of candidates' scripts, it is important the expectations are critically examined in the light of the data. To do this, the examiners must first formulate their expectations as clearly and as explicitly as possible. For instance, they might argue as follows.

The data will essentially be a two-way layout of responses with the rows corresponding to candidates and the columns to items. Since the examiners as yet know nothing about individual candidates, they may hold the rows to be exchangeable: i.e. that a particular data matrix and any row permutation of that matrix are equally likely. They might also hold the columns to be exchangeable either because the items were designed to be of equal difficulty or because, although the items were designed to be unequally difficult, the examiners do not know their order on the question paper. In practice, column exchangeability is unlikely to be reasonable, examiners usually design papers with a few easy questions at the beginning so that candidates are not disheartened early on. Moreover, questions are commonly grouped according to subject area. Some restricted version of column exchangeability may be reasonable, nonetheless. The point that matters is that some of the examiners' expectations can often be summarised by exchangeability or symmetry conditions.

Now exchangeability conditions have important implications for the form that one's subjective probability distribution may take. The classic result is De Finetti's Representation Theorem (De Finetti, 1937). He showed essentially that, if one considers an infinite sequence of 0-1 variables to be exchangeable (i.e. if attention is focussed on a finite subsequence, that subsequence and all permutations of it are considered equally likely), then one's probability for a particular finite subsequence must be given by a mixture of Bernoulli sequences. Put precisely, one's probability for the subsequence, 01001...01, in which there are r 1's and (n-r) 0's must have the form:

$$\Pr(01001\ldots01) = \int_0^1 \theta^r(1-\theta)^{(n-r)}p(\theta)d\theta. \qquad (*)$$

The precise density p(.) is not given by the exchangeability conditions, but the same p(.) applies whatever finite subsequence is considered. Furthermore, if a finite subsequence is observed, p(.) should be updated through Bayes' Theorem and this updated density used to form mixtures as in (*) to predict further subsequences.

Expression (*) may be interpreted naively as saying that, as a Bayesian believing in this exchangeability, one must use a Bernoulli model with unknown parameter $\theta$ and express one's beliefs in the value of $\theta$ through the prior distribution p(.). But this interpretation fails for a Bayesian because it suggests that the parameter $\theta$ has some physical interpretation, here the probability of a 1 at any given position in the sequence. A more satisfactory interpretation is that exchangeability and coherence imply that various relations must hold between the probabilities Pr(...) that model one's beliefs about the various subsequences. The functional form (*) ensures that these relations do hold. The indeterminacy of p(.) provides the degree of freedom that is left before the probabilities Pr(...) are completely determined. The parameter $\theta$ has no physical interpretation: it is purely a technical device to ensure that exchangeability holds. (See, e.g., Dawid, 1982.) In particular circumstances, one might construct an appropriate p(.) by considering what one would expect the mean number of 1's to be in a sequence and how confident one was in this by specifying a variance. A beta distribution could be fitted to these values to serve as p(.). Any subsequent analysis would, of course, include a check on the sensitivity of the conclusions to the particular choice of p(.).

Recently, much has been done to extend De Finetti's Theorem to exchangeable situations other than infinite sequences of 0-1 variables

(Aldous, 1985; Diaconis and Freedman, 1982; and Goldstein 1986a,b). Always the conclusion is the same. Exchangeability and similar symmetry requirements imply that beliefs should be modelled as mixtures of probability models. The parameters in these models act simply as 'indices' so that the mixture may be taken and exchangeability ensured. Other than that, parameters have no physical meaning. Much work still has to be done, but there is little doubt that it will lead to the same conclusion. Unfortunately, the two-way layout is one of the situations that still has to be fully investigated. So the comments and interpretations that may be made must be general rather than specific. None the less, some work has been done (e.g. Aldous, 1981). Perhaps most important in this context is that conditions leading to mixtures of Rasch models have been identified (Lauritzen, 1982).

From the above it may be seen that the general interpretation of a 'latent trait model' in the examination context would be the following. The examiners have some expectations about the data. These expectations, expressed as exchangeability conditions, demand that they represent their beliefs as a mixture of particular probability models. The parameters in these models, which are known as latent traits within the classical theory, have no physical significance; they are simply technical devices to ensure the exchangeability. If the examiners are to judge candidates fairly, it is important that their expectations are reasonable. Their reasonableness may be tested simply by fitting the model to the data and checking its goodness of fit: the Criticism phase of Bayesian analysis (Box, 1980). If the model fits, then they have reasonable expectations; but they should take care to note any outliers. Outliers may indicate either candidates who perform atypically and so will need careful, individual consideration or items which are atypical, perhaps easily misunderstood by certain categories of candidates.

Latent trait models are therefore important tools with which examiners may check their expectations. However, that is all they are. Parameters within the models do not have 'true' values and to estimate them is nonsense. Certainly estimates of 'ability' parameters do not provide measures of a candidate's performance. So when judging an individual candidate's performance, examiners should consider which items he got correct, not an estimated 'ability' parameter.

## THE FORM OF CANDIDATE RESPONSES

The above discussion assumes that candidates should say which answer to an item they believe to be correct. De Finetti (1965) suggested that, since they are unlikely to be certain in their beliefs, they should respond with their probability of the correctness of each possible answer. So that they are encouraged to reveal their beliefs honestly, he notes that they should know that their responses will be assessed by a proper scoring rule.

What can be said about this scheme? The first point to note is that its adoption would in no way invalidate the discussion of the previous section. One would need to adopt exchangeability conditions appropriate to a two-way layout in which the responses could be any value between 0 and 1, rather than being limited to two possible values: but that is all. The same interpretation of latent trait parameters, etc. would still hold.

Also, since candidates are allowed more possible responses, one might hope that the examiners can judge their performances better. But this will only be true if the candidates are fluent in the language in which they have to respond: that of uncertainty. Unfortunately few school leavers are. While Lindley (1984) is undoubtedly right that one of our most pressing needs in society is to train more people to appreciate uncertainty and coherent ways of reacting to it. I cannot but help feel that it is forlorn to hope that more than a minority of school leavers will understand the purpose of probability assessment sufficiently to answer items meaningfully. Part of the reason that multiple choice tests were introduced into the British public examination system was to allow candidates' knowledge and ability within a subject to be assessed, even though they might have poor powers of expression in English. Would weak candidates be better able to understand how to express their knowledge through probabilities? However, for the right candidature, knowing they had to sit multiple choice tests in this form might be an excellent incentive to learn to represent their strengths of belief as probabilities and to behave coherently. Indeed, in the Open University some courses on risk and professional judgement do use such tests and find many pedagogic advantages from doing so.

It is perhaps worth remarking at this point that, while it is clear to me that probabilities undoubtedly provide the most suitable framework in which an individual may think about his uncertainty (French, 1986), it is not clear to me that they necessarily are the most suitable medium in which he may communicate his uncertainty to another. Probabilities are subjective. They are part of an individual's thought processes. They are not part of a public language. Although they may have a role to play in communication, that is not their primary purpose and there is no reason to suppose that they are a particularly efficient means of communication. So asking candidates to reveal their knowledge and uncertainty by responding in terms of probabilities may not be quite such the ideal that it appears.

However, suppose that it is. Moreover, suppose that the candidates understand what is required of them and try to the best of their ability to answer in probabilities. Then there is still a difficulty. They will differ in their skills as probability assessors. Some will be better calibrated than others. This might not be a problem if it were possible to assess their calibration separately from assessing their performance in the subject examined. But such is unlikely to be the case. The fifty or so items asked in a typical test will not provide sufficient evidence to separate calibration effects from the substantive part of the candidate's performance, particularly since it is known that calibration in such tests depends on item difficulty (Lichtenstein et al, 1982). In short, one candidate might be graded higher than another in, say, biology simply because he was the better probability assessor. To be fair, whatever the style of questioning, there is always the problem that a candidate may be disproportionately rewarded for his examination technique: but, nevertheless, I do feel it is a significant issue here.

Apart from these practical difficulties in asking for probability responses, there is also a conceptual difficulty, at least there is in the scheme suggested by De Finetti. Who does the scoring rule belong to? De Finetti suggests that to encourage the candidate to state his probabilities honestly, he must be trained so that maximising a proper scoring rule becomes an end in itself. The rule must become his utility function. To ensure this, the rule must be used to give the aggregate mark on the test. But the aggregate mark on a component is a representation of the examiner's judgement of the quality of performance on that component (French, 1981, 1985). So the rule must also belong to the examiner: it represents one of his judgements. Why should it? Perhaps items in the test fall into a number of contexturally distinct areas and the examiner feels that a sound performance in a few of these areas is more worthy than a more mediocre performance in all of them. Modelling such judgements is unlikely to lead to a proper scoring rule. The examiner may not judge all the wrong answers in an item to be equally serious. Since scoring rules apply in circumstances in which one of a set of mutually exclusive events must happen, it is again not clear that the examiner's judgements would lead to a proper scoring rule.

However, this conceptual difficulty may disappear if one asks why the candidates need encouraging to give honest probabilities. When candidates are asked to write essays, examiners seldom consider that the candidates might deliberately or subconsciously misrepresent their knowledge. Why should they when responding in probabilities? Perhaps there is no need to tell candidates precisely what scoring rule will be used . . . or perhaps a Bayesian's awareness of the problems inherent in eliciting probability responses is pointing to problems that are also inherent when responses are elicited in other forms.

## REFERENCES


Aldous, D. J., 1981, Representations for partially exchangeable arrays of random variables, J. Mult. Anal., 11:581.

Aldous, D. J., 1985, Exchangeability and related topics, in "Ecole d'Ete de Probabilites de Saint-Flour XIII - 1983", D. J. Aldous, I. A. Ibragimov, J. Jacod, eds., Springer Verlag.

Box, G. E. P., 1980, Sampling and Bayes' inference in scientific modelling and robustness, J. Roy. Statist. Soc., A143:383.

Christie, T. and Forrest, G. M., 1981, "Defining Public Examination Standards", Macmillan.

Dawid, A. P., 1982, Intersubjective statistical models, in "Exchangeability in Probability and Statistics", G. Koch and F. Spizzichino, eds.

De Finetti, B., 1937, Foresight: its logical laws, its subjective sources (in French), English translation in "Studies in Subjective Probability", H. E. Kyburg and H. E. Smokler, eds., Wiley, 1964.

De Finetti, B., 1965, Methods for discriminating levels of partial knowledge concerning a test item, Brit. J. Math. Statist. Psych., 18:87.

Diaconis P. and Freedman, D., 1982, Partial exchangeability and sufficiency, to appear in Sankhya.

French, S., 1981, Measurement theory and examinations, Brit. J. Math. Statist. Psych., 34:38.

French, S., 1985, The weighting of examination components, The Statistician, 34:265.

French, S., 1986, "Decision Theory: An Introduction to the Mathematics of Rationality", Ellis Horwood.

French, S., Slater, J. B., Vassiloglou, M. and Willmott, A. S., 1986a, Descriptive and normative techniques in examination assessment, occasional publication, Oxford University Delegacy of Local Examinations (in press).

French, S., Slater, J. B., Vassiloglou, M. and Willmott, A. S., 1986b, The role of descriptive and normative techniques in examination assessment, submitted to a special issue of the Brit. J. Educ. Psych. on 'Evaluation and Assessment'.

Goldstein, H., 1979, Consequences of using the Rasch model for educational assessment, Brit. Educ. Res. J., 5:211.

Goldstein, M., 1986a, Exchangeable belief structures, J. Amer. Statist. Assoc. (to appear).

Goldstein, M., 1986b, Can we build a subjectivist statistical package? This volume.

Lauritzen, S. L., 1982, "Statistical Models as Extremal Families", Aalborg University Press.

Lindley, D. V., 1984, The next 50 years, J. Roy. Statist. Soc., A147:359.

Lord, F. M. and Novick, M. R., 1968, "Statistical Theories of Mental Test Scores", Addison-Wesley.

Mathews, J. C., 1985, "Examinations: A Commentary", George Allen and Unwin.
Vassiloglou, M., 1982, M.A. (Econ) Thesis, University of Manchester.
Vassiloglou, M., 1984, Some multiattribute models in examination assessment, Brit. J. Math. Statist. Psych., 37:216.
Vassiloglou, M., 1986, PhD. Thesis, University of Manchester.
Weiss, D. J. and Davison, M. L., 1981, Test theory and methods, Ann. Rev. Psych., 32:629.
Wood, R., 1978, Fitting the Rasch model: a heady tail, Brit. J. Math. Statist. Psych., 31:27.
Wright, B. D., 1977, Misunderstanding the Rasch model, J. Educ. Measur., 14:219.

# DYNAMIC INFERENCE ON SURVIVAL FUNCTIONS

Dani Gamerman

Department of Statistics
University of Warwick
Coventry CV4 7AL
England

ABSTRACT

The inference of survival functions based on information from censored observations is considered. The hazard function is assumed to be piecewise constant along intervals. The parameters are updated via a Bayesian conjugate analysis and information is passed through intervals via dynamic relations of the parameters. Inference is then made for the survival function of an individual (from the same population) conditioned on the observed data. Comparison with the product limit estimator, tools to criticise a model and some numerical examples are also provided.

## 1. INTRODUCTION

In this paper, I consider the statistical analysis of survival functions with no specific parametric family assumption. This problem has received a great deal of attention from both medical statistics and reliability areas and has often been referred to as nonparametric estimation. Here both words are to be avoided. The former because the Bayesian analysis pursued here gives meaning to the parameters modelling the sampling distribution and the latter because the problem is shown to be more of prediction than estimation. The distinction of this paper is in its use of a dynamic approach which filters the information collected up to a time passing it to future times.

The problem considered is that of a sample $Y=(Y_1,\ldots,Y_m)$ drawn from a population and interest lies in making inference about the populational survival (or reliability) function $S(u)=P(Y>u)$. The product-limit(PL) estimator (Kaplan & Meier, 1958) is obtained by estimating conditional probabilities at failure times by the observed conditional frequencies leading to

$$\hat{S}(u) = \prod_{i:y_i<u} \left[ 1 - \frac{d_i}{r_i} \right] \tag{1}$$

where $y_i$ is the $i^{th}$ ordered failure time and $d_i$ and $r_i$ are the number of observations that fail in $y_i$ and the number of observations that are known to be alive just prior to $y_i$, respectively. Susarla & Van Ryzin (1976) porposed a Bayesian analysis in which the survival function itself is treated as parameter and, after assuming a Dirichlet process prior, its posterior distribution can be obtained and inference made. This estimation approach to inference on survival functions is widely used in Bayesian statistics (Martz & Waller, 1982; Mashhoudy, 1985). It seems, however, that the survival function is a characteristic far more related to the individuals of the population than to the parameters of the model entertained. So, after observing $\underline{Y}=\underline{y}$ say, one should be obtaining

$$S(u|\underline{Y}=\underline{y}) = \int S(u|\underline{\theta},\underline{Y}=\underline{y})p(\underline{\theta}|\underline{Y}=\underline{y})d\underline{\theta} \tag{2}$$

for a new individual rather than the posterior distributions of $S(u|\underline{\theta})$ as functions of $\underline{\theta}$, the parameter of the model, for all u.

Both Susarla & Van Ryzin and the PL estimators have jumps at the observed failure times, although the former is smoother than the piecewise constant PL estimator, and they are dependent on a time factorisation at these points. In order to avoid these problems, Kalbfleisch & Prentice (1973) put forward a model for the observations that partitions the interval $(0,\infty)$ in intervals $I_1=(t_0,t_1)$, $I_i=(t_{i-1},t_i]$, $i=2,\ldots,n-1$ and $I_n=(t_{n-1},\infty)$, $(0=t_0<t_1<\ldots<t_{n-1})$, having constant hazards $\lambda_1,\ldots,\lambda_n$, respectively. This model is adopted here. In addition, the hazard function is generally expected to be smooth and therefore information gathered in an interval should exert some influence in the adjacent intervals. The dynamic approach enables the flow of information through intervals and smoothness assumptions on the hazard function can be implemented via dynamic relations between the $\lambda$'s.

## 2. OUTLINE OF THE ANALYSIS

Censoring is the main feature of reliability and survival data. In the former, data is often collected from industrial experiments that are designed to stop after the $k^{th}$ observed failure or after a certain amount of time (Barlow & Wu, 1981). In medical data, random censoring occurs due to loss to follow-up of patients being discharged from hospital, moving to other places and others.

In all those cases and indeed in the case of complete sample, the likelihood is

$$\prod_{i=1}^{n} \lambda_i^{d_i} e^{-\lambda_i a_i} \tag{3}$$

where $d_i$ and $a_i$ are the number of failures observed in $I_i$ and total observed time through $I_i$. This likelihood is a product of likelihood factors for each $\lambda_i$ (based on information collected in $I_i$) conditioned on prior information.

The analysis is such that information is passed sequentially through intervals. Also, the $\lambda$'s are assumed to be (marginally) Gamma distributed so that after updated they remain Gamma distributed due to the form of the likelihood. Let $[\lambda_{i-1}|D_{i-1}]\sim G(\alpha_{i-1},\gamma_{i-1})$ where $D_i$ is the information collected up to the end of interval $I_i$. The (Gamma) distribution for $\lambda_i$ is constructed in such a way that it retains the mean of $\lambda_{i-1}$ reflecting one's expectations about the smoothness of the sampling distribution in the absence of any other relevant information but has a larger variance to account for extra uncertainty as the analysis evolves in time. This implies that $[\lambda_i|D_{i-1}]\sim G(c_i\alpha_{i-1},c_i\gamma_{i-1})$, $c_i<1$ and after updating through the likelihood, one gets $\alpha_i = c_i\alpha_{i-1}+d_i$ and $\gamma_i=c_i\gamma_{i-1}+a_i$ and the cycle can restart for next interval. The values of the c's should take into account things like interval lengths and plausible expressions are suggested in the sequel.

The survival function for an individual with failure time X drawn from the same population is

$$S(x|D_n)=S(x|X>t_{i-1},D_n)\prod_{j=1}^{i-1} S(t_j|X>t_{j-1},D_n),\ x\in I_i \tag{4}$$

$$=\int_0^\infty e^{-\lambda_i(x-t_{i-1})}p(\lambda_i|D_n,X>t_{i-1})d\lambda_i \prod_{j=1}^{i-1}\int_0^\infty e^{-\lambda_j(t_j-t_{j-1})}p(\lambda_j|D_n,X>t_{j-1})d\lambda_j$$

because the distribution of X does not depend on $D_n$, the total information obtained from observing Y. The distributions for $[\lambda_i|D_n,X>t_{i-1}]$ have not been specified yet but are provided in section 5.

## 3. MODEL CRITICISM

A particular model can be assessed by its marginal likelihood. This is obtained after integrating out the parameters from the likelihood and gives the relative likelihoods of different entertained models by comparison. It can be readily obtained after integrating (3) with respect to the prior distributions for $[\lambda_i|D_{i-1}]$ giving

$$\prod_{i=1}^{n}\left\{(c_i\gamma_{i-1}+a_i)^{-d_i}\left[1+\frac{a_i}{c_i\gamma_{i-1}}\right]^{-a_i}\prod_{j=1}^{d_i}(c_i\alpha_{i-1}+j-1)\right\} \tag{5}$$

This is the main tool to criticise a model although some graphical comparisons could help. One could be interested in monitoring the smoothness of the survival of the population in which case a plot of the prediction obtained from different models is useful. Also, in some special cases, agreement with the data itself can be checked and this is related to some model assumptions.

## 4. COMPARISON WITH THE PL ESTIMATOR

In order to make this comparison, one has to assume that no initial information is available, there is no passage of information through different intervals and that the intervals $I_i$, $i=1,\ldots,n$ are determined by the ordered failure times. Those assumptions are implicit in the derivation of the PL estimator. They imply that the only relevant information for $\lambda_i$ is in its likelihood so that $[\lambda_i|D_n,X>t_{i-1}]\sim[\lambda_i|D_i-D_{i-1}]\sim G(d_i,a_i)$ since $c_i\to 0$. The survival function for $x\in I_i$ is

$$\left[1+\frac{x-y_{i-1}}{a_i}\right]^{-d_i}\prod_{j=1}^{i-1}\left[1+\frac{b_j}{a_j}\right]^{-d_j} \tag{6}$$

where $b_j$ = *length* $(I_j)$. This is a strictly decreasing continuous function with piecewise continuous first derivative. For the sake of clarity suppose further that all censored times coincide with any one of the (uncensored) failure times, so that $a_i = b_i r_i$. In moderate and large samples, each of the multiplying factors in (6) can be approximated by the leading terms in its Taylor expansion for $r_k^{-1}$ around 0, $k \le i$. This gives

$$\left[1-\frac{d_i}{r_i}\left(\frac{x-y_{i-1}}{b_i}\right)\right]\prod_{j=1}^{i-1}\left[1-\frac{d_j}{r_j}\right] \tag{7}$$

Comparing (7) with (1), it can be seen that this non-informative approach is approximately equal to the PL estimate for $x \to y_i$ from the right, for all i. Elsewhere, it is an approximately piecewise linear function joining these points whereas the PL estimate change by jumps at failure times. The approximation (7) gets poorer as k approaches n (see figure 2) and as sample size decreases.

These non-informative models provide mainly a smooth continuous version of the PL estimate which has merit on its own. They can not, however, be compared with other ones in the terms of section 3 due to the extremeness of its prior assumptions. Dynamic models avoid those problems while offering a wider choice to the modeller.

## 5. INFERENCE FOR THE SURVIVAL FUNCTION

As was previously said, the distribution of $[\lambda_i | D_n, X>t_{i-1}]$, for all i are needed for the evaluaiton of the predictive survival function. A specific stochastic model is necessary to establish the relations between the $\lambda$'s. To do this directly via a joint distribution for the $\lambda$'s would impose unnecessary numerical complications and a simple alternative is proposed.

One solution can be obtained using the structure of dynamic survival analysis developed by Gamerman (1985). This analysis is designed to be used with covariates and the study of random samples can be treated as a special case when the only covariate takes the value 1 for all observations. There, the system parameter $\eta$ undergoes a random walk $\eta_i=\eta_{i-1}+w_i$ where the w's are independent errors with zero mean and variances $W_i$, $i=1,\ldots,n$, respectively. The $\eta$'s are only partially defined through their means and variances that are obtained by relating $\eta$ to log $\lambda$. This transformation is used to minimise the effect of skewness of the Gamma distribution in the evolution. A linear approximation as in West and Harrison (1986) can be used first to construct $(\eta_{i-1}|D_{i-1}]\sim[m_{i-1},P_{i-1}]$ where $m_{i-1} = \ln(\alpha_{i-1}/\gamma_{i-1})$ and $P_{i-1} = \alpha_{i-1}^{-1}$ and then, after evolution, back to $[\lambda_i|D_{i-1}]\sim G[(P_{i-1}+W_i)^{-1},(P_{i-1}+W_i)^{-1}\exp(-m_{i-1})]$ implying the values of $c_i = P_{i-1}/(P_{i-1}+W_i) = (1+\alpha_{i-1}W_i)^{-1}$. Recursive smoothing is then used to obtain $[\eta_i|D_n]\sim[m_i^n,P_i^n]$ with

$$\begin{aligned} m_i^n &= m_i+c_{i+1}(m_{i+1}^n-m_i) \\ P_i^n &= P_i-c_{i+1}^2(P_i+W_{i+1}-P_{i+1}^n). \end{aligned} \tag{8}$$

This gives $[\lambda_i|D_n]\sim G(\alpha_i^n,\gamma_i^n)$ with $\alpha_i^n = (P_i^n)^{-1}$ and $\gamma_i^n = (P_i^n)^{-1}\exp(-m_i^n)$ and those are in fact the distributions that are used in the applications since the contributions of the events $[X>t_{i-1}]$ are negligible. They can be replaced in (4) giving, after integration,

$$S(x|D_n) \doteq \left[1 + \frac{x-t_{i-1}}{\gamma_i^n}\right]^{-\alpha_i^n} \prod_{j=1}^{i-1} \left[1+ \frac{b_j}{\gamma_j^n}\right]^{-\alpha_j^n}, \text{ for } x \in I_i \qquad (9)$$

## 6. NUMERICAL EXAMPLES

In all the examples, analysis starts with a vague prior $[\lambda_0|D_0] \sim G(\delta,\delta)$ with $\delta = 10^{-3}$. This implies $V[\lambda_0|D_0] = 10^3$ representing ignorance as to the value of $\lambda_0$. Also the values of $W_i$ for the evolution are taken as proportional to $b_i$. This is in line with an equivalent continuous-time model for $\eta$ having a Brownian motion process (Cox & Miller, 1965, pg. 206).

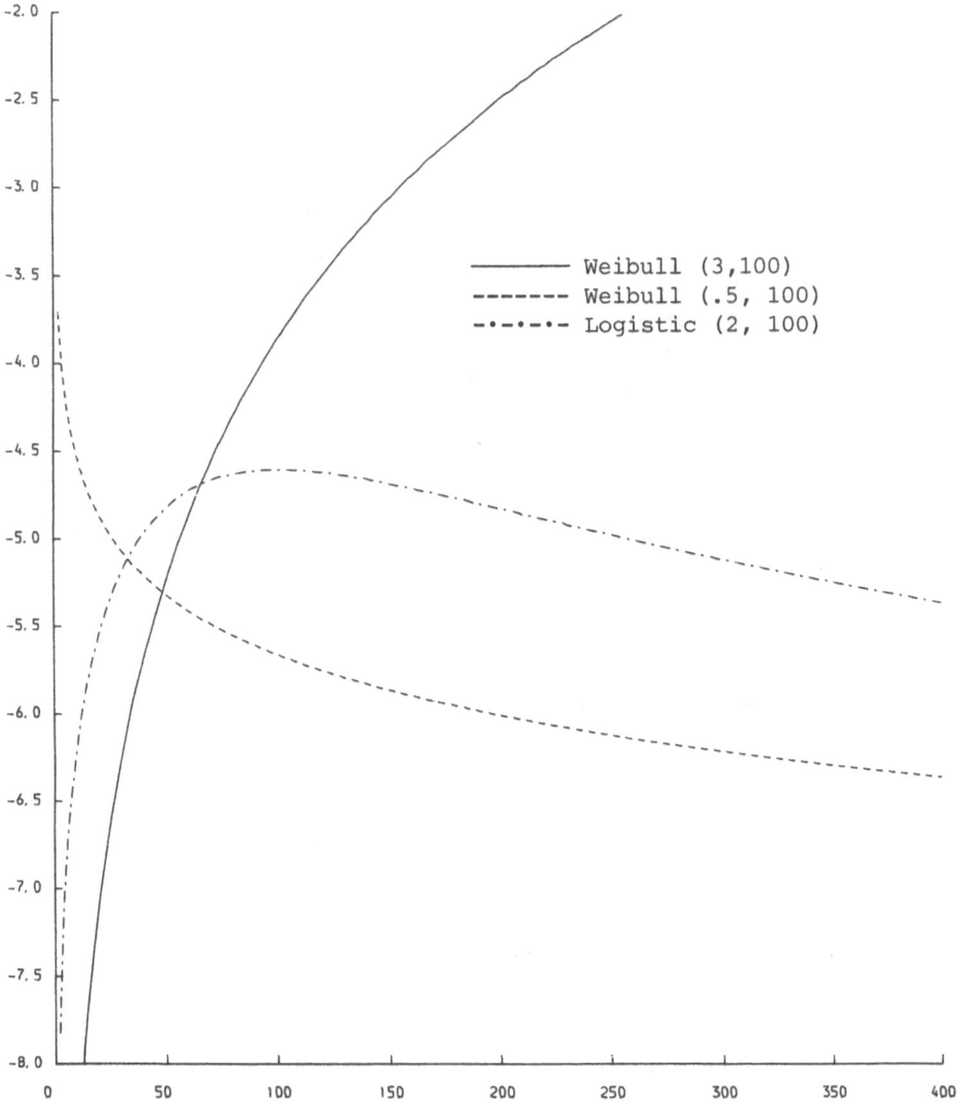


Figure 1. Hazard functions (in the log scale)

A sample of size 30 was generated from the Weibull (3,100), Weibull (.5,100) and logistic (2,100) distributions. These have hazard functions

$$.03(\log 2)\left[\frac{t}{100}\right]^2, \quad .005(\log 2)\left[\frac{t}{100}\right]^{-.5} \quad \text{and} \quad .02\,\frac{(t/100)}{1+(t/100)^2}, \; t>0$$

respectively, shown in Figure 1. Their common median (and scale parameter) is 100 and they represent quite different failure patterns. For each sample, a dynamic, a non-informative and the PL survival functions were obtained and are shown in Figure 2, along with the respective generating survival functions. The comparisons made in section 3 can be best appreciated in the figures. It is clear that although the non-informative model smooths the PL estimator, a proper smooth solution can only be obtained through a dynamic model. The values for $W_i$ ($.2b_i$, $.1b_i$ and $.05b_i$, respectively) were set on an illustrative basis and can be changed at the modeller's will giving more flexibility to the inference. The marginal likelihood can be called to assess model choices with respect to factors like interval lengths and values of the W's.

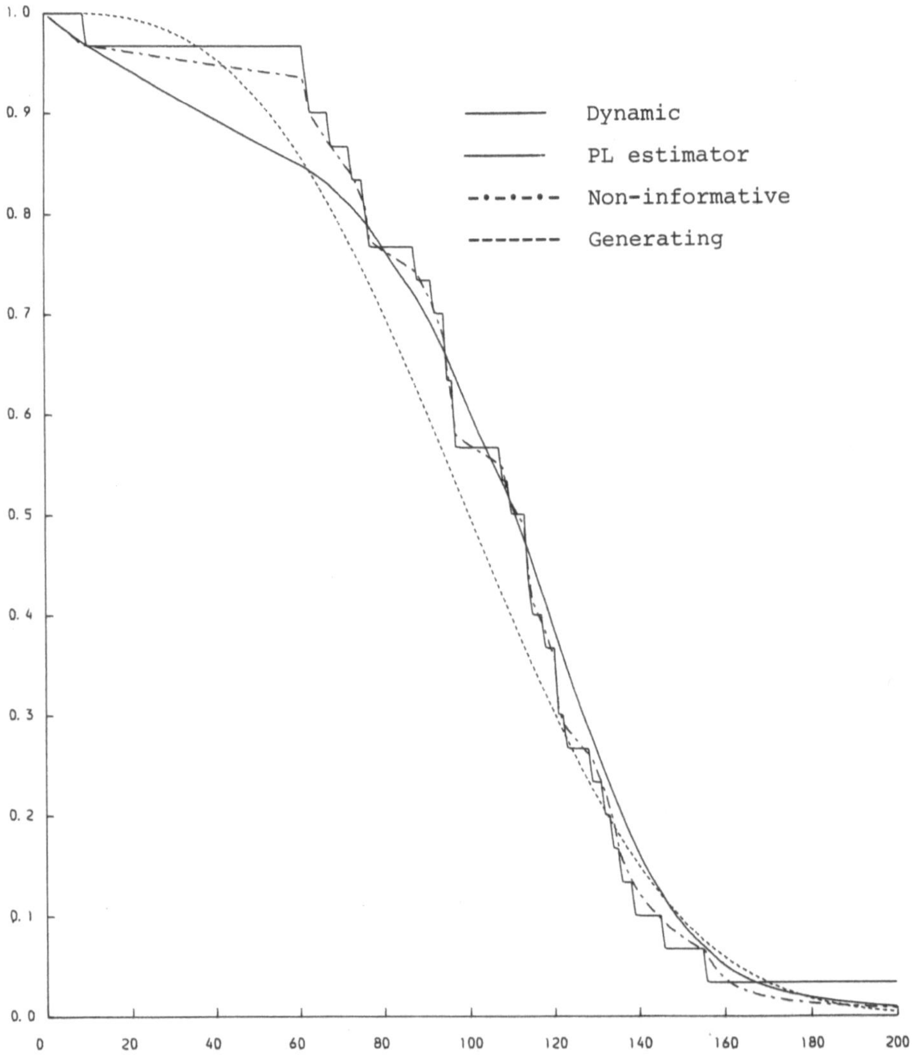


Figure 2(i). Survival functions for the Weibull (3,100).

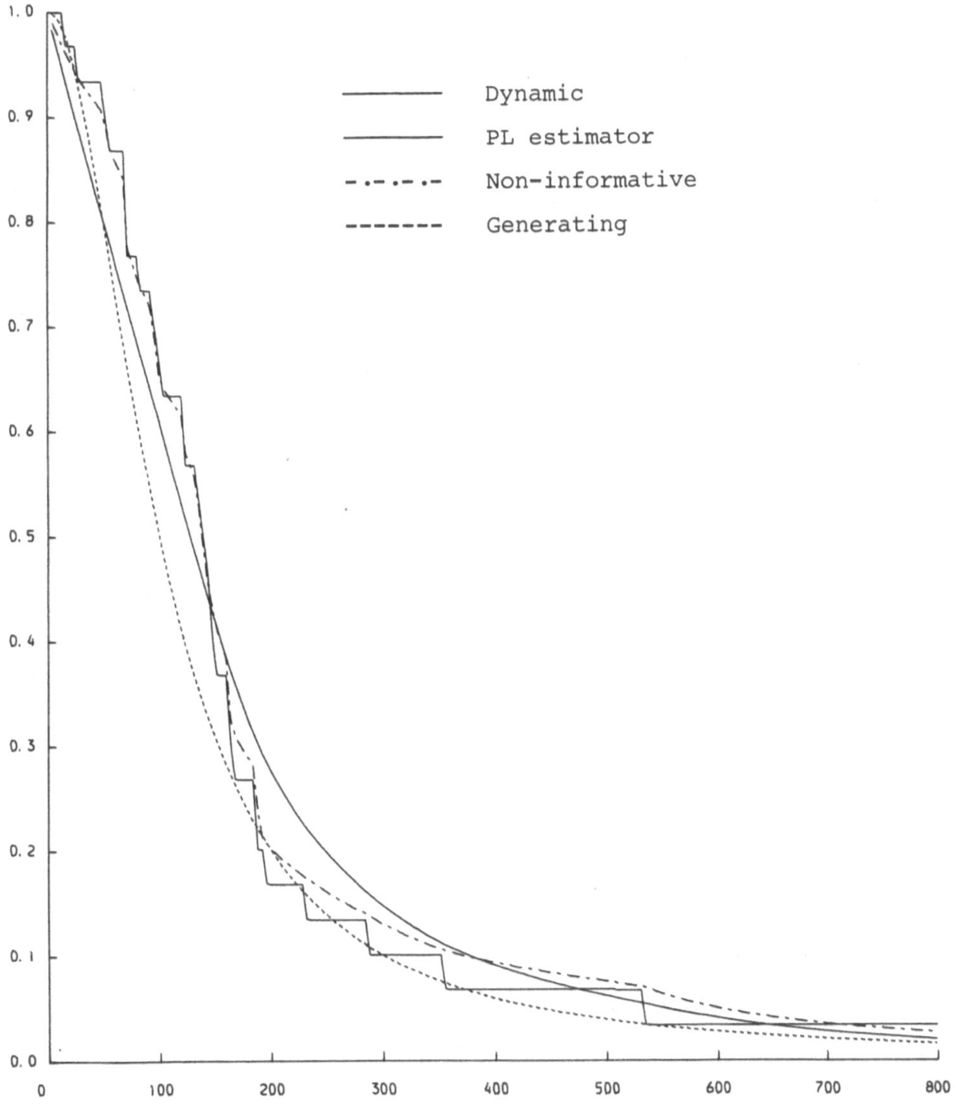


Figure 2(ii). Survival functions for the Logistic (2,100).

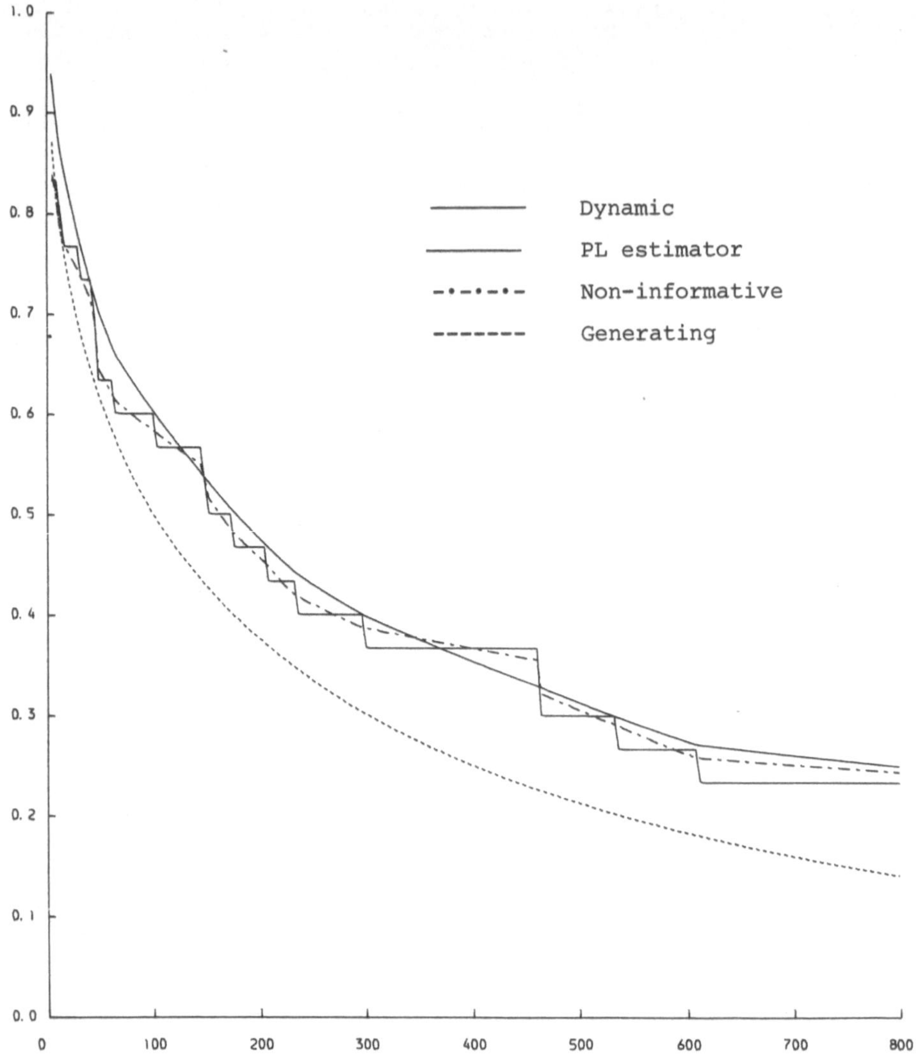


Figure 2(iii). Survival functions for the Weibull (.5,100).

The leukemia data from Gehan (1965) can be used to illustrate this point. They consist of remission times of patients in two groups (treatment and control) of equal size with heavy censoring in the treatment group (12 censored out of 21 patients). Each group is analysed here independently of the other. For each of them, a number of combinations of interval divisions and values for W are used and the corresponding marginal likelihood for each model are given in Tables 1 and 2.

It seems that a finely divided grid of points gives a better fit at least for this set of data. Also there is some preference manifested for the values $W_i = 0$ corresponding to $c_i = 1$. This supports the model with no evolution for $\lambda$, that is, exponential distribution. In this special case, there is no need for a dynamic analysis since $\lambda_i = \lambda$ and $[\lambda|D_n] \sim G(d,a)$ where $d = \sum_{i=1}^{n} d_i$, $a = \sum_{i=1}^{n} a_i$ and the predictive survival for time t is $[1+(t/a)]^{-d}$. The value otherwise obtained with a dynamic analysis

$$\left[1+\frac{t-t_{i-1}}{a}\right]^{-d} \prod_{j=1}^{i-1}\left[1+\frac{b_j}{a}\right]^{-d} , \quad t \in I_i$$

is close to that one specially for small values of the b's.

**Table 1**

Marginal likelihoods for treatment group

Data :6,6,6,6*,7,9*,10,10*,11*,13,16,17*,19*,20*,22,23,25*,32*,32*,34*,35*

| $W_i$ / $b_i$ | Interval division | | |
|---|---|---|---|
| | {uncensored times} | {5,10,15,20,25,30,35} | {1,2,3,...,34,35} |
| 0 | -48.11 | -49.26 | -49.26 |
| .05 | -48.71 | -49.90 | -49.75 |
| .10 | -49.13 | -50.29 | -50.07 |
| .20 | -49.78 | -50.85 | -50.54 |

*- censored

**Table 2**

Marginal likelihoods for control group

Data : 1,1,2,2,3,4,4,5,5,8,8,8,8,11,11,12,12,15,17,22,23

| $W_i$ / $b_i$ | Interval division | | |
|---|---|---|---|
| | {uncensored times} | {4,8,12,16,20,24} | {1,2,3,...,22,23} |
| 0 | -73.86 | -73.86 | -73.86 |
| .05 | -74.92 | -74.68 | -74.54 |
| .10 | -75.68 | -75.25 | -75.11 |
| .20 | -76.82 | -76.09 | -76.01 |

REFERENCES

Barlow, R. E. & Wu, A. S., 1981, Preposterior analysis of Bayes estimators of mean life, Biometrika, 68, 403-410.

Cox, D. R. & Miller, H. D., 1985, "The Theory of Stochastic Processes," London: Methuen.

Gamerman, D., 1985, Dynamic Bayesian models for survival data, Warwick Research Report No. 75.
Gehan, E. A., 1965, A generalized Wilcoxon test for comparing arbitrarily single-censored samples, Biometrika, 52, 203-224.
Kalbfleisch, J. D. & Prentice, R. L., 1973, Marginal likelihoods based on Cox's regression and life model, Biometrika, 60, 267-278.
Kaplan, E. L. & Meier, P., 1958, Nonparametric estimation from incomplete observations, J. Amer. Statist. Assoc., 53, 457-481.
Martz, H. F. and Waller, R. A., 1982, "Bayesian Reliability Analysis," New York: John Wiley.
Mashhoudy, H., 1985, Bayesian reliability growth analysis, Unpublished Ph.D. thesis, University of Warwick.
Susarla, V. & Van Ryzin, J., 1976, Nonparametric Bayesian estimation of survival curves from incomplete observations, J. Amer. Statist. Assoc., 71, 897-902.
West, M. & Harrison, P. J., 1986, Monitoring and adaptation in Bayesian forecasting models, J. Amer. Statist. Assoc. to appear.

# THE ROLE OF PROBABILITY AND STATISTICS IN PHYSICS

Gerhard Gerlich

Institut für Mathematische Physik
Technische Universität Braunschweig
Mendelssohnstr. 3, D-3300 Braunschweig

## INTRODUCTION

There are extensive fields in physics where probability theory and statistics are of great importance. It cannot be the intention of this lecture to describe the role of probability and statistics in the whole field of statistical physics.

Many parts of physics have got special statistical fields: statistical mechanics (ergodic theory, quantum statistics), statistical (magneto)hydrodynamics (theory of turbulence), statistical plasmaphysics (kinetic theory), statistical optics (partial coherence). Though all these fields are my specialities, I do not intend to talk about the success or failures of the probabilistic and statistical descriptions in these fields of physics.

The aim of my lecture is to show you that probability theory and statistics provide us with the essential mathematical structures in those fields of physics which do not have the attribute "statistical": axiomatic classical mechanics and axiomatic quantum theory. For the latter, it seems to be evident that probability and statistics play an important role as all students must learn the "statistical" or "probabilistic interpretation" of quantum theory. Nonetheless, as a student I had difficulty in connecting the mathematical structures of quantum theory on the one side and probability theory and statistics on the other side.

I suppose that it could be useful to tell you the story how I started my investigations of the mathematical foundations of quantum theory. I was concerned with the theory of the hydromagnetic dynamo (generation of the magnetic field of the earth or the sun). Very soon one could see that modern theories of the dynamo make an essential use of the statistical theory of turbulence. It is wellknown that the statistical theory of turbulence is very poor in successful mathematical technics whereas quantum theory has a lot of them giving accurate and excellent descriptions of very precise measurements. Therefore I told myself that it should only be neccessary to find out the mathematical statistical structure of quantum theory which is independent of the special physical situation. Then one should have a successful mathematical structure for the statistical theory of turbulence. I think that I found this mathematical structure. The result is a new system of axioms (laws) for classical and quantum mechanics containing both fields as special cases. The new aspect of this system is that one adds three statistical axioms to the usual (mechanical or quantum mechanical) axioms (for instance to the Newtonian axioms).

But what were the results for the statistical theory of turbulence? They are a great disappointment: the mathematical statistical structure of quantum theory is useless for the statistical theory of turbulence. At a first glance the cause sounds rather paradoxical: the probabilistic or statistical structure of quantum theory is more general than the conventional measure theoretic probability theory. The models of quantum theory give more precise probability distributions than the conventional probability theory. Using this structure in the statistical theory of turbulence is like using a steam hammer to crack a nut. In this connection the following historical remark could be of interest. The mathematically rigorously founded probability theory was published by A. N. Kolmogoroff in 1933 after the famous book of J. v. Neumann in 1932. Perhaps it is worthwhile to notice that, though Lebesgue measure was known for many years, a general "abstract" measure theory for $\sigma$-finite measures was established in those years. I suppose that this theory was first written down by E. Hopf in his famous book "Ergodentheorie"(1937). Thus the rigorously founded "classical" probability theory followed the "nonclassical" quantum theory. Therefore the following observation has a rather simple explanation. In the book of J. v. Neumann you can find only a few passages where measure theoretic concepts were used (the separability of the $L^2$- spaces, equivalence of the Heisenberg and Schrödinger picture of quantum theory). His intention was to give an algebraic and geometric formulation of the mathematical foundations of quantum theory.

My approach to the mathematical foundations of quantum theory is formulated with the measure theoretic probability theory, though the resulting structure is a bit more general. The generalization concerns a formula which is very closely related to Bayes' formula the thematic connecting link of this symposium.

## THE STATISTICAL AXIOMS

Let us begin with the three statistical axioms:

(A1) The statements of physics are statements about spaces of events. These statements about events are formulated with probability measures for <u>pairs of events</u>: q(A,B) is the probability observing the event B if one knows the event A. q(A,B) is called transition probability.

For the axiom (A2) one needs some technical definitions. They are not difficult, but awkward. I hope that in this lecture a very simplified definition is sufficient.

(D) We call a model for a physical experiment a unitary pair if it is possible to write the formula of the transition probability in the form

$$q(A,B) = \frac{\langle U(\chi_A) | P_B U(\chi_A) \rangle}{\langle U(\chi_A) | U(\chi_A) \rangle}$$

where U is a unitary map of the natural Hilbert spaces.

$\chi_A$: characteristic function of a set A, $\nu$ measure on a $\sigma$-algebra $\underline{A}$ over X, $\mu$ measure on a $\sigma$-algebra $\underline{B}$ over Y, natural Hilbert spaces $L^2(X,\nu)$, $L^2(Y,\mu)$, $P_B\psi|_y = \chi_B(y)\psi(y)$.

(A2) General probability distributions are given by (convex linear combinations of) the transition probability of a unitary pair.

(A3) The time dependence of unitary pairs is given by

$$i\hbar \frac{\partial U_t}{\partial t} = H_t \circ U_t$$

$H_t$ being (essential) self-adjoint operators defined on the natural Hilbert space in the decision.

The first axiom summarizes two empirical facts. Every physical theory is finally tested by reading numbers of a scale with error bounds. We call this the observation of events in the decision, where events are elements of a class of subsets of a certain set; in this example the observed event is a certain interval of $R^1$ (real line). The value of such measured numbers is meaningless if one does not know how the experiment was performed. In particular, this knowledge could be given by an observed or assumed event (in the condition). An event A one knows to predict the event B. You could consider the motion of a car on an inclined plane or the movement of the planets. The measurements of the space and velocity coordinates alone are not physics. Physics begins to predict the values at a later instant of time (with error bounds) with a model. Statements of measured values with error bounds are typical probability statements. For instance, with probability o.999 the measured value should be found in a certain interval performing the experiment the same way. This means that you have to choose the same event in the condition.

Spaces of events are sets X respectively Y with $\sigma$-algebras $\underline{A}$ respectively $\underline{B}$. Elements of $\sigma$-algebras are usually called measureable sets or, in the measure theoretic probability theory, events. In this sense, in (A1), very conventional concepts of the measure theoretic probability theory founded by Kolmogoroff are used. Only the concept of the transition probability q(A,B) is introduced as a fundamental concept and not as a derived concept.

In conventional probability theory, with two spaces of events $(X,\underline{A})$ and $(Y,\underline{B})$ one constructs a new common space of events $(X \times Y, \underline{A} \otimes \underline{B})$. $X \times Y$ is the cartesian product of the sets,

$\underline{A} \otimes \underline{B}$ the product σ-algebra generated by the sets $A \times B$ with $A \varepsilon \underline{A}$, $B \varepsilon \underline{B}$. The events A are replaced by $A \times Y$, the events B are replaced by $X \times B$. With a probability measure P on $\underline{A} \otimes \underline{B}$ one calculates the transition probability as a conditional probability

$$q_{cl}(A,B) = \frac{P(A \times B)}{P(A \times Y)}$$ (This gives Bayes' formula).

For $A \cap A' = \emptyset$ you get

$$q_{cl}(A \cup A', B) = \lambda_1 q_{cl}(A,B) + \lambda_2 q_{cl}(A',B) \quad \text{with} \quad \lambda_1 + \lambda_2 = 1,\ 0 \le \lambda_1, \lambda_2 \le 1.$$

(CL) The transition probability of a union of disjoint events in the condition is a convex linear combination of the individual classical transition probabilities.

I hope that you will allow me a short digression. Then I can touch the fields of physics I had originally excluded in my introduction. I would like you to remember that in some sense you can read the theory of stochastic processes as a theory of hidden variables.

Often one can find the following description of a stochastic process (1st picture): A stochastic process is not an ordinary function of time such that you have for an arbitrary set of n time values (for all n) the n function values

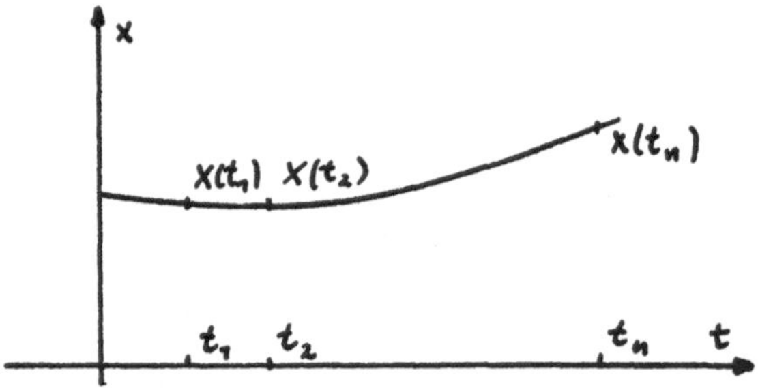


but you only know the probability distributions for the function values. You have the probability that the values x are elements of certain sets (I call them windows) at an arbitrary set of n time values.

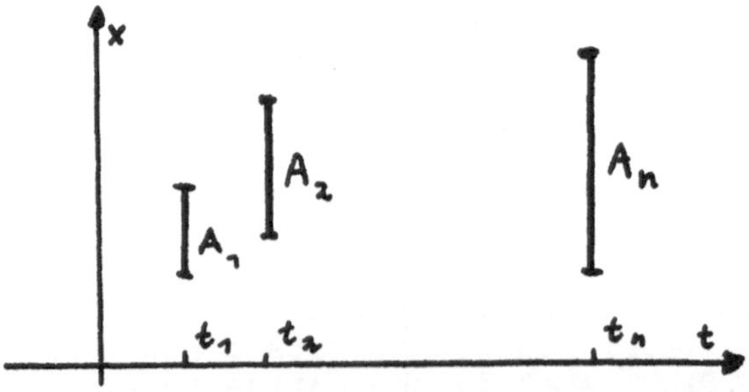

For instance you can write with densities

$$P(t_1,A_1;t_2,A_2;\cdots;t_n,A_n) = \int\limits_{A_1}\int\limits_{A_2}\cdots\int\limits_{A_n} f(t_1,x_1,t_2,x_2,\dots,t_n,x_n)\,dx_1dx_2\cdots dx_n\,.$$

What has produced these probabilities? This gives us the second picture of a stochastic process (Kolmogoroff's definition). We assume that the probability distribution is produced by a set of curves (paths) meeting the windows with a certain probability distribution. We index each curve with the parameter $\omega$ . The value of this parameter is usually unknown. I call $\omega$ the hidden variable of the stochastic process.

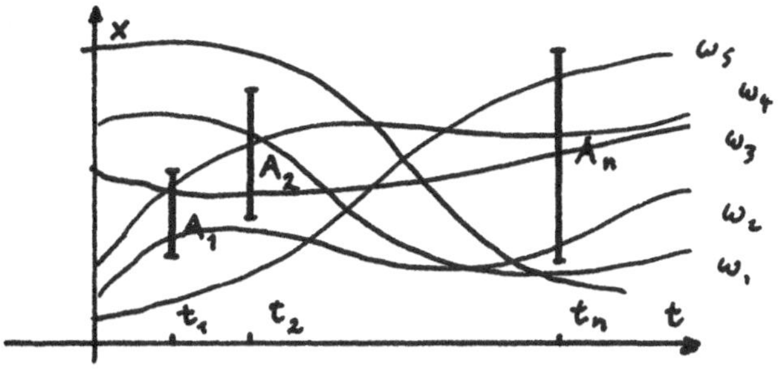


In this picture a stochastic process is a function depending on two variables: the time $t$ and the hidden variable $\omega$ , for which a probability distribution is given. We write for the function values $x_t(\omega)$. It is a wellknown fact:

KOLMOGOROFF'S THEOREM

Under certain mathematical assumptions both pictures are equivalent.


Résumé:
If you use Kolmogoroff's definition of a stochastic process, you use a certain theory of hidden variables, which are elements of <u>one</u> probability space $\Omega$.


A theory of stochastic processes (with differential paths!) is hidden behind the following catchwords:

Liouville equation, (n-time) BBGKY-hierarchy, classical Zwanzig formalism, Mori formalism, microscopical density correlation function, response function, van Hove's scattering formula, test particle diffusion, microscopical fluctuations, statistical theory of turbulence (E. Hopf), ergodic theory, Vlasov equation, Klimontovich formalism, Rostoker formalism, generalized Stratonovich method (best method for systems governed by a system of stochastic differential equations for short prediction times).

Let $x_t^i(\omega)$ be a family of stochastic processes on <u>one</u> probability space $(\Omega,\underline{A},P)$. Then you can take as a rather general formula for a transition probability

$$\bar{q}_{Cl}(A,B) = \frac{\sum_{i,j} w_{ij} P(x_{t_1}^{i^{-1}}(A) \cap x_{t_2}^{j^{-1}}(B))}{\sum_i w_i x_{t_1}^{i^{-1}}(A)}$$

with $0 \leq w_{ij} \leq 1$, $w_i = \sum_j w_{ij}$, $\sum_i w_i = 1$.

Also with this formula you can prove (CL). For the transition probability of (A2) this formula is wrong:

$$A \cap A' = \emptyset, \quad A \cup A' = X, \quad B \cap B' = \emptyset, \quad B \cup B' = Y$$

$$\mu(B) = \mu(B') = \nu(A) = \nu(A') = \frac{1}{2}$$

$$U(\chi_A) = \cos\varphi \, \chi_B + \sin\varphi \, \chi_{B'}$$

$$U(\chi_{A'}) = -\sin\varphi \, \chi_B + \cos\varphi \, \chi_{B'}$$

$$q(A,B) = \cos^2\varphi, \quad q(A',B) = \sin^2\varphi, \quad q(A,B') = \sin^2\varphi$$

$$q(A \cup A', B) = \frac{1}{2} - \frac{1}{2}\sin 2\varphi \ (= 0 \text{ for } \varphi = \frac{\pi}{4})$$

$$\lambda_1 q(A,B) + \lambda_2 q(A',B) = \lambda_1 \cos^2\varphi + \lambda_2 \sin^2\varphi \ \ (= \frac{1}{2} \text{ for } \varphi = \frac{\pi}{4})$$

Now we have our non-existence statement about quantum theories with hidden variables:

The conventional theory of stochastic processes is not sufficient to give the formulas for the quantum theoretic transition probabilities.

The precise predictions of the quantum theoretic models are the problem, not certain uncertainty relations!

Our axiom (A2) suggests a formulation of Dirac's superposition principle:

(AQSP) The unitary maps of the unitary pairs can be written as integral transformations of the natural Hilbert spaces of the events.

One should notice that our formulation of Dirac's superposition principle does not contain the assumption that all operators of the mathematical model can be written as integral transformations which was criticized by J. v. Neumann in Dirac's representation of the mathematical structure of quantum theory. Restricting this property to the operators of the unitary pairs I consider this assumption justifiable.

Let us go back to the first axiom. It contains an algebraic structure which can be worked out. Unlike classical probability theory, the pairs of events (A,B) are not automatically identified with the cartesian products $A \times B$. Usually incompatible events are disjoint sets. Therefore one could ask how one could define the incompatibility of pairs of events. Such a situation is given in the physical praxis if an experimental physicist says that he cannot verify the experiment of his colleague. Excluding events in the decision are meaningful only if the events in the condition are equal (meaningful comparison):

(MC) Pairs of events are comparable iff the events in the condition are equal.

With this, in the cartesian product of the σ-algebras $\underline{A} \times \underline{B}$ one can define a partial ordering $\trianglelefteq$ setting

(PO) $(A,B) \trianglelefteq (A',B')$ iff $A=A'$ and $B \cap B'=B$,

an orthocomplementation by

(POC) $(A,B)^{\perp} = (A, \complement B)$ ,

and an equivalence relation for the impossible and sure pairs of events

$$(A,B) \sim (A',B') \Leftrightarrow (B=B'=\emptyset \text{ or } B=B'=Y) \text{ or } (A=A' \text{ and } B=B').$$

With $\vee$ as the supremum and $\wedge$ as the infimum this poset is an orthocomplemented, orthomodular, quasimodular, not modular, not distributive lattice. Usually these are the general properties which were listed for a "quantum logic", the lattice of all closed linear subspaces of an infinite dimensional Hilbert space. But this is not the same lattice (the covering law is missing).

## THE PHYSICAL AXIOMS

If one intends to reproduce the conventional theories of classical mechanics and quantum theory, one has to treat these fields with different formulations of the next axioms. These axioms fix the spaces of events for particles, thus the natural Hilbert spaces and the self-adjoint operators H.

(AQS1)
Schrödinger's equation without and with external fields
(AQS2)

(AQD1)
Dirac's eqation without and with external fields
(AQD2)

(AMN1)
Newton's equation without and with external fields
(AMN2) (1st and 2nd axiom)

(AQS1) The space of events of the space measurements of a free particle is $R^3$ with the Lebesgue measure $\mu_L$ on the Lebesgue σ-algebra $\underline{A}_L$. The self-adjoint operator of the time evolution for this system is given by

$$H = -\frac{\hbar^2}{2m}\Delta\ , \quad \Delta = \frac{\partial^2}{\partial x^2} + \frac{\partial^2}{\partial y^2} + \frac{\partial^2}{\partial z^2}$$

defined on a suitable dense linear subspace of the natural Hilbert space $L^2(\mu_L, R^3)$.

(AQS2) The operator of the time evolution of a particle of charge Ze and mass m in an external electromagnetic field with the potential $(\underline{A}, -V)$ is given by

$$H = \frac{1}{2m}(-i\hbar\nabla - Ze\underline{A})\cdot(-i\hbar\nabla - Ze\underline{A}) + ZeV$$

defined on a suitable dense linear subspace of the natural Hilbert space defined in (AQS1)(with the same space of events).

(AQD1) The space of events of the space measurements of a free electron is $R^3$ with the Lebesgue measure $\mu_L$ on the Lebesgue $\sigma$-algebra $\underline{A}_L$. The operator of the time evolution for this system is given by

$$H = m_o c^2 \beta - ic\hbar\underline{\alpha}\cdot\nabla$$

defined on a suitable dense linear subspace of the $C^4$-valued natural Hilbert space.

(AQD2) The operator of the time evolution of an electron in an external electromagnetic field $(\underline{A}, -V)$ is given by

$$H = m_o c^2 \beta - ic\hbar\underline{\alpha}\cdot\nabla + ec\underline{\alpha}\cdot\underline{A} - eV$$

defined on a suitable dense linear subspace of the natural Hilbert space defined in (AQD1)(with the same space of events).

(AMN1) The space of events of the space and velocity measurements of a free particle is $R^6$ with the Lebesgue measure $\mu_L$ on the Lebesgue $\sigma$-algebra $\underline{A}_L$. The self-adjoint operator of the time evolution is given by

$$H = -i\hbar\underline{v}\cdot\nabla\ , \quad \underline{v}\cdot\nabla = v_x\frac{\partial}{\partial x} + v_y\frac{\partial}{\partial y} + v_z\frac{\partial}{\partial z}$$

defined on a suitable dense linear subspace of the natural Hilbert space $L^2(\mu_L, R^6)$.

(AMN2) The operator of the time evolution in an external acceleration field is given by

$$H = -i\frac{\hbar}{2}\frac{\partial b_t^k(r,v)}{\partial v^k} - i\hbar v^k\frac{\partial}{\partial r^k} - i\hbar b_t^k(r,v)\frac{\partial}{\partial v^k}$$

defined on a suitable dense linear subspace of the natural Hilbert space defined in (AMN1) (with the same space of events).

## GENERALIZED QUANTUM THEORIES

Systems of axioms should give hints for possible generalizations and appreciable modifications of a theory. Let us look at the system of axioms:

```
                    (A1)
                     |
                    (A2)
                     |
                    (A3)
           ____/     |     \____
(AMN1)              (AQS1)              (AQD1)
  |                   |                   |
(AMN2)              (AQS2)              (AQD2)
```

If you take the statistical structure seriously, the very general three statistical axioms should be left unchanged, only the last two lines are candidates for a change of the mathematical structure. This picture suggests looking for new axioms (A4) and (A5) containing all different fields in a single formulation. This is really possible taking the spaces of events from (AMN1) and adding the operators H. With this formulation one gets a generalization of quantum mechanics containing classical mechanics or vice versa a generalization of classical mechanics containing quantum mechanics. One gets field equations for the pilot waves.

We hope that these generalized Schrödinger or Dirac equations could solve some mathematical problems involved with the standard apparatus of quantum theory.

This generalization is suggested by the probabilistic and statistical structure of the mathematical foundations of quantum mechanics <u>and</u> classical mechanics.

## REFERENCES

Emmerich, A., Gerlich, G., and Kagermann, H., 1978, Particle Motion in Stochastic Force Fields, <u>Physica</u>, 92A:362.

Gerlich, G., 1977, "Eine neue Einführung in die statistischen und mathematischen Grundlagen der Quantentheorie," Vieweg, Braunschweig.

Gerlich, G., 1982, "Manuskripte zur Vorlesung Quantentheorie I, and II," TU Braunschweig.

Gerlich, G., 1985, A System of Axioms for Quantum Theory and Something about Hidden Variables, preprint, Institut f. Math. Physik, TU Braunschweig, submitted for publ.

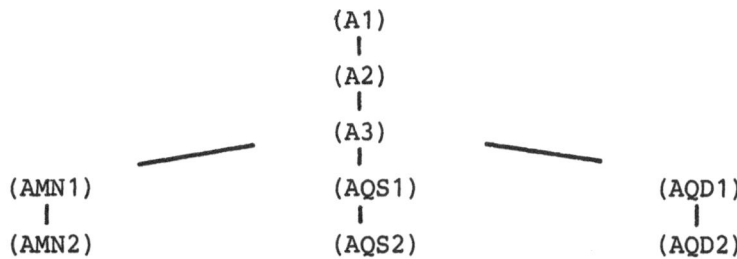

# CAN WE BUILD A SUBJECTIVIST STATISTICAL PACKAGE?

Michael Goldstein

Department of Statistics
University of Hull
England

## 1. INTRODUCTION

This paper concerns the practical implementation of subjectivist theory, and in particular the conceptions of Professor de Finetti, in the form of a statistical package. We begin by briefly considering certain features which distinguish de Finetti's approach from that of standard Bayesian methodologies.

Subjectivist theory can be characterised as the examination of the reasonableness of our modes of thought. In this study "Everything is essentially the fruit of a thorough examination of the subject matter, carried out in an unprejudiced manner, with the aim of rooting out nonsense." What is important is "...the systematic and constant concentration on the unity of the whole, avoiding piecemeal tinkering about, which is inconsistent with the whole; this yields, in itself, something new. "(Both these quotations are from de Finetti (1974:preface)).

This spirit is embodied in the content of de Finetti's work. As an example, he makes expectation, or prevision, fundamental instead of probability, because, once we can free ourselves from historical conventions, there are many advantages, practical, logical and philosophical in this choice. For example, we can now make directly those expectation statements that we require without exhaustive consideration of limiting partitions. De Finetti repeatedly emphasises the need to remain within the bounds of realism. He expresses this as a fundamental requirement as follows. "The fact is the possibility of expressing all that can legitimately be said by arguing solely in terms of the events (and random quantities) whose prevision is known. That is to say, without leaving the linear ambit determined by the latter, without imagining already present a probability distribution over larger ambits, those in which the extension is possible, albeit in an infinite number of ways. The criterion lies in the commitment to systematically exploiting this fact; the commitment considered as the expression of a fundamental methodological need in the theory of probability (at least in the conception which we here maintain). All this is not usually emphasised." (de Finetti (1974)).

As a second example, statistical models are constructed not in terms of unobservable (and ultimately undefinable) parameters, but instead through the notion of exchangeability, so that any such model can be explicated purely in terms of simple, verifiable statements of uncertainty about observable quantities. As de Finetti (1975:p.221) writes: "If we step out of this ambit, we not only find ourselves unable to reach out to something more concrete, but we tumble into an abyss, an illusory and metaphysical kingdom, peopled by Platonic shadows."

As a final example, in the Bayesian approach prior probabilities are transformed into "posterior" probabilities by Bayes theorem. However Bayes theorem actually evalutes "conditional" probabilities. Conditional beliefs are those based on "assumed knowledge", and are expressed as bets to be made now but to become operative only if certain events actually occur. Posterior beliefs are based on acquired knowledge, and are expressed as bets made after certain events are seen to occur, at terms which then seem fair. Logically, these are different concepts. De Finetti (1972:p.194) summarises the interpretation of conditioning as follows. "What emerges is this: only the predictive interpretation (according to which H is a proposition assumed, not acquired) is free of inextricable perplexities."

Each of the above quotations relates essentially to the difference in spirit between a full subjectivist formulation and the Bayesian implementation. In most Bayesian analyses it does not seem to matter whether beliefs are elicited in terms of previsions or probabilities, whether we view probability models as constructed from exchangeability arguments or from genuine beliefs in unknown parameters, whether we are dealing with conditional probabilities or posterior probabilities and so forth. This is because the language and ideas of "belief revision" are being used, in the main, to describe and support the process of "data analysis". However Bayesian methods appear to be tackling a quite different problem, namely how should you "reasonably" modify your beliefs in the light of (statistical) data.

To develop a subjectivist approach to the problems of learning from evidence, we must return to the roots of the theory and decide which elements are essential, which are peripheral and which are, possibly, wrong. This paper describes one such subjectivist approach, taking as a starting point the foundations set out in de Finetti (1974,1975). We shall concentrate upon general issues, as basic disagreements about the content and purpose of the theory can only be resolved when we view the whole structure in a unified manner. Thus we must clarify the substantitive content of the theory before we can describe the technical content of our methods.

The plan of the paper is as follows. In section (2), we suggest informal criteria for a subjectivist statistical package and set out various reasons why fully subjectivist approaches are needed. In section (3), we describe informally our approach to such a package, with particular emphasis on the role of exchangeability, the nature of inference and the organising principles for input and output. In section (4), we describe our first steps towards implementing these ideas. Finally, in section (5), we make very brief concluding comments relating to the problems and potential for the general development of subjectivist packages.

## 2. WHAT IS A SUBJECTIVIST STATISTICAL PACKAGE AND WHY SHOULD WE BUILD ONE?

### 2.1 What Is A Subjectivist Statistical Package?

A subjectivist package is primarily concerned with the judgements, however expressed, of an individual. The package is intended to facilitate the reasonable elicitation and modification (at least in part by "statistical" data) of beliefs by subjectivist principles. Let us identify some features which would distinguish a full subjectivist package from a package which applies subjectivist ideas in an informal fashion, for example as data-analytic tools. (Thus, we will not emphasise the features that both types of package might share, such as the fundamental matter of recognising that beliefs can be set in probabilistic form and constitute important inputs into the analysis.)

(1) Control over belief specification

All belief inputs should be genuine. We should not be forced to input hypothetical belief statements simply because the package demands more than we can meaningfully provide. Thus, we need control over the level of detail that is required by the program. The package should provide both automatic coherence checks and the identification of those aspects of our beliefs which require the most care and detail in specification.

(2) Specifications should relate to observable quantities

Beliefs should be specified for actual, not hypothetical, quantities. It should be possible (in practice, not merely in principle) to conduct the entire analysis in terms of such specifications for observable quantities. This means firstly that we will not consider "parameters" as primary quantities of interest and secondly that we are not forced into specifying beliefs about "underlying parametric models" solely in order to allow us to perform analyses upon observables.

(3) Output should relate to actual posterior beliefs

Just as the inputs to the program are (a portion of) initial beliefs, the output should relate to (a portion of) revised beliefs. The strongest possible relationship would be that the output was our actual revised beliefs. However, this is not usually possible (or even desirable!). Instead, we aim for output which has a useful and clearly defined relationship with such posterior belief. In particular, we should analyse "open" systems, for which we do not need to pretend that all considerations concerning all conceivable eventualities have been fully formulated a priori.

(4) Input and output should be governed by clear organising principles

A subjectivist package deals not with isolated belief inputs but with organised collections of such inputs. All that theory provides is the additional structure imposed by coherence. We require two organising principles to make this structure explicit. Firstly, we need to identify the kinds of belief input which will generate conclusions of interest. Secondly, we need to represent how the entire collection of beliefs is modified by the analysis, emphasising the most important features of the revision.

## 2.2 Why Do We Need Fully Subjectivist Packages?

(i) We are what we believe. Any help in examining our beliefs is valuable, from simple common-sense checks upon our ideas to an improved understanding of our whole reasoning process. Indeed it is because beliefs are fundamental that we must be scrupulous in our development. It is easy to make exaggerated or misleading claims for belief analyses, quite apart from the more insidious dangers involved in surrendering our reasoning to the computer.

(ii) Subjectivist ideas may offer both clear logical methods for integrating raw data into our belief systems, and also an efficient medium for the communication of such beliefs (and the basis for such beliefs) between individuals. In particular, in complicated situations, involving many sources of uncertainty, a theory which pays careful attention to the actual abilities of the individual to express beliefs (rather than treating this as a purely technical issue) may be expected to provide improved approaches to the handling of complexity.

(iii) Because the theory will be different, models within the theory will be different. The kinds of simplification which can be justified and exploited by the theory form the basis for a systematic approach to model formulation, specification and evaluation.

(iv) Theory suggests practice which, in turn, modifies theory. When theory directs us to perform taskswhich turn out to be unclear, useless or impossible, this provides a stimulus for re-examining the basis of the theory. The creation of computer software is a stringent application of this principle. Because we must describe all procedures in precise detail, we cannot treat the theory as a collection o f heuristic prescriptions. There is a danger, however, that theory may be arbitrarily frozen around whatever methods happen to be easy to program and so we must apply the same standards of stringency to the algorithm as we do to any other part of the theory. Indeed this is one of the fundamental issues in the subjectivist theory - to what extent can our modes of thought be reduced to formal routines?

We now turn to more speculative (but fundamental!) reasons for our enterprise. Subjectivist theory can be interpreted at two different levels. The first concerns an individual making a conscious effort to externalise various of his beliefs in numerical form. In this view, the beliefs are pre-existent and are given numerical values by some mysterious but conscious process. However, we can reverse this process by considering the numerical codings to be pre-existent, and our beliefs to be externalisations of these codings.

In describing such a structure, we must distinguish between process and implementation. The subjectivist theory is separate from the medium in which it is implemented. The theory describes the interface between an external reality and an internal representation of that reality, in a form which is (in principle) independent of both sides of the interface. (That is, we could rewrite the laws of physics without changing our theory. Similarly, we can switch from considering us, i.e. biological organisms, to it, i.e. a mechanical device such as a "machine intelligence", and though the implementation would be totally different, the theory would remain unchanged.) This internal/external interface establishes the logical parameters for understanding any "thought" process, and provides the ultimate subject matter of subjectivism. When we talk about, for example, exchangeability, we are considering one of the basic operations by which we cope with our environment, namely the simplification of giving many different objects the same "name". Most of our statements (for example, this one) can be represented strictly in terms of the exchangeability constructionsthat we will put forward (although to do so may be complicated in somewhat convoluted statements such as this one).

Thus, our ostensible statistical description of certain classes of situations allows us in principle to analyse the mechanisms of thought. Such study may find concrete application in the design of artificial intelligence systems, in which detailed numerical analyses (at levels both physically inaccessible and psychologically incomprehensible to users of the system) are converted into simple "verbal" summaries at the user interface. Here we have an explicit representation of beliefs as externalisations of pre-existent codings. Beyond this, however, the purpose of such investigation is self-knowledge. Fundamental questions as to the nature of and relationship between our conscious and subconscious processes and the interaction between these processes and external stimuli may be given precise algorithmic representations - at the least, we have a natural language in which to formulate these questions.

## 3. A FRAMEWORK FOR A SUBJECTIVIST STATISTICAL PACKAGE

We now describe our approach to the construction of a subjectivist statistical package, avoiding technicalities as much as possible. Instead, we summarise the various choices made, the reasons for these choices and the implications for managing the system. Clearly, a genuine subjectivist statistical package, as we have described it, is an ideal. I believe that it is an achievable ideal, but that is a long-term goal. We here provide a few steps towards this goal. Our concern is with the total structure rather than the individual operations of the system. Thus, we elicit beliefs in a form for which we can offer simple organising principles which are compatible with direct exchangeability constructions, which can be handled by our inferential principles, which in turn yield simple organising principles for the revisions of belief. It is the inter-relationship between organising principles (for input and output), exchangeability principles and inferential principles which constitutes the heart of the system. Although we can justify each principle individually, the ultimate justification for each ingredient lies in the role it plays in the overall structure.

### 3.1 Prevision Is the Fundamental Expression of Belief

Probability is too restrictive a concept to give us control over the level of detail in our belief specifications. The obvious alternative is to make expectation fundamental. (Probability specifications become expectations for the corresponding indicator functions.) De Finetti provides a full account as to how and why expectation, or prevision as he terms it, can and should be made the foundation of the subjectivist theory. Thus, we consider prevision (as operationally defined in de Finetti (1974, chapter 3)) to be the primitive quantity in terms of which we elicit beliefs. For any random (i.e. unknown to us) quantity X, we write P(X) for the prevision of X. (This may be thought of as expectation but elicited directly.) Our justification for this choice is that we can build a subjectivist system based on prevision, whereas we cannot satisfy the general requirements of section (2) by a system rooted in probability.

### 3.2 Previsions Should Be Organised into Inner Product Spaces

Collections of probabilities typically are organised into joint probability distributions. Is there an organising principle for prevision which is fundamentally different from that for probability? The difficulty with most possible organisations of beliefs is that they commit us to making far too many belief statements. Of course if a particular analysis genuinely requires a very large number of belief inputs, and we are both willing and able to provide these inputs, then there is no problem. However, for most situations, we may suspect that the majority of these inputs are largely irrelevant, and in any case are beyond our ability to specify.

Only if our method of organisation of previsions reflects the underlying structure of the subjectivist language will it, in general, generate interesting consequences. The essential property of prevision is linearity. The specification of previsions for some collection C = [X,Y,Z,...] of random quantities fixes previsions over the collection of all linear combinations of these quantities (and, in general, fixes no other previsions). Thus, it is natural to view C as generating a vector space L (where each element of C is a vector in L and linear combinations of vectors are the corresponding linear combinations of the elements in C).

Prevision is basic because linearity, i.e. adding quantities, is basic. We build multiplicative structure into L by defining an inner product over L, for each X,Y in L by (X,Y) = P(XY). We call any collection of previsions organised to define an inner product space with this inner product a BELIEF STRUCTURE. Thus, we form a belief structure by first specifying some collection C = [X,Y,Z,...] of random quantities, termed the BASE of the belief structure, and then specifying the values P(XY) for each pair of elements in C. (We restrict C to elements with finite squared prevision. We usually require that the constant is an element of C. Compare the geometric interpretation of prevision given in de Finetti (1974: Section 4.17).)

Different elements of C may refer to the same (mathematical) variable e.g. X and log(X) are different random quantities, as they are not linearly related. If we wish to specify beliefs relating to log (X) then we introduce log(X) into C. In many situations, we will introduce only a few such functions into the base of our belief structures. If we introduce all functions (and products) of all quantities of interest then the corresponding belief structure is equivalent to that given by the usual Bayesian prior specification. (The Bayesian specification concerns a probability measure over some space, while a belief structure is any subspace of the Hilbert space of square integrable functions over that space, with respect to the probability measure, under the usual product integral norm; further discussion in Goldstein (1984)). Of course, we could choose different methods of organisation for previsions. However, any such choice would be constrained by our requirements as to the use of the belief inputs, and, in particular, by the need for simple organising principles for output, a practical exchangeability principle and a satisfactory approach to the revision of belief.

### 3.3 Exchangeability Generates Statistical Models

Statistical models relate to quantities that carry the same "name" in a variety of situations (e.g. measurements of "height" on different individuals). Exchangeable beliefs for sequences are those which are not affected by permuting the order of the sequences. Exchangeable belief structures are those generated as follows.

Begin with a sequence C* = [X,Y,Z,...] of "names" (for example X,Y,Z,... might be "blood pressure", "log blood pressure", "temperature", etc.). This system of names is applied to a series of situations (e.g. the examination of a sequence of patients), to generate a sequence $B_1$, $B_2$, $B_3$,... of belief structures, one for each situation. The base of $B_i$ is [$X_i$,$Y_i$,$Z_i$,...], where $X_i$ is the value of X in situation i and so forth (e.g. $X_i$ might be the blood pressure of patient i). Exchangeability denotes the case where

(1) each belief structure is essentially the same i.e. for any X,Y in C*, and any i, ($X_i$,$Y_i$) = (X,Y), (a constant for all i).

(2) the relationship between each pair of belief structures is essentially the same, i.e. for any X,Y in C* and any $i \neq j$ ($X_i$,$Y_j$) = (X,Y)*, (a constant for all $i \neq j$).

Specification of all the values (X,Y) and (X,Y)* uniquely specifies the belief structure B with base consisting of all the observable quantities under consideration (i.e. the base generated by all quantities of form $Z_i$, Z in C*, and any i). Thus, we say that a collection of belief structures is EXCHANGEABLE if it is generated in the above manner, with inner product satisfying constraints (1) and (2) above.

As a simple illustration, consider tossing coins. C* might be [1,H], where 1 is the unit constant and H is 1 for a head, 0 for a tail. $B_i$ has base [1,$H_i$], where $H_i$ is 1 if toss i is heads otherwise 0. Conditions (1) and (2) become: (1) we assign the same probability p that each individual toss will show heads; (2) we assign the same probability q that any two different tosses will show heads. We can make as detailed specifications as we like, by adding further quantities to C*. However, if the values p and

and q, and the above conditions (1) and (2) are all that we are prepared to specify, then we should be able to obtain and analyse the implied model, without having to pretend to the infinite number of further specifications required by the usual exchangeability construction.

The standard exchangeability results concern in principle an infinite number of exchangeable situations, i.e. where there is no limit to the number of cases to which exchangeability relates. If we are considering, in principle, an infinite number of belief structures, then we may make a similar construction. For each X in C, we construct the quantity X* which is the Cauchy limit of the partial sums $(X_1+\ldots+X_n)/n$, as n tends to infinity. (So, X* is an element of B*, the closure of the overall belief structure B.) We now form the belief structure M whose base is all the quantities X* for X in C. M functions as the underlying "model" for our beliefs. We formalise this by creating a sequence of mutually orthogonal belief structures M, $R_1$, $R_2$, $R_3$,... for which, for each i, $B_i$ is a subspace of $M+R_i$. Each $R_i$ has, in a natural sense, the same belief structure. So, instead of constructing, from a sequence of exchangeable probability specifications, a further probability measure, conditional on which the sequence is iid, we construct, from a sequence of exchangeable belief structures, a further belief structure "given which" all of the residual belief structures, $R_i$, are individually the same, and mutually uncorrelated. (The details with modifications for finite exchangeability, are given in Goldstein (1986b).). As all the spaces $R_i$ are mutually orthogonal, all the relationships that we have expressed between the various belief structures $B_i$ are "explained" by the common relationship of each structure to the "model" M.

For example, in our coin example, we form M with base [1,H*], where H* is the Cauchy limit of the quantities $P_n$, the proportion of heads in the first n tosses. We write each $H_i = H^*+r_i$, where the quantities $r_i$ are mutually uncorrelated with the same variance, and so forth. H* acts as our primitive concept of a "limiting frequency of heads", and our previsions for this quantity are determined by our previsions for the individual tosses (for example $P(H^{*2}) = q$). (Note that exchangeability generates large numbers of beliefs from consideration of simple situations (in this case two coin tosses), using our perception of symmetries to extend these assessments to all of the cases that we might consider. One of the problems with the probabilistic formulation is that we must explicitly consider all the situations because we must specify all of the joint probabilities.)

Exchangeable belief systems form our basic models. Collections of sequences of belief structures which are mutually co-exchangeable (a natural extension of our definition of exchangeability) give the general model structure. (Details in Goldstein (1986b)). The mutual orthogonality relations between the spaces $R_i$ are fundamental, beyond their intuitive interpretation, because of the way that beliefs are revised within our system. We now consider the basis of our inferential arguments.

### 3.4 Temporal Coherence

What explicit, checkable, justifiable conditions can we impose upon our beliefs which will yield systematic approaches to the revision of belief structures? Our conditions must be stringent enough to provide a satisfactory account of the revision of models generated through exchangeability arguments while not treating statistical models as if they possessed some logical validity external to our specifications. The conditions should be sufficiently reasonable that breaching these conditions would suggest the situation to be sufficiently untypical that no standard analysis would be likely to cover it. (The Bayesian coherence condition is not particularly reasonable; we are supposed to be able to anticipate our reactions to all conceivable eventualities, and we intend our anticipated reactions to become our actual reactions without further consideration.) Our framework is as follows.

We suppose that at each particular moment our current beliefs should be coherent. This has two implications for our current beliefs about our future beliefs. Firstly we should now believe that our future beliefs (as specified by us at the future time point) should be coherent (at that time). Secondly, our current beliefs about our future beliefs should be coherent when we treat our future specifications simply as random quantities.

Further, we interpret coherence as an argument that we should avoid (or at least be wary of) finite collections of actions with the property that we can provide some automatic mechanical rule which, if we followed it, would automatically lead us to lower loss. This is fairly clear when applied to actions which are all taken at the same time. It is more subtle for collections of acts taken at several time points. These subtleties are certainly worthy of careful investigation. However, as long as our coherence requirements are clear and generally useful, then they may be applied in most situations. In those special circumstances when they do not seem so compelling, we will not expect our methods to apply. We do not view coherence as representing a fundamental rationality requirement, but rather as an efficient way of identifying wide classes of situations in which we may exploit certain useful structural properties of the subjectivist language. The coherence conditions outlined in skeletal form above seem to be reasonable for most statistical applications. They can be (loosely) reformulated that there is no information that we would pay money not to receive (that is, in practice there are plenty of reasons to avoid receiving information but none of them are related to the usual rationality requirements that statisticians are concerned with).

The above conditions are applied as follows. Suppose that we now consider our prevision, $P(X)$, for a random quantity $X$. Suppose that at time $t$ we intend to consider this problem again, at which time we will announce a new value for the prevision of $X$. At the present moment we do not know what value we will assign for this prevision, so that it is for us a random quantity, $P_t(X)$ say. This quantity is a genuine posterior prevision. Such future assignments are random quantities which are currently of interest to us. Stating beliefs about our future beliefs is not logically different from belief specifications for any of the other quantities over which we have some measure of control but for which we are prepared to specify beliefs. Sometimes, we may find these beliefs hard to specify, but that is why we need guidance from the theory. Given the above coherence conditions, we can show that our present and future previsions for any bounded random quantity $X$ must satisfy the fundamental relation $P(P_t(X)) = P(X)$. (This result is derived in Goldstein (1983), although the nature of the temporal constraints is dealt with rather briefly. Note that $t$ can be a fixed or a random time - e.g. tomorrow at noon, after the experiment is finished, etc.).

The above relation is somewhat similar to the relation that the expectation of a conditional expectation is equal to the original expectation. However, the relationship for conditional expectations is fundamentally different, in that it refers exclusively to beliefs revised by conditioning. The relationship for $P_t(X)$ is derived without making any assumptions about the way that $P_t(X)$ will be evaluated, excepting our requirement that at the future time when it is assigned, this assignment is consistent with any other specifications made at that time. Our beliefs are not revised, even in principle, by determining some limiting partition and then conditioning on a single outcome. Just as previsions made at one time point obey a single basic property, namely linearity, beliefs specified at different time points satisfy a single temporal property, namely $P(P_t(X)) = P(X)$. To build an inferential system exploiting this property, we must first consider what we require of such a system.

## 3.5 Inferential Framework - Preliminary Comments

The Bayesian revision of probabilistic beliefs, exclusively by conditioning, is inadequate for our purposes, for two reasons. Firstly, we may only condition upon events. This forces us back into consideration of limiting partitions, and so into impossible problems both in the anticipation of possibilities and in belief specification over these possibilities. Secondly, conditional beliefs are inherently different from posterior beliefs. Conditional prevision on some conditioning event A cannot be considered to be the value we would express were we to learn that A has occurred and nothing else, as if we happened to receive no other relevant evidence at the time that we happened to learn A (which would itself be a subjective judgement), then this would itself be relevant to our revision of belief. Further, even if we could give some substantitive meaning to learning only A, we still could not equate conditional previsions given A with posterior previsions on learning "only A", as conditional previsions relate to "called-off" bets or penalties for which posterior considerations under the special case of learning "only A" have no particular relevance. (For example, how do we interpret likelihood-type probability statements made "conditional" on models built from exchangeability arguments?)

The alternative description of the inferential process is based on such properties as can be justified by careful temporal accounting. Relations of the type $P(P_t(X)) = P(X)$ place no logical constraints upon actual future beliefs, but instead concern current attitudes to future beliefs. At any time point we may make probabilistic statements about our future beliefs. At a future time point, all that such statements can offer is guidance as to what conclusions we may reasonably reach. Sometimes we will take such guidance to be strongly suggestive, while, at other times, all it may provide is preliminary guidance as to which data to collect and which features of the data to give close scrutiny, with no implication whatever as to final decisions.

As an introduction to our formulation we now discuss further the relationship between conditional and posterior prevision. DeFinetti (1974, Chapter 4) defines the conditional prevision of X given E in terms of the "called off penalty" version of the definition of prevision, i.e. $P(X|E)$ is the number x that we would choose if confronted with a certain kind of penalty if E occurs, with no penalty otherwise. As with all tight definitions of conditioning this refers to an assignment of beliefs now, before the occurrence or otherwise of E is established. We must explicitly construct links between conditional previsions and any prevision we might specify on observing E.

Thus suppose that before time t we shall certainly observe an event whose possible values can be represented by the partition $H = [H_1, \ldots H_k]$. We can show from the basic temporal relations of section (3.4) that, for any bounded random X, we may currently write $P_t(X) = P(X|H) + R$. Here $P(X|H)$ is the prevision of X conditional on the partion H, which can be written, replacing each $H_i$ by the indicator function for the corresponding event, as $P(X|H) = P(X|H_1)H_1 + \ldots + P(X|H_k)H_k$, and R is a random quantity satisfying $P(R) = P(R|H_1) = \ldots = P(R|H_k) = 0$, and for which var(R) is not greater than var(X) - var(P(X|H)). (Details in Goldstein (1986a).)

In the relation $P_t(X) = P(X|H) + R$, $P_t(X)$ is somewhat like a posterior prevision for X "having seen H". However, we index this by t rather than H because we can precisely define t, so that $P_t(X)$ is a well defined quantity which can be analysed by the methods of the theory, whereas we can give no substantive meaning to phrases like "having seen H and nothing else." (If we want to emphasise H, then we write t as t(H).) Further, the crux of the relation is that if t is <u>any</u> well defined time point by which H will certainly have been observed, then conditional and posterior beliefs will be so related.

Our beliefs as to what else we might observe or reflect on are expressed in the random component R.

The use that we make of such relationships depends upon the context. When the simplification is harmless, we can operate a deterministic inference system governed by conditioning. Indeed, we now have a logical justification for so doing in that this is acceptable as a subjective judgement when we make the further subjective judgement that the random components of our uncertainty equations are unimportant to the case at hand. We replace a theory which concerns "perfect" inferential systems (but for which assessment of the relevance of the inferential procedures to the matter at hand lies outside the theory) with an approach in which judgements as to the adequacy of the deterministic inferential system are incorporated into the body of the theory (and can thus be analysed by the methods of the theory). In most cases the non-deterministic elements of the inferential procedure will be just as important as the deterministic elements; the interplay between these two aspects constitutes the subjectivist analysis. However, before we can discuss this interplay we must describe the revision of beliefs for a general belief structure.

## 3.6 Belief Revision Is Organised by Self-Adjoint Operators

What features are common to every revision of beliefs over a belief structure, A? We must specify, at the future time t at which beliefs are to be revised, the new inner product over A, i.e. for every pair of elements X,Y in the base of A, the quantities $P_t(XY)$ will be specified. In a stochastic system of inference, we must describe the stochastic rules governing the structure of our collection of revisions of belief. We can establish, using the temporal constraints, that for any revision of beliefs, we can construct an associated BELIEF TRANSFORM T, where T is a bounded, self-adjoint linear operator on A. The belief transform summarises the revision of beliefs as follows.

Denote the ordered eigenvalues of T by $m_1$, $m_2$,... with corresponding eigenvectors $V_1$, $V_2$,..., normalising each $V_i$ to unit norm. The collection $V = [V_1, V_2, \ldots]$ imposes an orthogonal coordinate frame over A which expresses all of the effects of the belief revision. For example, the expected reduction in variance for each $V_i$ is $m_i$. Take any element X in A, with zero prevision and unit norm. Resolve X along each coordinate axis. Then the expected reduction in variance for X is simply the sum of the reductions in variance along each coordinate axis (i.e. the sum of terms $m_i(X,V_i)$), with similar expressions for revisions of covariance.

The belief transform gives a simple geometric picture of expected revisions of belief. Elements of A with large components in the directions corresponding to large eigenvalues are those about which we expect to learn a lot, elements lying substantially in subspaces spanned by small eigenvalues are those which we do not expect to learn much about. (Details in Goldstein (1981).) In this picture all secondary features have been eliminated. The belief transform expresses changes in beliefs, not the reasons for these changes. Thus, if we want to compare various different ways of collecting information (fixed or variable sample sizes, how many auxiliary variables, how much pre-testing or whatever), then although the notional sample spaces for the different approaches (even if we can construct them) may be very different, the belief transform for each approach will be directly comparable, and will summarise the effectiveness of each approach in modifying our beliefs. The belief transform plays a fundamental role by virtue of its relationship with the inferential and exchangeability principles for the system. We now discuss these links.

## 3.7 Probabilistic and Deterministic Components of Belief Transforms

We now describe the general decomposition of the belief transform into probabilistic and deterministic components. The deterministic component relates to explicit data-based revisions of belief. We term a "data structure" (with respect to time t) to be any belief structure D with the property that every one of its base elements will certainly be known by time t. There is no single belief specification D which can, in general, carry all of the information that we receive by t. Further, even if there was such a structure, we would, in general, have no interest in specifying beliefs to such an extreme level of detail. Instead, just as we select certain beliefs of primary interest by our choice of belief structure A, we then specify those aspects of our beliefs about the evidence which we wish to explicitly introduce into our analysis by our choice of D.

In order to relate D to A, we construct the combined belief structure C = A+D. The effect of this specification upon the uncertainties of the elements of A evaluated at time t may be represented in terms of the space [A/D], the space A ADJUSTED by D, where [A/D] is the orthogonal complement of D in the combined space C. (Essentially, we replace each X in A by the corresponding quantity $(X - P_D(X))$, where $P_D$ is the orthogonal projection operator into D. A simple interpretation of [A/D] is as a summary of the "residual" variability for each element of A after we subtract from each of the base elements of A the corresponding "linear Bayes rule" selected from the base of D. Details in Goldstein (1985b).) In a Bayesian specification, D corresponds to a full probability specification over some sample space, and $P_D(X)$ is the conditional prevision of X given the sample. Our choice of "quotient space" notation emphasises that the adjustment of one belief structure by another is analogous to such probabilistic conditioning. However, there is nothing fundamental about the process of conditioning. The arguments apply to <u>any</u> data structure, with conditioning replaced by orthogonal projection.

Each general belief adjustment [A/D] carries an associated belief transform $T_D = P_A P_D$ (where $P_A$, $P_D$ are orthogonal projections into A,D). The essential property of $T_D$ is that it splits the overall belief transform over A, that is we may write $T = T_D + T_R$, where $T_R$ is also a bounded self adjoint "belief transform" over A, and the overall belief transform is the sum of $T_D$ and $T_R$. To assess our change in information about a general element X in A, we see how much we expect to learn about X according to $T_D$ and $T_R$ separately and our overall expected change in beliefs about X is the sum of the two quantities. (A derivation is given in Goldstein (1986b), although emphasis is placed on changes in beliefs about the individual quantities).

We construct and compare the two transforms $T_D$ and $T_R$, either directly or by comparison of $T_D$ with direct evaluation of T. We term $T_D$ the AUTOMATIC BELIEF TRANSFORM induced by D, as it is explicitly determined by our previsions over A and D, and term $T_R$ the posterior belief transform complementary to D. Quantities in A which relate to large eigenvalues in $T_D$ but small values in $T_R$ are essentially those for which we expect our automatic procedures to extract most of the relevant information without any further reflection on our part. Those corresponding to large eigenvalues in $T_R$ but small eigenvalues in $T_D$ are those for which we judge posterior reflection to be important. We increase the relative importance of the eigenstructure of $T_D$ as compared to $T_R$ by adding new quantities into D, both by observing more and also by specifying in greater detail our beliefs. The automatic transform induced by each D expresses the efficiency of the corresponding automatic procedure.

How we specify the various belief transforms will depend upon the situation. Sometimes, we start with little idea about our global revisions of belief, but detailed ideas about how to specify the elements needed to evaluate $T_D$. Many standard statistical problems fall into this category, and interest may focus on the conversion of plausible automatic procedures into genuine posterior procedures. In many other problems, we begin with detailed notions as to our global revisions of belief but little quantitive idea as to the factors which tend to lead us to such beliefs. For example, a doctor routinely diagnosing patients might find it easy to describe the kinds of information that he expected from the examination, but might find it very difficult to break this information down into specific inferencesfrom particular diagnostic procedures, particularly as he would not expect the examination to develop in the same way for each patient. In such cases, we might try to identify those areas for which judgements can be reduced to an automatic routine, and those areas which appear to rely intrinsically on posterior judgements.

Our final step in describing the framework is to relate the belief transforms, and their separation, to the general statistical models which we have generated via exchangeability.

## 3.8 Belief Transforms and Exchangeable Systems

Suppose that we will observe a sample of individual cases from an exchangeable system of belief structures. We intend to revise our beliefs about further cases that we might observe. We want to treat the associated statistical model for the system as though it were an observable, rather than hypothetical, belief structure, revise our beliefs about the model using the methods for revising belief structures described above, and then use our representation theorem to revise beliefs about each further observable case (by writing each such case as the sum of the revised model space and the original, unrevised residual structure). In this way we could reduce a large number of difficult revisions of belief into a single, straightforward revision.

However, we must be careful in our formulation. Firstly, we will not simply observe a sequence of n individual cases "and nothing else". Secondly, at any moment we may abandon our belief in exchangeability for the sequence, at which point the model ceases to exist. We must determine those additional constraints that our beliefs must satisfy in order that the inferential arguments of the preceding sections may be applied to our models.

These conditions are as follows. Begin with the sequence of names $C^* = [X,Y,\ldots]$ which generate the exchangeable sequence, $B_1$, $B_2,\ldots$ . Create the new sequence $C^*_t = [P_t(X),P_t(Y),\ldots]$ which generates the sequence $B_{t1}$, $B_{t2},\ldots$ of belief structures, where the base of $B_{ti}$ is $[P_t(X_i),P_t(Y_i),\ldots]$ where $P_t(Z)$ is the prevision for Z that we will express at time t. (All assignments are made now, so that all future assessments are currently random quantities.) It can be shown that we are justified in using our model analysis to make statements about future observables if there is some N for which the collection of all structures $B_{ti}$, $i > N$, is an exchangeable system. This condition is a precise expression of the informal requirement that at present we cannot identify any particular subsequence of the future situations about which we already expect that, by time t, we will have received a different amount of information than any other subsequence. (We exclude the first N situations as these will form part of our sampling procedure).

Orthogonality between the residual structures and the model structure is crucial to these results. For example, if we consider the adjustment of $B_i$ by observation on $B_{(n)} = B_1+\ldots+B_n$, (the belief structure corresponding to the first n situations), where $i > n$, then just as we may write $B_i$ as a subspace of $M + R_i$, we can write the adjusted space $[B_i/B_{(n)}]$, as the corresponding subspace of $[M/B_{(n)}] + R_i$. (Notice that this is just what we want. We adjust a future observable by adjusting the model and then applying the representation theorem).

The key quantity is the belief transform. We must evaluate transforms over belief structures corresponding to future observables, based on observation of structures exchangeable with the structures of interest. Given the above conditions on our beliefs, we can decompose such a transform into the sum of a transform over the model structure, plus an (essentially random) transform over the residual spaces. The model analysis has a very simple form. For example if we construct the automatic belief transform over M induced by the data structure $B_{(n)}$, then it is essentially the same for each value of n (i.e. the eigenvectors are the same for all n, and the eigenvalues for different n are related by simple formulae). If we look instead at the automatic belief transform over $B_i$ induced by $B_{(n)}$, $i > n$, then we find that it is essentially the same as the transform over M (the eigenvectors are the same, and the eigenvalues are related by simple formulae). Thus the model analysis preserves all the information for the analysis of future observables.

To summarise our development, the argument is as follows. Working with belief structures gives us automatic access to intuitive, easily programmed and interpreted summaries of our revisions of belief in the form of belief transforms. For exchangeable structures these transforms have much extra structure that we can easily exploit to understand in detail the effect of increasing quantities of information upon our beliefs. These transforms can be considered to operate upon simple models of our system. These models are directly constructed from our statements about observables. We have sound logical reasons for relating our hypothetical model analysis to our actual posterior beliefs about future observables. In this way, all the features of a proper subjectivist system work together. (The results referred to in this section will be reported more fully elsewhere.)

## 4. IMPLEMENTATION

Having been awarded a three year grant (by the Science and Engineering Research Council) to produce a sample program, we decided to concentrate on certain limited aspects of the theory in order to have a useful working module by the end of the period (we being myself and David Wooff, who is converting the general theory to matrix form and writing the computer implementation). Thus our program focuses upon the analysis of exchangeable and co-exchangeable belief structures by means of direct evaluation of the automatic belief transforms for the structure. (A brief discussion of the program, with examples, is given in Goldstein (1987).)

To illustrate the type of questions we might address, suppose that there are some basic quantities of interest (e.g. effectiveness of various drugs for individual patients). We want good "estimates" for certain effects, and we can list a further collection of possible covariates. We do not want to use them all, for reasons of cost, difficulties in specifying all the required beliefs, desire for simplicity etc.. Thus we want guidance as to how adding or deleting these quantities from our formal structure changes our state of information concerning the primary quantities of interest, given a sample of n upon whichever elements we decide to include in our analysis. We collect the primary quantities of interest into an exchangeable belief system A. We view A as a subspace of the larger exchangeable belief structure B corresponding to all the quantities under consideration. The automatic transform over B for a sample of n induces an automatic transform, $T_A$, over A, which summaries the information provided by observation of the sample of elements from B. We compare the eigenstructure of $T_A$ for varying choices of B and n, until we find a satisfactory choice.

The simplest choice criterion is the trace of $T_A$ which is a scale free measure of information gained over A. We might, for example, proceed in a stepwise fashion, adding terms to B one at a time to stepwise maximise the trace. (This approach, for a collection of m possible candidates for B, reduces the number of belief specifications from order $m^2$ down to order m). Alternately, we may decompose a large system $B_i$ into orthogonal components, each of which makes a "separate" contribution to the trace of $T_A$, for various choices of n. We can then see how the quantities in B are combining to provide the information that we receive, and how this is affected by sample size. We then select those combinations of elements in B which seem to be most informative.

As time permits, we are adding various further facilities; general analysis of co-exchangeable structures; data diagnostics (highly informative components of the belief transform modify our beliefs while apparently uninformative components form the diagnostic system, rather like residual plots); basic model specification (corresponding to constraints upon belief specifications beyond those imposed by coherence). Although hopefully this program will provide useful and interesting output in its own right, it is not an end in itself. Rather, it is a module within a larger system, namely (at least the skeleton of) a fully articulated subjectivist system.

In our implementation, we have concentrated on the automatic transform. Partly, this is for simplicity - we can easily extract rich output from this transform. However, partly this is because given a fully developed system for analysing this transform, we can "bootstrap" ourselves into constructing the random components of such transforms, in many situations. This is because the random components will tend to be informative precisely in those situations where we have experience of making similar judgements. For example, we may place our procedure online, monitoring actual posterior judgements and using these judgements as the base of a further exchangeable system. The automatic transform for this system directly reduces the random component of the original transform for future judgements of a similar nature. By relating our theory directly to actual posterior judgements, these judgements themselves become available to us as raw material to which we can apply the full subjectivist methodology.

Given the caveat that currently our program is very much in an embryo form for which the usermust provide the "subjectivist environment", we would welcome enquiries from people interested in trying it.

## 5. CONCLUDING COMMENTS

There are various technical problems involved in the construction of subjectivist statistical systems. However, for the most part these problems can be overcome by careful analysis. More intractable are those questions relating to the objectives of the analysis.

We have argued that our beliefs are worth serious study, and that it is intrinsically worthwhile to step back from our activities and reflect systematically upon the reasoning which underpins our lives. However, while many people might informally agree with such a proposition (while reserving a natural scepticism as to whether this is actually possible), there are few individuals to whom such an activity would appear to have direct professional relevance. To the extent that any group uses Bayesian methods, then such use is always subordinate to some specific application. It often seems that belief examination methodology is being used to examine everything except our beliefs. However, subjectivist theory has enormous and currently untapped potential for integrating information into our belief systems, in a clear and logically justifiable fashion (quite apart from its philosophical and mathematical appeal). We have described one possible attempt to realise some of this potential, and argued that this approach is not arbitrary, but rather is directed by our view of the intrinsic nature of the subjectivist language. These are simply first steps - an exploration of possibilities demonstrating that there are alternative ways to establish the foundations of our subject, that these choices have important practical consequences and that foundations and implementation can and should be developed in a unified manner.

REFERENCES


De Finetti, B. (1972) Probability, Induction and Statistics, London:Wiley.

De Finetti, B. (1974, 1975) Theory of Probability, Vol. 1,2, London:Wiley.

Goldstein, M. (1981) Revising Previsions: A Geometric Interpretation. J. Roy. Statist. Soc. B. 43, 105-130.

Goldstein, M. (1983) The Prevision of a Prevision. J. Amer. Statist. Assoc. 78, 817-819.

Goldstein, M. (1984) Belief Structures. Mathematics Research Center, University of Wisconsin-Madison, Technical Report 2738.

Goldstein, M. (1985a) Temporal Coherence, in: Bayesian Statistics 2, ed. Bernardo, J.H. et al, North Holland .

Goldstein, M. (1985b) Adjusted Belief Structures, Mathematics Research Center, University of Wisconsin-Madison, Technical Report 2804.

Goldstein, M. (1986a) Separating Beliefs, in: Bayesian Inference and Decision Techniques: Essays in honour of Bruno de Finetti, ed. Goel, P.K. and Zellner, A., North Holland .

Goldstein, M. (1986b) Exchangeable Belief Structures, J. Amer. Statist. Assoc., to appear.

Goldstein M. (1987) Systematic Analysis of Limited Belief Specifications, The Statistician, to appear.

Kyburg, H.E. and Smokler, H.E. (editors) (1964) Studies in Subjective Probability, London:Wiley.

LIFE TIME DISTRIBUTIONS AND

STOCHASTIC DYNAMICAL SYSTEMS

Johan Grasman

Department of Mathematics
University of Utrecht
Utrecht, The Netherlands

## 1. INTRODUCTION

In reliability analysis the Weibull distribution and other standard distribution functions have successfully been used to describe the probability of failure of a system as a function of its age, operational time or some other measure of its life time. One is led to choose such distributions because of their fit to experimental data. For systems with extended life times one is confronted with the question how to estimate life time distributions beforehand. The method of accelerated life testing deals with this problem, see Viertl (1983). One has to extrapolate from life time distributions of systems under stress by constructing appropriate acceleration functions. Because of the uncertainty in the validity of the extrapolation, it is worth to use available information about the system in the process of estimating life times under normal and under stress conditions.

In our presentation we analyse stochastic dynamical systems and try to extract from their statistical properties an approximation for the shape of the life time distribution. The inverse Gaussian distribution, used in reliability, illustrates this strategy, see Martz and Waller (1982). It is the distribution of first passage time for a Brownian motion process with drift. It describes failure due to wearout and chance. The stochastic differential equation for this process P(t) reads

$$dP = a\,dt + r\,dW, \qquad P(0) = 0, \tag{1.1}$$

where W(t) is a Wiener process. The first passage time distribution at P = 1 is

$$f(t;a,r) = \frac{1}{r\sqrt{2\pi t^3}} \exp\left\{-\frac{(1-at)^2}{2r^2 t}\right\}, \tag{1.2}$$

which is equivalent with the inverse Gaussian distribution.

## 2. STOCHASTIC FORCING OF SYSTEMS WITH UNKNOWN INTERNAL DYNAMICS

When a system is put under stress, it is indeed so that wearout fastens and life time decreases as argued in the introduction. In order to let accelerated life tests gain in predictive power, we will model more closely the way stress is acting upon the system. Let us assume that in its normal operation mode the system is stable in the sense that, if the system is perturbed, it returns to its operation mode. Let this restoring force have a relaxation time constant b. Stress means that at frequent time instances the system is pushed against this restoring force. Assuming that these external perturbations have a stochastic nature and are of a type known as white noise, we arrive at an Ornstein-Uhlenbeck process for modeling the response of the system:

$$dX = -bX\,dt + s\,dW, \tag{2.1}$$

where W(t) denotes a Wiener process. The state variable X can be seen as the deflection in the direction of the eigenvector related with the real eigenvalue -b of the system (for oscillatory damping the meaning is slightly different). The size of the stress is measured by s. Let for $|X| = L$ the deflection be so large that it significantly influences the life time of the system. Then assuming that the life time is proportional with the first passage time of arriving at $X = \pm L$, we may concentrate ourselves on the solution of this well-known problem in the theory of stochastic processes. Let us first apply a scaling such that $L = 1$. Then for (2.1) with

$$X(0) = x, \qquad |x| < 1 \tag{2.2}$$

we may study the first passage at $X = \pm 1$. The Laplace transform of the distribution function $f(t;x)$ of the first passage time $T(x)$ can be expressed in terms of parabolic cylinder functions, see e.g. Capocelli and Ricciardi (1971). However, this expression is too complicated for the practical use in accelerated life testing. It suffices to find the first statistical moments $T_i(x)$ of this distribution:

$$T_i(x) = \int_0^\infty t^i f(t;x)\,dt. \tag{2.3}$$

The expected value of T(x) satisfies Dynkin's equation

$$\tfrac{1}{2}s^2 \frac{d^2T_1}{dx^2} - bx\frac{dT_1}{dx} = -1, \qquad T_1(\pm 1) = 0$$

or

$$T_1(x) = \frac{-2}{s^2}\int_{-1}^{x}\int_{0}^{y} \exp\left\{\frac{b(y^2-z^2)}{s^2}\right\} dz\,dy, \tag{2.4}$$

see Gardiner (1983). For the second moment we obtain

$$\tfrac{1}{2}s^2 \frac{d^2T_2}{dx^2} - bx\frac{dT_2}{dx} = -2T_1(x), \qquad T_2(\pm 1) = 0$$

or

$$T_2(x) = \frac{-4}{s^2}\int_{-1}^{x}\int_{0}^{y} T_1(z)\exp\left\{\frac{b(y^2-z^2)}{s^2}\right\} dz\,dy. \tag{2.5}$$

Because of the scaling the stress parameter s will be small compared with unity and so we may expand (2.4) and (2.5) with respect to s giving

$$E\{T(0)\} = T_1(0) = P, \tag{2.6a}$$

$$\mathrm{Var}\{T(0)\} = T_2(0) - T_1^2(0) \approx P^2 - R \tag{2.6b}$$

with

$$P = \frac{s}{2b}\sqrt{\frac{\pi}{b}}\,e^{b/s^2}, \qquad R = \frac{2s}{b^2}\sqrt{\frac{\pi}{b}}\,e^{b/s^2}. \tag{2.7}$$

Since R is small with respect to $P^2$, one may approximate the life time function by an exponential distribution with $\lambda = 1/P$. However, in order not to loose information that is contained in R, we propose to use the Gamma distribution with

$$\lambda = 1/P \quad \text{and} \quad \alpha = 1 + R/P^2. \tag{2.8}$$

Let in experiments on accelerated life tests the physical stress be measured by $\sigma$. Assuming that $\sigma$ is proportional with the mathematical stress parameter s, we can estimate the life time under normal conditions as follows. For given $\sigma_{exp}$ the mean and variance of the life time of N samples in the experiment under stress are computed or a Gamma distribution is fitted to the outcome of the experiment. Then by using (2.6)-(2.8) we derive the values for b and $s_{exp}$. Since it is assumed that

$$s_{norm} = \frac{s_{exp}}{\sigma_{exp}} \sigma_{norm}, \tag{2.9}$$

we can compute the Gamma distribution approximating the life time distribution under normal conditions.

The first passage time problem for the parameter P of the introduction and the analogous problem for the state variable X above, can be combined:

$$T = \min(T_p, T_x) \tag{2.10}$$

with T having a mixed distribution

$$f(t) = q\, f_p(t) + (1-q) f_x(t),$$

$$q = \mathrm{Prob}(T_p < T_x) = \int_0^\infty F_p(s)\{1-F_x(s)\}ds.$$

However, it is expected that changes in P will affect the dynamics of X, like the change in stiffness of a spring will influence its deflection. A correct modelling of the interaction between system parameters and system variables requires a better knowledge of the internal dynamics.

## 3. THE FIRST PASSAGE TIME PROBLEM FOR SYSTEMS WITH CHANGING PARAMETERS

In this section we give the general formulation of the class of dynamical systems for which the first passage time problem can be analyzed. We assume that the change of the parameters is described by a Brownian motion process with a drift depending on the parameter values only. Then the system is written as

$$dX_i = F_i(X,P)dt + \sum_{k=1}^{m} s_{ik}(X,P)dW_k, \qquad i = 1,\dots,m, \tag{3.1a}$$

$$dP_j = G_j(P)dt + \sum_{l=1}^{n} r_{jl}(P)dW_l, \qquad j = 1,\dots,n, \tag{3.1b}$$

where $W_k(t)$ and $W_l(t)$ denote independent Wiener processes. The diffusion coefficients $s_{ik}$ and $r_{jl}$ may vanish in some of the equations. Moreover, forcing by coloured noise may be included, see Grasman (1985). Since P changes slowly compared with X, $G_j$ and $r_{jl}$ will be small. However, introduction of multi-time scales will not be necessary, as we may employ the smallness of coefficients in the asymptotic analysis of Dynkin's equation. If $G_j$ and $r_{jl}$ are allowed to depend on X, an averaging technique has to be applied, see Freidlin and Wentzell (1984). Since this brings about a considerable complication in the analysis, we exclude this possibility in our present investigations. For (3.1) with starting values within a domain $\Omega$ of the x,p-space we consider the time T(x,p) of reaching the boundary $\partial\Omega$. The statistical moments of T(x,p) satisfy a recurrent system of partial differential equations:

$$LT_q = -qT_{q-1}(x,p) \quad \text{in } \Omega \tag{3.2a}$$

$$T_q = 0 \text{ at } \partial\Omega, \quad q = 1,2,\ldots, \tag{3.2b}$$

where L denotes the elliptic partial differential operator

$$\tfrac{1}{2}\sum_{i,k=1}^{m} s_{ik}^2 \frac{\partial^2}{\partial x_i \partial x_k} + \tfrac{1}{2}\sum_{j,l=1}^{n} r_{jl}^2 \frac{\partial^2}{\partial p_j \partial p_l} + \sum_{i=1}^{m} F_i \frac{\partial}{\partial x_i} + \sum_{j=1}^{n} G_j \frac{\partial}{\partial p_j},$$

see Gardiner (1983). In the next section we will work out this problem in a particular example.

## 4. RELIABILITY OF A SWITCH

A switch is assumed to satisfy the differential equation

$$\frac{d^2z}{dt^2} + c\,\frac{dz}{dt} + \frac{dV}{dz} = F(t), \tag{4.1}$$

where parameter c is sufficiently large and

$$V(z) = z^4 - z^2. \tag{4.2}$$

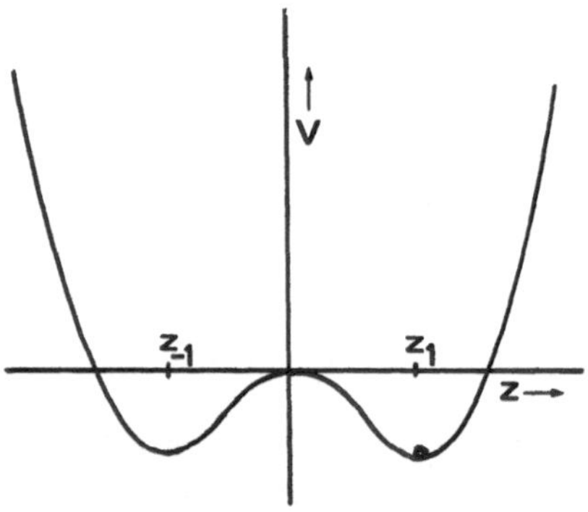


Fig. 1. A point mass in the potential field V(z) describing the dynamics of the switch with two stable positions $z_{-1}$ and $z_1$.

In fig. 1 it is observed that the system has only two stable stationary states. By the external force F(t) it may be put from one position in the other. The reliability of the switch depends on the parameter c and on the effect of external unintended perturbations upon the system. In order to investigate these factors we formulate the first passage time problem of the switch starting in position $z_1$ with random perturbations (white noise) for slowly decreasing c (stiffness). In the form (3.1) the system (4.1) becomes

$$dX_1 = X_2\, dt, \qquad X_1(0) = z_1, \tag{4.2a}$$
$$dX_2 = \{-PX_2 - V'(X_1)\}dt + s\, dW_2, \qquad X_2(0) = 0, \tag{4.2b}$$
$$dP = -a dt + r\, dW_1, \qquad P(0) = c_0. \tag{4.2c}$$

The domain $\Omega$ is bounded by $x_1 = 0$ (switch changes spontaneously from position) and $p = c_{min}$ (switch does not pass the quality requirments), see fig. 2. Since in this problem Eq. (3.2a) is parabolic, the boundary condition differs from (3.2b): $T_q = 0$ for $p = c_{min}$ and for $x_1 = 0$ with $x_2 < 0$.

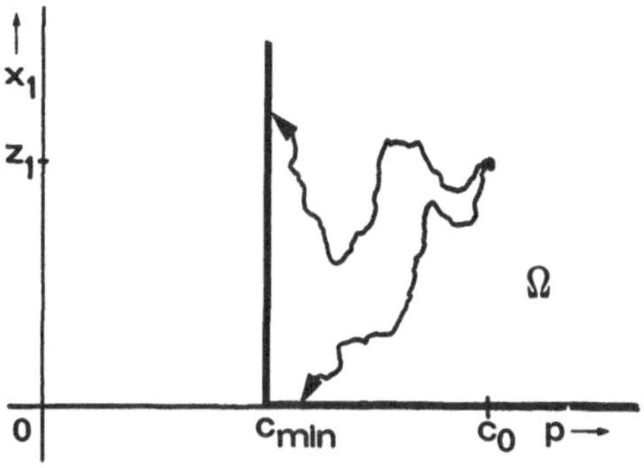


Fig. 2. The domain $\Omega$ for the first passage time problem (4.2).

In the present formulation we can also solve the problem of finding the right criterion to decide about replacement of the switch in order that exit through $x_1 = 0$ (failure) has a probability of $\alpha$ or less. Let the system have initial values

$$\{X(0), P(0)\} = (x, p)$$

Then the probability $u(x,p)$ of exit through $x_1 = 0$ satisfies

$$Lu = 0 \quad \text{in} \quad \Omega,$$
$$u = \begin{cases} 1 \text{ at } x_1 = 0,\ x_2 < 0, \\ 0 \text{ at } p = c_{min}. \end{cases}$$

Consequently, the probability of failure depends only on $c_{min}$ if one starts with a new switch in position 1:

$$\alpha(c_{min}) = u(z_1, 0, c_0),$$

see fig. 3. From this relation one derives the value $c_{min}$ at which the switch should be replaced such that the probability of failure is $\alpha$ or less.

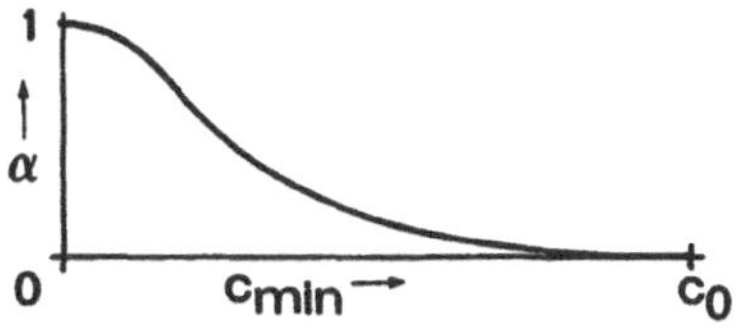


Fig. 3. Probability of failure as a function of the minimal stiffness

## REFERENCES


Capocelli, R.M., and Ricciardi, L.M., 1971, Diffusion approximation and first passage time problem for a model neuron, Kybernetik 8:214-233.

Freidlin, M.I., and Wentzell, A.D., 1984, "Random Perturbations of Dynamical Systems", Springer-Verlag, New York.

Gardiner, C.W., 1983, "Handbook of Stochastic Methods for Physics, Chemistry and the Natural Sciences", Springer-Verlag, Berlin.

Grasman, J., 1985, Estimates of large failure times from the theory of stochasticly perturbed dynamical systems with and without feedback, in "The Road-Vehicle-System and Related Mathematics", H. Neunzert, ed., B.G. Teubner, Stuttgart.

Martz, H.F., and Waller, R.A., 1982, "Bayesian Reliability Analysis", Wiley, New York.

Viertl, R., 1983, Nonlinear acceleration functions in life testing, Meth. of Operations Res. 47:115-122.

# CALIBRATION OF A RADIATION DETECTOR: CHROMOSOME DOSIMETRY FOR NEUTRONS

Peter G. Groer

Center for Epidemiologic Research
Oak Ridge Assoc. Univs.
Oak Ridge, TN 37831

Carlos A. De B. Pereira

Operations Research Center
U. C. Berkeley
Berkeley, CA 94720

## ABSTRACT

Calibrative densities for the unknown neutron dose $D_f$ of an individual accidentally exposed to high doses (>20 rad) of neutrons are derived. These densities incorporate prior dose information (e.g., from a dosimeter reading, or from dose reconstruction efforts), information from an *in vitro* calibration experiment with neutrons of the same energy, and information from the number of dicentric chromosome aberrations $y_f$ observed shortly (< 4 weeks) after exposure in a sample of $n_f$ lymphocytes from the exposed individual. If Y, the number of dicentric aberrations induced by a known neutron dose d in n lymphocytes is assumed to be Poisson distributed ($Y \sim Po(n\alpha d)$) and if $D_f$ and the parameter $\alpha$ are assumed to have gamma priors it is possible to give an analytic solution for the calibrative density $f(d_f|\mathfrak{D})$. $\mathfrak{D}$ consists of the calibration data and the observed aberrations in a sample of lymphocytes from the exposed individual. This density characterizes the remaining uncertainty about $D_f$ after consideration of the prior information about $D_f$ and $\alpha$ and of the data $\mathfrak{D}$.

## INTRODUCTION

After accidental exposures to low or high LET radiation it is desirable to obtain dose estimates for the accident victims. Estimation of doses is also mandated by regulations dependent on the severity of the accident. Personal dosimeters and dose reconstruction by health physicists can provide initial information about the magnitude of the radiation doses. In this paper we take the view that this prior dose information should be combined with data on chromosome aberrations observed in a sample of lymphocytes from the accident victim(s) to reduce the uncertainty about the radiation doses to which the accident victims were exposed. Data on chromosome aberrations can be used to make inferences about radiation doses with the help of data from *in vitro* calibration experiments. In such experiments $n_i (i=1,2,\ldots N)$ lymphocytes are exposed *in vitro* to several fixed doses $d_i$ and the resulting aberrations $y_i$ are scored under the microscope. The resulting reduction in the dose uncertainty helps physicians to decide if and how the victim(s) should be treated.

## DERIVATION OF THE CALIBRATIVE DENSITY

We are interested in obtaining an expression for the calibrative density[1] $f(d_f|\mathfrak{D})$ . $\mathfrak{D}$ stands for all the data and consists of the following observed events:

$$\mathfrak{D} = \left\{\{Y_1=y_1|D_1=d_1,n_1\},\ldots\{Y_N=y_N|D_N=d_N,n_N\},\ \{Y_f=y_f|n_f\}\right\}$$

The first $N$ events correspond to the data from the "controlled" calibration experiment. In this experiment $n_i$ cells are exposed to a neutron dose $d_i$ which is accurately controlled by the experimenter. In the $n_i$ cells so exposed $y_i$ chromosome aberrations are observed. The last event consists of the observation of $y_f$ chromosome aberrations in $n_f$ cells of the accident victim who was exposed to an unknown neutron dose $D_f$ . The subscript $f$ is mnemonic for "future" and indicates that $y_f$ is observed after the calibration experiment has been performed.

Derivation of the calibrative density for $D_f$ , $f(d_f|\mathfrak{D})$ involves expressing this density in terms of other densities and probabilities using the rules of probability. First, we will give a general derivation without a specific model and priors. This derivation involves a model parameter $\alpha$ which will later be identified as the rate at which chromosome aberrations are produced after neutron irradiation.

$$f(d_f|\mathfrak{D}) = \int_0^\infty f(d_f,\alpha|\mathfrak{D})\, d\alpha$$

$$= \int_0^\infty f(d_f|\alpha,y_f,\mathfrak{D}')\, f(\alpha|y_f,\mathfrak{D}')\, d\alpha \qquad (1)$$

Where $\mathfrak{D}' = \mathfrak{D} - \{y_f|n_f\}$ stands for the calibration data and $\{y_f|n_f\}$ is abbreviated by $y_f$ in Equ. (1). Since conditional on $\alpha$, $D_f$ is independent of $\mathfrak{D}'$, we can rewrite the last line of Equ. (1):

$$f(d_f|\mathfrak{D}) = \int_0^\infty f(d_f|\alpha,y_f)\, f(\alpha|y_f,\mathfrak{D}')\, d\alpha$$

$$= \int_0^\infty \frac{p(y_f|d_f,\alpha)\, f(d_f|\alpha)}{p(y_f|\alpha)} \cdot f(\alpha|y_f,\mathfrak{D}')\, d\alpha$$

$$= \int_0^\infty \frac{p(y_f|d_f,\alpha)\, f(d_f)}{p(y_f|\alpha)} \cdot \frac{p(y_f|\alpha,\mathfrak{D}')\ \ f(\alpha|\mathfrak{D}')}{p(y_f|\mathfrak{D}')}\, d\alpha$$

$$\propto \quad f(d_f) \int_0^\infty p(y_f|d_f,\alpha)\, f(\alpha|\mathfrak{D}')\, d\alpha \qquad (2)$$

Equ. (2) states that the calibrative density is proportional to the prior density for $D_f$ and the predictive density for a future number of chromosome aberrations $y_f$. The predictive distribution incorporates, of course, the information from the calibration experiment through the posterior distribution $f(\alpha|\mathfrak{D}')$. For the derivation of Equ. (2) we assumed that the prior for $D_f$ does not depend on $\alpha$. We proceed now to insert the appropriate model and the prior distributions used into Equ. (2) to obtain the special form of $f(d_f|\mathfrak{D})$ which applies to dose estimation after neutron irradiation.

In the literature on cytogenetic dosimetry (see e.g. Ref. 2) we found that a Poisson model for $Y$, the number of chromosome aberrations induced, is used for all types - high or low LET - of ionizing radiation. In the case of neutron exposure this model is:

$$Y|(\alpha,n,d) \sim Po(n\alpha d) \qquad (3)$$

In words the Poisson mean is proportional to the neutron dose $d$ delivered to the $n$ cells. This simple model neglects the background rate of chromosome aberrations. Since the background frequency is in the range of 1 to 2 per thousand cells,[3] it can be neglected for doses of 20 rads and more. We judged gamma priors for $\alpha$ and $D_f$: $\alpha \sim Ga(a,b)$, $D_f \sim Ga(A,B)$ . The gamma family of distributions is rich enough to be able to express a multitude of judgements about the uncertainty surrounding $\alpha$ and $D_f$ and in addition it provides for the Poisson model the usual mathematical conveniences of a family of conjugate prior distributions.

Inserting the Poisson model for Y and the priors for $\alpha$ and $D_f$ into Equ. (2) one finds:

$$f(d_f|\mathcal{D}) \propto d_f^{\mathcal{B}-1} (d_f + \mathcal{A})^{-\mathcal{b}} \exp(-Ad_f) \tag{4}$$

where $\mathcal{B} = B + y_f$ , $\mathcal{A} = \mathcal{a}/n_f$ with

$$\mathcal{a} = a + \sum_{i=1}^{N} n_i d_i \quad \text{and} \quad \mathcal{b} = \sum_{i=1}^{N} y_i + y_f + b \quad .$$

The mode of $f(d_f|\mathcal{D})$ occurs at

$$d_M = \left[-\mathcal{C} + \sqrt{(\mathcal{C}^2 + 4A\mathcal{A}(\mathcal{B}-1)}\right]/2A$$

with $\mathcal{C} = A\mathcal{A} + \mathcal{b} + 1 - \mathcal{B}$

In the following section we will use Equ. (4) for a particular calibration experiment and show graphs of $f(d_f|\mathcal{D})$ for different observed $y_f$ and $n_f$ .

EXAMPLE

The example is based on a hypothetical radiation accident with a $^{210}$Po-Be neutron source. Readout of the neutron dosimeter worn by the victim and subsequent calculations by Bayesian health physicists based on calibration data for the neutron dosimeter and the geometry of the accident yielded a calibrative density $f(d_f)$ for the unknown dose $D_f$ to the lymphocytes of the victim which could be approximated by a $Ga(A = .1, B = 10)$ . This gamma density becomes the prior for the subsequent analysis of dicentric chromosome aberrations in a sample of lymphocytes from the victim. In $n_f = 104$ metaphases a cytogenetic technician scored $y_f = 64$ dicentric aberrations shortly after the accident. Based on calibration experiments with other types of radiation the prior distribution for $\alpha$ was judged to be $Ga(a = 1000 , b = 10)$ . As stated already earlier $Y \sim Po(n\alpha d)$ is the statistical model. This model is thought to be valid for a wide range of neutron doses.[2] If this model, the gamma priors for $\alpha$ and

TABLE 1. Calibration data for Po-Be Neutrons (from reference 2)

| dose $d_i$ (rad) | Cells $n_i$ | Dicentrics $y_i$ |
|---|---|---|
| 50 | 269 | 109 |
| 75 | 78 | 47 |
| 100 | 115 | 94 |
| 150 | 90 | 114 |
| 200 | 84 | 138 |
| 250 | 59 | 125 |
| 300 | 37 | 97 |

$D_f$ and the calibration data[2] shown in Table 1 are used one obtains the calibrative density shown in Fig. 1 from Equ. (2). Fig. 1 shows also the prior density for $D_f$ . Both densities are divided by $f(d_M|\mathcal{D})$ where $d_M$ is the modal dose. Fig. 2 shows $f(d_f|\mathcal{D})$ for $y_f = 8$ dicentrics in $n_f = 13$ metaphases. This density is clearly wider than $f(d_f|\mathcal{D})$ shown in Fig. 1. In practice the calibrative densities could be updated sequentially and scoring of metaphases could stop whenever the physician is satisfied by the obtained precision.

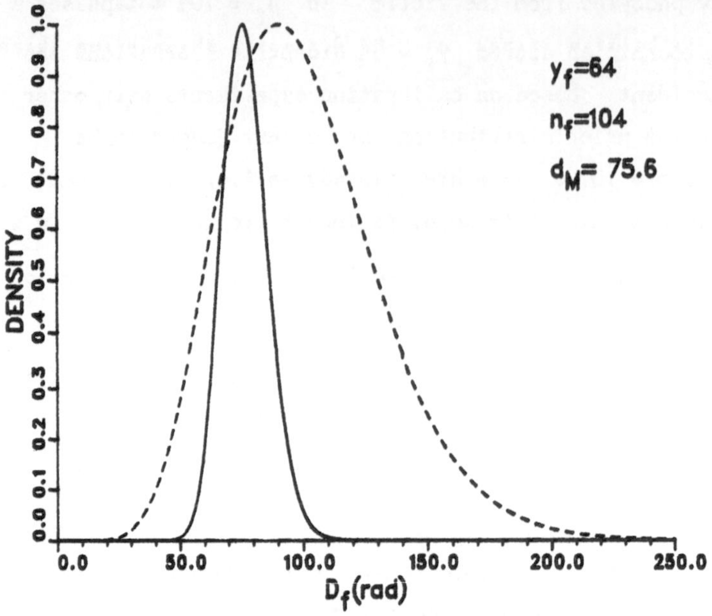


Fig. 1 : Prior and calibrative density for $D_f$ with $y_f = 64$ and $n_f = 104$.

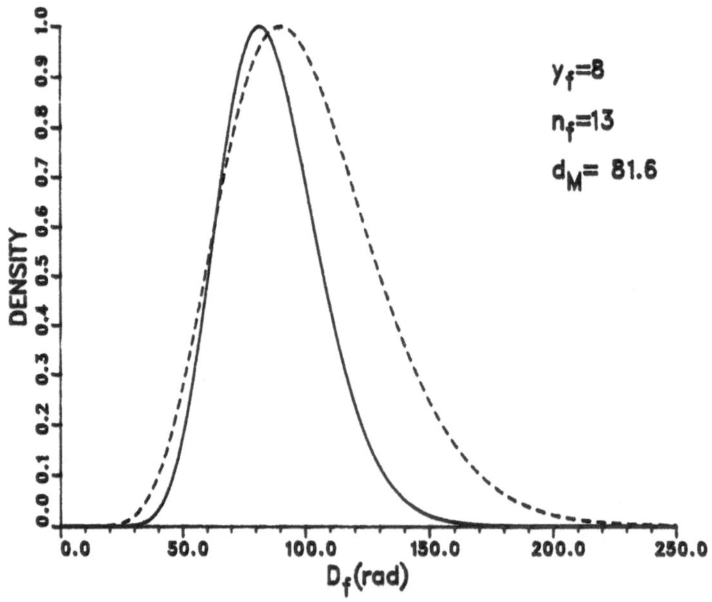


Fig. 2 : Prior and calibrative density for $D_f$ with $y_f = 8$ and $n_f = 13$.

## SUMMARY

We derived calibrative densities for the unknown neutron dose $D_f$ of a hypothetical accident victim and pointed out that it is possible to obtain analytic solutions for this type of radiation if gamma priors and the "public" Poisson model from the cytogenetic literature are used. Our calculations neglected the small background frequency and are therefore only valid for doses which evidence a much greater number of aberrations. Incorporation of a background rate $\alpha_0$ into the analysis would extend the results to lower doses. This extension is presently under investigation. Another extension involves exposure to so-called low LET radiation like $\gamma$- and X-rays. For this case the public model[2] is $Y \sim Po[n(\alpha_0+\alpha d+\beta d^2)]$ and $f(d_f|\mathscr{D})$ cannot be given in closed form.

It is standard practice to estimate doses for accident victims by deterministic procedures using a maximum likelihood estimate for the model parameters. In our example this would give $\hat{d}_f = y_f/(n_f\hat{\alpha})$ as our estimate of the neutron dose. With this procedure the uncertainty about $D_f$ cannot be specified and other information about $D_f$ from a personal dosimeter or from dose reconstruction efforts cannot be incorporated.

## ACKNOWLEDGEMENTS

We wish to thank Drs. Michael Bender, Gayle Littlefield and Julian Preston for teaching us the essentials of cytogenetic dosimetry. Without their knowledge, encouragement and patience this work would not have been possible. Support for this work was provided by the National Cancer Institute under interagency agreement 40-849-85, NCI No. Y01-CP-50512 with the U.S. Department of Energy and by contract No. DE-AC05-760R00033 between the U.S. Department of Energy Office of Energy Research and Oak Ridge Assoc. Univs.

## REFERENCES

1. Aitchison, J., and Dunsmore, I.R., *Statistical Prediction Analysis*, Cambridge University Press, Cambridge (1975).

2. Edwards, A.A., Lloyd, D.C., and Purrot, R.J., 1979, "Radiation induced chromosome aberrations and the Poisson distribution," *Rad. and Environ. Biophys.*, 16:80.
3. Bender, M.A., Preston, R.J., Leonard, R.C., and Shelby, M.D., 1986, *Chromosomal aberration analysis in peripheral lymphocytes of a large control population*, unpublished manuscript, Brookhaven National Laboratory, Upton, New York.

# ON SOME BAYES AND EMPIRICAL BAYES SELECTION PROCEDURES

Shanti S. Gupta and TaChen Liang

Department of Statistics
Purdue University
West Lafayette, IN

## 1. INTRODUCTION

A common problem faced by an experimenter is one of comparing several populations (processes, treatments). Suppose that there are $k(\geq 2)$ populations $\pi_1, \ldots, \pi_k$ and for each i, $\pi_i$ is characterized by the value of a parameter of interest, say $\theta_i$. The classical approach to this problem is to test the homogeneity hypothesis $H_0: \theta_1 = \ldots = \theta_k$. However, the classical tests of homogeneity are inadequate in the sense that they do not answer a frequently encountered experimenter's question, namely, how to identify the "best" population or how to select the more promising (worthwhile) subset of the populations for further experimentation. These problems are known as ranking and selection problems. The formulation of ranking and selection procedures has been accomplished generally using either the indifference zone approach (see Bechhofer (1954)) or the subset selection approach (see Gupta (1956, 1965)). A discussion of their differences and various modifications that have taken place since then can be found in Gupta and Panchapakesan (1979).

In many situations, an experimenter may have some prior information about the parameters of interest, and he would like to use this information to make an appropriate decision. In this sense, the classical ranking and selection procedures may seem conservative if the prior information is not taken into consideration. If the information at hand can be quantified into a single prior distribution, one would like to apply a Bayes procedure since it achieves the minimum of Bayes risks among a class of decision procedures. Some contributions to ranking and selection problems using Bayesian approach have been made by Deely and Gupta (1968), Bickel and Yahav (1977), Chernoff and Yahav (1977), Goel and Rubin (1977), Gupta and Hsu (1978), Miescke (1979), Gupta and Hsiao (1981), Gupta and Miescke (1984), Gutpa and Yang (1985), and Berger and Deely (1986).

Now, consider a situation in which one is repeatedly dealing with the same selection problem independently. In such instances, it is reasonable to formulate the component problem in the sequence as a Bayes decision problem with respect to an unknown prior distribution on the parameter space, and then, use the accumulated observations to improve the decision at each stage. This is the empirical Bayes approach due to Robbins (1956, 1964, and 1983). Empirical Bayes procedures have been derived for subset selection goals by Deely (1965). Recently, Gupta and Hsiao (1983) and Gupta and Liang (1984, 1986) have studied some selection problems using the empirical Bayes approach. Many such empirical Bayes procedures have been shown to be asymptotically optimal in the sense that the risk for the n-th decision problem converges to the optimal Bayes risk which would have been obtained if the prior distribution was fully known and the Bayes procedure with respect to this prior distribution was used.

In the present paper, we describe selection and ranking procedures using prior distributions or using the information contained in the past data. Section 2 of this paper deals with the problem of selecting the best population through Bayesian approach. An essentially complete class is obtained for a class of reasonable loss functions. We also discuss Bayes-P* selection procedures which are better than the classical subset selection procedures in terms of the size of selected subset. In Section 3, we set up a general formulation of the empirical Bayes framework for selection and ranking problems. Two selection problems dealing with binomial and uniform populations are discussed in detail.

## 2. BAYESIAN APPROACH

### 2.1 Notations and Formulation of the Selection Problem

Let $\theta_i \in \Theta \subset \mathbb{R}$ denote the unknown characteristic of interest associated with population $\pi_i$, $i = 1,..,k$. Let $X_1,\ldots,X_k$ be random variables representing the k populations $\pi_i$, $i = 1,\ldots,k$, respectively, with $X_k$ having the probability density function (or probability frequency function in discrete case) $f_i(x|\theta_i)$. In many cases, $X_i$ is a sufficient statistic for $\theta_i$. It is assumed that given $\underset{\sim}{\theta} = (\theta_1,\ldots,\theta_k)$, $\underset{\sim}{X} = (X_1,\ldots,X_k)$ have a joint probability density function $\underset{\sim}{f}(\underset{\sim}{x}|\underset{\sim}{\theta}) = \prod_{i=1}^{k} f_i(x_i|\theta_i)$, where $\underset{\sim}{x} = (x_1,\ldots,x_k)$. Let $\theta_{[1]} \leq \theta_{[2]} \leq \ldots \leq \theta_{[k]}$ denote the ordered values of $\theta_i$'s and let $\pi_{[i]}$ denote the unknown population associated with $\theta_{[i]}$. The population $\pi_{[k]}$ will be called the best population. If there are more than one population satisfying this condition, we arbitrarily tag one of them and call it the best one. Also let $\Omega = \{\underset{\sim}{\theta}|\theta_i \in \Theta, i = 1,..,k\}$ denote the parameter space and let $G(\cdot)$ denote a prior distribution on $\underset{\sim}{\theta}$ over $\Omega$.

Let $\mathcal{A}$ be the action space consisting of all the $2^k - 1$ nonempty subsets of the set $\{1,\ldots,k\}$. When action S is taken, we mean that population $\pi_i$ is included in the selected subset if $i \in S$. For each $\underset{\sim}{\theta} \in \Omega$ and $S \in \mathcal{A}$, let $L(\underset{\sim}{\theta},S)$ denote the loss incurred when $\underset{\sim}{\theta}$ is the true state of nature and the action S is taken. A decision procedure d is defined to be a mapping from $\mathcal{X} \times \mathcal{A}$ into $[0,1]$, where $\mathcal{X}$ is the sample space of $\underset{\sim}{X} = (X_1,\ldots,X_k)$.

Let D be the set of all decision procedures $d(\underset{\sim}{x},S)$. For each $d \in D$, let $B(d,G)$ denote the associated Bayes risk. Then, $B(G) = \inf_{d\in D} B(d,G)$ is the minimum Bayes risk. An optimal decision procedure, denoted by $d_G$, is obtained if $d_G$ has the property that

$$B(d_G,G) = B(G). \tag{2.1}$$

Such a procedure is called Bayes with respect to G. Under some regularity conditions,

$$(2.2) \qquad B(d,G) = \int_{\mathcal{X}} \sum_{S \in \mathcal{A}} d(\underset{\sim}{x},S) \int_{\Omega} L(\underset{\sim}{\theta},S)\, \underset{\sim}{f}(\underset{\sim}{x}|\underset{\sim}{\theta})dG(\underset{\sim}{\theta})d\underset{\sim}{x}.$$

Now let

$$(2.3) \qquad r_G(\underset{\sim}{x},S) = \int_{\Omega} L(\underset{\sim}{\theta},S)\, \underset{\sim}{f}(\underset{\sim}{x}|\underset{\sim}{\theta})dG(\underset{\sim}{\theta}),$$

$$(2.4) \qquad A_G(\underset{\sim}{x}) = \{S \in \mathcal{A} | r_G(\underset{\sim}{x},S) = \min_{S' \in \mathcal{A}} r_G(\underset{\sim}{x},S')\}.$$

Then, a sufficient condition for (2.1) is that $d_G$ satisfies

$$(2.5) \qquad \sum_{S \in A_G(\underset{\sim}{x})} d_G(\underset{\sim}{x},S) = 1.$$

## 2.2 An Essentially Complete Class of Decision Procedures

In this subsection, we consider a class of loss functions possessing the following properties:

Let H denote the group of all permutations of the components of a k-component vector.

Definition 2.1: A loss function L has property T if

(a) $L(\underset{\sim}{\theta},S) = L(h\underset{\sim}{\theta},hS)$ for all $\underset{\sim}{\theta} \in \Omega$, $S \in \mathcal{A}$ and $h \in H$, and

(b) $L(\underset{\sim}{\theta},S') \leq L(\underset{\sim}{\theta},S)$ if the following holds for each pair $(i,j)$ with $\theta_i \leq \theta_j$: $i \in S$, $j \notin S$ and $S' = (S - \{i\}) \cup \{j\}$.

The property (a) assures the invariance under permutation and property (b) assures the monotonicity of the loss function. In many situations, a loss function satisfying these assumptions seems quite natural.

We now let $x_{(1)} \leq x_{(2)} \leq \cdots \leq x_{(k)}$ denote the ordered observations. Here the subscript $(i)$ can be viewed as the (unknown) index of the population associated with the observation $x_{(i)}$. For each $j = 1, \ldots, k$, let $S_j = \{(k),\ldots,(k - j + 1)\}$, and the remaining subsets $S_j$ be associated one-to-one with $j = k + 1,\ldots,2^k - 1$, arbitrarily. Also, let $\mathcal{A}_m = \{S \in \mathcal{A} \| S| = m\}$, $m = 1,\ldots,k$, and $D_1 = \{d \in D | \sum_{j=1}^{k} d(\underset{\sim}{x},S_j) = 1 \text{ for all } \underset{\sim}{x} \in \mathcal{X}\}$.

Theorem 2.1: Suppose that $f_i(x_i|\theta_i) = f(x_i|\theta_i)$, $i = 1,\ldots,k$, where the pdf $f(x|\theta)$ possesses the monotone likelihood ratio (MLR) property, and the prior distribution G is symmetric on $\Omega$. Also, suppose that the loss function has property T. Then,

(a) for each $m = 1,\ldots,k$, $r_G(\underset{\sim}{x},S_m) \leq r_G(\underset{\sim}{x},S)$ for all $S \in \mathcal{A}_{k-m+1}$, $\underset{\sim}{x} \in \mathcal{X}$, and

(b) $D_1$ is an essentially complete class in D.

Proof: The proof for part (a) is analogous to that of Theorem 3.3 of Gupta and Yang (1985). For part (b), let d be any decision procedure in D. Consider the decision procedure d* defined as: for $\underset{\sim}{x} \in \mathcal{X}$,

$$d^*(\underset{\sim}{x},S_m) = \sum_{S \in \mathcal{A}_{k-m+1}} d(\underset{\sim}{x},S),\ m = 1,\dots,k;$$

$$d^*(\underset{\sim}{x},S) = 0,\ S \neq S_m,\ m = 1,\dots,k.$$

Then, $d^* \in D_1$. Also, by part (a) and (2.2), one can see that $B(d^*,G) \leq B(d,G)$, which completes the proof.

Let $A'_G(\underset{\sim}{x}) = \{S_j | 1 \leq j \leq k,\ r_G(\underset{\sim}{x},S_j) = \min_{1 \leq i \leq k} r_G(\underset{\sim}{x},S_i)$. Then, under the condition of Theorem 2.1, any Bayes procedure $d_G$ satisfies $\sum_{S_j \in A'_G(\underset{\sim}{x})} d_G(\underset{\sim}{x},S_j) = 1$ for all $\underset{\sim}{x} \in \mathcal{X}$.

## 2.3 Bayes-P* Selection Procedures

A selection procedure $\underset{\sim}{\psi} = (\psi_1,\dots,\psi_k)$ is defined to be a mapping from $\mathcal{X}$ to $[0,1]^k$, where $\psi_i(\underset{\sim}{x})$: $\mathcal{X} \to [0,1]$ is the probability that $\pi_i$ is included in the selected subset when $\underset{\sim}{X} = \underset{\sim}{x}$ is observed. A correct selection (CS) is defined to be the selection of any subset that includes the best population.

In the decision-theoretic approach, a Bayes decision (selection) procedure always provides a decision with the minimum risk under a certain loss. However, in practive, one always has the difficulty in figuring out what the loss may be and the Bayesian result is quite sensitive to the loss used; in this sense, a Bayes procedure does not mean that its quality is good enough to pass a certain level. For guaranteeing the quality of a decision (selection) procedure one would like to have a "quality control" criterion about the class of all possible decision (selection) procedures. That is, any procedure with lower quality will be removed, even though it might be the cheapest one under some losses. Analogous to the classical subset selection approach, Gupta and Yang (1985) set up a control criterion using the Bayesian approach. Let

$$(2.6)\qquad p_i(\underset{\sim}{x}) = P(\pi_i \text{ is the best } | \underset{\sim}{X} = \underset{\sim}{x}) = P(\theta_i \text{ is the largest } | \underset{\sim}{X} = \underset{\sim}{x})$$

be the posterior probability that population $\pi_i$ is the best population when $\underset{\sim}{X} = \underset{\sim}{x}$ is observed. Then, for selection procedure $\underset{\sim}{\psi}$, the posterior probability of a correct selection given $\underset{\sim}{X} = \underset{\sim}{x}$ is

$$(2.7)\qquad P(CS|\underset{\sim}{\psi},\underset{\sim}{X} = \underset{\sim}{x}) = \sum_{i=1}^{k} \psi_i(\underset{\sim}{x})p_i(\underset{\sim}{x})$$

Definition 2.2: Given a number P*, $k^{-1} < P^* < 1$, and a prior G on $\Omega$, we say a selection procedure $\underset{\sim}{\psi}$ satisfies the PP*-condition (posterior P*-condition) if

(a) $\psi_i(\underset{\sim}{x}) = 1$ for at least some i, $1 \leq i \leq k$, and

(b) $P(CS|\underset{\sim}{\psi},\underset{\sim}{X} = \underset{\sim}{x}) \geq P^*$ for all $\underset{\sim}{x} \in \mathcal{X}$.

Note that $\sum_{i=1}^{k} p_i(\underset{\sim}{x}) = 1$ for all $\underset{\sim}{x} \in \mathcal{X}$; hence this kind of selection procedures always exist. We let $C = C(P^*)$ be the class of all selection procedures satisfying the PP*-condition.

Let $p_{[1]}(\underset{\sim}{x}) \leq \ldots \leq p_{[k]}(\underset{\sim}{x})$ be the ordered $p_i(\underset{\sim}{x})$'s and let $\pi_{(i)}$ be the population associated with $p_{[i]}(\underset{\sim}{x})$, $i = 1,\ldots,k$. Then a selection procedure $\underset{\sim}{\psi}$ can be completely specified by $\{\psi_{(1)},\ldots,\psi_{(k)}\}$, where

$$\text{(2.8)} \qquad \psi_{(i)}(\underset{\sim}{x}) = P(\pi_{(i)} \text{ is selected } | \underset{\sim}{\psi}, \underset{\sim}{X} = \underset{\sim}{x}\}, \ i = 1,\ldots,k.$$

For a given number $P^*$, $k^{-1} < P^* < 1$, and an observation $\underset{\sim}{X} = \underset{\sim}{x}$, let $j = \max\{m | \sum_{i=m}^{k} p_{[i]}(\underset{\sim}{x}) \geq P^*\}$. Gupta and Yang (1985) proposed a selection procedure $\underset{\sim}{\psi}^G = (\psi_1^G,\ldots,\psi_k^G)$ defined as below:

$$\psi_{(k)}^G(\underset{\sim}{x}) = 1, \text{ and for } 1 \leq i \leq k-1,$$

$$\psi_{(i)}^G(\underset{\sim}{x}) = \begin{cases} 1 & \text{if } i > j, \\ \lambda & \text{if } i = j, \\ 0 & \text{if } i < j, \end{cases}$$

where the constant $\lambda$ is determined so that

$$\lambda p_{[j]}(\underset{\sim}{x}) + \sum_{m=j+1}^{k} p_{[m]}(\underset{\sim}{x}) = P^*.$$

It is clear that $\underset{\sim}{\psi}^G \in C$. In the following, optimality of this selection procedure is investigated.

<u>Definition 2.3</u>: A selection procedure $\underset{\sim}{\psi}$ is called ordered if for every $\underset{\sim}{x} \in \mathcal{X}$, $x_i \leq x_j$ implies $\psi_i(\underset{\sim}{x}) \leq \psi_j(\underset{\sim}{x})$. It is called monotone or just if for every $i = 1,\ldots,k$, and $\underset{\sim}{x}, \underset{\sim}{y} \in \mathcal{X}$, $\psi_i(\underset{\sim}{x}) \leq \psi_i(\underset{\sim}{y})$ whenever $x_i \leq y_i$, $x_j \geq y_j$ for any $j \neq i$.

Sufficient conditions for $\underset{\sim}{\psi}^G$ to be ordered and monotone are given below:

<u>Theorem 2.2</u>: Let $G(\underset{\sim}{\theta}|\underset{\sim}{x})$ be the posterior cdf of $\underset{\sim}{\theta}$, given $\underset{\sim}{X} = \underset{\sim}{x}$. Let $G(\underset{\sim}{\theta}|\underset{\sim}{x})$ be absolutely continuous and have the generalized stochastic increasing property, that is:

(1) $G(\underset{\sim}{\theta}|\underset{\sim}{x}) = \prod_{i=1}^{k} G_i(\theta_i|\underset{\sim}{x})$, $G_i(\cdot|\underset{\sim}{x})$ = posterior cdf of $\theta_i$.

(2) $G_i(t|\underset{\sim}{x}) \geq G_j(t|\underset{\sim}{x})$ for any $t$, whenever $x_i \leq x_j$.

Then, $\underset{\sim}{\psi}^G$ is ordered and monotone.

Gupta and Yang (1985) also investigated some optimal behavior of this procedure through the decision-theoretic approach over a class of loss functions.

Definition 2.4: A loss function L has proporty T' if

(a) L has property T, and

(b) $L(\underset{\sim}{\theta},S) \leq L(\underset{\sim}{\theta},S')$ if $S \subset S'$.

Theorem 2.3: Under the assumption of Theorem 2.2, the selection procedure $\underset{\sim}{\psi}^G$ is Bayes in C provided that the loss function has property T'.

Gupta and Yang (1985) investigated the computation of $p_i(\underset{\sim}{x})$ for the "normal model" by using normal and non-informative priors. Berger and Deely (1986) have considered another selection problem, and given a more detailed discussion about the computation of $p_i(\underset{\sim}{x})$ under several different priors.

## 3. EMPIRICAL BAYES APPROACH

In this section, we continue with the general setup of Section 2. However, we assume only the existence of prior distribution G on $\Omega$, and the form of G is unknown or partially known. In Section 3.1, we consider decision procedures for general loss functions. In Sections 3.2 and 3.3, empirical Bayes selection procedures are concerned.

### 3.1 Formulation and Summary of the Empirical Bayes Selection Problems

For each i, $i = 1,\ldots,k$, let $X_{ij}$ denote the random observation from $\pi_i$ at stage j. Let $\Theta_{ij}$ denote the random characteristic of $\pi_i$ at stage j. Conditional on $\Theta_{ij} = \theta_{ij}$, $X_{ij}|\theta_{ij}$ has the pdf (or pf in discrete case) $f_i(x|\theta_{ij})$. Let $\underset{\sim}{X}_j = (X_{1j},\ldots,X_{kj})$ and $\underset{\sim}{\theta}_j = (\theta_{1j},\ldots,\theta_{kj})$. Suppose that independent observations $\underset{\sim}{X}_1,\ldots,\underset{\sim}{X}_n$ are available and $\underset{\sim}{\theta}_j$, $1 \leq j \leq n$, have the same distribution G for all j, though $\underset{\sim}{\theta}_j$ are not observable. We also let $\underset{\sim}{X} = (X_1,\ldots,X_k)$ denote the present random observation.

Consider an empirical Bayes decision procedure $d_n$. Let $B(d_n,G)$ be the Bayes risk associated with the decision procedure $d_n$. Then

$$B(d_n,G) = \int_{\Omega} E \int_{\mathcal{X}} \sum_{S \in \mathcal{A}} d_n((\underset{\sim}{x};\underset{\sim}{X}_1,\ldots,\underset{\sim}{X}_n),S)L(\underset{\sim}{\theta},S)f(\underset{\sim}{x}|\underset{\sim}{\theta})d\underset{\sim}{x}dG(\underset{\sim}{\theta}),$$

where $d_n((\underset{\sim}{x};\underset{\sim}{X}_1,\ldots,\underset{\sim}{X}_n),S)$ $(\equiv d_n(\underset{\sim}{x},S))$ is the probability of selecting the subset S when $(\underset{\sim}{x};\underset{\sim}{X}_i,\ldots,\underset{\sim}{X}_n)$ is observed, and the expectation E is taken with respect to $(\underset{\sim}{X}_1,\ldots,\underset{\sim}{X}_n)$. Note that $B(d_n,G) - B(G) \geq 0$, since $B(G)$ is the minimum Bayes risk. This nonnegative difference may be used as a measure of the optimality of the decision procedure $d_n$.

Definition 3.1: A sequence of decision procedures $\{d_n\}_{n=1}^{\infty}$ is said to be asymptotically optimal relative to the prior distribution G if $B(d_n,G) \to B(G)$ as $n \to \infty$.

Let $\underset{\sim}{L}(\theta) = \max_{S \in \mathcal{A}} |L(\underset{\sim}{\theta},S)|$ and assume that $\int L(\underset{\sim}{\theta})dG(\underset{\sim}{\theta}) < \infty$. Following Robbins (1964), one can see that a sufficient condition for the sequence $\{d_n\}$ to be asymptotically optimal is that $d_n(\underset{\sim}{x},S) \xrightarrow{P} d_G(\underset{\sim}{x},S)$ for all $\underset{\sim}{x} \in \mathcal{X}$ and $S \in \mathcal{A}$, where "$\xrightarrow{P}$" means convergence in probability (with respect to $(\underset{\sim}{X}_1,\ldots,\underset{\sim}{X}_n)$).

Let $G_n$ be a distribution function on the parameter space $\Omega$. Suppose $G_n$ is a function of $(\underset{\sim}{X}_1,\ldots,\underset{\sim}{X}_n)$ such that $P\{\lim_{n\to\infty} G_n(\underset{\sim}{\theta}) = G(\underset{\sim}{\theta})$ for every continuous point $\underset{\sim}{\theta}$ of $G\} = 1$, where the probability is with respect to $(\underset{\sim}{X}_1,\ldots,\underset{\sim}{X}_n)$. Let the loss function $L(\underset{\sim}{\theta},S)$ and the density $f(\underset{\sim}{x}|\underset{\sim}{\theta})$ be such that $L(\underset{\sim}{\theta},S)f(\underset{\sim}{x}|\underset{\sim}{\theta})$ is bounded and continuous in $\underset{\sim}{\theta}$ for every $S \in \mathcal{A}$. Then $\{d_{G_n}\}$ is asymptotically optimal with respect to G if $\int_\Omega L(\underset{\sim}{\theta})dG(\underset{\sim}{\theta}) < \infty$, where $d_{G_n}$ is a Bayes procedure with respect to the distribution $G_n$.

To find $G_n$, we may assume G to be a member of some parametric family $\Gamma$ with unknown hyperparameters, say $\underset{\sim}{\lambda} = (\lambda_1,\ldots,\lambda_k)$. Suppose now an estimator $\underset{\sim}{\lambda}_n = (\lambda_{1n},\ldots,\lambda_{kn})$ depending on the previous observations $(\underset{\sim}{X}_1,\ldots,\underset{\sim}{X}_n)$ can be found such that $G_n$ converges to G with probability one. Note that $G_n$ is also a member in $\Gamma$. We then follow the typically Bayesian analysis and derive the Bayes procedure $d_{G_n}$ with respect to the estimated prior distribution $G_n$. Then, according to the result of Deely (1965), the sequence of empirical Bayes procedures $\{d_{G_n}\}$ is asymptotically optimal. This approach is referred to as parametric empirical Bayes. Deely (1965) has derived the empirical Bayes procedures through the parametric empirical Bayes approach in several special cases among which are (a) normal-normal, (b) normal-uniform, (c) binomial-beta, and (d) Poisson-gamma.

In another approach, called nonparametric empirical Bayes, one just assumes that $\underset{\sim}{\theta}_j$, j = 1,2,..., are independently and identically distributed; however, the form of the prior distribution G on $\Omega$ is completely unknown. In this situation, one may represent the Bayes procedure in terms of the unknown prior and then use the data to estimate the Bayes procedure directly. This approach has been used by Van Ryzin and Susarla (1977), Gupta and Hsiao (1983), and Gupta and Liang (1984, 1986), among others.

In the following sections, we consider some selection problems with underlying populations having binomial or uniform distributions. We will use the nonparametric empirical Bayes approach.

## 3.2 Empirical Bayes Procedures Related to Binomial Populations

In this section, two selection problems related to binomial populations are discussed: selecting the best among k binomial populations and selecting populations better than a standard or a control. For each i, the observations $X_i$ can be viewed as the number of successes among N independent trials taken from $\pi_i$, and the parameter $\theta_i$ as the probability of a success for each trial in $\pi_i$. Then $X_i|\theta_i$ has probability function $f_i(x|\theta_i) =$

$\binom{N}{x} \theta_i^x (1 - \theta_i)^{N-x}$, $x = 0,1,\ldots,N$. We let $G_i(\cdot)$ denote the prior distribution of $\theta_i$ and assume that $G(\underset{\sim}{\theta}) = \prod_{i=1}^{k} G_i(\theta_i)$.

3.2.1 Selecting the Best Binomial Population. Gupta and Liang (1986) considered the loss function

$$(3.1) \qquad L(\underset{\sim}{\theta},\{i\}) = \theta_{[k]} - \theta_i$$

for the problem of selecting the largest binomial parameter $\theta_{[k]}$ among k binomial populations.

Let $f_i(x) = \int_0^1 f_i(x|\theta)\, dG_i(\theta)$, $W_i(x) = \int_0^1 \theta f_i(x|\theta)\, dG_i(\theta)$ and $\varphi_i(x) = W_i(x)/f_i(x)$. Then, from (3.1), following a straightforward computation, a randomized Bayes selection procedure, say $\underset{\sim}{\psi}^B = (\psi_1^B,\ldots,\psi_k^B)$, is given below:

$$(3.2) \qquad \psi_i^B(\underset{\sim}{x}) = \begin{cases} |S(\underset{\sim}{x})|^{-1} & \text{if } i \in S(\underset{\sim}{x}), \\ 0 & \text{otherwise,} \end{cases}$$

where

$$(3.3) \qquad S(\underset{\sim}{x}) = \{i | \varphi_i(x_i) = \max_{1 \leq j \leq k} \varphi_j(x_j)\}.$$

Here, $\psi_i^B(\underset{\sim}{x})$ is the probability of selecting $\pi_i$ as the best population given $\underset{\sim}{X} = \underset{\sim}{x}$.

Note that $\varphi_i(x)$ is the Bayes estimator of the parameter $\theta_i$ under the squared error loss given $X_i = x$. One can see that $\varphi_i(x)$ is increasing in x for $i = 1,\ldots,k$ and hence $\underset{\sim}{\psi}^B$ is a monotone selection procedure.

Due to the surprising quirk that $\varphi_i(x)$ cannot be consistently estimated in the usual empirical Bayes sense (see Robbins (1964) and Samuel (1963)), an idea of Robbins in setting up the empirical Bayes framework for binomial populations is used below.

For each i, $i = 1,\ldots,k$, at stage j, consider $N + 1$ independent trials from $\pi_i$. Let $X_{ij}$ and $Y_{ij}$, respectively, stand for the number of successes in the first N trials and the last trial. Let $\underset{\sim}{Z}_j = ((X_{1j},Y_{1j}),\ldots,(X_{kj},Y_{kj}))$ denote the observations at the jth stage, $j = 1,\ldots,n$. We also let $\underset{\sim}{X}_{n+1} = \underset{\sim}{X} = (X_1,\ldots,X_k)$ denote the present observations.

By the monotonicity of the estimators $\varphi_i(x)$, $1 \leq i \leq k$, in terms of the Bayes risk, one can see that all monotone procedures form an essentially complete class in the set of all selection procedures. In view of this fact, it is reasonable to require that the appropriate empirical Bayes procedures possess the above mentioned monotone property. For this purpose, we first need to have some monotone empirical Bayes estimators for $\varphi_i(x)$, $1 \leq i \leq k$.

For each $x = 0, 1, \ldots, N$, and $n = 1, 2, \ldots$, define

$$(3.4) \qquad f_{in}(x) = \frac{1}{n} \sum_{j=1}^{n} I_{\{x\}} (X_{ij}) + n^{-1},$$

$$(3.5) \qquad W_{in}(x) = \frac{1}{n} \sum_{j=1}^{n} Y_{ij} I_{\{x\}} (X_{ij}) + n^{-1},$$

Also, let $V_{ij} = X_{ij} + Y_{ij}$, $j = 1,2,\ldots$ Define

$$(3.6) \qquad \tilde{W}_{in}(x) = \{[\frac{x+1}{n(N+1)} \sum_{j=1}^{n} I_{\{x+1\}}(V_{ij})] \wedge [\frac{1}{n} \sum_{j=1}^{n} I_{\{x\}}(X_{ij})]\} + n^{-1},$$

where $a \wedge b = \min\{a,b\}$. Let

$$(3.7) \qquad \varphi_{in}(x) = W_{in}(x)/f_{in}(x),$$

$$(3.8) \qquad \tilde{\varphi}_{in}(x) = \tilde{W}_{in}(x)/f_{in}(x),$$

and for each $0 \leq x \leq N$, define

$$(3.9) \qquad \varphi^*_{in}(x) = \max_{0 \leq s \leq x} \min_{s \leq t \leq N} \{\sum_{y=s}^{t} \varphi_{in}(y)/(t - s + 1)\},$$

$$(3.10) \qquad \tilde{\varphi}^*_{in}(x) = \max_{0 \leq s \leq x} \min_{s \leq t \leq N} \{\sum_{y=s}^{t} \tilde{\varphi}_{in}(y)/(t - s + 1)\}.$$

By (3.9) and (3.10), one can see that both $\varphi^*_{in}(x)$ and $\tilde{\varphi}^*_{in}(x)$ are increasing in x. Gupta and Liang (1986) proposed $\varphi^*_{in}(x)$ (or $\tilde{\varphi}^*_{in}(x)$) as an estimator of $\varphi_i(x)$. They also proposed two empirical Bayes selection procedures, say $\underset{\sim}{\psi}^*_n = (\psi^*_{1n}, \ldots, \psi^*_{kn})$, and $\tilde{\underset{\sim}{\psi}}_n = (\tilde{\psi}_{1n}, \ldots, \tilde{\psi}_{kn})$, which are given below, respectively:

$$(3.11) \qquad \psi^*_{in}(\underset{\sim}{x}) = \begin{cases} |S^*_n(\underset{\sim}{x})|^{-1} & \text{if } i \in S^*_n(\underset{\sim}{x}), \\ 0 & \text{otherwise,} \end{cases}$$

where

$$(3.12) \qquad S^*_n(\underset{\sim}{x}) = \{i | \varphi^*_{in}(x_i) = \max_{1 \leq j \leq k} \varphi^*_{jn}(x_j)\},$$

$$(3.13) \qquad \tilde{\psi}_{in}(\underset{\sim}{x}) = \begin{cases} |\tilde{S}_n(\underset{\sim}{x})|^{-1} & \text{if } i \in \tilde{S}_n(\underset{\sim}{x}), \\ 0 & \text{otherwise,} \end{cases}$$

where

(3.14) $$\tilde{S}_n(\underset{\sim}{x}) = \{i \mid \tilde{\varphi}^*_{in}(x_i) = \max_{1\le j\le k} \tilde{\varphi}^*_{jn}(x_j)\}$$

It is easy to verify that $\underset{\sim}{\psi}^*_n$ and $\tilde{\underset{\sim}{\psi}}_n$ are both monotone selection procedures.

Without ambiguity, we still use $B(\underset{\sim}{\psi},G)$ to denote the Bayes risk associated with the selection procedure $\underset{\sim}{\psi}$ when G is the true prior distribution.

Gupta and Liang (1986) proved that the two sequences of selection procedures $\{\underset{\sim}{\psi}^*_n\}$ and $\{\tilde{\underset{\sim}{\psi}}_n\}$ have the following asymptotically optimal property:

$$B(\underset{\sim}{\psi}^*_n,G) - B(\underset{\sim}{\psi}^B,G) \le O(\exp(-c_1 n)),$$

$$B(\tilde{\underset{\sim}{\psi}}_n,G) - B(\underset{\sim}{\psi}^B,G) \le O(\exp(-c_2 n)),$$

for some positive constants $c_1$ and $c_2$.

3.2.2 Selecting Populations Better Than A Control. Let $\theta_0 \in (0,1)$ denote a control parameter. Population $\pi_i$ is said to be good if $\theta_i \ge \theta_0$ and bad if $\theta_i < \theta_0$. Gupta and Liang (1984) considered the loss function

(3.15) $$L(\underset{\sim}{\theta},S) = \sum_{i\in S} (\theta_0 - \theta_i) I_{(0,\theta_0)}(\theta_i) + \sum_{i\notin S} (\theta_i - \theta_0) I_{(\theta_0,1)}(\theta_i),$$

for the problem of selecting (excluding) all good (bad) populations. The value of the control parameter $\theta_0$ is either known or unknown. When $\theta_0$ is unknown, a sample from the control population, say $\pi_0$, is needed. To be consistent with the notation used in earlier sections, we assume $\theta_0$ is known. We note that Gupta and Liang (1984) have studied the case when $\theta_0$ is unknown.

For the loss function (3.15), a nonrandomized Bayes selection procedure $\underset{\sim}{\alpha}^B = (\alpha_1^B,\ldots,\alpha_k^B)$ is given by

(3.16) $$\alpha_i^B(\underset{\sim}{x}) = \begin{cases} 1 & \text{if } \varphi_i(x_i) \ge \theta_0, \\ 0 & \text{otherwise,} \end{cases}$$

where $\alpha_i^B(\underset{\sim}{x})$ is the probability of selecting $\pi_i$ as a good population given $\underset{\sim}{X} = \underset{\sim}{x}$.

Note that $\underset{\sim}{\alpha}^B$ is also a monotone selection procedure. Hence, based on the estimators $\varphi^*_{in}(x)$ and $\tilde{\varphi}^*_{in}(x)$, two intuitive empirical Bayes procedures, say $\underset{\sim}{\alpha}^*_n = (\alpha^*_{1n},\ldots,\alpha^*_{kn})$ and $\tilde{\underset{\sim}{\alpha}}_n = (\tilde{\alpha}_{1n},\ldots,\tilde{\alpha}_{kn})$ can be obtained where

(3.17) $$\alpha^*_{in}(\underset{\sim}{x}) = \begin{cases} 1 & \text{if } \varphi^*_{in}(x_i) \ge \theta_0, \\ 0 & \text{otherwise;} \end{cases}$$

$$(3.18) \qquad \tilde{\alpha}_{in}(\underset{\sim}{x}) = \begin{cases} 1 & \text{if } \tilde{\varphi}^*_{in}(x_i) \geq \theta_0, \\ 0 & \text{otherwise.} \end{cases}$$

As before, one can show that these two sequences of selection procedures $\{\underset{\sim}{\alpha}^*_n\}$ and $\{\tilde{\underset{\sim}{\alpha}}_n\}$ have the following asymptotically optimal property:

$$B(\underset{\sim}{\alpha}^*_n, G) - B(\underset{\sim}{\alpha}^B, G) \leq O(\exp(-c_3 n)),$$

$$B(\tilde{\underset{\sim}{\alpha}}_n, G) - B(\underset{\sim}{\alpha}^B, G) \leq O(\exp(-c_4 n)),$$

for some positive constants $c_3$ and $c_4$.

## 3.3 Empirical Bayes Procedures Related to Uniform Populations

In this section, we assume that the random variables $X_i$, $1 \leq i \leq k$, have uniform distributions $U(0,\theta_i)$, $\theta_i > 0$ and unknown. The parameter space is $\Omega = \{\underset{\sim}{\theta} | \theta_i > 0, 1 \leq i \leq k\}$. It is also assumed that the prior distribution $G$ on $\Omega$ has the form $G(\underset{\sim}{\theta}) = \prod_{i=1}^{k} G_i(\theta_i)$, where $G_i(\cdot)$ is a distribution on $(0,\infty)$, $i = 1,\ldots,k$.

Let $\theta_0 > 0$ be a known control parameter. Gupta and Hsiao (1983) considered the problem of selecting populations better than the standard using the loss function

$$(3.19) \quad L(\underset{\sim}{\theta},S) = L_1 \sum_{i \notin S} (\theta_i - \theta_0) I_{(\theta_0,\infty)}(\theta_i) + L_2 \sum_{i \in S} (\theta_0 - \theta_i) I_{(0,\theta_0)}(\theta_i),$$

where $L_i$, $i = 1, 2$, are positive and known.

Let $m_i(x)$ be the marginal pdf of $X_i$ and $M_i(x)$ be the marginal distribution of $X_i$. Then we have

$$(3.20) \qquad m_i(x) = \int_x^\infty \frac{1}{\theta} \, dG_i(\theta) \text{ for } x > 0,$$

$$(3.21) \qquad M_i(x) = \int_0^x \int_t^\infty \frac{1}{\theta} \, dG_i(\theta) dt = x m_i(x) + G_i(x).$$

Note that the marginal pdf $m_i(x)$ is continuous and decreasing in x.

By a direct computation, a Bayes procedure $\underset{\sim}{\psi}^B = (\psi_1^B,\ldots,\psi_k^B)$ for this selection problem is given by

$$(3.22) \qquad \psi_i^B(\underset{\sim}{x}) = \begin{cases} 1 & \text{if } (x_i \geq \theta_0) \text{ or } (x_i < \theta_0 \text{ and } \Delta_{iG}(x_i) \geq 0), \\ 0 & \text{otherwise,} \end{cases}$$

where

$$(3.23)\quad \Delta_{iG}(x_i) \equiv L_2 m_i(x_i)(x_i - \theta_o) + L_2[M_i(\theta_o) - M_i(x_i)] + L_1[1 - M_i(\theta_o)].$$

Since $m_i(x)$, $1 \le i \le k$ are decreasing in x, one can see that $\Delta_{iG}(x)$, $1 \le i \le k$, are increasing in x for $x < \theta_o$; and hence, the Bayes procedure $\psi^B$ has the monotone property.

To derive an empirical Bayes procedure, we first need to have some estimators, say $m_{in}(x)$ and $M_{in}(x)$, for $m_i(x)$ and $M_i(x)$, respectively. Due to the decreasing property of $m_i(x)$, we require that the estimators $m_{in}(x)$, $n = 1, 2, \ldots$, possess the same property. Once an estimator $m_{in}(x)$ is obtained, we let

$$(3.24)\qquad M_{in}(x) = \int_0^x m_{in}(y)dy,$$

$$(3.25)\qquad \Delta_{in}(x) = L_2 m_{in}(x)(x - \theta_o) + L_2[M_{in}(\theta_o) - M_{in}(x)] + L_1[1 - M_{in}(\theta_o)].$$

Then, an empirical Bayes procedure $\psi_n = (\psi_{1n}, \ldots, \psi_{kn})$ can be given as follows:

$$(3.26)\qquad \psi_{in}(x) = \begin{cases} 1 & \text{if } (x_i \ge \theta_o) \text{ or } (x_i < \theta_o \text{ and } \Delta_{in}(x_i) \ge 0), \\ 0 & \text{otherwise.} \end{cases}$$

This empirical Bayes procedure $\psi_n$ is a monotone procedure if $m_{in}(x)$, $1 \le i \le k$, are decreasing in x. We use the method of Grenander (1956) to obtain such an estimator having the decreasing property.

Let $X^n_{i(1)} \le X^n_{i(2)} \le \cdots \le X^n_{i(n)}$ be the ordered observations of the first n observations taken from $\pi_i$. Let $F_{in}$ be the empirical distribution based on $X_{i1}, \ldots, X_{in}$. For each j, $1 \le j \le n$, let

$$(3.27)\qquad \beta_{ij} = \min_{s \le j-1} \max_{t \ge j} \frac{F_{in}(X^n_{i(t)}) - F_{in}(X^n_{i(s)})}{X^n_{i(t)} - X^n_{i(s)}},$$

when $X^n_{i(0)} \equiv 0$, and define

$$(3.28)\qquad m_{in}(x) = \begin{cases} 0 & \text{for } x \le 0, \\ \beta_{ij} & \text{for } X^n_{i(j-1)} < x \le X^n_{i(j)}, \\ 0 & \text{for } x > X^n_{i(n)}. \end{cases}$$

From (3.27) and (3.28), one can see that the estimator $m_{in}(x)$ is decreasing in x. Thus, the empirical Bayes procedures $\psi_n$ defined by (3.24 - 3.28) is a monotone procedure. It is known that both estimators $M_{in}(x)$ and $m_{in}(x)$ have strong consistency property. Hence, $\Delta_{in}(x)$ is a strongly consistent estimator of $\Delta_{iG}(x)$. Then by Theorem 2.1 of Gupta and Hsiao (1983), the sequence of empirical Bayes procedures $\{\psi_n\}$ is asymptotically optimal provided $\int_0^\infty \theta dG_i(\theta) < \infty$ for each $i = 1, \ldots, k$.

ACKNOWLEDGEMENT

This research was partially supported by the Office of Naval Research Contract N00014-84-C-0167 and NSF Grant DMS-8606964 at Purdue University.

REFERENCES

Bechhofer, R. E., 1954, A single-sample multiple decision procedure for ranking means of normal populations with known variances, Ann. Math. Statist., 25:16-39.

Berger, J., and Deely, J. J., 1986, A Bayesian approach to ranking and selection of related means with alternatives to AOV methodology, Technical Report #86-8, Department of Statistics, Purdue University, West Lafayette, Indiana.

Bickel, P. J., and Yahav, J. A., 1977, On selecting a subset of good populations, Statistical Decision Theory and Related Topics-II (Eds. S. S. Gupta and D. S. Moore), Academic, New York, 37-55.

Chernoff, H., and Yahav, J. A., 1977, A subset selection problem employing a new criterion. Statistical Decision Theory and Related Topics-II (Eds. S. S. Gupta and D. S. Moore), Academic, New York, 93-119.

Deely, J. J., 1965, Multiple decision procedures from an empirical Bayes approach, Ph.D. Thesis (Mimeo. Ser. No. 45), Department of Statistics, Purdue University, West Lafayette, Indiana.

Deely, J. J., and Gupta, S. S., 1968, On the properties of subset selection procedures, Sankhyā, A30:37-50.

Goel, P. K., and Rubin, H. 1977, On selecting a subset containing the best population--A Bayesian approach, Ann. Statist., 5:969-983.

Grenander, U., 1956, On the theory of mortality measurement, Part II. Skand. Akt., 39:125-153.

Gupta, S. S., 1956, On a decision rule for a problem in ranking means, Ph.D. Thesis (Mimeo. Ser. No. 150), Inst. of Statist., University of North Carolina, Chapel Hill.

Gupta, S. S., 1965, On some multiple decision (selection and ranking) rules, Technometrics, 7:225-245.

Gupta, S. S., and Hsiao, P., 1981, On Γ-minimax, minimax, and Bayes procedures for selecting populations close to a control, Sankhyā, B43:291-318.

Gupta, S. S., and Hsiao, P., 1983, Empirical Bayes rules for selecting good populations, J. Statist. Plan. Infer., 8:87-101.

Gupta, S. S., and Hsu, J. C., 1978, On the performance of some subset selection procedures, Commun. Statist.-Simula. Computa., B7(6):561-591.

Gupta, S. S., and Liang, T., 1984, Empirical Bayes rules for selecting good binomial populations, to appear in The Proceedings of the Symposium on Adaptive Statistical Procedures and Related Topics.

Gupta, S. S., and Liang, T., 1986, Empirical Bayes rules for selecting the best binomial population, to appear in Statistical Decision Theory and Related Topics-IV (Eds. S. S. Gupta and J. O. Berger).

Gupta, S. S., and Miescke, K., 1984, On two-stage Bayes selection procedures, Sankhyā, B46:123-134.

Gupta, S. S., and Panchapakesan, S., 1979, "Multiple Decision Procedures", Wiley, New York.

Gupta, S. S., and Yang, H. M., 1985, Bayes-P* subset selection procedures for the best population, J. Statist. Plan. Infer., 12:213-233.

Miescke, K., 1979, Bayesian subset selection for additive and linear loss functions, Commun. Statist.-Theor. Meth., A8(12):1205-1226.

Robbins, H., 1956, An empirical Bayes approach to statistics, Proc. Third Berkeley Symp. Math. Probab., University of California Press, 155-163.

Robbins, H., 1964, The empirical Bayes approach to statistical decision problems, Ann. Math. Statist., 35:1-19.

Robbins, H., 1983, Some thoughts on empirical Bayes estimation, Ann. Statist., 11:713-723.

Samuel, E., 1963, An empirical Bayes approach to the testing of certain parametric hypothesis, Ann. Math. Statist., 34:1370-1385.
Van Ryzin, J., and Susarla, V., 1977, On the empirical Bayes approach to multiple decision problems, Ann. Statist., 5:172-181.

BAYESIAN ASPECTS IN THE THEORY OF COMPARISON

OF STATISTICAL EXPERIMENTS

Herbert Heyer

Mathematisches Institut der Universität
Auf der Morgenstelle 10
74 Tübingen
Deutsche Bundesrepublik

INTRODUCTION

The purpose of this paper is to present some aspects of the theory of comparison of experiments where in an efficient way the Bayesian concept of apriori knowledge can be applied in order to enrich the motivation and the understanding of the mathematical analysis involved. There are various kinds of comparisons of experiments based on orderings of decision functions and their risks. Of particular interest in the applications are the Bayesian orderings introduced by De Groot and elaborated by Feldman. See [4], [3] and also [10], [11]. In the following we shall adopt the comparison invented by Blackwell and generalized by LeCam. Although this comparison is rather strong it has proved to be an important tool in asymptotic decision theory. Basic knowledge of the Blackwell-LeCam theory can be obtained from the text books [7] and [13]. We recall a few key notions. An experiment is determined by three data: a list of possible outcomes (the sample space $(E,\mathfrak{A})$), a collection of possible explaining theories (the parameter set I), and a correspondence which to every explaining theory associates the random mechanism governing the random outcome (a mapping $i \to P_i$ from I into the set $\mathfrak{M}^1(E,\mathfrak{A})$ of probability measures on $(E,\mathfrak{A})$). We shall consider experiments $\mathfrak{E} = (E,\mathfrak{A},\{P_i : i\in I\})$ with fixed parameter set I. Since there is no explicit definition of the information contained in an experiment, we content ourselves with the comparison of information whenever two experiments $\mathfrak{E}$ and $\mathfrak{F} = (F,\mathfrak{B},\{Q_i : i\in I\})$ are given. The question then arises how much information gets lost, under the worst possible circumstances, if the experiment $\mathfrak{E}$ receives preference with respect to the experiment $\mathfrak{F}$. In order to measure this loss of information LeCam in 1964 introduced the notion of deficiency $\rho(\mathfrak{E},\mathfrak{F})$ between the experiments $\mathfrak{E}$ and $\mathfrak{F}$. In the case of a subexperiment $\mathfrak{F}$ of $\mathfrak{E}$ which is defined by a subalgebra $\mathfrak{B}$ of $\mathfrak{A}$ the relationship $\rho(\mathfrak{F},\mathfrak{E}) = o$ means that $\mathfrak{B}$ is sufficient for $\mathfrak{E}$.

A rather abstract setting for the Bayesian model has been proposed in [1]. In order not to lose the readers among our fellow statisticians we shall work in moderate generality. The first encounter with the notion of minimal Bayes risk will be produced in Section 1 where the comparison of binary experiments appears in the context of the Neyman-Pearson theory.

Section 2 is devoted to the generalized comparison. In particular we describe the comparison of Bayes risks in the case of k-decision problems and finite parameter sets. In Section 3 we study Bayesian sufficiency and compare it with classical sufficiency. Section 4 contains properties of totally informative and totally noninformative experiments. They play an important role in the Bayesian comparison of Section 5 where we report on some work of Torgersen [15] of 1976 which was the basis of our presentation in Chapter X of [7]. Finally we include in Section 6 a few results on the comparison of powers of experiments. In the special case of finite experiments the asymptotic behavior of the corresponding minimal Bayes risks can be related to some early work of Chernoff [2]. Recent progress concerning the statistical information contained in additional observations is due to Helgeland [5] and Mammen [8].

## 1. COMPARISON OF BINARY EXPERIMENTS

Let $\mathfrak{E} = (E, \mathfrak{A}, \{P_1, P_2\})$ denote a binary experiment. By $\mathfrak{M}^{(1)}(\mathfrak{E})$ we abbreviate the set of all $\mathfrak{A}$-measurable functions on E with values in $[0,1]$. The <u>error function</u> $g_{\mathfrak{E}}$ of $\mathfrak{E}$ is defined by

$$g_{\mathfrak{E}}(\alpha) := \inf_{t \in \mathfrak{M}^{(1)}(\mathfrak{E})} \left[(1-\alpha)\int t\,dP_1 + \alpha \int t\,dP_2\right]$$

for all $\alpha \in [0,1]$, where the right side of the equality equals the <u>minimal Bayes risk</u> corresponding to the apriori measure $(1-\alpha)\varepsilon_1 + \alpha\varepsilon_2$. We note that the above inf is attained exactly for the Neyman-Pearson test

$$t^* := 1_{\left[\frac{dP_2}{dP_1} > \frac{1-\alpha}{\alpha}\right]}$$

The following well-known result motivates the desired order relation between experiments.

1.1 <u>Proposition</u>. For binary experiments $\mathfrak{E}$ and $\mathfrak{F} = (F, \mathfrak{B}, \{Q_1, Q_2\})$ with error functions $g_{\mathfrak{E}}$ and $g_{\mathfrak{F}}$ resp. and for a tolerance value $\varepsilon \geq 0$ the following statements are equivalent:

(i) $g_{\mathfrak{E}} \leq g_{\mathfrak{F}} + \frac{\varepsilon}{2}$.

(ii) For every $s \in \mathfrak{M}^{(1)}(\mathfrak{F})$ there exists a $t \in \mathfrak{M}^{(1)}(\mathfrak{E})$ such that

$$\begin{cases} \int t\,dP_1 \leq \int s\,dQ_1 + \frac{\varepsilon}{2} \text{ and} \\ \int t\,dP_2 \geq \int s\,dQ_2 - \frac{\varepsilon}{2}. \end{cases}$$

In presence of statement (i) we say that $\mathfrak{E}$ is $\varepsilon$-<u>deficient</u> with respect to $\mathfrak{F}$ and abbreviate by $\mathfrak{E} >_{\varepsilon} \mathfrak{F}$. One observes that $> := >_0$ is an order relation giving rise to an equivalence relation $\sim$ between binary experiments and leading to the notion of types of experiments. More generally one introduces the <u>deficiency</u>

$$\rho(\mathfrak{E},\mathfrak{F}) := \inf\{\varepsilon \geq 0 : \mathfrak{E} >_{\varepsilon} \mathfrak{F}\}$$

and the pseudo distance

$$\Delta(\mathfrak{E},\mathfrak{F}) := \rho(\mathfrak{E},\mathfrak{F}) \vee \rho(\mathfrak{F},\mathfrak{E})$$

between experiments $\mathfrak{E}$ and $\mathfrak{F}$, and one obtains that

$$\mathfrak{E} \sim \mathfrak{F} \iff \Delta(\mathfrak{E},\mathfrak{F}) = 0.$$

## 2. GENERALIZED COMPARISON

Let $\mathfrak{E} = (E, \mathfrak{A}, \{P_i : i \in I\})$ and $\mathfrak{F} = (F, \mathfrak{B}, \{Q_i : i \in I\})$ be two

experiments with arbitrary but fixed parameter set I. We denote by $(D, \mathfrak{D})$ an arbitrary decision space, by $\mathfrak{D}(\mathfrak{E}) = \text{Stoch}((E, \mathfrak{A}), (D, \mathfrak{D}))$ the set of randomized decision functions for $\mathfrak{E}$, and by $\mathfrak{V}$ the set of bounded measurable loss functions on I×D. For given $V \in \mathfrak{V}$ and $\delta \in \mathfrak{D}(\mathfrak{E})$ the risk function associated with $\mathfrak{E}$ is the mapping

$$i \to R_{\delta}^{V}(i) := \int V(i,t)\delta(P_i)(dt)$$

on I. Finally we are given a tolerance function $i \to \varepsilon(i)$ on I.

<u>Definition</u>. $\mathfrak{E}$ is said to be $\varepsilon$-<u>deficient</u> with respect to $\mathfrak{F}$ ($\mathfrak{E} >_{\varepsilon} \mathfrak{F}$) if for all decision spaces $(D, \mathfrak{D})$ with $|D| < \infty$, all $V \in \mathfrak{V}$, given $\sigma \in \mathfrak{D}(\mathfrak{F})$ there exists $\delta \in \mathfrak{D}(\mathfrak{E})$ such that

$$R_{\delta}^{V} \leq R_{\sigma}^{V} + \varepsilon ||V||.$$

The notion of deficiency $\rho$ and the related pseudo metric $\Delta$ are defined as in the binary case above. In the case of $|D| = k$ one talks about $\varepsilon$-<u>deficiency for</u> k-<u>decision problems</u>.

2.1 <u>Theorem</u> (LeCam 1964). We make the following standard hypotheses: Let

(a) $\{P_i : i \in I\}$ be dominated and let

(b) $(F, \mathfrak{B})$ be a standard Borel space.

Then the subsequent statements are equivalent:

(i) $\mathfrak{E} >_{\varepsilon} \mathfrak{F}$

(ii) For every $\sigma \in \mathfrak{D}(\mathfrak{F})$ there exists a $\delta \in \mathfrak{D}(\mathfrak{E})$ such that $||\delta(P_i) - \sigma(Q_i)|| \leq \varepsilon(i)$ for all $i \in I$.

(iii) There exists an $N \in \text{Stoch}((E, \mathfrak{A}), (F, \mathfrak{B}))$ satisfying $||N(P_i) - Q_i|| \leq \varepsilon(i)$ for all $i \in I$.

Moreover

$$\rho(\mathfrak{E},\mathfrak{F}) = \inf_N \sup_i ||N(P_i) - Q_i||,$$

where N runs through $\text{Stoch}((E, \mathfrak{A}), (F, \mathfrak{B}))$ and i through I.

2.2 Remark. Under the additional assumptions $|I| = m$ and $|D| = k$ the $\varepsilon$-deficiency of $\mathfrak{E}$ with respect to $\mathfrak{F}$ is equivalent to the comparison of Bayes risks:

(iv) Given an apriori probability measure $\Lambda$ on I with $\Lambda(\{i\})>0$ for all $i\in I$, for each $V\in\mathcal{V}$ and each $\sigma\in\mathfrak{D}(\mathfrak{F})$ there exists a $\delta\in\mathfrak{D}(\mathfrak{E})$ such that

$$\sum_{i=1}^{m} R_\delta^V(i)\Lambda(\{i\}) \leq \sum_{i=1}^{m} R_\sigma^V(i)\Lambda(\{i\}) + \sum_{i=1}^{m} (\varepsilon(i)\sup_t|V(i,t)|)\Lambda(\{i\}).$$

Since the implication (i) ⇒ (iv) is trivial, it remains to prove the implication (iii) ⇒ (ii). This, however, is done following an idea of Torgersen's of 1970 by applying the minimax theorem to the mapping

$$(V,\delta) \to \sum_{i=1}^{m} [R_\delta^V(i)-R_\sigma^V(i)- \frac{1}{2}\varepsilon(i)\max_t|V(i,t)|]\Lambda(di)$$

which is concave in V and convex in $\delta$.

## 3. BAYESIAN SUFFICIENCY

Let $\mathfrak{F} := (E,\mathfrak{B},\{Res_{\mathfrak{B}} P_i:i\in I\})$ be a subexperiment of $\mathfrak{E}$ defined by a sub-$\sigma$-algebra $\mathfrak{B}$ of $\mathfrak{A}$. Then a specialization of Theorem 2.1 yields the equivalence of the following two statements:

(i) $\rho(\mathfrak{F},\mathfrak{E}) = 0$ ($\Longleftrightarrow \Delta(\mathfrak{E},\mathfrak{F}) = 0$)

(ii) $\mathfrak{B}$ is sufficient (in the sense of Halmos and Savage) for $\mathfrak{E}$, i.e. there exists a conditional probability on $\mathfrak{A}$ given $\mathfrak{B}$ which is independent of i in I.

For further characterizations and generalizations of sufficiency the reader is referred to [6] and [7], § 22. In [6] the notion of Bayesian sufficiency has been mentionned.

Intuitively Bayesian sufficiency says that given any prior probability $\Lambda\in\mathcal{M}^1(I,\mathfrak{I})$ the posterior probability on I given $\mathfrak{A}$ is the same as the posterior probability on I given $\mathfrak{B}$. Mathematically we start with a stochastic kernel $P\in Stoch((I,\mathfrak{I}),(E,\mathfrak{A}))$ and we put $P_i(\cdot) := P(i,\cdot)$ for all $i\in I$. Given $\Lambda\in\mathcal{M}^1(I,\mathfrak{I})$ we define the mixture $P\circ\Lambda\in\mathcal{M}^1(E\times I,\mathfrak{A}\otimes\mathfrak{I})$ by

$$P\circ\Lambda(A\times C) := \int_C P_i(A)\Lambda(di)$$

for all $A\times C\in\mathfrak{A}\otimes\mathfrak{I}$. In what follows we shall work with the projected $\sigma$-algebras $E\times\mathfrak{I} = \{E\times C:C\in\mathfrak{I}\}$, $\mathfrak{A}\times I$, $\mathfrak{B}\times I$ and similarly with the lifted functions $f^*$ on $E\times I$ defined by

$$f^*(x,i) := f(x)$$

for all $(x,i)\in E\times I$ whenever f is a function on E.

Definition. $\mathfrak{B}$ is B-sufficient for $\mathfrak{E}$ if for all $C\in\mathfrak{I}$

$$E_{P\circ\Lambda}(1_{E\times C}|\mathfrak{B}\times I) = E_{P\circ\Lambda}(1_{E\times C}|\mathfrak{A}\times I).$$

3.1 Propositon. The following statements are equivalent:

(i) $\mathfrak{B}$ is B-sufficient for $\mathfrak{E}$.

(ii) $\mathfrak{A} \times I$ and $E \times \mathfrak{I}$ are conditionally independent given $\mathfrak{B} \times I$ on $(E\times I, \mathfrak{A} \otimes \mathfrak{I}, P\circ\Lambda)$ for all $\Lambda \in \mathcal{M}^1(I, \mathfrak{I})$.

(iii) For every bounded $\mathfrak{A}$-measurable f on E there exists a $\mathfrak{B}$-measurable g on E such that

$$g^* = E_{P\circ\Lambda}(f^* | \mathfrak{B} \otimes \mathfrak{I})$$

for all $\Lambda \in \mathcal{M}^1(I, \mathfrak{I})$.

For the definition of conditional independence and a proof of the equivalences see [9].

3.2 Theorem. Under the standard hypotheses of Theorem 2.1 the following statements are equivalent:

(i) $\mathfrak{B}$ is sufficient for $\mathfrak{E}$.

(ii) $\mathfrak{B}$ is B-sufficient for $\mathfrak{E}$.

For the proof of (i) $\Rightarrow$ (ii) one takes a bounded $\mathfrak{A}$-measurable function f on E and notes that by (i) there exists a version of $E_{P_i}(f | \mathfrak{B})$ which is independent of $i \in I$. But then for all $i \in I$

$$E_{P_i}(f) = E_{P_i}(g),$$

where for all $\Lambda \in \mathcal{M}^1(I, \mathfrak{I})$, $g^*$ is a version of $E_{P\circ\Lambda}(f^* | \mathfrak{B} \times I)$. Proposition 3.1 yields (ii).

In order to show that (ii) $\Rightarrow$ (i) we let $\Lambda := \alpha\varepsilon_{i_1} + (1-\alpha)\varepsilon_{i_2}$ with $i_1, i_2 \in I$ and $\alpha \in [0,1]$. For each bounded $\mathfrak{A}$-measurable function f on E it follows from (ii) that $E_{P\circ\Lambda}(f^* | \mathfrak{B} \times I)$ is a version of $E_{P\circ\Lambda}(f^* | \mathfrak{I} \otimes \mathfrak{B})$. Thus for $i \in \{i_1, i_2\}$ and all $B \in \mathfrak{B}$

$$\int_B f dP_i = \int_B E_{P\circ\Lambda}(f^* | \mathfrak{B} \times I) dP_i.$$

Hence $E_{P\circ\Lambda}(f^* | \mathfrak{B} \times I)$ is a version of $E_{P_i}(f | \mathfrak{B})$ for $i \in \{i_1, i_2\}$ which means that $\mathfrak{B}$ is pairwise sufficient for $\mathfrak{E}$. Since $\{P_i : i \in I\}$ is assumed to be dominated, this implies (i).

3.3 Remark. If the family $\{P_i : i \in I\}$ defining the experiment $\mathfrak{E}$ is not dominated, then the crucial implication (ii) $\Rightarrow$ (i) is not true even if the $\sigma$-algebras $\mathfrak{A}$ and $\mathfrak{B}$ are countably generated. If, however, the latter is the case, then (ii) appears to be equivalent to

(iii) For every $\Lambda \in \mathcal{M}^1(I, \mathfrak{I})$ there exists a $C_\Lambda \in \mathfrak{I}$ with $\Lambda(C_\Lambda) = 1$ such that $\mathfrak{B}$ is sufficient for $\mathfrak{E}_{C_\Lambda} := (E, \mathfrak{A}, \{P_i : i \in C_\Lambda\})$.

For a proof of this statement see [12].

## 4. EXTREMELY INFORMATIVE EXPERIMENTS

An experiment $\mathcal{E}$ is said to be totally informative if for each pair $(i,j) \in I \times I$, $i \neq j$ we have $P_i \perp P_j$. Clearly any two such experiments are equivalent; the class or type will be denoted by $\mathcal{E}_a$. $\mathcal{E}_a$ is the experiment which consists of observing the underlying parameter i. We note that

4.1 $\rho(\mathcal{E}_a, \mathcal{E}) = 0$,

but for dominated $\mathcal{E}$ and uncountable I

4.2 $\Delta(\mathcal{E}, \mathcal{E}_a) = \rho(\mathcal{E}, \mathcal{E}_a) = 2$.

An experiment $\mathcal{E}$ is said to be totally uninformative if $P_i = P$ independent of $i \in I$. Again $\mathcal{E}$ defines a class $\mathcal{E}_n$, for which

4.3 $\rho(\mathcal{E}, \mathcal{E}_n) = 0$

holds.

Obviously $\mathcal{E}_a > \mathcal{E} > \mathcal{E}_n$, and it appears well motivated to introduce the information numbers

$$\rho_a(\mathcal{E}) := \rho(\mathcal{E}, \mathcal{E}_a)$$

and

$$\rho_n(\mathcal{E}) := \rho(\mathcal{E}_n, \mathcal{E}).$$

4.4 Statistical interpretation

From Theorem 2.1 follows that

$$\rho_a(\mathcal{E}) = 2 \min_N \sup_i N(P_i)(\complement\{i\}),$$

and $\frac{1}{2}\rho_a(\mathcal{E})$ is the minimax risk in the problem of estimating i on the basis of $\mathcal{E}$ when the loss is 0 or 1 according to success or failure of the estimator.

Analoguously we obtain

$$\rho_n(\mathcal{E}) = \min_Q \sup_i \| P_i - Q \|,$$

and $\rho_n(\mathcal{E})$ is the minimax risk in the estimation problem where no observations are available and the loss is given by the statistical distance.

## 5. BAYESIAN DEFICIENCY

In this section we want to discuss the Bayesian comparison of experiments in greater generality. Let $\mathcal{E}$ be an experiment $(E, \mathfrak{A}, \{P_i : i \in I\})$ with measurable parameter space $(I, \mathfrak{I})$. By $\mathcal{M}^1_f(I, \mathfrak{I})$ and $\mathcal{M}^1_c(I, \mathfrak{I})$ we denote the spaces of probability measures of $(I, \mathfrak{I})$ with finite and countable support respectively.

Given the data $(D, \mathfrak{B})$ and $V \in \mathcal{V}$ we introduce

Definition for any $\Lambda \in \mathcal{M}^1(I, \mathfrak{I})$ the Bayes risk

$$r_{\mathcal{E}}(D, V, \delta, \Lambda) := \int R^V_\delta(i)\Lambda(di)$$

incurred by the choice of a decision function $\delta \in \mathfrak{D}(\mathfrak{E})$, and the <u>minimal Bayes risk</u>

$$r_{\mathfrak{E}}(D,V,\Lambda) := \inf_{\delta} r_{\mathfrak{E}}(D,V,\delta,\Lambda).$$

5.1 <u>Theorem</u> (Bayes criterion). Under the standard hypotheses of Theorem 2.1

$$\rho(\mathfrak{E},\mathfrak{F}) = \sup_{\substack{|D|<\infty \\ |V|\leq 1 \\ \Lambda\in \mathcal{M}_f^1(I,\mathfrak{I})}} (r_{\mathfrak{E}}(D,V,\Lambda) - r_{\mathfrak{F}}(D,V,\Lambda)).$$

For the proof one notes that for every $\varepsilon>\rho(\mathfrak{E},\mathfrak{F})$, given $\sigma\in\mathfrak{D}(\mathfrak{F})$ there exists a $\delta\in\mathfrak{D}(\mathfrak{E})$ such that

$$R_{\delta}^{V} \leq R_{\sigma}^{V} + \varepsilon||V||$$

implies

$$r_{\mathfrak{E}}(D,V,\Lambda) \leq r_{\mathfrak{F}}(D,V,\sigma,\Lambda) + \varepsilon\,||V||.$$

But then

$$\rho(\mathfrak{E},\mathfrak{F}) \geq \sup_{\sigma} \frac{1}{||V||}(r_{\mathfrak{E}}(D,V,\Lambda)-r_{\mathfrak{F}}(D,V,\sigma,\Lambda))$$

$$= \frac{1}{||V||}(r_{\mathfrak{E}}(D,V,\Lambda)-r_{\mathfrak{F}}(D,V,\Lambda)).$$

5.2 <u>Specialization</u> to o-1 estimation as related to $\rho_a(\mathfrak{E})$ yields that for all $\Lambda\in\mathcal{M}_c^1(I,\mathfrak{I})$

$$r(\mathfrak{E}|\Lambda) := r_{\mathfrak{E}}(D,V,\Lambda) = 1-||\bigvee_{i\in I}\Lambda(\{i\})P_i||,$$

in particular, for $|I|=2$

$$r(\mathfrak{E}|\Lambda) = ||\Lambda(\{1\})P_1 \wedge \Lambda(\{2\})P_2||.$$

Moreover, $\Lambda \to r(\mathfrak{E}|\Lambda)$ determines the type of $\mathfrak{E}$. From minimax theory follows that

$$\frac{1}{2}\rho_a(\mathfrak{E}) = \sup_{\Lambda\in\mathcal{M}_c^1(I,\mathfrak{I})} r(\mathfrak{E}|\Lambda).$$

5.3 <u>Remark.</u> Torgersen in [15] studies also for a given $\Lambda\in\mathcal{M}_c^1(I,\mathfrak{I})$ the $\Lambda$-<u>weighted deficiency</u> of $\mathfrak{E}$ with respect to $\mathfrak{F}$ defined by

$$\rho(\mathfrak{E},\mathfrak{F}|\Lambda) := \inf\{\sum_{i\in I}\Lambda(\{i\})\varepsilon(i) : \mathfrak{E} >_{\varepsilon} \mathfrak{F}\}.$$

One immediately notes that

$$\sup_{\Lambda\in\mathcal{M}_f^1(I,\mathfrak{I})} \rho(\mathfrak{E},\mathfrak{F}|\Lambda) = \rho(\mathfrak{E},\mathfrak{F}).$$

## 6. APPLICATION TO POWERS OF EXPERIMENTS

In general the deficiency between two experiments is difficult to calculate. For translation experiments it suffices to evaluate the infimum of the representation

$$\rho(\mathcal{E},\mathcal{F}) = \inf_N \sup_i ||N(P_i) - Q_i||$$

within the set of invariant randomized decision functions. See [7], Chapter VIII. For efficient computations one still needs more specialization, f.e. to the n-th power

$$\mathcal{E}^n := (E^n, \mathfrak{A}^{\otimes n}, \{P_i^{\otimes n} : i \in I\})$$

of an experiment $\mathcal{E} = (E, \mathfrak{A}, \{P_i : i \in I\})$ which gives rise to the quantification of the statistical information contained in additional oberservations. For powers of $\mathcal{E}$ we have $\mathcal{E}^n < \mathcal{E}^m$ whenever $n \leq m$. The question arises of how much more informative than $\mathcal{E}^n$ is $\mathcal{E}^m$ or, in other words, of what is $\rho(\mathcal{E}^n, \mathcal{E}^m)$. Knowing this quantity can be useful in planning replicated experiments in case the decision problem is not fully determined. We shall now motivate the method of attacking the problem of estimating $\rho(\mathcal{E}^n, \mathcal{E}^m)$, give some history and develop a few recent results.

6.1 Let $K(\mathcal{E})$ be the cost of performing $\mathcal{E}$ and V a loss function. Then the (global) risk function under the decision function $\delta \in \mathfrak{D}(\mathcal{E})$ is

$$R_{\mathcal{E}} := R_\delta^V + K(\mathcal{E}).$$

Let $||V|| \leq 1$. We prefer $\mathcal{E}^n$ to $\mathcal{E}^{n+1}$ if

$$\rho(\mathcal{E}^n, \mathcal{E}^{n+1}) \leq K(\mathcal{E}^{n+1}) - K(\mathcal{E}^n)$$

and $\mathcal{E}^{n+1}$ to $\mathcal{E}^n$ if

$$\rho(\mathcal{E}^n, \mathcal{E}^{n+1}) \geq K(\mathcal{E}^{n+1}) - K(\mathcal{E}^n).$$

$\mathcal{E}^n$ is better than $\mathcal{E}^m$ means that to every risk function $R_{\mathcal{E}^m}$ there exists a risk function $R_{\mathcal{E}^n}$ such that

$$R_{\mathcal{E}^n} \leq R_{\mathcal{E}^m}.$$

Already in 1972 Torgersen established for normal experiments $\mathcal{E} = (\mathbb{R}, \mathfrak{B}, \{N(a,\sigma^2) : a \in \mathbb{R}\})$, where $\sigma^2 > o$ is known, the asymptotic equivalence

$$\rho(\mathcal{E}^n, \mathcal{E}^{n+1}) \sim \frac{1}{n}\sqrt{\frac{2}{\pi e}}.$$

If $K(\mathcal{E}^n) := k_o + nk_1$, then $n_o := \sqrt{\frac{2}{\pi e}}/k_1$ is the optimal sample size.

Intuition suggests that $\mathcal{E}^n$ becomes more and more informative as n increases, i.e. that one additional observation becomes more and more unimportant. In fact, for $|I| < \infty$ one obtains

$$\lim_{n\to\infty} \rho_a(\mathcal{E}^n) = o.$$

Moreover, one has

6.2 (Torgersen 1981). There exists a constant $C(\mathcal{E})\in[o,1]$ (independent of the special decision problem) such that for any apriori measure $\Lambda\in \mathcal{M}^1(I,\mathfrak{I})$ with $\Lambda(\{i\})>o$ for all $i\in I$,

$$\lim_{n\to\infty} r(\mathcal{E}^n|\Lambda)^{\frac{1}{n}} = \lim_{n\to\infty} \rho_a(\mathcal{E}^n)^{\frac{1}{n}} = C(\mathcal{E}),$$

where

$$C(\mathcal{E}) = \max_{i\neq j} \inf_{t\in]o,1[} \int dP_i^{1-t} dP_j^t = \inf_{t\in]o,1[} H(\mathcal{E}|t).$$

This last relationship between the constant $C(\mathcal{E})$ and the Hellinger transform $H(\mathcal{E}|t)$ is due, in the binary case, to Chernoff [2].

6.3 (Helgeland 1982). Let I be a compact subset of the parameter set of a one-dimensional exponential family, and let I contain a nondegenerate interval. Let $(r_n)_{n\geq 1}$ be a sequence of real numbers satisfying $1\leq r_n\leq n^\beta$ for some $\beta<1$. Then

$$\sqrt{\frac{2}{\pi e}} \leq \underline{\lim}_n \frac{n}{r_n} \Delta(\mathcal{E}^n, \mathcal{E}^{n+r_n})$$

$$\leq \overline{\lim}_n \frac{n}{r_n} \Delta(\mathcal{E}^n, \mathcal{E}^{n+r_n}) \leq 2\sqrt{\frac{2}{\pi e}} .$$

The lower bound has been extended by Mammen [8] to experiments $\mathcal{E}$ which can be locally approximated in a point of their parameter set by a Gaussian experiment $\mathcal{G}$ : For all $(r_n)_{n\geq 1}$ with $r_n = o(n)$

$$\underline{\lim}_n \frac{n}{r_n} \Delta(\mathcal{E}^n, \mathcal{E}^{n+r_n}) \geq \lim \frac{n}{r_n} \Delta(\mathcal{G}^n, \mathcal{G}^{n+r_n}).$$

6.4 (Mammen 1986). Let $\mathcal{E}$ be an experiment which is finite-dimensional in the sense of Dacunha-Castelle. Then there exists a constant C depending only on the dimension of $\mathcal{E}$ such that for all n and r

$$\Delta(\mathcal{E}^n, \mathcal{E}^{n+r}) \leq C\frac{r}{n}.$$

A few additional explanations seem to be in order. Let $d(i,j): = H(P_i,P_j)$ denote the Hellinger pseudo metric on I. The <u>dimension</u> of I is the smallest natural number n such that for every $\delta>o$ every subset of I of diameter $\delta$ can be covered by $2^n$ sets of diameter $\delta/2$.

The method of proof of the above inequality consists in applying a stochastic kernel N which describes the following chance mechanism: Estimate i by an estimator $\hat{i}$ which depends only on some of the ovservations available. Then observe a random variable with distribution $P_{\hat{i}}$ and mix this random variable randomly under the remaining observations. Then

$$\sup_i || N(P_i^n) - P_i^{n+1} || = o(\tfrac{1}{n}).$$

## REFERENCES


[1] Ph. Caillot, F. Martin, Le modèle bayésien, Ann. Inst. Henri Poincaré Section B, Vol. VIII, no 2 (1973), 19-40.

[2] H. Chernoff, A measure of asymptotic efficiency for tests of a hypothesis based on the sum of observations, Ann. Math. Statist. 23 (1952), 493-507.

[3] D. Feldman, Some properties of Bayesian ordering of experiments, Ann. Math. Statist. 43 (1972), 1428-1440.

[4] P. K. Goel, M. H. DeGroot, Comparison of experiments and information measures, The Annals of Statistics 7 (1979), 1066-1077.

[5] J. Helgeland, Additional observations and statistical information in the case of 1-parameter exponential distributions, Z. Wahrscheinlichkeitstheorie verw. Gebiete 59 (1982), 77-100.

[6] H. Heyer, Information-type measures and sufficiency, Symposia Mathematica Vol. XXV (1981), 25-54.

[7] H. Heyer, "Theory of Statistical Experiments", Springer Series in Statistics, New York-Heidelberg-Berlin (1982).

[8] E. Mammen, The statistical information contained in additional observations, The Annals of Statistics 14 (1986), 665-678.

[9] F. Martin, J.-L. Petit, M. Littaye, Indépendence conditionelle dans le modèle statistique bayésien, Ann. Inst. Henri Poincaré Section B, Vol. IX, no 1 (1973), 19-40.

[10] L. Piccinato, On the comparison among decisions from the Bayesian viewpoint, Metron 32 (1974), 269-298.

[11] L. Piccinato, On the orderings of decision functions, Symposia Mathematica Vol. XXV (1981), 61-71.

[12] K. K. Roy, R.V. Ramamoorthi, Relationship between Bayes, classical and decision theoretic sufficiency, Tech. Report No 30/78, Stat. Math. Division, Indian Statistical Institute, Calcutta (1978).

[13] H. Strasser, "Mathematical Theory of Statistics", De Gruyter Studies in Mathematics Vol. 7, Berlin-New York (1985).

[14] E. N. Torgersen, Comparison of statistical experiments, Scand. J. Statist. 3 (1976), 186-208.

[15] E. N. Torgersen, Deviations from total information and from total ignorance as measures of information, Statistical Research Report No 3 (1976), Institute of Mathematics, University of Oslo.

[16] E. N. Torgersen, Measures of information based on comparison with total information and with total ignorance, The Annals of Statistics 9 (1981), 638-657.

MAXIMAL SEMIGROUPS AND THE SUPPORT OF GAUSS - SEMIGROUPS

Joachim Hilgert

Technische Hochschule Darmstadt

D-6100 Darmstadt, FRG

The purpose of this note is to describe a connection between the theory of probability measures on Lie groups and the Lie theory of semigroups. The objects under consideration will be one parameter semigroups of probability measures on Lie groups and their supports. We start by giving the basic definitions.

Let G be a connected topological group. A family $(\mu_t)_{t>o}$ of probability measures on G is called a *Gauss-semigroup*, if no $\mu_t$ is a point measure, $\mu_{t+s} = \mu_t * \mu_s$ with the usual convolution and $\lim_{t\to o} t^{-1}\mu_t(G\setminus U)=0$ for every open neighborhood U of the identity in G. The Gauss-semigroup $(\mu_t)_{t>o}$ is called absolutely continuous if each $\mu_t$ is absolutely continuous with respect to a, once and for ever fixed, left Haar measure on G.

If G is a Lie group we can associate with any Gauss-semigroup $(\mu_t)_{t>o}$ an *infinitesimal generator* N of the form:

$$N = \sum_{i=1}^{n} a_i x_i + \sum_{i=1}^{r} x_i^2$$

where $\{x_1, 1\leqq i\leqq n\}$ is a basis of the Lie algebra L(G) of G, viewed as left invariant first order differential operators on G. The pair $(M,x_o)$, where M is the Lie algebra generated by $\{x_1,\dots,x_r\}$ and $x_o = \sum_{i=1}^{n} a_i x_i$ is called the *carrier* of $(\mu_t)_{t>o}$. We have (cf.[Si82]):

THEOREM 1. *Let G be a Lie group and $(\mu_t)_{t>o}$ be a Gauss-semigroup on G with carrier $(M,x_o)$. Then*

(i) Supp $\mu_t = (\bigcup_{n=1}^{\infty} (G_M \exp n^{-1}tx_o)^n)^-$, *where* Supp $\mu_t$ *is the support of the measure* $\mu_t$ *and* $G_M$ *is the analytic subgroup of* G *with Lie algebra* M.

(ii) (Supp $\mu_t$)(Supp $\mu_s$) $\subset$ Supp $\mu_{t+s}$ *for all* s,t>0. □

It is clear from this theorem that the sets $S_{\mu,a} = (\bigcup_{t>a} \text{Supp } \mu_t)^-$ are semigroups for any $a\geqq 0$. The semigroups $S_{\mu,a}$ will in general not contain the identity and hence are not suited too well to the Lie theory of semigroups which studies subsemigroups of Lie groups via their tangent object at the identity (see below for the precise definitions). But $S_\mu = S_{\mu,o}$ does contain the identity and it will be this semigroup we will concentrate on.

Let L be the Lie algebra in L(G) generated by $\{x_o, x_1,\dots,x_r\}$ and $G_L$ the corresponding analytic subgroup of G. We will call a Gauss-semigroup *generating* if $G = G_L$. By [Si82] Theorem 2, we know that $\mu_t(G\setminus G_L) = 0$ for all $t>0$. Thus, for the purpose of studying the support behaviour of Gauss-semigroups, it is no serious loss of generality to assume that $G = G_L$.

It is necessary to have some control over the interior points $\mathrm{int}(S_\mu)$ of $S_\mu$ in order to apply the techniques developed in [HHL85] and [La86]. We find:

LEMMA 2. *Let* $(\mu_t)_{t>o}$ *be a Gauss-semigroup with carrier* $(M,x_o)$ *then we have*

(i) *The interior* $\mathrm{int}(S_\mu)$ *of* $S_\mu$ *is dense in* $S_\mu$.

(ii) $S_\mu$ *is equal to the closed subsemigroup* $\overline{S}$ *of* G *generated by* exp(M) *and* $\exp(\mathbb{R}^+x_o)$.

Proof. Note first that Theorem 1 implies that $\exp(\mathbb{R}^+x_o)$ is contained in $S_\mu$. Moreover $G_M$ is contained in $S_\mu$ as well. In fact, let $y\varepsilon M$ then $(\exp(n^{-1}y)\exp((mn)^{-1}x_o))^n \subset \mathrm{Supp}(\mu_{1/m}) \subset S_\mu$ so that the Trotter product formula shows $\exp(y) \subset \overline{S}_\mu = S_\mu$ for y small enough. Therefore $S_\mu$ contains a neighborhood of the identity in $G_M$ and hence all of $G_M$. But by [JS72] the semigroup S generated by exp(M) and $\exp(\mathbb{R}^+x_o)$ satisfies $(\mathrm{int}(S))^- = \overline{S}$ since M and $x_o$ generate L(G) by our assumptions. Note finally that Theorem 1 shows that $S_\mu \subset \overline{S} = (\mathrm{int}(S))^- \subset (\mathrm{int}(S_\mu))^- \subset S_\mu$. □

Lemma 2 allows us to conclude that $S_\mu$ is contained in some *maximal subsemigroup* $S_{max}$ of G unless $S_\mu = G$ (cf[La86]). Here by maximality we mean that $S_{max}$ is no group and $S_{max}$ and G are the only subsemigroups of G containing $S_{max}$.

Now suppose that $(\mu_t)_{t>o}$ is a generating Gauss-semigroup and $S_\mu$ is contained in a maximal semigroup $S_{max}$ which is proper, i.e. $S_{max} \neq G$. Recall from Lemma 2 that $G_M$ is contained in $S_{max}$. This implies that $\exp(\mathbb{R}x_o)$ can not be contained in the group of units $H = S_{max} \cap S_{max}^{-1}$ of $S_{max}$. Now suppose that H is normal in G then L(H) is a subalgebra of L(G) which contains M and is $\mathrm{ad}(x_o)$-invariant. Thus the following remark, taken from [Si82], shows that $(\mu_t)_{t>o}$ can not be absolutely continuous.

REMARK 3. *A Gauss-semigroup is absolutely continuous if and only if the only* $\mathrm{ad}(x_o)$ *- invariant subalgebra of* L(G) *containing* M *is all of* L(G). □

We collect the obtained information in

PROPOSITION 4. *If* $(\mu_t)_{t>o}$ *is a generating absolutely continuous Gauss-*

*semigroup and* $S_\mu$ *is a proper semigroup contained in a maximal semigroup* $S_{max}$ *then* $H = S_{max} \cap S_{max}^{-1}$ *can not be not be normal in* G. □

Maximal subsemigroups of Lie groups may look very different and the theory describing them is by no means complete, but there are large classes of groups where they can be handled quite well (cf.[La86],[Hi86a]). The way these semigroups are described is typical for the Lie theory of semigroups in so far as it proceeds via their tangent object.

Given a closed subsemigroup S of a Lie group G we define the *tangent cone* L(S) of S by $L(S) = \{x \in L(G): \exp(\mathbb{R}^+x) \subset S\}$. It turns out (cf[HL83]) that L(S) is a closed convex cone satisfying

$$e^{ad(x)}L(S) = L(S) \quad \text{for all} \quad x \in L(S) \cap -L(S).$$

A closed subsemigroup S of a Lie group G is called a *halfspace semigroup* if L(S) is a halfspace. We give some examples:

The subsemigroup $\mathbb{R}^+$ of non-negative real numbers in $\mathbb{R}$ is a halfspace semigroup in $\mathbb{R}$. Let $Aff^+$ be the group of real 2 x 2 - matrices of the form

$$\left\{ \begin{bmatrix} a & b \\ 0 & 1 \end{bmatrix} : a > 0 \right\}$$

and

$$Aff^{++} = \left\{ \begin{bmatrix} a & b \\ 0 & 1 \end{bmatrix} : a > 0,\ b \geq 0 \right\}.$$

Then $Aff^{++}$ is a halfspace subsemigroup of $Aff^+$ .

Let $Sl(2,\mathbb{R})^\sim$ be the simply connected covering group of $Sl(2,\mathbb{R})$ and $\Omega^+$ be the closed subsemigroup of $Sl(2,\mathbb{R})^\sim$ generated by $\exp(\mathbb{R}^+u)$, $\exp(\mathbb{R}h)$ and $\exp(\mathbb{R}p)$ where

$$u = \begin{bmatrix} 0 & -1 \\ 1 & 0 \end{bmatrix}, \quad h = \begin{bmatrix} 1 & 0 \\ 0 & -1 \end{bmatrix}, \quad p = \begin{bmatrix} 0 & 1 \\ 0 & 0 \end{bmatrix}$$

in $sl(2,\mathbb{R})$. Then $\Omega^+$ is a halfspace semigroup in $Sl(2,\mathbb{R})^\sim$ (cf[HH85a]).

Note that for any half space in a Lie algebra bounded by a subalgebra there is a closed halfspace semigroup in the simply connected group corresponding to the Lie algebra whose tangent wedge is just the halfspace we started with (cf[Hi86b], [La86]). Moreover this halfspace semigroup is maximal and its group of units is the analytic subgroup corresponding to the hyperplane contained in the halfspace.

We are now ready to prove a converse to Proposition 4 in the case that G is simply connected:

PROPOSITION 5. *Suppose that* G *is simply connected and let* $(\mu_t)_{t>0}$ *be a generating Gauss-semigroup which is not absolutely continuous. Then* $S_\mu$ *is contained in a halfspace semigroup* $S_{max}$ *whose group of units* $S_{max} \cap S_{max}^{-1}$ *is a closed normal subgroup in* G *of codimension* 1.

Proof. Note first that by hypothesis there exists an $ad(x_o)$-invariant subalgebra of $L(G)$ containing $M$ which is not all of $L(G)$. Let $P$ be such an algebra of maximal dimension. We claim that $P$ must be a hyperplane in $L(G)$. In fact, suppose that $codim(P) > 1$ then $P + \mathbb{R}x_o$ is a subalgebra of $L(G)$ containing $M$ which is $ad(x_o)^{\sim}$ invariant, but not all of $L(G)$. But on the other hand we assumed $M$ and $x_o$ to generate the whole algebra which contradicts our earlier statement. Thus $P$ is a hyperplane and by the argument given above it cannot contain $x_o$. Therefore the $ad(x_o)$- invariance of $P$ shows that $P$ is an ideal in $L(G)$. Let $G_P$ be the analytic subgroup of $G$ corresponding to $P$ then $G_P$ is the group of units of a maximal halfspace semigroup $S_{max}$ containing also $\exp(\mathbb{R}^+x_o)$. Since $P$ is an ideal we know that $G_P$ is normal in $G$. Finally we note that $S_{max}$ contains $\exp(P)$, hence $\exp(M)$, so that Lemma 2 implies that $S_{max}$ contains $S_\mu$. □

Of course one wonders how serious the assumption in Proposition 5 that $G^{\sim}$ be simply connected is. Let $G$ be the simply connected covering group of $G$ and $\varphi: G^{\sim} \to G$ be the covering morphism. If $(\sigma_t)_{t>o}$ is a Gauss-semigroup on $G$ with infinitesimal generator $N$ then $(\varphi\sigma_t)_{t>o}$, consisting of the image measures, is the Gauss-semigroup on $G$ with infinitesimal generator $N$. Let $Exp: L(G) \to G^{\sim}$ be the exponential function for $G^{\sim}$. Then $S_\sigma$ is the closed subsemigroup of $G^{\sim}$ generated by $Exp(M)$ and $Exp(\mathbb{R}^+x_o)$ by Lemma 2. Therefore we get $\varphi(int(S_\sigma))$ is open dense in $S_{\varphi\sigma}$ again by Lemma 2. Thus practically all the information on the support of Gauss-semigroups we can expect to obtain via the Lie theory of semigroups, we can already get from the simply connected case.

Proposition 4 and 5 have some immediate consequences. For instance, Proposition 5 says that any generating Gauss-semigroup on $Sl(2,\mathbb{R})^{\sim}$ is absolutely continuous and, since the absolute continuity of a Gauss-semigroup depends only on its infinitesimal generator, the same is true for $Sl(2,\mathbb{R})$. On the other hand Proposition 4 shows that any generating absolutely continuous Gauss-semigroup on a nilpotent Lie group satisfies $S_\mu = G$ the group of units of maximal semigroups in nilpotent Lie groups contains the commutator subgroup (cf[HHL85]). Of course all of this, and more, is well known (cf[Mc84],[McW83]), but the methods given above are quite general so any kind of information one has on the maximal subsemigroups of a Lie group will yield some information on the support of Gauss-semigroups on this group.

Note that for any subsemigroup $S$ of $G$ containing the identity there is a largest normal subgroup contained in $S$ (cf[La86]). It is denoted by $Core(S)$. The core of a closed semigroup $S$ is closed, so it makes sense to talk about the *reduced pair* $(G_R,S_R)$ where $G_R = G/Core(S)$ and $S_R = S/Core(S)$. If $S$ is a closed halfspace semigroup then we have a complete description of $(G_R,S_R)$:

<u>THEOREM 6</u>. (cf.[Po77]). *Let $S$ be a closed halfspace semigroup in a connected Lie group $G$. Then for the reduced pair $(G_R,S_R)$ one of the following cases occurs:*

(i) $(G_R,S_R)$ *is topologically isomorphic to* $(R,R^+)$

(ii) $(G_R,S_R)$ *is topologically isomorphic to* $(Aff^+,Aff^{++})$

(iii) $(G_R,S_R)$ *is topologically isomorphic to* $(Sl(2,R)^{\sim},\Omega^+)$. □

Theorem 6 tells us that the group of units of a closed halfspace semigroup S is normal if and only if the reduced pair $(G_R, S_R)$ is equal to $(R, R^+)$.

Thus if we, for some reason, know that any maximal semigroup S in G has to be a halfspace semigroup with reduced pair $(G_R, S_R) \equiv (R, R^+)$ then Proposition 4 tells us that for any absolutely continuous Gauss-semigroup $(\mu_t)_{t>o}$ the semigroup $S_\mu$ has to be all of G.

In this context we recall the following theorem from [La86]:

THEOREM 7. *Let G be a Lie group such that G/Rad(G) is compact, where Rad(G) is the radical of G. If S is a maximal subsemigroup of G with non-empty interior, then S is a halfspace semigroup containing every semisimple analytic subgroup and for the reduced pair $(G_R, S_R)$ one of the following two cases occurs*

(i) $(G_R, S_R)$ *is topologically isomorphic to* $(R, R^+)$

(ii) $(G_R, S_R)$ *is topologically isomorphic to* $(Aff^+, Aff^{++})$. □

From this we derive

COROLLARY 8. *Let G be a Lie group such that Rad(G) is nilpotent and G/Rad(G) is compact, then for every absolutely continuous Gauss-semigroup $(\mu_t)_{t>o}$ we have $S_\mu = G$.*

Proof. It remains to show that case (ii) of Theorem 7 cannot occur. To this end note that the conjugate of a semisimple analytic subgroup is again semisimple so that the subgroup of G generated by all semisimple analytic subgroups of G is a normal subgroup and, by Theorem 7, contained in the core of any maximal semigroup. Thus $G_R$ is nilpotent which excludes case (ii) of Theorem 7. □

COROLLARY 9. *Let G be a Lie group such that L(Rad(G)) = Rad(L(G)) carries the structure of a complex Lie algebra and G/Rad(G) is compact, then for every absolutely continuous Gauss-semigroup $(\mu_t)_{t>o}$ we have $S_\mu = G$.*

Proof. As in Corollary 8 we see that any Levi complement of G is contained the core C of an arbitrary maximal semigroup with nonempty interior. Thus $G_R = G/C \cong Rad(G)/(Rad(G) \cap C)$ and $G_R$ contains a halfspace semigroup. Taking the inverse image in Rad(G) this shows that Rad(G) contains a halfspace semigroup. If we look at the tangent cone of this semigroup it follows from [HH85b] that it contains the commutator algebra of Rad(L(G)) because of the complex structure. Thus $Rad(G) \cap C$ contains the commutator subgroup of Rad(G) so that $G_R$ is abelian which again excludes case (ii) of Theorem 7. □

Let us draw a short resumé of what has been said in this note: The supports of the measures in a Gauss-semigroup give rise to subsemigroups of the Lie groups involved. These semigroups can be studied by methods from the Lie theory of semigroups. The results will in general not be results on the supports of the single measures but on the semigroups one associates to them. In special cases, however, as in the case of decreasing supports it is possible to derive results on the supports of the single measures.

It has not been my intention to give a polished exposition of all the results that can be obtained using the methods indicated, but rather I wanted to explain the methods themselves. It is clear that one can construct many examples along these lines and it seems reasonable to believe that many related results could be obtained without a lot of extra effort.

REFERENCES


[Bou72] N. Bourbaki, "Groupes et Algébres de Lie II, III," Hermann, Paris (1972).

[He77] H. Heyer, "Probability Measures on Locally Compact Groups," Springer, Berlin (1977).

[HH85a] J. Hilgert and K. H. Hofmann, Old and New on Sl(2), Manus. Math. 54:17 (1985).

[HH85b] ---, Lie Semialgebras are real Phenomena, Math. Ann. 270:97 (1985).

[HHL85] J. Hilgert, K. H. Hofmann and J. D. Lawson, Controllability of Systems on Nilpotent Lie Groups, Beitr. Alg. Geom. 20:185 (1985)

[Hi86a] J. Hilgert, Maximal Semigroups and Controllability in Products of Lie Groups, THD Preprint 971 (1986).

[Hi86b] ---, Infinitesimally Generated Subsemigroups of Lie Groups, submitted (1986).

[HL83] K. H. Hofmann and J. D. Lawson, Foundations of Lie Semigroups, in: LNM 998:128 (1983).

[Hoch65] G. Hochschild, "The Structure of Lie Groups," Holden Day, San Francisco (1965).

[JS72] V. Jurdjevic and H. Sussmann, Control Systems on Lie Groups, J. Diff. Equ. 12:313 (1972).

[La86] J. D. Lawson, Maximal Subsemigroups of Lie Groups that are Total, submitted (1986).

[Mc84] M. McCrudden, On the Supports of Absolutely Continuous Gauss Measures on Connected Lie Groups, Mh. Math. 98:295 (1984).

[McW83] M. McCrudden and R. M. Wood, On the Supports of Absolutely Continuous Gauss Measures on SL(2,R), in: LNM 1064 (1983).

[Po77] D. Poguntke, Well-Bounded Semigroups in Connected Groups, Semigroup Forum 15:159 (1977).

[Si82] E. Siebert, Absolute Continuity, Singularity, and Supports of Gauss Semigroups on a Lie Group, Mh. Math. 93:239 (1982).

# SUFFICIENCY COMPLETENESS PRINCIPLE

Julián de la Horra

Dep. Matemáticas, Univ. Autónoma Madrid
28049-Madrid
Spain

## 1. INTRODUCTION

Different sensible principles have been considered in Statistics: likelihood, weak conditionality, weak sufficiency...Many papers have been writen on these topics, after the pioneer work by Birnbaum (1962). Berger and Wolpert (1984) is a good reference for the study of different principles and relations between them. These principles generally work on rather vague concepts: Evidence (Birnbaum (1962)), Inference patterns (Dawid (1977)),... We shall work on the somewhat more concrete concept of generalized estimator or inferences (Eaton (1982), De la Horra (1987)).

In Section 2 we state the sufficiency completeness principle (stronger than the weak sufficiency principle). Its relation with posterior distributions is researched (Jeffreys' rule is compatible with the sufficiency completeness principle). In Section 3, the relation with other principles is studied.

## 2. THE SUFFICIENCY COMPLETENESS PRINCIPLE

Let $E=(X,\theta,\{P_\theta\}_{\theta\in\Theta})$ be an experiment, consisting of a realization of the random variable X (taking values on the sample space $\mathcal{X}$, where $\mathcal{X}$ is a Borel set of $\mathbb{R}^k$), with distribution given by the probability measure $P_\theta$, for $\theta\in\Theta$ (Borel set of $\mathbb{R}^p$). If the family $\{P_\theta\}_{\theta\in\Theta}$ is absolutely continuous with respect to a $\sigma$-finite measure $\mu$, we shall denote their densities by $\{f_\theta\}_{\theta\in\Theta}$.

<u>Definition 2.1</u>

A <u>generalized estimator or inference</u> is a function H mapping the sample space $\mathcal{X}$ into the set of probability measures on $\Theta$.

Point estimators are generalized estimators: they assign a degenerate distribution to each $x\in\mathcal{X}$. In this case, H(x) will denote the only mass point. Generalized estimators are closely related to the work by Dawid (1977) on Inference patterns based on E and x. Also of interest are the studies by Blyth (1970) and Gatsonis (1984).

We shall consider principles as tests which a given generalized estimator must undergo. We shall start with the weak sufficiency principle (Dawid (1977)) applied to generalized estimators:

Weak Sufficiency Principle (WSP)

A generalized estimator H satisfies the WSP when for all sufficient statistic T:

$$H(x)=g(T(x)),\ a.s.-\{P_\theta\}_{\theta\in\Theta}$$

Therefore, this principle demands that an estimator be a function of any sufficient statistic, but it does not use all the information contained in the concept of sufficiency. An example can add ligth to this question: let $\dot{x}=(y_1,\dots,y_n)$ be a sample, where $y_1,\dots,y_n$ are i. i. d. observations from a distribution $N(\theta,\sigma)$ ($\sigma$ known). $\bar{y}$ is a minimal sufficient statistic. Let H be an estimator defined as:

$$H(x)=\begin{cases}10 & \text{if } \bar{y}\ (8,\ 12)\\ \bar{y} & \text{otherwise}\end{cases}$$

H satisfies the WSP. But, what reason is there for reaching the same conclusion, for all the samples with $\bar{y}$ (8, 12)? There is no reason for doing so (by invoking sufficiency). I think that different conclusions must be reached if different $\bar{y}$'s are obtained. This is the idea leading to the following principle:

Sufficiency Completeness Principle (SCP)

A generalized estimator H satisfies the SCP when:

$$H(x)=g(T(x))\ a.s.-\{P_\theta\}_{\theta\in\Theta} \text{ if and only if T is sufficient.}$$

More generally, a statistical procedure giving generalized estimators satisfies the SCP, when all these estimators satisfy the principle. Clearly, the SCP is stronger than the WSP. We shall next see that posterior distributions verify the SCP.

Theorem 2.1.

Let H(x) be the posterior distribution obtained from the experiment $E=(X,\ \theta,\ \{f_\theta\}_{\theta\in\Theta})$ and the prior distribution with density $g(\theta)>0$, for all $\theta\in\Theta$. Then, H satisfies the sufficiency completeness principle.

Proof: It is obvious, by observing that this theorem states (with other words) the equivalence between classical and Bayesian definitions of the sufficiency. See De Groot (1970, p. 156) □

We remark that the application of Jeffreys' rule is compatible with the SCP. This interesting property does not hold true for the likelihood principle (see Berger and Wolpert (1984, p. 20)). Of interest could be additional research to find what principles are needed for posterior distributions to be the only sensible estimators.

There is an easy characterization for estimators satisfiing the SCP:

Lemma 2.1.

Let $E=(X,\ \theta,\ \{P_\theta\}_{\theta\in\Theta})$ be an experiment and let $T_0$ be a minimal sufficient statistic. A generalized estimator H satisfies the sufficiency completeness principle if and only if H is a function of $T_0$ a.s.-$\{P_\theta\}_{\theta\in\Theta}$

and H distinguishes values of $T_0$ a.s.-$\{P_\theta\}_{\theta\in\Theta}$.

Proof: The proof is easy and is omitted □

## 3. RELATION WITH OTHER PRINCIPLES

Next, we shall study the relation between the SCP and other principles (namely, likelihood and weak conditionality).

a) The likelihood principle (LP) does not imply the SCP, as example 3.1 below shows:

Example 3.1

Let $x=(y_1,\ldots,y_n)$ be a sample, where $y_1,\ldots,y_n$ are i. i. d. observations from the distribution with $\lambda$-density ($\lambda$ being the Lebesgue measure):

$$f_\theta(y)=\theta(1-\theta)^{-1}\exp\left(\frac{2\theta-1}{1-\theta}\log y\right) I_{(0,1]}(y) \quad , \text{ for } \theta\in\Theta=[1/2,\ 1]$$

$T_0(x)=\Sigma_{i=1}^{n}\log y_i$ is a minimal sufficient statistic. The maximum likelihood estimator is:

$$H(x)=\begin{cases} n/(n-\Sigma_{i=1}^{n}\log y_i) & \text{if this amount is greater than } 1/2 \\ 1/2 & \text{otherwise} \end{cases}$$

The inequality $n/(n-\Sigma_{i=1}^{n}\log y_i)\leq 1/2$ is equivalent to $\Pi_{i=1}^{n} y_i\leq 1/e^n$ . Thus, for all $x=(y_1,\ldots,y_n)$ such that $\Pi_{i=1}^{n} y_i\leq 1/e^n$ , $H(x)=1/2$, but $T_0(x)$ takes different values. H does not distinguish values of $T_0$ , and therefore, H does not verify the SCP (from the lemma 2.1). Of course, H verifies the LP.

b) As a consequence, the weak conditionality principle (WCP) does not imply the SCP (from the fact that the LP is equivalent to the WCP and the WSP; see Birnbaum (1962) and Berger and Wolpert (1984, p. 27)).

c) On the other hand, the SCP does not imply the LP. This is proved by observing that Jeffreys' rule is compatible with the SCP, but is not compatible with the LP.

d) As a consequence, the SCP does not imply the WCP.

## REFERENCES


Berger, J. O. and Wolpert, R. L., 1984, "The Likelihood Principle", Lecture Notes, Vol. 6, Institute of Mathematical Statistics, Hayward, California.

Birnbaum, A., 1962, On the foundations of statistical inference , *J. Amer. Statist. Assoc.* 57, 269-306.

Blyth, C. R., 1970, On the inference and decision models in statistics, *Ann. Math. Statist.* 41, 1034-1058.

Dawid, A. P., 1977, Conformity of inference patterns , in: Recent Developments in Statistics, J. R. Barra et al. (eds.), North-Holland, Amsterdam.

De Groot, M. H., 1970, "Optimal Statistical Decisions", Mc Graw-Hill, New York.

De la Horra, J., 1987, Generalized estimators: A Bayesian decision-theoretic view , *Statist. & Dec.* (to appear).

Eaton, M. L., 1982, A method for evaluating improper prior distribu-

tions, in: Statistical Decision Theory and Related Topics, Vol. 1. S. S. Gupta and J. O. Berger (eds.), Academic Press, New York.
Gatsonis, C. A., 1984, "Deriving posterior distributions for a location parameter: A decision-theoretic approach", Ann. Statist. 12, 958-970.

# ON THE INTERPRETATION OF HYPOTHESIS TESTS FOLLOWING NEYMAN AND PEARSON*

David Johnstone

Faculty of Economics
University of Sydney

## 1. HYPOTHESIS TESTS AS METHODS FOR DECISION

### Neyman and Pearson

To begin with, Neyman and Pearson agreed with Fisher that the result in a hypothesis test is a measure of evidence. In their first joint paper, which was published in 1928, they declared that the level of significance (P-level) attained in a likelihood ratio test is a measure of *evidence* against the null hypothesis [1928, pp.28-9], and that a hypothesis test is a method with which to "accept" or "reject" the null hypothesis "..with a greater or less degree of confidence" or certainty [1928, pp.1, 67]. This was clearly an inferential interpretation.[1] However, in 1933, in their celebrated joint paper to the Royal Society, they introduced the notion that a hypothesis test is simply a "rule of behavior", *i.e.* a rule for *decision* between alternate courses of action. This was Neyman's idea. Pearson, it seems, retained the position that a hypothesis test is a legitimate method for *inference*. Indeed, in a paper of his own published in 1955, Pearson agreed explicitly with Fisher that a test is a "means for learning" [1955, p.206]. Furthermore, in this same paper, he disclaimed any association with Neyman's idea that tests are methods merely for "inductive behavior" or decision [1955, pp.206-7]. Moreover, there seems a commitment to inference throughout Pearson's writing. For example, consider his statement in a joint paper with Clopper [1934, pp.404-5] that some confidence interval $(p_1,p_2)$ with "confidence coefficient" $1-\alpha$ entails a degree of confidence or certainty ($\text{probability}_1$)[2] of $1-\alpha$ that the true paramater $\theta$ lies in the interval $p_1$ to $p_2$, *i.e.* $prob_1(\theta\epsilon(p_1,p_2))=1-\alpha$. More recently, commenting in a paper by Barnard *et.al.* [1962, p.363], Pearson explained clearly that the

*I have benefited from the comments of H.E. Kyburg, D.V. Lindley, J.W. Pratt and participants in the Popper Seminar at the London School of Economics.

[1]Specifically, Neyman and Pearson [1928, p.4] state that our *confidence* in hypothesis $A$ depends on the likelihood $f(\Sigma/A)$ of the sample $\Sigma$ under $A$, or moreover on the likelihood ratio $f(\Sigma/A)/f(\Sigma/B)$, although [p.67] the "..confidence with which we form a judgement" cannot be based entirely on the likelihood ratio, or "..any single numerical criterion ...because there will nearly always be present certain *a priori* conditions and limitations which cannot be expressed in exact terms." These remarks were apparently Bayesian, for to speak of the likelihood ratio and *subjective* prior information determining the confidence we place in hypotheses presumes (implies) both Bayes' theorem and a subjective $\text{probability}_1$ interpretation of "degree of confidence".

[2]The abbreviations "$\text{probability}_1$" (degree of certainty) and "$\text{probability}_2$" (relative frequency) are from Carnap [1962, pp.23-5].

"action" or "decision" in a hypothesis test may be *cognitive* (inferential) rather than *behavioral*. His example is one of "deciding that for the moment there is not enough evidence to justify the claim that some "difference" has been established".

## Neyman's Logic

The result in a hypothesis test is to either "reject $h_o$" or "accept $h_o$". Neyman declared that the terms "reject $h_o$" and "accept $h_o$" designate *courses of action A* and *B*, not states of inference or belief:

> The terms "accepting" and "rejecting" a statistical hypothesis are very convenient and are well established. It is important, however, to keep their exact meaning in mind and to discard various additional implications which may be suggested by intuition. Thus, to accept a hypothesis *H* means only to decide to take action *A* rather than action *B*. This does not mean that we necessarily believe that the hypothesis *H* is true. Also, if the application of a rule of inductive behavior [a hypothesis test] "rejects" *H*, this means only that the rule prescribes action *B* and does not imply that we believe that *H* is false. [1950, pp.259-60]

Thus, for Neyman, the result in a hypothesis test is a *decision* between the alternate courses of action *A* and *B*. These are labelled "reject $h_o$" and "accept $h_o$", yet there is strictly *no inference* about the hypothesis $h_o$, express or implied. Thus, Neyman's logic is a logic merely for *decision*. Inference is specifically denied.

Decision *without inference* (behavioralism) has no apparent logic. Moreover, it seems that decision presumes (requires) inference. This is of course the Bayesian position. To quote Lindley:

> Whilst it is possible to make inferences without considering decisions, the implementation of decision-making requires an earlier calculation of the appropriate inference, $p(\theta/x)$. [1977, pp.51-2]

For example, a judge decides that the accused should hang only once he believes, or comes to the *inference*, that the accused is quite certainly guilty. Neyman, however, denied inference. His logic is that if we decide (in each case) to "reject $h_o$" or "accept $h_o$" with a hypothesis test (or "rule of behavior") with low error frequencies $(\alpha,\beta)$, then we will not in any case learn whether the hypothesis $h_o$ is true (probably) or false, but we will rest assured that in "the long run" our decisions will mostly be correct. Consider, for example, his statement below, from a paper written with Pearson:

> Without hoping to know whether each separate hypothesis is true or false, we may search for rules to govern our behavior with regard to them, in following which we ensure that, in the long run of experience, we shall not be too often wrong. Here, for example, would be such a "rule of behavior": to decide whether a hypothesis, *H*, of a given type be rejected or not, calculate a specified character, *x*, of the observed facts; if $x>x_o$ reject H, if $x \leq x_o$ accept *H*. Such a rule tells us nothing as to whether in a particular case *H* is true when $x \leq x_o$ or false when $x>x_o$. But it may often be proved that if we behave according to such a rule, then in the long run we shall reject *H* when it is true not more, say, than once in a hundred times, and in addition we may have evidence that we shall reject *H* sufficiently often when it is false. [1933, p.142]

Here Neyman explained that if we use *good* hypothesis tests, *i.e.* tests with low error frequencies, then in "the long run" we will *usually* make the right decision. He did not say that we will make the right decision in any particular *single case*, or that there is a high probability[1] that we will make the right decision in any single case. To the contrary, Neyman was a *frequentist* in the strictest sense. He maintained that we can not be concerned with any single case. On his account, the best that we can do is to control the *frequencies* $(\alpha,\beta)$ with which we make errors in a sequence of tests described as the "long run"; *cf.* de Finetti [1972, p.172]. To wit:

It would be nice if something could be done to guard against errors in each particular case. However, as long as the postulate is maintained that the observations are subject to variations affected by chance (in the sense of frequentist theory of probability), all that appears possible to do is to control the frequencies of errors in a sequence of situations $(s_n)$, whether similar, or very different. [1971, p.13]

Neyman's Philosophy

Neyman was a *deductivist*, *i.e.* one who maintains that there is no reasonable method for inductive inference. Perhaps his strongest statement is one he made with Pearson in 1933. Here he claimed that no test which makes use of the probability calculus, which surely any reasonable test must, can provide evidence for or against any particular hypothesis:

> We are inclined to think that as far as a particular hypothesis is concerned, no test based upon the theory of probability can by itself provide any valuable evidence of the truth or falsehood of that hypothesis. [1933, pp.141-2]

Thus, for Neyman, there is no method with which we can say that the result in a hypothesis test constitutes *evidence* (in any degree) for or against any particular hypothesis. This includes the methods of inverse probability$_1$ (such as those of Carnap, Jeffreys, Savage and de Finetti) and the methods of likelihood inference (such as those of Barnard, Hacking and Edwards) as well as tests of significance following R.A. Fisher. Neyman abandoned the methods of inverse probability$_1$ generically. These methods all employ *Bayes' theorem*, which Neyman [1941, p.378] thought logical but not often useful. Only if the probability distribution *a priori* is a frequency (probability$_2$) distribution derived theoretically would Neyman use Bayes' theorem. But theoretical priors are hard to find. And Neyman abhorred both subjective priors, and priors based on rules such as "Bayes' postulate", *i.e.* the Laplacean "principle of insufficient reason". Hence, he decided with Pearson to abandon Bayes' theorem altogether; *e.g.* Neyman and Pearson [1928, p.69].

Despite his belief that there is no reasonable method for inference, Neyman was hardly despondent. He maintained that the role of statistics, and science in general, is to enable *decision* between alternate courses of action [1957, p.16], and that the methods he developed with Pearson enable decision *without recourse to inference*. Specifically, these methods enable "inductive behavior", which is decision (behavior) such that in the "long run" the courses of action decided upon are *most often* for the best [Neyman (1961) p.148]. Thus, Neyman had no need for inference. He professed concern merely with results in actual practical decisions, specifically results in the "long run". In this respect, he was a sort of *instrumentalist*. Indeed, he was especially close to the mathematician and philosopher C.S. Peirce, who anticipated his concern with results on average in the "long run".

## 2. HYPOTHESIS TESTS AS METHODS FOR INFERENCE

Neyman's Practice

During all the years from 1933 until his death in 1981, Neyman insisted more and more that a hypothesis test is simply a "rule of behavior", *i.e.* a method for *decision*, but not inference. However, in practice, he was less than convincing. Both Kempthorne [1976, p.773] and Rosenkrantz [1977, p.191] have suggested that Neyman's practice seemed distinctly *Fisherian*, *i.e.* "evidential", rather than "behavioristic". It is not difficult to find signs of inference in Neyman's practice. Let me cite one or two examples:

(i) In 1936, in a joint paper with Tokarska, Neyman said that in practice the decision in a hypothesis test rests on *evidence* gained in that test for or against the hypothesis tested. For example, the decision to accept a consignment of electric lamp bulbs rests on *evidence* against the hypothesis that the lamp bulbs in that consignment are generally defective:

The consignment is accepted when the trial of a sample of lamps furnishes sufficient evidence for the rejection of the hypothesis tested. [1936, p.239]

(ii) In a more recent paper, Neyman [1969, pp.1060-1] reported the results from one study not as decisions to "reject $h_o$" or "accept $h_o$", but as levels of significance (P-levels) in the manner following Fisher. He had conceded already that in practice the procedure is not to "reject $h_o$" or "accept $h_o$" in accord with some *critical* level of significance $\alpha$, but to assess the "significance" of the sample, presumably in the sense due to Fisher, using the benchmarks which Fisher established (*e.g.* 1%, 5% *etc.*):

> In practice, ..it is likely that the level of significance will not be fixed so rigidly. In most cases we should probably consider a result significant whenever it is above the 0.01 level. In cases when the criterion ($t$ or $z$) is in between 1% and 5% points, we should probably remain in some doubt, and finally, if the criterion is below 5%, we should judge the result as non-significant. [1935, p.229].

If the result in a hypothesis test is merely a *decision* between two courses of action, it seems hardly sensible to think of that result, in some sense, as more or less *significant*. Moreover, "significance" (like support) is a continuous measure, between 0 and 1, whereas a decision is simply one way or the other; *cf.* Johnstone [1987].

<u>Neyman's Theory</u>

In the English translation of his *Theory of Probability*, de Finetti [1974a, p.14] claimed that Neyman's solution to the problem of scientific inference, whereby the "logical formulation" (Bayes' theorem and subjective probability) is eschewed, resorts sometimes to "particular tricks (which are sometimes rather contrived)". More specifically, de Finetti intimated that the behavioristic doctrine attributable to Neyman is semantic humbug, meretricious, and even deceitful:

> In order to be able to provide 'conclusions'- but without being able to state that they are *certain*, because they are undoubtedly not so, and not wanting to say that they are *probable*, because this would involve admitting subjective probability - a search is made for words that appear to be expressing something meaningful, it is then made clear that they do not, in fact, mean what they say, and then, finally, a strenuous attempt is made to get people to believe that it is wise to act as if the words did, in fact, have some meaning (though what it is heaven only knows!). [1975, p.200]

This was strong criticism, but not without reason. Neyman's writing on the interpretation of hypothesis tests is equivocal, if not evasive and tendentious. Let me support this claim with reference to specific passages:

(i) In 1975, at the Annual Meeting of the Institute of Mathematical Statistics, Neyman took part withas the result in a hypothesis test is a decision between alternate courses of action. Similarly, a distinction was made between nce?". During this discussion, Kempthorne [1976, pp.764-7, 768, 770] distinguished conceptually between tests interpreted as methods for *decision*, which he labelled "hypothesis tests", and tests interpreted as methods for *inference*, which he labelled "significance tests".[3] In his terms, the result

[3]This distinction was neither new nor controversial, tracing at least to Barnard [1949, p.116]. Indeed, Tukey [1960, p.433] distinguished between "tests of significance (conclusions)" and "tests of hypotheses (decisions)" in the same terms exactly as Kempthorne. It matters not of course which tests are labelled which. The need is merely to distinguish semantically between tests interpreted as methods for *inference* and tests interpreted as *decision rules*. It is sensible, however, to attach the term "hypothesis test" to tests interpreted as decision rules, for this was the term which Neyman himself introduced to distinguish his tests from those of Fisher.

in a significance test (the level of significance $P$) is a measure of *evidence*, whereas the result in a hypothesis test is a *decision* between alternate courses of action. Similarly, a distinction was made between *decision* and *conclusion*, where, following Tukey [1960, p.425], conclusion means inference or belief. But Neyman would not admit any such distinction, not even conceptually. He responded:

> [I am] not aware of a conceptual difference between a "test of a statistical hypothesis" and a "test of significance" and [I use] these terms interchangeably. [1976, p.737]
>
> A similar remark applies to the use of the words "decision" or "conclusion". It seems to me that at our discussion these particular words were used to designate only something like a final outcome of complicated analysis involving several tests of different hypotheses. In my own way of speaking, I do not hesitate to use the words "decision" or "conclusion" every time they come handy. [p.750]

This is abhorrent. It is one thing to deny any logic or program for inference, but quite another to deny inference as a *concept* distinct from decision. Moreover, if there is no conceptual distinction between inference and decision, what sense does it make to write, as Neyman did consistently, that hypothesis tests are methods for decision rather than inference.

(ii) Neyman maintained that to "accept $h_o$" is merely to choose action $A$, and to "reject $h_o$" is merely to choose action $B$. But he failed to identify these actions $A$ and $B$. Tukey [1960, pp.424-5] suggested that to "accept $h_o$" is to act (in the present situation) *as if* the hypothesis $h_o$ is true, and to "reject $h_o$" is to act *as if* the hypothesis $h_o$ is false. This well known interpretation was due initially to Neyman, but Neyman was by no means as definite as Tukey. For example, consider his remarks below concerning the interpretation of confidence intervals:

> ..after observing the values of the $x$'s in a case where the $\theta$'s are unknown and calculating [the confidence limits] $\underline{\theta}(E')$ and $\overline{\theta}(E')$, we may *decide* to behave as if we actually knew that the true value $\theta_1$ of $\theta_1$ were between $\underline{\theta}(E')$ and $\overline{\theta}(E')$. This is done as a result of our *decision* and has nothing to do with 'reasoning' or 'conclusion'. ... The above process is also devoid of any 'belief' concerning the value $\theta_1$ of $\theta_1$. Occasionally we do not behave in accordance with our beliefs. Such, for example, is the case when we take out an accident insurance policy while preparing for a vacation trip. In doing so, we surely act against our firm belief that there will be no accident; otherwise we would probably stay at home. [1941, pp.379-80]

Here Neyman said that given the confidence interval $f(X)$, we *may* decide to act as if we know or believe that the interval $f(X)$ includes the true $\theta$. However, then he added that it can be reasonable to act in a way which is inconsistent with propositions which we firmly believe. This seems effectively an escape clause, and thus it is hard to know how Neyman would act given the confidence interval $f(X)$, or given the result in a hypothesis test. He might act in a way which is consistent with that result $f(X)$ being true, and then again he might not. Unlike Tukey, he seemed not to commit himself either way.[4]

(iii) Neyman maintained that the term "accept $h_o$" ("reject $h_a$") is merely a label for a particular course of action. However, in his discussion with Kempthorne *et.al.* [1976, p.749], he admitted that rather than the label "accept $h_o$" he preferred the locution "no evidence against $h_o$ is found". Surely this expression can not be merely a label for a course of action. Moreover, on any reasonable interpretation, the expression "no evidence against $h_o$

---

[4]Wisely so perhaps, for to act *as if* $h_o$ is true, or *as if* $h_o$ is false, precludes any repetition of the experiment. Moreover, if we *know* that $h_o$ is true, or false, we can not rationally run another experiment. Thus, the 'act as if' interpretation of hypothesis tests is incompatible with the doctrine of experimental repetition; *cf.* de Finetti [1972, pp.176-7].

is found" is a statement of a *measure of evidence*, which is of course a statement of inference.

(iv) On careful inspection, there is scope for an inferential interpretation in much of Neyman's writing, including some of the most unlikely passages. For example, consider the passage below:

> ..the theory of testing hypotheses has no claim of any contribution to the "inductive reasoning." ...The application of the Bayes' formula and of its consequences, as advocated by Jeffreys, when the problem treated does not contain in its conditions the probabilities *a priori* interpretable in terms of frequencies, does not lead to results having any clear frequency interpretation. Instead it measures the "intensity of our belief". On the other hand, the theory of testing statistical hypotheses predicts relative frequencies - in so far as it is possible to do so - but does not claim to measure directly the confidence. [1942, p.301]

Here Neyman states that his theory for hypothesis tests "does not claim to measure directly the confidence". Thus, he denies any direct measure of confidence in the hypothesis tested, but not an *indirect* or elliptical (intuitive) measure. If strictly a behavioralist, he would deny *any* measure of confidence whatever, direct or otherwise. That is, he would not have qualified his disclaimer with the adverb "directly". By including this qualifier, Neyman admits frequency based evidential interpretations such as Birnbaum [1977, pp.24-5], whereby the result in a hypothesis test, being described by a triple of the form *(reject $h_1$ for $h_2$, $\alpha$, $\beta$)* or *(reject $h_2$ for $h_1$, $\alpha$, $\beta$)*, is interpreted on an intuitive notion of evidence (known as the "confidence concept") as a measure of evidence against $h_1$ for $h_2$, or against $h_2$ for $h_1$.

(v) Neyman found respite in equivocacy, especially in "difficult" company. Consider, for example, his remarks below, taken from his contribution to the meeting of the International Statistical Institute in 1963:

> ...an experiment will be performed yielding a value, say *x*, of *X* and the statistician will be faced with the necessity of taking a practical step, which I shall describe as "concluding step." The description and, probably, also the essence of the concluding step varies greatly from one section of our literature to the next. Some of us speak of "inference" regarding θ. Some others prefer the term "decision." I wish to emphasize that for purposes of the present discussion these differences are not material and that the term "concluding step" is meant to designate the final outcome of the work that the statistician does with reference to his problem in instance $I_n$. Thus, this term is meant to apply to the decision, say, to institute or not to institute an immunization campaign using a given vaccine, to the decision to treat two galaxies as forming a physical system or an optical pair and, equally, to Professor Barnard's "inference that $y_{10}$ is $N(1.047,1)$." [1963, p.929]

Here Neyman acknowledges the inference (conclusion) mentioned by Barnard that the variate *y* is distributed $N(1.047,1)$. Professor Barnard has always distinguished inference from decision. For example, in a well known discussion with Savage *et.al.*, he wrote:

> Before continuing with the discussion of statistical inference it is necessary to refer to a topic which, in my opinion, is distinct from inference, namely, decision-making. That there is a distinction seems to me to follow from the fact that in decision-making it is necessary to specify a goal to be aimed at in the result of the decision, whereas inferences can be made without reference to any such goal. [Barnard (1962) p.40]

In Barnard's terms, the "inference that *y* is $N(1.047,1)$" means something like *$N(1.047,1)$ is the most probable, credible or best supported hypothesis, e.g.* perhaps 1.047 is the value of $\mu$ with maximum likelihood $f(X/\mu)$. Faced with this interpretation, there was little chance that Neyman would succeed in describing Barnard's inference as a decision. But rather than admit inference, he played down any distinction between inference and decision, and invented (without definition) the vacuous catchall "concluding step" to cover both inference and decision alike. This merely confused (suppressed) the issue.

(vi) In his paper on "inductive behavior", Neyman [1957, p.16] wrote that the result in a hypothesis test is a decision, always a decision. However, he allowed that this result might be the decision "to assume a particular attitude towards the various sets of hypotheses mentioned". Thus, the result in a hypothesis test might be to "assume the attitude" that the hypothesis $h_o$ is true, or that the hypothesis $h_o$ is false. But what does it mean to "assume the attitude" that $h_o$ is true (or that $h_o$ is false)? Does it mean to believe that $h_o$ is true, or that $h_o$ is probably true? Typically, Neyman didn't say. But it would seem that to "assume the attitude" that the hypothesis $h_o$ is true is *not to* believe that $h_o$ is true, but merely to *deem* that $h_o$ is true for the purpose of decision; decision between courses of action, the best of which depends on whether $h_o$ is in fact true. For example, to "assume the attitude" that the Salk vaccine is effective against polio is merely to deem the vaccine effective, thereby enabling a decision to recommend the vaccine. This interpretation is tenable if there happens to be a decision *pending* between courses of action of which the best depends on whether $h_o$ is in fact true. But this is not always to be. Indeed, Neyman [1957, p.16] conceded that whilst research is often "..for purposes of some immediate practical action (*e.g.* should one use the Salk vaccine against polio?)", there is also research purely "..for the sake of scientific curiosity (*e.g.* does our Universe expand?)". However, if the hypothesis $h_o$ is tested purely out of curiosity, it seems pointless to assume any particular attitude toward $h_o$ other than an epistemic or cognitive attitude, *e.g.* the attitude that the hypothesis $h_o$ is true (probably) or false, which is of course a conclusion or inference.

(vii) In relation to a test comparing the effects of two types of insulin, Neyman [and Pearson (1936) p.204] maintained that the scientist must make the decision to report either that "I can detect no indication that the cheaper insulin is worse than the more expensive one..." or alternatively that "The cheaper insulin seems to be of inferior quality...". Extensibly, these are statements of *inference*. Hence, if the scientist follows Neyman's instructions, surely he must admit inference (either inference or sham). Mercifully, Erich Lehmann, a leading statistician in Neyman's school, put matters straight. He explained that the result in a hypothesis test is very often an *inference*, and thus not a decision in any ordinary literal sense:

> Frequently it is a question of providing a convenient summary of the data or indicating what information is available concerning the unknown parameter or distribution. This information will be used for guidance in various considerations but will not provide the sole basis for any specific decision. In such cases the emphasis is on the inference rather than on the decision aspect of the problem, although formally it can still be considered a decision problem if the inferential statement itself is interpreted as the decision to be taken. [1959, pp.4-5][5]

## Neyman's Logic for Inference

Notwithstanding his insistent deductivism, there are threads in Neyman's opus of a logic for inference. In one passage, Neyman condones inference explicitly, with something of a logic attached. Specifically, he suggests that if the *power* of the test concerned is uniformly high, say 0.95 or better, then it is reasonable intuitively (indirectly) to interpret the result "accept $h_o$" as logical *confirmation* (his word) of the hypothesis $h_o$, and as a basis for

[5]Birnbaum [1977, pp.25-6] explained that to interpret (dress) an *inference* as a decision, we need merely preface that inference with the words "decide that". For example, he suggested that we may "'decide that' a certain hypothesis is true or supported by strong evidence."

confidence in that hypothesis [1955, pp.40-1]. This is clearly an inferential or evidential interpretation, a slip of the pen perhaps amid Neyman's behavioralist strictures.

Further evidence of inference in Neyman's writing, although more subtle, concerns his interpretation of $\text{probabilities}_2$, specifically the error probabilities $(\alpha,\beta)$. If the probabilities $(\alpha,\beta)$ are interpreted as $\text{probabilities}_1$ in the single case, then the result in a hypothesis test constitutes an inference.[6] Of course, Neyman liked to say that $(\alpha,\beta)$ are not $\text{probabilities}_1$ but strictly $\text{probabilities}_2$, *e.g.* [1963, p.929-30]. However, despite his frequentist testimony, he tended (like most) to treat $\text{probabilities}_2$ as $\text{probabilities}_1$ in the single case, at least intuitively. For example:

(i) On appeal to Bernoulli's law, Neyman [1955, p.18] declared that it is almost certain, *i.e.* the $\text{probability}_1$ is near 1, that in the "long run" the error frequencies $(\alpha,\beta)$ will pertain, at least approximately. Yet Bernoulli's law entails only that the $\textit{probability}_2$ is near 1 that those frequencies will pertain. Thus, Neyman interpreted Bernoulli's $\text{probability}_2$ as a $\text{probability}_1$ in the single case, *i.e.* in the single "long run". This has been noted by Hacking [1965, p.105] and Seidenfeld [1979, pp.65-6].

(ii) Neyman liked to speak of the probabilities $(\alpha,\beta)$ in terms of "chance". Specifically, he often referred to the power of a test, *i.e.* $1-\beta(h)$, as the *chance* in that test of detecting the alternative hypothesis $h$, if in fact $h$ is true [*e.g.* (1935) p.227; (1977) p.107]. However, unlike the term probability, which Neyman defined very carefully, the word "chance" is not defined in any of Neyman's writing. Some philosophers and statisticians have defined "chance" strictly as relative frequency ($\text{probability}_2$). But more often, "chance" is interpreted in the sense of Popper's word "propensity", which entails both $\text{probability}_2$ and $\text{probability}_1$ in the single case. In *The Grammar of Science*, Karl Pearson [1892, pp.174-5] defined chance as both degree of belief or certainty ($\text{probability}_1$), which he called "subjective chance", and relative frequency ($\text{probability}_2$), which he called "objective chance". On this inter pretation, which is not uncommon, to say that the "chance" that a die will turn up an ace is 1/6 is to say both that the $\text{probability}_2$ of an ace is 1/6 and that the $\text{probability}_1$ of an ace in the single case is 1/6. That is, the word chance has definite $\textit{probability}_1$ connotations, at least in the context of mechanical "games of chance", where $\text{probabilities}_2$ are natural $\text{probabilities}_1$. Hence, it is interesting that Neyman likened hypothesis tests, and systems of confidence intervals, to gambling mechanisms. For example:

> ...the situation of a statistician who decided to use the 95 per cent confidence intervals is exactly the same as that of a gambler participating in a game with probability of winning equal to 0.95. [1963, p.930]

Suppose we interpret the chance $1-\beta(h)$ of detecting the alternative $h$ as both a relative frequency and a $\text{probability}_1$ in the single case.[7] After all, this is the way we interpret the chance that a die will turn up an ace, or the chance that a chocolate wheel will win us a prize. Moreover, if Neyman intended that "chance" be interpreted strictly and unequivocally as relative frequency ($\text{probability}_2$) he ought not to have spoken of the chance of detecting alternatives $h$ in any particular single case (test). Alternatively, he might have kept to the less equivocal term "frequency", or even "probability", which he defined strictly as $\text{probability}_2$.

---

[6]Specifically, if the probability $\alpha$ is interpreted as a $\text{probability}_1$, then the result $t(X)$ in a hypothesis test of size $\alpha$ entails a confidence interval $f(X)$ such that $prob_1(\theta\epsilon f(X))=1-\alpha$, where $\theta$ is the unknown paramater and $X$ is the sample observed.

[7]Note that Giere, who interprets the probabilities $(\alpha,\beta)$ in N-P theory as propensities in the single case, refers to $(\alpha,\beta)$ as measures of "chance"; *e.g.* [1976, p.84].

GIERE, R.N. [1976] 'Empirical Probability, Objective Statistical Methods and Scientific Inquiry' in Harper, W.L. and Hooker, C.A. (eds.) *Foundations of Probability Theory, Statistical Inference, and Statistical Theories of Science* Vol.2: 63-101 (Dordrecht; D.Reidel).

GOOD, I.J. [1976] 'The Bayesian Influence, Or How to Sweep Subjectivism Under the Carpet' in Harper, W.L. and Hooker, C.A. (eds.) *Foundations of Probability Theory, Statistical Inference, and Statistical Theories of Science* Vol.2: 125-74 (Dordrecht; D.Reidel).

HACKING, I. [1965] *Logic of Statistical Inference* (Cambridge University Press).

JOHNSTONE, D.J. [1987] 'Tests of Significance following R.A. Fisher' *The British Journal for the Philosophy of Science*, In press.

KEMPTHORNE, O. [1976] 'Of What Use are Tests of Significance and Tests of Statistical Hypothesis', *Communications in Statistics - Theory and Methods* A5 8: 763-77.

LEHMANN, E.L. [1959] *Testing Statistical Hypotheses* (New York; Wiley).

LINDLEY, D.V. [1977] 'The Distinction between Inference and Decision' *Synthese* 36: 51-8.

NEYMAN, J. [1935] 'Complex Experiments' *Journal of the Royal Statistical Society* (Supplement) 2: 235-42; reprinted in *A Selection of Early Statistical Papers of J. Neyman*: 225-32 (University of California Press, 1967).

NEYMAN, J. [1941] 'Fiducial Argument and the Theory of Confidence Intervals' *Biometrika* 32: 128-50; reprinted in *A Selection of Early Statistical Papers of J. Neyman*: 375-94 (University of California Press, 1967).

NEYMAN, J. [1942] 'Basic Ideas and Some Recent Results of the Theory of Testing Statistical Hypotheses' *Journal of the Royal Statistical Society* 105: 292-327.

NEYMAN, J. [1950] *First Course in Probability and Statistics* (New York; Henry Holt).

NEYMAN, J. [1955] 'The Problem of Inductive Inference' *Communications on Pure and Applied Mathematics* 8: 13-45.

NEYMAN, J. [1957] '"Inductive Behavior" as a Basic Concept of the Philosophy of Science' *Review of the International Statistical Institute* 25: 7-22.

NEYMAN, J. [1961] 'Silver Jubilee of My Dispute with Fisher' *Journal of the Operations Research Society of Japan* 3: 145-54.

NEYMAN, J. [1963] 'Fiducial Probability' *Bulletin of the International Statistical Association* 40: 919-39.

NEYMAN, J. [1969] 'Statistical Problems in Science. The Symmetric Test of a Composite Hypothesis' *Journal of the American Statistical Association* 64: 1154-71.

NEYMAN, J. [1971] 'Foundations of Behavioristic Statistics' in Godambe, V.P. and Sprott, D.A. (eds.) *Foundations of Statistical Inference*: 1-19 (Toronto; Holt, Rinehart and Winston).

NEYMAN, J. [1976] 'Tests of Statistical Hypotheses and Their Use on Studies of Natural Phenomena' *Communications on Statistics - Theory and Methods* A5 8: 737-51.

NEYMAN, J. [1977] 'Frequentist Probability and Frequentist Statistics' *Synthese* 36: 97-131.

NEYMAN, J. and TOKARSKA, B. [1936] 'Errors of the Second Kind in Testing 'Student's' Hypothesis' *Journal of the American Statistical Association* 31: 318-26; reprinted in *A Selection of Early Statistical Papers of J. Neyman*: 238-45 (University of California Press, 1967).

NEYMAN, J. and PEARSON, E.S. [1928] 'On the Use and Interpretation of Certain Test Criteria for Purposes of Statistical Inference' Part I *Biometrika* A 20: 175-240; reprinted in *J. Neyman and E.S. Pearson Joint Statistical Papers*: 1-67 (Cambridge University Press, 1967).

NEYMAN, J. and PEARSON, E.S. [1933] 'On the Problem of the Most Efficient Tests of Statistical Hypotheses' *Philosophical Transactions Royal Society* A 231: 289-337; reprinted in *J. Neyman and E.S. Pearson Joint Statistical Papers*; 140-85 (Cambridge University Press, 1967).

NEYMAN, J. and PEARSON, E.S. [1936] 'Contributions to the Theory of Testing Statistical Hypotheses' *Statistical Research Memoirs* 1: 1-37; reprinted in *J. Neyman and E.S. Pearson Joint Statistical Papers*: 203-39 (Cambridge University Press, 1967).

PEARSON, E.S. [1955] 'Statistical Concepts in their Relation to Reality' *Journal of the Royal Statistical Society* B 17: 204-7.

PEARSON, K. [1892] *The Grammar of Science* (London; Scott).

SEIDENFELD, T. [1979] *Philosophical Problems of Statistical Inference: Learning from R.A. Fisher* (Dordrecht; D. Reidel).

TUKEY, J.W. [1960] 'Conclusions vs Decisions' *Technometrics* 2: 423-33.

# DE FINETTI'S METHODS OF ELICITATION

Joseph B. Kadane
Department of Statistics
Carnegie-Mellon University
Pittsburgh, PA 15213
U.S.A.

Robert L. Winkler
Fuqua School of Business
Duke University
Durham, NC 27706
U.S.A.

## INTRODUCTION

De Finetti (1974) uses payoffs through promissory notes, bets, or scoring rules in the elicitation of an expert's probabilities and introduces his "hypothesis of rigidity" to argue that as long as the payoffs are small, nonlinearities in the expert's utility function can be ignored for practical purposes. In an analysis considering not just the elicitation-related payoffs, but all uncertainties related to the expert's fortune, we find that the hypothesis of rigidity is not sufficient to eliminate the impact of the utility function in probability elicitation. We propose an "extended hypothesis of rigidity" that adds an extra condition to de Finetti's hypothesis. The extra assumption is that, ignoring elicitation-related payoffs, the fortune of the expert is independent of the events for which probabilities are being elicited.

The purpose of the paper, then, is to investigate the implications of de Finetti's hypothesis of rigidity and our extended hypothesis of rigidity. We focus specifically on de Finetti's method of eliciting probabilities in terms of price ratios but note that similar results can be derived for a variety of elicitation methods used by de Finetti and others (Kadane and Winkler, 1986). First, we show that the original hypothesis of rigidity is not sufficient to provide price ratios equal to the expert's odds ratios. The extended hypothesis of rigidity is presented and shown to be sufficient in this sense. Next, we take a slightly deeper look at the relationship between price ratios and odds ratios by considering a second-order analysis that reveals some systematic shifts in elicited probabilities. We then ask when the extended hypothesis of rigidity might be justified and find that it seems quite fragile, and we close with a brief summary of our conclusions.

## ELICITATION AND THE HYPOTHESIS OF RIGIDITY

To make everything as simple as possible, we consider the elicitation of a probability by an expert for a single event A. Let N be a promissory note that pays r if A occurs and nothing otherwise, where $r > 0$. De Finetti's approach (de Finetti, 1974) implies that if p is the largest price the expert will pay for N (i.e., the price that makes the expert indifferent between buying N and not buying it, then the ratio $p/(r-p)$

equals the expert's odds in favor of A. The expert's probability for A is therefore p/r.

Suppose that the densities $g(f|A)$ and $g(f|\bar{A})$ represent the probability distributions of the expert's fortune f given A and its complement $\bar{A}$, respectively, without the promissory note N. Then if U denotes the expert's utility function for f, the expert's expected utility without N is

$$E[U(f)] = \pi \int U(f)g(f|A)df + (1-\pi) \int U(f)g(f|\bar{A})df, \tag{1}$$

where $\pi$ represents the expert's probability that A will occur. If the expert buys N at price p, the expected utility becomes

$$E[U(f)|N] = \pi \int U(f-p+r)g(f|A)df + (1-\pi) \int U(f-p)g(f|\bar{A})df. \tag{2}$$

The expert's indifference price for N is the value of p for which

$$E[U(f)] = E[U(f)|N]. \tag{3}$$

Equating (1) with (2) as required by (3) yields

$$\pi \int [U(f-p+r)-U(f)]g(f|A)df = (1-\pi) \int [U(f)-U(f-p)]g(f|\bar{A})df, \tag{4}$$

which simplifies to

$$\pi/(1-\pi) = [p/(r-p)]c, \tag{5}$$

where

$$c = \frac{\int \{[U(f)-U(f-p)]/p\}g(f|\bar{A})df}{\int \{[U(f-p+r)-U(f)]/(r-p)\}g(f|A)df} . \tag{6}$$

From (5), we see that the price ratio $p/(r-p)$ equals the expert's odds ratio $\pi/(1-\pi)$ iff

$$c = 1. \tag{7}$$

As expected, (7) is satisfied when U is linear. The case of primary interest here, however, is that of nonlinear utility, since the main purpose of the hypothesis of rigidity is apparently to enable us to ignore utility considerations in the elicitation of probabilities. De Finetti states that his hypothesis of rigidity

> ... is acceptable ... provided the amounts in question are "not too large." Of course, the proviso has a relative and approximate meaning relative to you, to your fortune and temperament (in precise terms, to the degree of convexity of your utility function U); approximate, because, in effect, we are substituting in place of the segment of the curve U which is of interest, the tangent at the starting point. (de Finetti, 1974, p. 80)

To see what happens when the amounts are not large, we consider the limit of (6) as r approaches zero (implying that p and r-p also approach zero):

$$\lim_{r\to 0} c = c_0 = \frac{\int U'(f)g(f|\bar{A})df}{\int U'(f)g(f|A)df} , \tag{8}$$

where the prime is used to denote differentiation. For small amounts, then, c is approximately equal to the right-hand-side of (8).

If U is not linear, then $c_0$ clearly depends on $g(f|A)$ and $g(f|\bar{A})$. For a simple example, suppose that $g(f|A)$ and $g(f|\bar{A})$ are degenerate, placing probability one at $f_1$ and $f_2$, respectively, and

$$U(f) = -e^{-df}$$

with $d > 0$ (risk-averse exponential utility). In this situation we have

$$c_0 = \frac{U'(f_2)}{U'(f_1)} = e^{d(f_1-f_2)}.$$

Thus, $c_0$ is a strictly increasing function of $f_1-f_2$. For any fixed $d > 0$, however small (i.e., however weak the risk aversion), $c_0$ can differ considerably from one if $|f_1-f_2|$ is large.

A sufficient condition for $c_0=1$ is

$$g(f|A) = g(f|\bar{A}) \quad \text{for all } f. \tag{9}$$

Since de Finetti apparently does not rely on linear utility to get $c_0=1$, he must be assuming that (9) holds, although we have found no explicit discussion of this assumption in his writing. Of course, to assume (9) is to assume that, apart from the possible purchase of the promissory note N, no part of the expert's fortune is contingent on whether or not A occurs. This assumption is close to Ramsey's (1931) assumption of ethical neutrality. We call the combination of both assumptions, de Finetti's hypothesis of rigidity <u>and</u> $g(f|A) = g(f|\bar{A})$ for all f, the extended hypothesis of rigidity.

As illustrated by the above example, violations of this extended hypothesis of rigidity associated with the dependence of an expert's fortune on the events of interest [violations of (9)] can cause substantial differences between the expert's odds ratio and price ratio. Clearly de Finetti's claim that "the hypothesis of rigidity ... is acceptable in practice ... provided the amounts in question are not 'too large'" (de Finetti, 1974, p. 80) should be clarified in view of (6) and (8). If U is not linear, it may not be sufficient for r (and hence p) to be small, because the entire range of values of f implied by $g(f|A)$ and $g(f|\bar{A})$ is relevant. This range reflects all of the many uncertainties related to the expert's fortune.

## SECOND-ORDER ANALYSIS

The limiting analysis in the previous section leading to (9) as a condition of interest shows what happens as the stakes approach zero. For finite stakes, approximating $U(f-p+r)$ and $U(f-p)$ in (2) by a first-order expansion gives equivalent results: c is approximately equal to $c_0$. In this section we take a deeper look at the relationship between odds ratios and price ratios by considering the impact of including second-order terms in the expansion. This yields

$$U(f-p+r) = U(f) + (r-p)U'(f) + [(r-p)^2/2]U''(f) + O(r^3) \tag{10}$$

and

$$U(f-p) = U(f) - pU'(f) + (p^2/2)U''(f) + O(r^3). \tag{11}$$

Substituting (10) and (11) in (2) and equating (2) with (1) gives

$$\pi(r-p) \int U'(f)g(f|A)df + [\pi(r-p)^2/2] \int U''(f)g(f|A)df + O(r^3)$$
$$= (1-\pi)p \int U'(f)g(f|\bar{A})df - [(1-\pi)p^2/2] \int U''(f)g(f|\bar{A})df + O(r^3),$$

which reduces to

$$\pi/(1-\pi) = [p/(r-p)]c, \tag{12}$$

where

$$c = \frac{\int U'(f)g(f|\bar{A})df - (p/2) \int U''(f)g(f|\bar{A})df + O(r^2)}{\int U'(f)g(f|A)df + [(r-p)/2] \int U''(f)g(f|A)df + O(r^2)} . \tag{13}$$

Now the ratio $p/(r-p)$ equals the expert's odds if $c=1$. Let $r \to 0$ (so a fortiori $p \to 0$ and $r-p \to 0$), so that $p/r$ has a limit $s$, with $0 \le s \le 1$. Expanding $c$ in a Taylor series in $r$ around zero, we find it has the form

$$c = c_0 + rc_1 + O(r^2), \tag{14}$$

with

$$c_0 = \frac{\int U'(f)g(f|\bar{A})df}{\int U'(f)g(f|A)df} \tag{15}$$

and

$$c_1 = \frac{[\int U'(f)g(f|A)df][(-s/2) \int U''(f)g(f|\bar{A})df] - [\int U'(f)g(f|\bar{A})df][(1-s)/2][\int U''(f)g(f|A)df]}{[\int U'(f)g(f|A)df]^2} . \tag{16}$$

Note that $c_0$, the zeroth-order term of the expansion of $c$, is the limiting value of $c$ as given by (8).

The first-order term $c_1$ can be rewritten as follows:

$$c_1 = -\frac{c_0}{2}\left[ s \frac{\int U''(f)g(f|\bar{A})df}{\int U'(f)g(f|\bar{A})df} + (1-s) \frac{\int U''(f)g(f|A)df}{\int U'(f)g(f|A)df} \right] . \tag{17}$$

Let

$$w(f) = -U''(f)/U'(f) \tag{18}$$

be the Pratt-Arrow risk-aversion function (Pratt, 1964). Also, define

$$g_A(f) = U'(f)g(f|A)/ \int U'(f)g(f|A)df, \tag{19}$$

$$g_{\bar{A}}(f) = U'(f)g(f|\bar{A})/ \int U'(f)g(f|\bar{A})df, \tag{20}$$

and

$$h(f) = sg_{\bar{A}}(f) + (1-s)g_A(f). \tag{21}$$

Since we assume that $U'(f) > 0$ (the expert prefers more to less), $g_A(f)$ and $g_{\bar{A}}(f)$ are probability densities, as is the convex combination $h(f)$.

Using (18)-(21), we can rewrite (16) as

$$c_1 = (c_0/2)E_h(w) \tag{22}$$

and (14) as

$$c = c_0 + r(c_0/2)E_h(w) + O(r^2). \tag{23}$$

From (23), we see that (9) may not be sufficient for c=1. In fact, if U' > 0 (more is preferred to less) and U" < 0 (the expert is strictly risk averse), then independence of N and the rest of the expert's prospects, as given by (9), implies that $c_1 > 0$. Then there is a region of values of r, close to zero, for which c > 1. In this region, a risk-averse expert satisfying (9) will understate the odds in favor of A. That is, the expert's odds in favor of A are greater than the odds implied if the ratio p/(r-p) is taken at face value and the impact of nonlinear utility is ignored. On the other hand, a strictly risk-taking expert will have U" > 0, implying that $c_1 < 0$, so that in a region of values of r near zero, p/(r-p) overstates the expert's odds in favor of A.

Thus, working with finite stakes can lead to systematic shifts in the elicited probabilities. The extent of the shifts depends on how much the expert's utility function deviates from linearity, on the distributions $g(f|A)$ and $g(f|\bar{A})$, and on the magnitude of the stakes. Even if (9), the extra assumption in the extended hypothesis of rigidity, is satisfied, c can differ from one, although the discrepancy between c and one is reduced as the stakes become smaller. If (9) is <u>not</u> satisfied, then all bets are off regarding how close c is to one.

## WHEN IS THE EXTENDED HYPOTHESIS OF RIGIDITY JUSTIFIED?

Essentially, the extended hypothesis of rigidity requires two conditions. First, as in de Finetti's original hypothesis of rigidity, the stakes in the elicitation-related bets, scores, or other payoffs must be small. Second, the extension of the hypothesis of rigidity requires that aside from elicitation-related payoffs, the expert's fortune should be independent of the events for which probabilities are being elicited. The first condition can be controlled somewhat in the design of the elicitation procedure, keeping in mind that an acceptable size for the stakes depends on the perceived degree of nonlinearity of the expert's utility function and on the desired degree of accuracy in terms of potential deviations of the elicited probability from the expert's judgments about the events. We will set aside more detailed questions about "how small is small" and concentrate here on the second condition, which is the primary focus of this paper.

One way of viewing the problem posed here is that de Finetti does not ask whether, or to what extent, the expert is already making bets on the very stochastic events for which the expert's probabilities are to be elicited. In many instances, the experts concerning certain events are likely to already have significant stakes relating to these events. These stakes may be difficult to untangle, but nonetheless bear an important weight in the further bets the expert might make. Without studying these, with no "conflict of interest" statement, de Finetti-style elicitations could make serious errors. This is admittedly speculation, but in many cases the impact of violations of (9) seems likely to be much greater than the impact of violations of de Finetti's original hypothesis of rigidity. In this sense, the extended hypothesis of rigidity is more fragile than the original hypothesis, and elicitation procedures should be reexamined carefully with this extended hypothesis in mind.

## CONCLUSION

De Finetti (1974, p. 79) sees himself as being in the Ramsey-Savage tradition of expected utility maximization, which is surely correct. However, his proposed simplification via the hypothesis of rigidity is not innocuous and does not necessarily allow the utility aspects of choice to be suppressed. De Finetti may well have been aware of this problem. Although he seems not to have addressed the issue directly, a broad interpretation of "everyday affairs" in the following quote to include not just elicitation-related payoffs but also other stakes would make it relevant to the concerns discussed in this paper.

> Essentially, our assumption amounts to accepting as practically valid the hypothesis of rigidity with respect to risk: in other words, the identity of monetary value and utility within the limits of "everyday affairs" ... actually, it seems safe to say that under the heading of "everyday affairs" one can consider all those transactions whose outcome has no relevant effect on the fortune of an individual (or firm, etc.), in the sense that it does not give rise to substantial improvements in the situation, nor to losses of a serious nature. (de Finetti, 1974, p. 82)

In any event, it would be sounder to maintain the full decision-theoretic structure in the analysis from the beginning. The problem impacts not just de Finetti's method of eliciting probabilities in terms of price ratios, but other elicitation methods as well. In Kadane and Winkler (1986), we explore further the separation of probability elicitation from utilities and indicate that probability elicitation procedures need to be reassessed in view of possible utility-related complications.

## ACKNOWLEDGMENTS

This research was sponsored by the Office of Naval Research (Kadane) under Contract N00014-82-0622 and by the National Science Foundation (Winkler) under Grants PRA8413106, ATM8507495, and IST8600788. Conversations with Gary Chamberlain and Teddy Seidenfeld and comments (particularly those of Morris DeGroot) at a seminar given by Kadane at the Statistics Department of Carnegie-Mellon University have been very helpful.

## REFERENCES

de Finetti, B., 1974, "Theory of Probability," Vol. 1, Wiley, New York.

Kadane, J.B., and Winkler, R.L., 1986, Separating probability elicitation from utilities, unpublished manuscript.

Pratt, J.W., 1964, Risk aversion in the small and in the large, Econometrica, 32: 122-136.

Ramsey, F.P., 1931, "The Foundations of Mathematics and Other Logical Essays," Kegan Paul, London.

# BAYESIAN ESTIMATION OF DESIGN FLOODS UNDER REGIONAL AND SUBJECTIVE PRIOR INFORMATION

Robert Kirnbauer, Sylvia Schnatter and Dieter Gutknecht

Institut fuer Hydraulik, Gewaesserkunde und Wasserwirtschaft
Technische Universitaet Wien

## INTRODUCTION

Usually design floods are estimated as certain quantiles of a cumulative distribution function (CDF) fitted to a sample of yearly maxima of floods observed at a gauging station. If such observations do not exist at a site where a hydraulic structure is to the built, the hydrologist can collect flood data for several years during the planning phase. This small new sample, however, will not be sufficient for estimating the design flood by means of common flood statistics but it fits as one source of information to be combined with some prior information within the BAYESIAN estimation procedure. Prior information can be taken from long time flood records observed at stations of the same region to be incorporated in a data-based a-priori density function of the parameters.

In other cases a non-data-based a-priori density function can be derived from interviews with experts in hydrology concerning their estimation of the statistical properties of floods at the project site.

If the short flood record accidentally is taken from a period of extremely wet or dry years, sample information can be "put in the right place". This can be done by examining the longer records and modifying the likelihood function following a method proposed by Bardossy (1982).

In the sequel for each of the three above mentioned estimation procedures an example is given using data from Upper Austrian streams and taking advantage of a method proposed by Cunnane and Nash (1971). This method yields not only a point estimate of the design flood but its whole CDF. Under the hydrological point of view this procedure is better than calculating a point estimate because the final security of a hydraulic structure not only depends on the probability of non exceedance of the design flood but also on the accuracy of its estimation.

The integration procedures had to be performed numerically because no natural conjugate a-priori probability density function (PDF) of the likelihood function was used. For this purpose a computer algorithm provided by Schnatter (1982) could be applied.

## FLOOD STATISTICS UNDER THE BAYESIAN POINT OF VIEW

### Common Flood Statistics

In common flood statistics a PDF $f(x/\underline{p})$ is fitted to a sample of yearly maxima by calculating a point estimate $\underline{p}^*$ of the parameter vector $\underline{p}$. The design value then is found by choosing a particular recurrence interval T and solving equation (1) with respect to $x_T$

$$1-\frac{1}{T}=\int_{-\infty}^{x_T} f(x|\underline{p}^*)\,dx \tag{1}$$

Using a two parameter PDF this solution can be noted as

$$x_T=\mu+k_T.\sigma=\mu+k_T.\mu.c_v \tag{2}$$

with $k_T$ being the frequency factor, dependent only on the recurrence interval T and the type of the distribution, and $\mu$ and $c_v$ being estimates of the mean and coefficient of variation derived from the sample.

### Incorporation of Parameter Uncertainty

Within the Bayesian framework the parameters are assumed to be random variables, the distribution of which can be derived from two sources of information: The sample information is contained in the sample vector $\underline{X}$ = $(x_1,x_2,\ldots x_i\ldots x_N)^T$ and quantified by the likelihood function:

$$L(\mu,c_v|\underline{X})=\prod_{i=1}^{N} f(x_i|\mu,c_v)\ . \tag{3}$$

The second source of information is the a-priori PDF $f'(\mu,c_v|I_P)$ of the parameters $\mu$ and $c_v$, given some prior information $I_p$. The combination of those two kinds of information yields the a-posteriori PDF $f''(\mu,c_v|\underline{X},I_P)$ of the parameters $\mu$ and $c_v$.

$$f''(\mu,c_v|\underline{X},I_p)=\frac{L(\mu,c_v|\underline{X})\cdot f'(\mu,c_v|I_p)}{\int_\mu\int_{c_v} L(\mu,c_v|\underline{X})\cdot f'(\mu,c_v|I_p)\,dc_v\cdot d\mu}\ . \tag{4}$$

It contains both the sample information and the prior information. If the a-priori PDF is a natural conjugate of the likelihood function (see e.g. Zellner, 1971) these calculations can be done in closed form, and the a-priori PDF can be called a "convenient prior" (Vicéns et al., 1974).

### Evaluation of the Cumulative Distribution Function of $x_T$

It would to some exent be contradictory to the Bayesian way of thinking to calculate a point estimate of the quantile $x_T$ by means of the a-posteriori PDF and thus ignoring the randomness of the parameters. Therefore a method proposed by Cunnane and Nash (1971) was used to derive the cumulative distribution function of the quantile $x_T$.

If we assume any value $x_j$ as a fixed flood discharge, the probability of the quantile $x_T$ being less than or equal to the value $x_j$ can be found by the following considerations: $k_T$ be the frequency factor of the chosen distribution of the floods of the recurrence interval T, then in the $(\mu,c_v)$-plane

$$c_v(j)=\frac{x_j-\mu}{k_T\cdot\mu} \tag{5}$$

$c_v(j)$ forms a hyperbola. Every $(\mu, c_v)$-combination between the origin and the hyperbola represents a flood less than $x_j$. Hence the probability of $x_T$ being less than or equal to $x_j$ can be found by integrating the a-posteriori PDF over the region between the origin and the hyperbola.

$$P(x_T \leq x_j) = \int_{-\infty}^{\infty} [\int_0^{c_v(j)} f''(\mu, c_v | \underline{X}, I_p)\, dc_v]\, d\mu . \tag{6}$$

Variation of $x_j$ yields the CDF of $x_T$.

## REAL WORLD APPLICATION OF THE ESTIMATION PROCEDURE

### The Sample

Five yearly maxima of floods at the Upper Austrian river Krumme Steyrling at the gauging station Molln, catchment area $A_E = 129,4\ km^2$, were taken as the sample. As a longer record of 57 years existed at the gauging station Molln it was possible to test the performance of the Bayesian estimation procedure.

In order to eliminate the influence of the catchment size, especially with regard to the planned combination of sample and regional information, the flood discharge values $HQ_1(m^3/s)$ of 1973 to 1977 were divided by the catchment area $A_E$ so that the sample vector resulted in $\underline{X} = (0.723, 0.420, 0.702, 0.231, 0.835)^T$.

### Probability Density Function and Likelihood Function.

The double exponential or Gumbel-I distribution was chosen as the model distribution. Its PDF is given by

$$f(x|\mu, c_v) = \alpha . exp((-\alpha(x-u) - exp(-\alpha .(x-u))) \tag{7a}$$

$$\alpha = \frac{\pi}{\mu . c_v \sqrt{6}} \quad ; \quad u = \mu - \frac{\gamma_E}{\alpha} . \tag{7b}$$

From the PDF the likelihood function can be derived

$$l(\mu, c_v | \underline{X}) = exp(N . ln(\frac{\pi}{\mu . c_v \sqrt{6}}) - \sum_{i=1}^{N} y_i - \sum_{i=1}^{N} exp(-y_i)) \tag{8a}$$

$$y_i = \frac{\pi}{\mu . c_v \sqrt{6}} \cdot (x_i - \mu + \frac{\gamma_E . \mu . c_v . \sqrt{6}}{\pi}) \tag{8b}$$

$\gamma_E$ is Euler's constant ($\gamma_E = 0,5772157$).
No natural conjugate ("convenient prior") of this likelihood function was found, so that there were no mathematical restrictions influencing the choice of the a-priori PDF, and only hydrological aspects had to be considered.

The integrations in equations (4) and (6), however, had to be performed numerically, taking advantage of an integration procedure provided by Schnatter (1982). It is based on a two dimensional Gaussian integration and automatically restricts the integration area from an infinite to a finite region considering requirements of accuracy.

### Data-Based Regional Prior Information

Investigations reported in Kirnbauer (1981) showed that regional prior information yields suitable results only if the data are taken from a region which is similar to the catchment where the sample is taken from with respect to its hydrological conditions. Thus the data were taken from longer flood records at twelve gauges in surrounding catchments in Upper Austria. In spite of the homogeneity of the region considerable variability appeard especially with respect to the mean flood $\mu$ (not so with respect to the coefficient of variation $c_v$). Therefore investigations had to be performed to find a morphological parameter in order to reduce the variability. Many attempts failed: Yearly mean precipitation, sums of precipitation causing the floods, land use, fall of slopes and rivers did not differ enough to explain different mean values of the flood records. The density of streams (GD) remained as the parameter with the strongest influence on the mean flood. This influence was quantified in the form of a regression model (see fig. 1)

$$\ln \hat{\mu} = a_2 \cdot GD^2 + a_1 \cdot GD + a_0 \tag{9a}$$

$$\ln \mu = 0{,}113 . GD^2 + 0{,}476 . GD - 1{,}434 + \varepsilon_1 \tag{9b}$$

which allowed to estimate the expected mean flood at the gauge "Molln" with the stream density in the catchment upstream of the gauge (GD=1,13 km/km$^2$): $\ln \hat{\mu}^* = -0{,}752$, $\hat{\mu}^* = 0{,}471 m^3/(s.km^2)$. The standard deviation was estimated from the (M=12) residuals of the regression model:

$$\varepsilon_{1k} = \ln \hat{\mu}_k - \ln \mu_k \tag{10a}$$

$$\sigma_{\varepsilon 1} = \left( \frac{1}{M-2} \sum_{k=1}^{M} \varepsilon_{1k}^2 \right)^{1/2} . \tag{10b}$$

Due to the logarithmic form of the regression model the mean flood at Molln is distributed log-normal with mean $\hat{\mu}^*$ and standard deviation $\sigma_{\varepsilon 1}$.

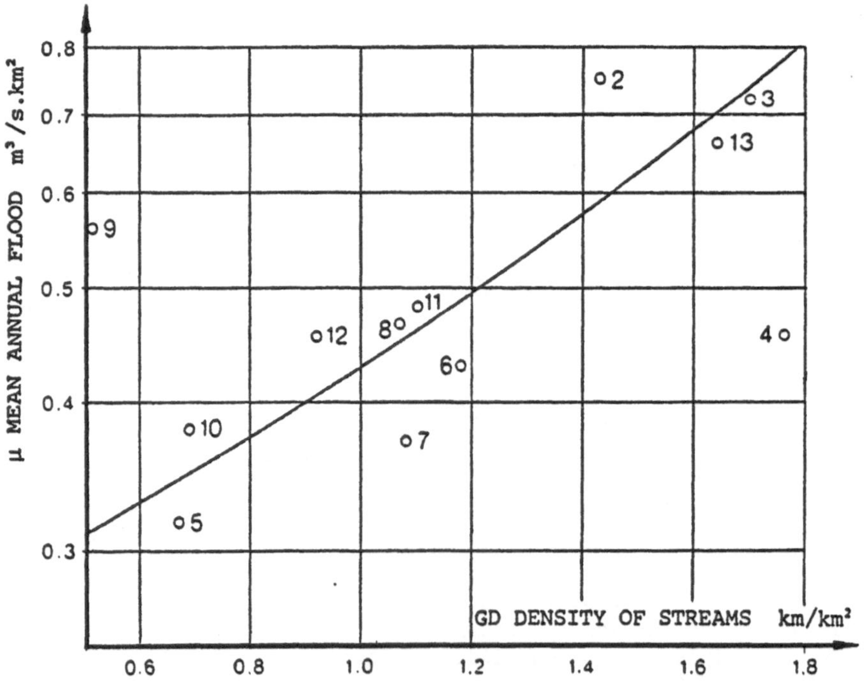


Fig. 1. Data-based regional prior information: mean value plotted versus morphological parameter.

This distribution is a marginal distribution of the a-priori PDF

$$\varphi_1(\mu|I_p) = LN(\hat{\mu}^*; \sigma_{\varepsilon 1}) = LN(0{,}471; 0{,}129) \tag{11a}$$

$$\varphi_1(\mu|I_p) = \frac{1}{\mu} \cdot \frac{1}{\sigma_{\varepsilon 1} \cdot \sqrt{2\pi}} \exp\left(-\frac{1}{2\sigma_{\varepsilon 1}^2}(\ln\mu - \ln\hat{\mu}^*)^2\right). \tag{11b}$$

For the coefficient of variation $c_v$ it was impossible to find a regionalisation parameter, and thus the marginal distribution of $c_v$ was directly estimated from the M $c_v$-values of the regional flood records. The distribution was assumed to be log-normal with mean and standard deviation estimated from the regional sample:

$$\hat{c}_v^* = \frac{1}{M}\sum_{k=1}^{M} \ln c_{vk} \tag{12a}$$

$$\sigma_{\varepsilon 2} = \left(\frac{1}{M-2}\sum(\ln c_{vk} - \hat{c}_v^*)^2\right)^{1/2}. \tag{12b}$$

The marginal ditribution of $c_v$, therefore, has density

$$\varphi_2(c_v|I_p) = LN(\hat{c}_v^*; \sigma_{\varepsilon 2}) = LN(0{,}468; 0{,}164) \tag{13a}$$

$$\varphi_2(c_v|I_p) = \frac{1}{c_v} \cdot \frac{1}{\sigma_{\varepsilon 2} \cdot \sqrt{2\pi}} \cdot \exp\left(-\frac{1}{2\sigma_{\varepsilon 2}^2} \cdot (\ln c_v - \ln\hat{c}_v^*)^2\right). \tag{13b}$$

As the $c_v$-values appeared to be independent from the residuals $\varepsilon_1$ of the regression model (9b) the a-priori PDF of $\mu$ and $c_v$ was found by multiplying the marginal distributions (see next page).

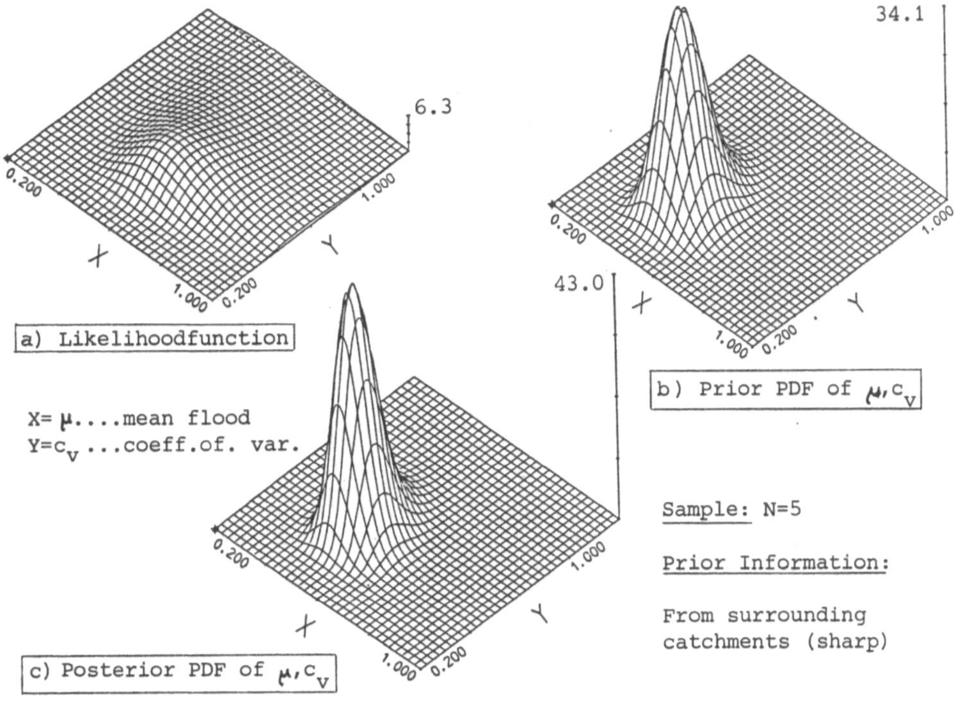


Fig. 2. Bayes' estimation. Distributions of parameters.

$$f'(\mu, c_v | I_p) = \varphi_1(\mu | I_p) \cdot \varphi_2(c_v | I_p). \qquad (14)$$

Likelihood function, a-priori PDF and the (not yet normalized) a-posteriori PDF are shown in fig. 2. The small information content of the small sample and the relatively sharp regional information can be observed in this figure. Likelihood function and a-priori PDF were combined to achieve the a-posteriori PDF following equation (4) and the cumulative distribution function of the design flood $x_T$=$HQ_{100}$ with recurrence interval of T=100 years was calculated due to equation (6).

Non-Data-Based Subjective Prior Information

In order to test the performance of subjective prior information five experts in hydrology were interviewed about their opinion concerning floods at the "project site" Molln. Preliminary interviews showed that the hydrologists were not accustomed to the Bayesian way of thinking so that they could not give a numerical assessment about the distribution parameters mean and coefficient of variation and their PDF. Therefore, the interviews had to concentrate on values the experts were familiar with. Those values were quantiles of the flood distribution of 10 or 1 percent probability of exceedance respectively (recurrence intervals $T_1$ = 10 or $T_2$= 100). They were asked for the most likely, the highest probable and the lowest probable value of the above mentioned quantiles.

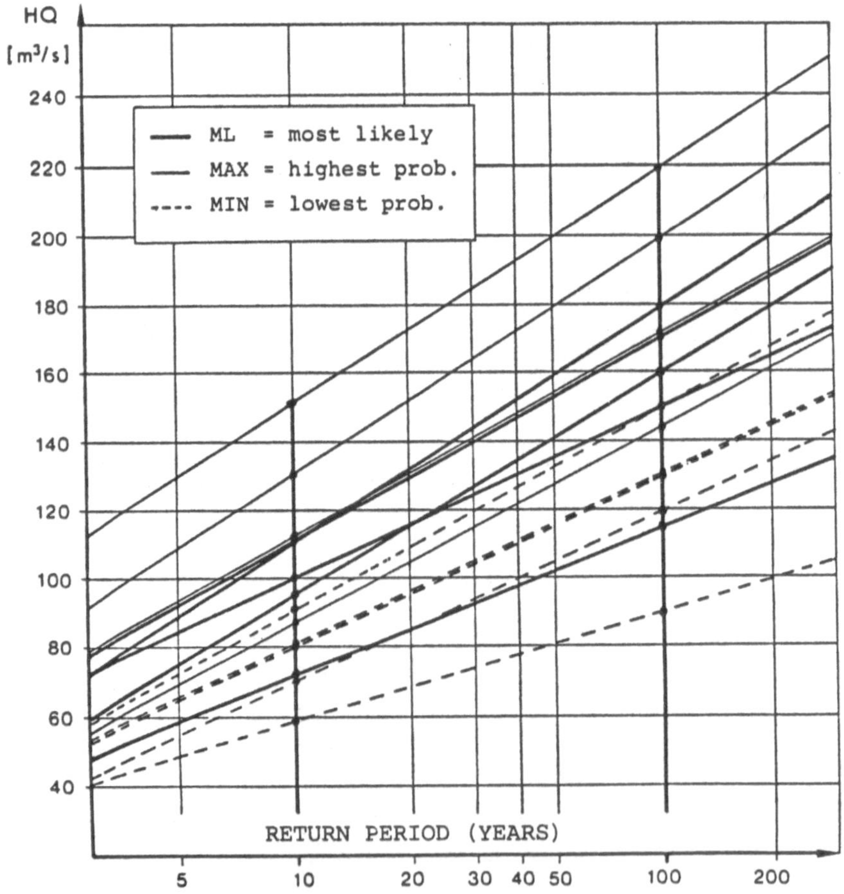


Fig. 3. The experts' assessments about design floods $HQ_{10}$ and $HQ_{100}$

The result of those interviews can be characterized by the latin proverb "QUOT CAPITA TOT SENTENTIAE". As shown in fig. 3 flood values, judged by one expert to be the "highest probable", were lower than those estimated as "most likely" by another. Therefore the discrimination into the three categories was abandoned, and each value was used equally, as if 15 experts would have been interviewed about $HQ_{10}$ and $HQ_{100}$.

Thus two distributions resulted from the interviews: One for $HQ_{10}$ and one for $HQ_{100}$. With the index x standing for the expert's statements about $HQ_{10}$ and y for $HQ_{100}$ a joint normal distribution in x and y was fitted (with the parameters $\mu_x, \sigma_x, \mu_y, \sigma_y, \varrho_{x,y}$ estimated by the method of moments from the two times 15 statements).

$$g(x,y)=\frac{1}{\sigma_x\cdot\sigma_y\cdot 2\pi\sqrt{1-\varrho_{xy}^2}}\cdot exp\left(-\frac{1}{2(1-\varrho_{xy}^2)}\left(\left(\frac{x-\mu_x}{\sigma_x}\right)^2-2\varrho_{xy}\left(\frac{x-\mu_x}{\sigma_x}\right)\cdot\left(\frac{y-\mu_y}{\sigma_y}\right)+\right.\right.$$
$$\left.\left.+\left(\frac{y-\mu_y}{\sigma_y}\right)^2\right)\right) \qquad (15a)$$

$$\mu_x = 0,7575\,;\ \sigma_x = 0,1977\,;\ \mu_y = 1,1999\,;\ \sigma_y = 0,2759\,;\ \varrho_{xy} = 0,9874. \qquad (15b)$$

This distribution implicitly contains the subjective a-priori PDF in and $c_v$ if we consider that

$$x=\mu+k_{10}\cdot\mu\cdot c_v \qquad (16a)$$

$$y=\mu+k_{100}\cdot\mu\cdot c_v \qquad (16b)$$

(with $k_{10}$ = 1.30455 and $k_{100}$ = 3.13667 being the frequency factors of the standardized Gumbel-I distribution). The equations (16) are the transformation equations to transform the joint density of x and y to the a-priori PDF, which results in

$$f'(\mu,c_v|I_p)=\mu\cdot(k_{100}-k_{10})\cdot g\left(\mu+k_{10}\cdot\mu\cdot c_v,\ \mu+k_{100}\cdot\mu\cdot c_v\right) \qquad (17)$$

with g(...) given in equations (15).

The calculation of the cumulative distribution function of the design flood followed equations (4) and (6).

## Modified Likelihood Function

It is a well known phenomenon, that big floods can occur in a series of consecutive years, or that during several years no remarkable flood can be observed. If the sample at the project site was taken from such a period, the design flood would be over- or underestimated respectively.

Past experience has shown that the floods in a whole region are rather homogeneous, either big or small. Therefore the relative magnitude of the sample floods can be estimated from the corresponding floods in surrounding catchments. A distribution of regional frequency factors can be derived from the regional information gauges (their number be M, their index j) for each of the N years the sample values are taken from, utilizing mean and standard deviation of the longer records at the regional information gauges.

For the year i of the sample a set of M frequency factors can be calculated as follows

$$k_{yji}=\frac{y_{ji}-\mu_{yj}}{\sigma_{yj}}\,,\qquad j=1\,(1)\,M \qquad (18)$$

with $y_{ji}$ being the flood at gauge j observed in the year i, and $\mu_{yj}$ and $\sigma_{yj}$ being mean and standard deviation of the (long) flood record at gauge j. Then a PDF can be fitted to that set of frequency factors, with mean and standard deviation according to equation (19)

$$\bar{k}_{yi} = \frac{1}{M} \sum_{j=1}^{M} k_{yji} \tag{19a}$$

$$s_{yi} = \left( \frac{\sum_{j=1}^{M} (k_{yji} - \bar{k}_{yi})^2}{M-1} \right)^{1/2} . \tag{19b}$$

This PDF can be assumed to be a measure of probability of the unknown real magnitude of the frequency factor $k_{xi}$ at the project site. From this consideration a modified likelihood function can be derived. If we assume $k_{xi}$ being normally distributed with mean $\hat{k}_{xi} = \bar{k}_{yi}$ and standard deviation $\hat{s}_{xi} = s_{yi}$ and with sample element $x_i$, then the modified likelihood function appears to be

$$L(\mu, c_v | \underline{X}) = \prod_{i=1}^{N} \frac{1}{\hat{s}_{xi}\sqrt{2\pi}} \exp\left(- \frac{(k_{xi} - \hat{k}_{xi})^2}{2\, \hat{s}_{xi}^2}\right) \tag{20a}$$

$$k_{xi} = \frac{x_i - \mu}{\mu \cdot c_v} . \tag{20b}$$

This likelihood function has to be combined with the data-based regional a-priori PDF (equation (14)), and the distribution of the design flood can be calculated following equations (4) and (6).

Cumulative Distribution Function of the Design Flood

The cumulative distribution functions of the design flood $x_T$ (with T=100 years) are plotted in fig. 4 in normal probability paper. The following numbers in the circles correspond to the line-numbers in fig. 4. The different sources of information leading to the respective line are

① Sample information alone, estimation procedure due to common flood statistics: point estimate of parameters by method of moments, calculation of the expected value of $x_T$ and estimation of the confidence intervals following Kaczmarek (1957).
② Sample Information (equation (8)) combined with a subjective a-priori PDF (equation (17)) by Bayes' theorem (equation (4)) and CDF of $x_T$ calculated following equation (6).
③ Sample information (equation (8)) combined with data-based regional a-priori PDF (equation (14)) by Bayes' theorem (equation (4)) and CDF of $x_T$ calculated following equation (6).
④ Modified sample information (equation (20) combined with data based a-priori PDF (equation (14)) by Bayes' theorem (equation (4)) and CDF of $x_T$ calculated following equation (6).
⑤ "true" CDF of $x_T$, estimated by common flood statistics from the whole flood record of 57 annual maxima.

It can been seen from fig. 4 that sample information alone gives just vague information about the design flood. (No civil engineer should dare to design e.g. a dam based on a design value derived from such a distribution.)

If the sample information is combined with subjective prior information given by five experts there is much information yield, though the experts' opinion tends towards overestimating the design value. Maybe they are biased by including a factor of safety in their opinion.

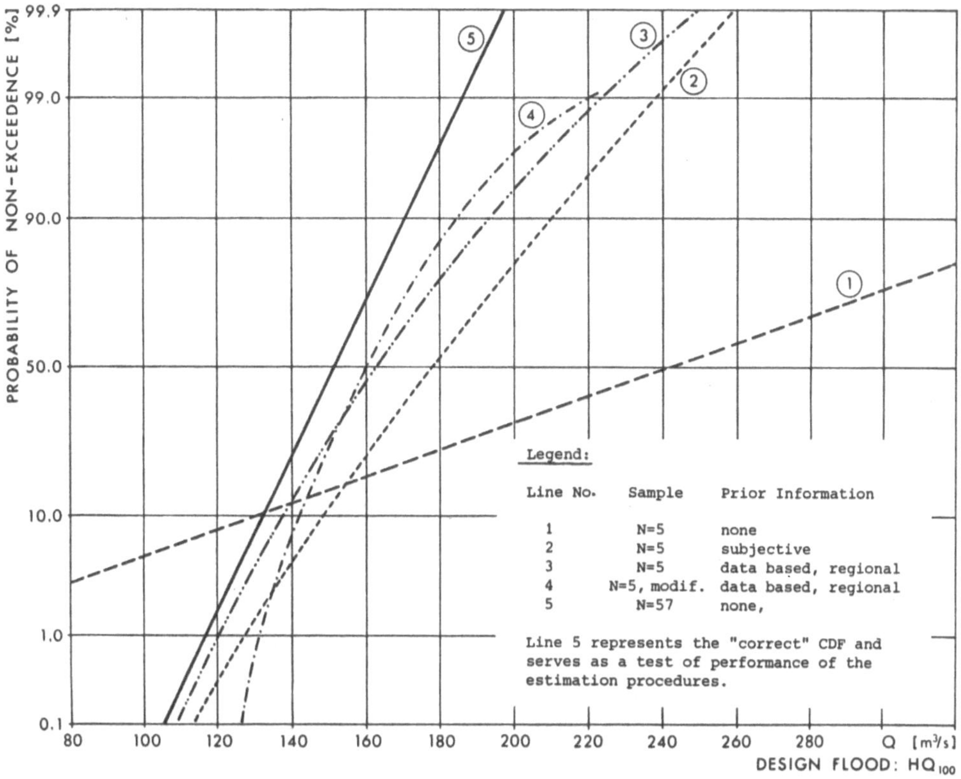


Fig. 4. Cumulative distribution functions of the design flood $HQ_{100}$ due to different sources of information.

The use of data-based regional prior information combined with sample information to a large extent compensates the lack of information contained in the small sample.

If the hydrological characteristics of the years the sample is taken from are considered in addition to the regional information, there is some more information yield.

The cumulative distribution function of the design value estimated from a relatively large sample of 57 elements can be looked upon as the standard of comparison. It can be used to test the performance of the estimation procedures.

## CONLUSIONS

The Bayesian way of thinking meets many requirements of hydrology and of design problems in civil engineering. To consider a design value as a random variable makes it clear, that designing a structure is a decision problem and not merely the result of mathematical calculations. Every uncertainty should be taken into consideration within the decision procedure. From this point of view some more work has to be done in hydrology: There is a considerable amount of uncertainty hidden in the data the hydrologist is compelled to calculate with. The influence of this uncertainty on the accuracy of design values should be made evident.

## REFERENCES


Bardossy, A., 1982, Private communication.

Cunnane, C. and Nash, J.E., 1971, Bayesian estimation of frequency of hydrological events, in: "Proceedings of the Warsaw Symposium. Mathematical Models in Hydrology, July 1971", IAHS-AISH Publication No. 100, Vol. 1, pp. 47-55.

Kaczmarek, Z., 1957, Efficiency of the estimation of floods with a given return period, in: "General Assembly of Toronto 1957", UGGI-IAHS Publication No.45, Vol. 3, pp. 144-159.

Kirnbauer, R., 1981, "Zur Ermittlung von Bemessungshochwaessern im Wasserbau", Wiener Mitteilungen Wasser-Abwasser-Gewaesser, Band 42, Wien.

Schnatter, S., 1982, "Numerische Methoden zur Ermittlung von Bemessungshochwasserverteilungen", Diplomarbeit, Technische Universitaet Wien.

Vicéns, G.J., Rodriguez-Iturbe, I. and Schaake jr., J.C., 1974, "A Bayesian Approach to Hydrologic Time Series Modeling", Ralph M. Pearson Laboratory for Water Resources and Hydrodynamics, Massachusetts Institute of Technology, Report No. 181, Cambridge, Mass.

Zellner, A., 1971, "An Introduction to Bayesian Inference in Econometrics", John Wiley and Sons, New York.

BAYESIAN METHODS IN MULTIPERIOD

FINANCIAL DECISION MAKING

Peter Kischka

Universität Karlsruhe, FRG

## INTRODUCTION

Bayesian methods are applied in financial decision making in order to incorporate estimation risk and/or to incorporate subjective elements in the decision process. In both cases it is assumed, that the distributions of the rates of return depend on some parameter and there is a diffuse or an informative prior distribution for this parameter.From a decision theoretic point of view this approach is justified in Klein et al. (1978), the strong economic reasins to do so are demonstrated e.g. in Bawa et al. (1979), Kischka (1984).

Most work in this field is done with respect to single-period decision problems. Whereas the theory for the single period is well developed only few papers consider the multiperiod approach; special Bayesian aspects are considered in Winkler/Barry (1976), Kischka (1984), Jammernegg (1985). Of course, financial decision making is a multiperiod problem for every investor has the possibility to change his portfolio within his planning horizon. In this paper we show the implications of the Bayesian approach with respect to the first period decision. We compare this decision to the one of a so called classical investor who assumes the rates of return to be identical and independently distributed during the planning horizon. The analysis is done using expected utility maximization; the main result shows that whether a "Bayesian investor" acts more or less risky than a classical one depends on his relative risk aversion in the sense of Pratt (1964). A similiar problem in another context and with a linear utility function is examined in Tonks (1984).

## THE MODEL

We use the following notations

| | |
|---|---|
| $W_0$ | initial wealth |
| $T$ | planning horizon |
| $W_t$ | wealth at the end of period t |

| | |
|---|---|
| $R_t = (R_{1t},\ldots,R_{nt})$ | random vector denoting the rates of return of assets in period t, n being the number of assets available |
| $a_t = (a_{1t},\ldots,a_{nt})$ | portfolio chosen in period t; $a_{it}$ is the amount invested in asset i in period t |
| $P(\|\gamma)$ | given $\gamma$ $R_t$ is i.i.d. with $P(\|\gamma)$ |
| $\phi = \psi_0$ | prior distribution of the parameter $\gamma$ |
| $\psi_t(=\psi_t(r_1,\ldots,r_t)$ | posterior distribution of the parameter depending on the realizations $r_1,\ldots,r_t$ of rates of return |
| $P_t(=P_t(\psi_{t-1}))$ | predictive distribution in period t |
| U | concave utility function for wealth |

As usual we assume that the first asset is riskless

$R_{1t} \equiv r_{st}$.

Of course the amounts $a_{it}$ invested in period t sum up to total wealth available

$$\sum_{i=1}^{n} a_{it} = W_{t-1}.$$

Therefore starting with $W_{t-1}$ and investing the amounts $a_{it}$ we have at the end of period t the random wealth

$$\begin{aligned} W_t &= (1+r_{st})a_{1t} + \sum_{i=2}^{n}(1+R_{it})a_{it} \\ &= (1+r_{st})W_{t-1} + \sum_{i=2}^{n}(R_{it}-r_{st})a_{it}. \end{aligned}$$

The problem is to maximize expected wealth at the end of the planning horizon

$$\max E(U(W_T)).$$

We neglect the problem of intermediate consumption. The functional equations derived from this problem can be written as

$$\begin{aligned} \phi^B_{T-1}(W_{T-1},\psi_{T-1}) &= \max_{a_T} E_T(U(W_T)) \\ &\;\vdots \\ \phi^B_t(W_t,\psi_t) &= \max_{a_{t+1}} E_{t+1}(\phi^B_{t+1}(W_{t+1},\psi_{t+1})) \end{aligned}$$ [1]

1) $\psi_{t+1}$ is determined by $\psi_t$ and $R_{t+1}$; $W_{t+1}$ by $W_t, a_{t+1}$ and $R_{t+1}$ (see Kischka (1984), pp106).

The expectations $E_t$ are taken with respect to the distribution $P_t$ of $R_t$.

The Arrow-Pratt measure of absolute risk aversion is

$$A(W) = -\frac{U''(W)}{U'(W)},$$

and of relative risk-aversion it is

$$R(W) = W\,A(W).$$

There is strong empirical evidence for absolute risk aversion to be decreasing ($A'<0$); no such evidence is given for the sign of the derivative of relative risk aversion.

We consider the class $\mathcal{U}$ of utility functions exhibiting a linear tolerance function

$$\mathcal{U} = \{U|\quad (A(W))^{-1} = d + bW\}$$

This class contains logarithmic, exponential and power functions (see e.g. Bertsekas (1976) pp 89).

NO RESTRICTIONS FOR BORROWING AND SHORT-SELLING

In this section we assume that there are no restrictions for borrowing and short-selling, i.e. the investor can borrow money paying the interest rate $r_{st}$ or he can sell short the risky assets. Formally, we have: $a_{it} \in \mathbb{R}$. We compare the optimal first period decision of the Bayesian, given by

$$a_1^B := \arg\max_{a_1} E_1(\phi_1^B(W_1,\psi_1))$$

with the optimal decision $a_1^C$ of investor C(lassical) who assumes the rates of return to be i.i.d. according to $P_1$. I.e. in the first period both investors assume that the distribution of the rates of return is $P_1$, but Investor B(ayesian) will change this distribution according to realized returns.

Let $\phi_t^C(W_t)$ denote the derived utility functions for investor C:

$$\phi_{T-1}^C(W_{T-1}) = \max_{a_T} E_1(U(W_T))$$

$$\vdots$$

$$\phi_t^C(W_t) = \max_{a_{t+1}} E_1(\phi_{t+1}^C(W_{t+1}))$$

We have the following simple condition for the initial decisions to coincide

<u>Proposition 1</u>

Assume there is some constant $e > 0$ and a function f s.t.

$$\phi_1^B(W_1,\psi_1) = e\phi_1^C(W_1) + f(\psi_1)$$

Then: $a_1^C = a_1^B$

For utility functions $U \in \mathcal{U}$ the solution for investor C is given in Bertsekas (1976),pp 90. From this analysis it follows that for $U \in \mathcal{U}$ the tolerance functions of the derived utility function $\phi_t^B$ are independent of $\psi_t$. This can be used to prove

Proposition 2

For $U \in \mathcal{U}$ there is a vector-valued function $g_t$ such that the optimal portfolio $a_t^B$ depending on $W_{t-1}$ and $\psi_{t-1}$ is given as

$$a_t^B = a_t^B(W_{t-1},\psi_{t-1}) = g_t(\psi_{t-1})(d(\prod_{i=t+1}^{T}(1+r_{si}))^{-1} + b(1+r_{st})W_{t-1})$$

for $t=1,..,T-1$

$$a_T^B = a_T^B(W_{T-1},\psi_{T-1}) = g_T(\psi_{T-1})(d+b(1+r_{sT})W_{T-1})$$

where d and b are constants determining the function $U \in \mathcal{U}$.

The important point in proposition 2 is the independence of wealth of the first factor determining $a_t^B$. We apply this result to the special utility function $U(W) = \ln(W+d)$, $W>-d$. These functions are elements of $\mathcal{U}$ , with $b=1$, exhibiting decreasing absolute risk aversion while the relative risk aversion depends on the sign of d.

Corollary

For utility functions $U(W) = \ln(W+d)$ we have $a_1^B = a_1^C$, i.e. the Bayesian and the classical investor take the same initial decision.

Proof:

Inserting the solution of proposition 2 into the functional equation we get

$$\phi_t^B(W_t,\psi_t) = \ln(W_t + d(\prod_{i=t+1}^{T}(1+r_{si}))^{-1}) + f_t(\psi_t)$$

The first summand doesn't depend on the distribution of $R_{t+1}$. Therefore for the classical investor we have

$$\phi_t^C(W_t) = \ln(W_t + d(\prod_{i=t+1}^{T}(1+r_{si}))^{-1}) + \text{const.}$$

and therefore the corollary follows from proposition 1.

For other functions out of the class $\mathcal{U}$ a similiar result doesn't hold; e.g. in Kischka (1984), pp 123, it is shown that for an exponential utility function the Bayesian investor behaves less risky than the classical one.

## RESTRICTIONS FOR BORROWING AND SHORT-SELLING

The result of proposition 2 depends on the first order condition for the solution of the maximization problem. If we take into account restrictions for borrowing and short-selling there may be no interior solutions. Especially in the Bayesian case the predictive distributions may adjust in a way that it is optimal to invest all in the risky assets at time t having wealth $W_t$ while it would be not optimal to do so if the "old" distribution $P_1$ for the rates of return is applied.

This situation is more complicated than the unrestricted case. In order to simplify the analysis we make some additional assumptions:

| | |
|---|---|
| $n = 2$ | There are only one risky and one riskless asset |
| $P(\|\gamma)$ | is a two-point distribution with realizations z,v |
| $T = 2$ | There is only one possibility for portfolio revision |

To exclude trivial situations we assume $z<r_{si}<v$ for $i = 1,2$.

Furthermore we assume that there are only two parameters $\gamma,\gamma'$ (e.g. with the interpretation that the business conditions are good or bad).

If there is no borrowing and no short selling we have the restriction for the amount invested in the risky assets

$$0 \leq a_{21} \leq W_0, \quad 0 \leq a_{22} \leq W_1.$$

We assume that investor C chooses an interior solution in both periods[1]. Under these conditions it follows from Bertsekas (1976), p 94, that for $U\in\mathcal{U}$ the initial problem is

$$(*) \max_{a_{21}} E_1(U[(1+r_{s2})W_1]).$$

Now consider the decision problem at t=1 for investor B. There are two possible predictive distributions in period 2, which we denote by

$$P_2(\,|z),\ P_2(\,|v)$$

depending on the realization -z or v- of the risky asset in the first period.
E.g. it is possible that investor B invests nothing in the risky asset if z is realized in the first period, and he chooses an interior solution if v is realized.In this case one can show, that for logarithmic utility functions the initial decisions coincide.
In the following we make an assumption which implies that the learning effect is essentially compared to the assumptions of i.i.d. rates of return.

---

1) sufficient for this assumption is

$$zP_1(z) + vP_1(v) > \min\{r_{s1},r_{s2}\}$$

$$zU'(W_{t-1}(1+z))P_1(z) + vU'(W_{t-1}(1+v))P_1(v)$$

$$<r_{st}(U'(W_{t-1}(1+z)) + U'(W_{t-1}(1+v))) \quad (t=1,2)$$

We assume

(A)
$$zP_2(z|z) + vP_2(r|z) \leq r_{s2}$$
and
$$zU'(H(1+z))P_2(z|v) + vU'(H(1+v))P_2(v|v)$$
$$\geq r_{s2}(U'(H(1+z)) + U'(H(1+v))$$
with $H = (1+z)W_0$

The first assumption assures that investor B will invest nothing in the risky asset if in the first period z is realized. As mentioned above one can assume absolute risk aversion to be at least non-increasing; in this case the second assumption assures that total wealth will be invested in the risky asset if v is realized, for $(1+z)W_0$ is the minimal wealth possible at t = 1 and non-increasing risk aversion implies that the investor will not invest less in the risky asset if wealth increases.
With this assumption we have

$$\phi_1^B(W_1,\psi) = \begin{cases} U((1+r_{s2})W_1) & \psi=\psi(|z) \\ & \text{if} \\ U((1+v)W_1)P_2(v|v)+U((1+z)W_1)P_2(z|v) & \psi=\psi(|v) \end{cases}$$

The initial problem for investor B therefore is

$$(**)\qquad \max_{a_{21}} \{U((1+r_{s2})W_1)P_1(z) + U((1+v)W_1)P_1(v)P_2(v|v) + U((1+z)W_1)P_1(v)P_2(z|v)\}$$

<u>Proposition 3</u>

Assume relative risk aversion R of an utility function $U\varepsilon\mathcal{U}$ is smaller than 1 for all possible values of wealth. Then, for every prior distribution $\phi$ there exists some constant k<1, such that:

If $P(v|\gamma)>k$, $P(z|\gamma')>k$ and assumption (A) is fulfilled, then the optimal amount to be invested in the risky asset for (*) is smaller than for (**), i.e. $a_{21}^B > a_{21}^C$.

Proof:

Consider the following function of $\delta$ and $a_{21}$

$$H(\delta,a_{21}):=U[(1+r_{s2})((1+r_{s1})W_0+(z-r_{s1})a_{21})]P_1(z)$$

$$+ U[(1+\frac{v-r_{s2}}{1+r_{s2}}\delta)(1+r_{s2})((1+r_{s1})W_0 + (v-r_{s1})a_{21})]P_1(v)P_2(v|v)$$

$$+ U[(1+\frac{z-r_{s2}}{1+r_{s2}}\delta)(1+r_{s2})((1+r_{s1})W_0 + (v-r_{s1})a_{21})]P_1(v)P_2(z|v).$$

Maximization of H with respect to $a_{21}$ is equivalent to (*) for $\delta = 0$ and to (**) for $\delta = 1$.

In order to simplify the formulas we assume without loss of generality:

$r_{s1} = r_{s2} = 0.$

For given $\delta$ the necessary (and sufficient) condition for an interior maximum $a^*_{21}$ is

$$\frac{\partial}{\partial a_{21}} H(\delta, a^*_{21}) = 0 .$$

Because of the concavity of U we have $\frac{\partial^2 H}{a_{21}{}^2} < 0$

and therefore from the implicit differentiation theorem it follows

$$\frac{\partial^2 H}{\partial\delta\partial a_{21}} (\delta, a^*_{21}) > 0 \Rightarrow a^B_{21} > a^C_{21}$$

We have

$$\frac{\partial H}{\partial a_{21}} = U'(W_0 + za_{21})zP_1(z)$$
$$+ U'[\underbrace{(1+v\delta)(W_0+va_{21})}_{A}](1+v\delta)vP_1(v)P_2(v|v)$$
$$+ U'[\underbrace{(1+z\delta)(W_0+va_{21})}_{B}](1+z\delta)vP_1(v)P_2(z|v)$$

and

$$\frac{\partial^2 H}{\partial\delta\partial a_{21}} = U''[A]v(W_0+va_{21})(1+v\delta)vP_1(v)P_2(v|v) + U'[A]v^2P_1(v)P_2(v|v)$$
$$+ U''[B]z(W_0+va_{21})(1+z\delta)vP_1(v)P_2(z|v) + U'[B]z\ vP_1(v)P_2(z|v) > 0$$
$$\Longleftrightarrow \quad U'[A]vP_2(v|v)(1-R[A]) + U'[B]zP_2(z|v)(1-R[B]) > 0$$

R denotes relative risk-aversion as defined above. Since $z<r_{s1} = 0$ and $R<1$ this condition is satisfied for $P_2(v|v)$ greater some constant m.
From Bayes' theorem we have

$$P_2(v|v) = \frac{(P(v|\gamma))^2\phi(\gamma) + (P(v|\gamma'))^2\phi(\gamma')}{P(v|\gamma)\phi(\gamma) + P(v|\gamma')\phi(\gamma')} \text{ and}$$

$P(v|\gamma) \to 1,\ P(v|\gamma') \to 0 \Rightarrow P_2(v|v) \to 1.$

Therefore there exists some k, depending on $\phi$ ,s.t.

$P(v|\gamma) > k,\ P(v|\gamma') < 1-k \Rightarrow P_2(v|v) > m.$

From this proof it is obvious that for relative risk aversion $R > 1$ there exists some constant k as in proposition 3 implying

$$a_{21}^{B} < a_{21}^{C}.$$

Finally we will consider the case of 'total Bayesian learning' assuming

(T) $\quad P(v|\gamma) = 1 = P(z|\gamma')$ [1].

In this case Bayes' theorem implies for all prior distributions $\phi$: $P_2(v|v) = 1 = P_2(z|z)$.

Total learning therefore means that after observing the high or the low rate of return in the first period the investor is sure that this rate of return will occur in the second period.

Corollary

Assume (T). Then for every $U \varepsilon \mathcal{U}$ with relative risk aversion R and every prior distribution $\phi$ we have

$$R \gtreqless 1 \Rightarrow a_{21}^{B} \lesseqgtr a_{21}^{C}$$

The first condition has to be fulfilled for all possible values of wealth.

Proof:

Under assumption (T) we have $P_2(v|v) = 1$ and therefore the sign of $\frac{\partial^2 H}{\partial \delta \partial a_{21}}$ is determined by relative risk aversion only.

CONCLUSION

The Bayesian approach in multiperiod financial decision making avoids the strong classical assumption that investors will assume rates of return to be i.i.d. over the whole planning period. Furthermore -contrary to arbitrary Markovian approaches- no additional assumptios have to be made compared to the single-period case.

If there are no restrictions for borrowing and short-selling there are utility functions-expressing all types of empirically relevant risk aversion- s.t. there is no difference in the initial decision between a Bayesian investor and a classical one. The reason is, that new information can be totally exploited in a new period, since there are no financial restrictions.

---

1) Every distribution $P_1$ can be derived from this assumption by choosing an appropriate prior.

For the restricted case we have shown that relative risk aversion is decisive for the comparison of initial decisions. Loosely speaking, if the Bayesian assumes a relatively high probability for the rates of return depending on the parameter, then his relative risk aversion is the crucial point deciding whether he behaves more or less risky than a classical investor.

## REFERENCES


Bawa,S., Brown,J., Klein,W., 1979, "Estimation Risk and Optimal Portfolio Choice", North-Holland, Amsterdam.

Bertsekas, D.P., 1976, "Dynamic Programming and Stochastic Control", Academic Press, New York.

Jammernegg, W., 1985, A Dynamic Portfolio Model under Uncertainty: The Two-Point Distribution Case, in: "Optimal Control Theory and Economic Analysis", G.Feichtinger,ed., North-Holland, Amsterdam.

Kischka,P., 1984, "Bestimmung optimaler Portfolios bei Ungewißheit - Bayessche Verfahren in der Portfoliotheorie", Athenäum, Königstein.

Klein,W., Rafsky,L.C., Sibley,D.S., Willig,R.D.,1978, Decisions with Estimation Uncertainty, Econometrica 46.

Pratt,J.W., 1964, Risk Aversion in the Small and in the large, Econometrica 32.

Tonks,I., 1984, A Bayesian Approach to the Production of Information with a Linear Utility Function, Review of Economic Studies 51.

Winkler, R.L., Barry,B., 1975, A Bayesian Model for Portfolio Selection and Revision, The Journal of Finance 30.

# COGNITIVE REPRESENTATION

# OF INCOMPLETE KNOWLEDGE

Gernot D. Kleiter

Institut für Psychologie
Universität Salzburg
Austria

## 1. INTRODUCTION

### 1.1 Mapping Knowledge into Subjective Probabilities

When asked for the meaning of subjective probability, Bayesian statisticians refer to concepts like confidence, feeling of uncertainty, incomplete knowledge, partial knowledge, or degrees of belief. Usually these concepts are taken as primitives, i.e. their meaning is not interpreted within a theory. Often subjective probabilities are "defined" by betting behavior. But such "definitions" are in the spirit of Bridgman's operationalism and black box thinking. These paradigms were dominant around 1930 when the fundamentals of subjective probability theory were first developed by de Finetti.

During the last twenty years there has been extensive research on the representation and processing of knowledge. The work was done mainly within psychology and artificial intelligence. Do the recent cognitive models of knowledge representation provide a framework for the interpretation of subjective probabilities? Imagine the following thought-experiment: down-load the description of a person's (incomplete) knowledge about a given domain from the human brain onto a computer. The resulting representation in the computer is a frozen but "objective" snap-shot of a human knowledge base. Can we find a rule which maps such a knowledge structure into betting ratios and probability distributions? And furthermore, if such a rule can be found, what is the justification for calling the resulting probabilities "subjective"? At present we are far from realizing such an experiment and, perhaps, it can principally never be done. But there are many attempts to day to describe common-sense and expert knowledge in an objective way. Theories and models showing how such descriptions can be made are first steps in this direction.

The aim of the present paper is to review some of the work done on the cognitive representation of knowledge and to discusss its relation to the interpretation of subjective probability as incomplete knowledge. Subjective probabilities will be interpreted as summary descriptions of partial knowledge states. I will first discuss a fundamental difference between the way in which knowledge systems and in which Bayesian probability theory processes new data. Knowledge systems process new information in a constructive additive way, Bayesian probability theory process new information in a suppresive subtractive way.

## 1.2 Suppression of Incompatible Alternatives

An idea that always impressed me in de Finetti's work is his two stage model of induction: We begin the inductive process with a large set of possibilities, later each incoming piece of information logically suppresses some of these possibilities, namely those which are incompatible with the new information and which should thus be excluded from further analysis. This first stage reminds me of the work of a sculptor: he begins with a large square stone and uses his chisel to remove more and more of those pieces of the stone which are incompatible with his ideas.

The second stage of induction deals with the according probabilities. Given a possibility space together with a probability distribution defined on it; the removal of possibilities leads also to a loss of probability mass. To fit the new situation the probabilities must be re-standardized to sum up to one. De Finetti showed that this re-standardization - if done in a coherent way - "automatically" follows Bayes-Theorem (de Finetti, 1974, I, 141).

According to this model a learning system should start with a wide possibility space and reduce it according to the incoming experience. Only the very first prior probabilities really matter because at all the later stages they are completely determined logically.

## 1.3 Constructive Composition of Facts

The psychological processes seem to be reversed. We do not represent possibility spaces in our memory but store the actual experiences. We try to keep track of the facts instead. We do not work like a sculptor but like a painter who puts colours on a sheet of paper: the picture is created by adding new shapes and colours to those already existing. This is a constructive composition process and not a suppression of incompatible alternatives. Large possibility spaces impose unrealistically high processing loads on human memory and thought. Common-sense thinking is guided by positive knowledge, not by the "up until now not excluded possibilities".

We also do not normally store negative properties like "the hero is not bad", "ants are not big" (Graesser and Clark, 1985). Again, "... Storing negative expressions would rapidly clutter a data base because thousands of true negative expressions could be potentially stored in every..." knowledge structure (Graesser and Clark, 1985, 138). Young children have no concept of ambiguous information. "Their strategy is to make the best interpretation they can on the basis of prior assumptions and expectations." (Robinson and Robinson, 1982, 279)

An important advantage of storing facts instead of possibilities is that facts can be re-analysed in different models and under changed conditions. Multi-purpose data processing is a vital advantage for a complex system striving for multiple goals in a complex environment.

# 2. KNOWLEDGE REPRESENTATION

The most important concepts in cognitive science are "representation" and "process".

A cognitive representation is the mentally coded content of information. A taxonomy according to the representational format is given in Fig. 1.

The denotational formats work on the basis of static structures. The processes which operate on these structures are not part of the represented knowledge and are well separated from it. The declarative formats are the

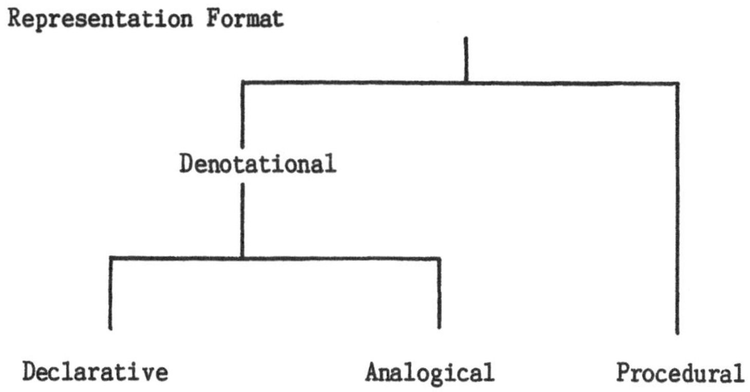


Fig. 1. Main types of representational formats.

most frequently investigated forms of knowledge representation. Typical examples are propositional systems, semantic networks, and frames. Analogical formats contain a homomorphism between the external objects and their internal representation. Mental images are the best examples. Procedural formats represent knowledge by dynamic structures or small "programs". The knowledge of how to ride a bicycle or how to tie shoe laces is stored in a procedural format.

A cognitive process is an activity which changes and manipulates the representations in a cognitive system. In the present context the distinction between processes which operate on the object-knowledge level and processes which operate on the meta-knowledge level will be important.

## 3. DECLARATIVE REPRESENTATION

### 3.1 Logical Systems

Famous philosophers like Boole or John Stuart Mill considered the laws of logic as fundamental for human reasoning. We don't agree today (e.g. Braine, 1978). But even if natural reasoning does not follow formal logic - wouldn't it be a first powerful approximation to model human thinking and knowledge representation on a logical system? No wonder that the first knowledge representation systems in artificial intelligence (AI) followed closely first order propositional calculus.

Logical data bases. Logical systems of knowledge representation were mainly developed for data bases. PROLOG (e.g. Warren, 1979; Clark and McCabe, 1979; Futo et al., 1978) is the best known example for a logical data base. In a logical data base knowledge is translated into a set of well formed formulas in predicate calculus, e.g.(Robinson, 1979, 105):

for all X, if X is human then X is fallible
Socrates is human
Aristotle is human
Socrates is Athenian

The user of a data base may interact with the system by asking questions like

Are there any Z, such that Z is Athenian and Z is fallible?

A question is treated as goal statement. The system answers the question by YES if the questioned formula can be shown to be a theorem of the data base.

Z = Socrates.

It's answer is NO SOLUTION if the query is not a theorem of the data base.

Aristotle is Athenian?

Such a system is an automated deductive inference device. Technically, the answer is often computed by a machine-orientated refutation procedure known as the resolution method (Robinson, 1979, 1982). It consists of three steps: the transformation of the original formulas into standardized clause forms, the main resolution process, and a final test.

Automated deductive inference devices provide only two principles to process incomplete knowledge: (i) If the missing information consist of not stated axioms then the set of possible YES-answers (derivable theorems) is reduced and the NO SOLUTION set grows larger; (ii) if there are missing elements within one of the axioms this leads to an error condition and a break down of the process. The explicit admittance of non-stated axioms (case i) is the most fundamental way to deal with incomplete knowledge. We will turn to this question next.

Note that the assignment of probabilities to quantified propositions is rather unusual. The only example I know in which probabilities are attached to expressions containing existential and universal quantifiers is Nilsson (1986). He considers the convex hull of a polyeder built by the truth values of the constituents and derives upper and lower bounds for coherent probabilities. His treatment is completely within the de Finetti's Fundamental Theorem (1974, I, p. 112; Kleiter, 1981, p.83).

Closed and open worlds. A fundamental decision is whether we explicitely allow for the lack of knowledge in the representation of a knowledge domain. If we declare a knowledge base to contain all relevant knowledge we say that we work under the closed world assumption, when we admit lack of knowlege we work under the open world assumption (Collins et al., 1975; Reiter, 1978). If a data base runs under the open world assumption and if an item cannot be found or a deduction cannot be proven, the output of the system is a DON'T KNOW; there may be more to know. A consequence of the closed world assumption is that "... certain answers are admitted as a result of failure to find a proof." (Reiter, 1978, 56)

In probability theory we work under the closed world assumption. This is a consequence of the basic event-structures to which probabilities are assigned; they are closed under Boolean operators or some other principle. A theory that was developed to process uncertainty partly relies on the assumption of complete knowledge.

From a psychological viewpoint the closed world assumption is realistic in many cases because it relieves memory from the problem of storing negative facts; the negative facts can easily be computed at the time when they are needed. "People do not store most things that are not true, for example, that Mexico has no king. Therefore, deciding that something is not true normally requires an inference." (Collins et al. 1975, 386/7) Decomposing a problem space and searching for closed subsets is an efficient strategy in problem solving. But the decision to close a possibility space is always critical. Only too often we feel safe under the closed world assumption but, at a later time, we painfully realize that we overlooked one or more possibilities. This is a well known phenomenon in the assessment and evaluation of technological and natural hazards.

## Learning, non-monotonic inference, and circumscription

> When I tell you that Tweety is a bird, you will conclude that Tweety can fly. When you later learn that Tweety is a penguin, you will withdraw your prior conclusion.

In this example additional knowledge invalidated the previous inference. This is commen-sense reasoning; but in a deductive system this is forbidden because we are not allowed to suppress old theorems in the light of new evidence.

Let q be a formula and let A and B be sets of formulas, then

if $A \vdash q$ and A is a subset of B then $B \vdash q$.

In the Tweety-example the original set of formulas (A = Tweety is a bird) is extended by an additional formula (Tweety is a penguin); but from the list "Tweety is a bird, Tweety is a penguin" we do not want to derive that Tweety can fly. Deductive systems are so 'permissive' that new data (axioms) cannot invalidate old theorems. The corpus of theorems, therefore, can only monotonically grow larger. In a deductive system learning leads to a blow-up of the set of inferences. The inability to modify or delete old inferences is such a devastating feature for any data base that the problem of non-monotonic reasoning was vigorously attacked by a number of mathematical logicians (e.g. McCarthy, 1980a, 1980b, 1986; McDermott and Doyle, 1980; Reiter, 1980; Ginsberg, 1984; Moore, 1985). Before starting the inference process tests can be performed to check whether all relevant aspects are normal or abnormal ("would the exceptions please stand up!"). Abnormal results may be written on a SOMETHING-WRONG-LIST. A circumscription (McCarthy, 1980a) declares that if the SOMETHING-WRONG-LIST is empty no exceptions are present and everything is just normal and o.k.

A circumscription is a rule that selects the minimal set of all known objects having a certain property as long as no exceptions are explicitly derivable. It draws a contour-line around the things which are allowed because they are not explicitly forbidden. Exceptions should be stated or derivable; as long as they are not you may safely trust you are in a normal world. McCarthy (1986) argues that circumscriptions are closely related to qualitative probabilities. Circumscriptions are "a very streamlined expression of probabilistic information when numerical probabilities, especially conditional probabilities, are unobtainable." (1986, 91)

The qualitative probabilities take on three values:

(i) Infinitesimally close to one.
Given Tweety's SOMETHING-WRONG-LIST for the aspect FLY is known to be empty, i.e., as long as we do not know that Tweety is a penguin, the probability that Tweety can fly is infinitesimally close to one.

(ii) Intermediate.
Given the SOMETHING-WRONG-LIST for the aspect FLY is not known the probability that Tweety can fly is intermediate.

(iii) Infinitesimally close to zero.
Given Tweety's SOMETHING-WRONG-LIST for aspect FLY contains IS A PENGUIN, the probability that Tweety can fly is infinitesimally close to zero.

McCarthy (1986) emphasized that his "circumscriptional probabilities" do not need a possibility space. In this respect, his formalization is, indeed, radically different from other proposals of probabilistic thinking. Circumscriptions are an interesting candidate for the cognitive representation of some forms of incomplete knowledge. They may prevent infinitely long testing procedures. Applied to domains where we do not have much experience to distinguish what is normal and what is abnormal it may lead to the error of declaring the SOMETHING-WRONG-LIST as being empty. This may happen in the case of rare events. A discussion of the relationship between circumscriptions and closed world assumptions if found in Lifschitz (1985).

Autoepistemic knowledge and meta-inference. Autoepistemic knowledge evolves from "reasoning about one's own knowledge or belief" (Moore, 1985, 78). Autoepistemic knowledge may be used to explain non-monotonic reasoning. Take the following (slightly adjusted) example of Moore:

I know, I don't have a younger brother.

How do I know that? My parents never explicitly assured me that I do not have a younger brother. I also do not infer my knowledge from other "object level" facts. But a highly plausible explanation of my knowledge is the following autoepistemic counterfactual:

If I did have a younger brother I would know about it.

The example demonstrates that we make inferences from lack of knowledge. Our common-sense model of memory assumes that we store positive information. We seem to apply an inference rule like:

From "someone should know q but his memory search fails to find q" infer "not q".

Psychologists have tried to answer the question how we decide that we don't know something without performing an extensive memory search. Norman described the first stage of such a process: "There appears to be a preliminary rapid, cursory search of the information presented to determine if anything at all is known about the query. If this rapid search fails, then the reason for the failure determines the type of response made to the question." (Norman, 1973, p.138)

When you watch a quiz on TV or listen to BBC's Brain of Britain you will observe two kinds of of don't know answers:

a. Slow and low-confidence don't-no-responses or just time-outs, e.g.
   What was Harry Truman's middle name? - Let me think for a while! Was is ... no, I don't know.
b. Fast and high-confidence don't knows, e.g. responses to
   What is Professor Viertl's telephone number? - I have absolutely no idea; why on earth would I know his phone number? (Such examples are due to George Mandler).

These two types of don't-know-responses were investigated in three reaction time experiments by Glucksberg & McCloskey (1981). The authors found good evidence in favor of a two stage decision process. In the first preliminary stage we search memory for stored information that may be relevant to the question. If we do not find relevant information we respond with a rapid don't know answer. If we find potentially relevant facts the second stage is entered in which these facts are evaluated in detail. If the evaluation in the second stage fails to provide a sufficient result a slow and low-confidence don't know response is given.

Gentner & Collins (1981) investigated the questions what we infere when someone else states that he does not know something. This is of importance when we want to apply expert knowledge. Take this example:

Fred H. Tschirley (1986), an expert in the toxicality of dioxin, in a recent Scientific American paper states that he does not know of any investigation in which it was shown that dioxin caused a chronic desease in humans?

Gentner & Collins showed that similar lack-of-knowledge statements in a written text constitute strong evidence against a fact in the reader. "The more important the assertation and the more expert the person who lacks knowledge, the more certain is the lack-of-knowledge inference, as measured by a decrease in the rated likelihood of the assertation." (1981, p.434) Autoepistemic processes save time by preventing exhaustive memory search and they save memory space by allowing reasonable inferences from lack-of-knowledge. A system with well developed autoepistemic strategies can thus handle incomplete knowledge without explicitely representing it.

Space limitations do not allow to discuss incomplete knowledge representation in primitive features and quantitative dimension. Some aspects are treated in Slovic and Phillamy (1974), Yates et al. (1978), Huber (1983), or Kühberger (1986).

### 3.2 Conceptual Prototypes

The representation of concepts is a traditional and important subject matter both in psychology and artificial intelligence. Often declarative knowledge is centered around conceptual entities with associated descriptions ("object-centered factorization of knowledge", Bobrow and Winograd, 1977). According to the classical view "all instances of a concept share common properties that are necessary and sufficient conditions for defining the concept." (Medin and Smith, 1984, 115) An example for the classical view is the definition of a BACHELOR as an adult unmarried male human.

In a series of experiments Rosch (1975) showed that the instances of a concept vary in their typicality. There are typical instances (a robin is a typical bird) and atypical instances (like the penguin Tweety). The typicality of instances predicts several experimental effects like the speed of categorization. Amstrong et al. (1983) demonstrated that typicality-effects are not restricted to natural concepts but are also found within integer-judgments. The integer 4 is a more typical even number than 18 and 7 is a more typical odd number than 501 (Amstrong et al., 1983, 276). Attempts were made to explain the typicality-effects by fuzzy set theory, especially by membership functions (Zadeh, 1982). But Osherson and Smith (1981, 1982) argued that the conceptual combination of two concepts does not follow the laws of fuzzy set operators.

A prototype is a summary representation (Medin and Smith, 1984, 117) of a concept consisting of its most typical attribute values. Often the most typical attribute values are defined by salient features or by averages. The instances of concepts may now be categorized by a critical distance measure; the similarity between the instances and the prototypes can be determined by simple feature counts or by complex linear discriminant functions. A summary representation is the result of an abstraction process, it should be applicable to all relevant test items, but it need not correspond to a particular instance (Smith and Medin, 1981, 132).

Typicality effects with well-defined concepts like even and odd numbers should warn us that human subjects may report superficial, unessential, or even misleading characteristics of their knowledge representations when they answer our questions. The typicality may easily be used as a heuristic for judging probabilities. The representativeness heuristic of Kahneman and Tversky (1972) demonstrates that the similarity between a data sample and its parent data generating process may lead human subjects to unreasonable probability assessments in important cases.

### 3.3 Frames and Scripts

"As you are walking through an unfamiliar house, you come to a normal interior-type door, open it, and walk through. At the moment that you open the door, your (entirely reasonable) expectations have already brought a 'room' frame to mind." (Kuipers, 1975, 154) The frame (Minsky, 1975) contains information about the standard arrangement of the walls, windows etc.. A frame is a predefined default description of a scenario. The default values are standard values suggested unless there is contradictory evidence. The function of a frame is to write its default values into the missing slots of an as yet incomplete individual scenario representation. Frames help to explain how we can understand stories and other forms of verbal communication. Especially in AI frames made it obvious, that language comprehension is not possible without an appreciable amount of world knowledge and about what is normal and usual. Psychologically, special effects may result from the interaction of the expected default values and actually observed values. Let me illustrate this by an example taken from Tversky and Kahneman (1983) investigation on the "conjunction fallacy":

> A health survey was conducted in a representative sample of adult males in British Columbia of all ages and occupations.
> Mr. F. and Mr. G. were both included in the sample. They were unrelated

and were selected by chance from the list of participants.

Which of the following statements is more probable?
A = Mr. F. has had one or more heart attacks.
B = Mr. G. has had one or more heart attacks and Mr. G. is over 55 years old.

In both examples many subjects judge B to be more probable than A. This is a violation of a fundamental principle of probability theory, Tversky and Kahneman call it the principle of extensionality:

if B is a subset of A then $p(B) \leq p(A)$

Because a conjunction of two events A and B is a subset of B it follows that

$p(A \text{ and } B) \leq p(B)$.

Intuitive probability judgments do not follow the extension rule but are trapped by the conjunction fallacy. The conjunction fallacy occurs if a default value is explicitly confirmed and emphasized.

Another example of the interaction between default values and actual information is our Tweety example. As we hear that Tweety is a penguin the default value CAN FLY is explicitly denied with the effect that the original inference is withdrawn by non-monotonic reasoning. In the case a hierarchy of low, intermediate and high level frames exists a "specialization-of slot is used to establish a property inheritance hierarchy among the frames, which in turn allows information about the parent frame to be inherited by its children, much like the ISA link in semantic net representations... " (Barr and Feigenbaum, 1981, 218).

Scripts are memory structures which store standard event sequences. They were introduced by Schank and Abelson (1977) and are in many respects similar to frames. More recently Schank (1982) dropped the assumption that scripts are homogenuous memory structures. Instead, he hypothesized a hierarchical memory structure from low level (like events) to high level (like situations) information. A basic function of memory is the prefiguring of the knowledge system for the interpretation and understanding of future information. It is a self-organizing growing structure.

Bobrow and Winograd (1977, 1979; Lehnert and Wilks, 1979) designed a complex knowledge representation language (KRL). It is a hybride of many representational formats and makes extensive use of prototypes, frames and procedures.

### 3.4 Network Systems

A network is represented by a finite labelled graph consisting of nodes and labelled links. Networks belong to the most popular systems used in AI and psychology to represent knowledge in long term memory. Psycholinguists like Chomsky or Fillmore had a strong influence on the semantic network models in psychology (Quillian, 1968; Anderson and Bower 1973; Rumelhart, Lindsay, and Norman, 1975; Anderson, 1976, 1983). A critical review of discrimination networks is given by Barsalou and Bower (1984). Often, the models were simulated by computer programs written in LISP. The main research interest was to model language comprehension. The encoding of information by semantic elements and relations is often called a propositional representation.

Fig. 2 represents a net-structure in the Anderson-Bower-style for the sentence "Last night in the city a ganster shot a policeman". The surface structure of the sentence is rewritte by elementary semantic relations like LOCATION and TIME (building the CONTEXT), SUBJECT and PREDICATE (building the FACT), RELATION and OBJECT (building the PREDICATE), MEMBERSHIP RELATION, SUBSET RELATION, QUANTIFIERS, INDIVIDUAL OBJECTS etc..

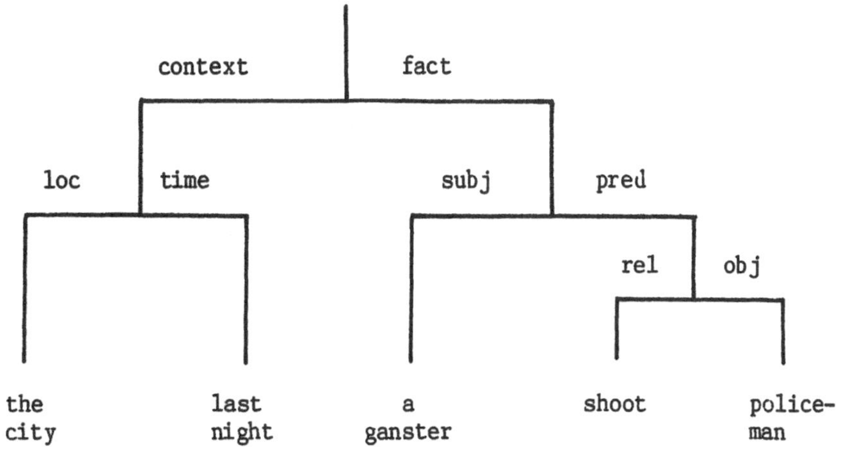


Fig. 2: A proposition tree corresponding to the proposition "Last night in the city a gangster shot a policeman", as represented in the Anderson and Bower (1973) model; loc=location, subj=subject, pred=predicate, rel=relation, obj=object.

In a Lindsay-Norman-Rumelhart-representation the structure of propositional representation is derived from the verb-structure. For example, the verb GIVE is characterized by three arguments: GIVE AGENT, RECIPIENT, OBJECT. An important link in a LNR-net is the ISA-link expressing the class membership relation between a concept and its SUPERORDINATE category; the HAS-link is used for the assignment of properties to concepts. The reverse direction of ISA indicates an EXAMPLE node or a SUBORDINATE category.

Incomplete knowledge in propositional networks has not been the subject of major concern. But the networks provide the simple possibility to introduce predicates like UNCERTAIN(X) or PROBABLE(X) and to add them, for example, to the CONTEXT description of a HAM-structure. The result is an explicit declarative representation of uncertainty about facts or events. Psychological investigations indicate that such a representation of incomplete knowledge seems to be an exception which is used only in rare cases.

In the case of missing nodes incomplete knowledge on low levels may be resolved by top-down property inheritance through the ISA-links. A similar principle may work in the bottom-up direction, i.e. from instances to categories. Missing links or link-labels may lead to a loss of availability. Spreading activation models introduce assumptions on the dynamic flow of information in the network and on various strength parameters (Collions and Loftus, 1975; Anderson, 1983) but there is not enough space to treat them here.

Graesser and Clark (1985) tried to model the essential cognitive structures and processes which underlie story comprehension. They focussed on the representation of general world knowledge and not on the problems of language comprehension. They employ "conceptual graph structures" (CGS) to represent all the knowledge a subject has on a concept. A CGS is a very rich structure, containing 160 nodes on the average. Incomplete knowledge may enter the theory in several ways:

1) A CGS may contain a description in a slot filler that specifies not just one but a distribution of acceptable values.

2) Uncertainty may be expressed by meta nodes. A meta node refers to an

embedded set of nodes and may comment on the embedded information. A substructure may thus explicitely be declared as more or less partial. Other kinds of propositonal attitutes may also be expressed that way.
Another form of uncertainty may enter the system via the story comprehension process:
3) The comprehension process leads to a sequential updating. The kind of updating and information integration depends on the matching process between the two conceptual knowledge structures in the working memory. Many shared nodes lead to many exact matches.

The number of matches in two consecutive conceptual knowledge structures may be used to explain base rate fallacies in diagnostic classification problems - one of the classical Kahneman and Tversky examples (Kahneman, Slovic, and Tversky, 1982). The problems are described and presented by short stories. The diagnostic alternatives are stated first, the specific case descriptions are given next. If the subjects classify the case description on the basis of its number of matching nodes in the conceptual knowledge structures the base rates will be ignored.

Extensive use of network systems was also made in AI and expert systems. PROSPECTOR (Duda et al., 1978; Duda et al., 1979) is an expert system designed to aid geologists in the diagnosis of the favorability of a region for the occurrence of an ore deposit. Its architecture is general, however, and may easily be exported to other domains. PROSPECTOR represents its knowledge in an inference network. The nodes of the network consist of assertations; there are two types of assertations, evidence and hypothesis. The network has a hierarchical tree-like structure. At the lowest level, the "leaves" of the tree, the user provided evidence is located, at the highest level, the "origin" of the tree, the top-hypothesis is located. The diagnosis of the system is inferred according to Bayesian or quasi-Bayesian principles. Before the system can start its routine work it must be supplied with prior probabilities and likelihood ratios. The prior probabilities are attached to the hypothesis nodes and the likelihood ratios are attached to the links connecting a lower level assertation to a higher level assertation. As a consequence, the effect of telling the system that a number of lower-level assertations are true, leads to an updating of the probabilities of all the directly or indirectly linked higher-level assertations. When the user inputs his pieces of evidence he may express his certainty for each piece on a scale from -5 (virtually certainly absent) to 5 (virtually certainly present). A value of 0 does not change the probabilities of the hypothesis and is equivalent to no information. The values -5 and 5 lead to a "full" Bayesian updating; the intermediate values (-4,-3,-2,-1,1,2,3,4) lead to a degraded updating only. The logic of "certainty factors" was adapted from Shortliffe's (1976) pioneering work on expert systems.

The pieces of evidence are processed under the assumption of conditional independence. There was a debate on the possibility to change the probabilities of hypothesis with incoming data in PROSPECTOR under the conditional independence assumption (Pednault et al., 1981; Glymour, 1985). An interesting proposal to rewrite the Bayes Theorem in a linear form was made by Kadesch (1986). He defines a "relevance parameter" as the difference of the two posteriori probabilities which result from the occurence of an event and from its complement. The concept is similar to what has been called "diagnosticity". The advantage of the given formulas is that they make transparent the impact of the data probabilities upon the posterior probabilities. Kadesch treats the case of multiple evidence in general, under the assumption of mutual independence of the pieces of evidence, and under the additional assumption of relevance independence. Pearl (1986) described two updating techniques for trees and general graph hierarchies; conditional independence is assumed throughout. The first technique performs the updating by a normalization phase. The second technique works by propagation-based updating where an impacted hypothesis transmits "messages" to its neighbors. The messages finally result in probability changes of the hypotheses. The propagation is analogous to a spreading activation process in a semantic network.

## 4. ANALOGICAL REPRESENTATION

How much do you know about Austrian geography? How sure are you that Innsbruck is north of Vienna? Answers to these questions are typically given on the basis of mental images and mental maps. Parts of the information stored in human memory are coded in such a way that they can easily be transformed into a visual format and be depicted on an "internal screen". Mental images are a subject of high interest in recent psychological research (Paivio, 1971; Shepard and Cooper, 1982; Kosslyn, 1980; Finke, 1986). To see how incomplete knowledge may be represented in mental images let's first have a look at visual perception.

A countryman of de Finetti, professor Maffei at the University of Pisa (Maffei, 1978) and the English professor Campbell at the Cambridge University (Campbell and Robson, 1968) performed a remarkable theoretical "transplantation": they took the frequency concept - which usually is defined in the time domain - and re-defined it in the space domain. The result was the <u>spacial frequency theory</u>. Spacial frequency is measured by the number of contrast maxima per degree in the visual field. The contrast - comparable to the amplitude of a wave - is defined by the ratio of intensities (maximum-minimum)/(maximum+minimum). A picture is processed by a set of spacial frequency filters. Each filter performs a Fourier analysis. Each filter is an operator which generates a <u>convolution</u> of the whole picture (Marr, 1982).

Harvey (1986) applied spacial frequency concepts to mental imagery. He showed how introspective intuitions about the representation of incomplete knowledge in images can be experimentally investigated. Mental imagery is perhaps the most natural medium to deal with incomplete knowledge. Mental pictures allow for different degrees of <u>resolution</u> and clarity they allow for <u>continuous mappings</u>, independent foreground and background handling, mental rotation, efficient symmetry handling, zooming-in, local modifications etc.. Although classes cannot be depicted directly (Kosslyn, 1984, 106) it is well known, that imagery processes are also used to represent abstract and non-pictorial information.

It may be suspected that numbers are sometimes represented on an imagery "number line". Shepard et al. (1975, 113) showed that the subjective spacing between numbers follows a logarithmic Weber-Fechner type of relationship: the subjective distance between 1 and 2 is larger than that between 9 and 10. This may lead to confusion in direct numerical assessment techniques: A logarithmic utility scale obtained from a direct rating by numbers may tell us more about the cognitive representation of numbers than about the utilities. The same effect may distort direct probability estimates.

Another example of analogical representations is the solution of linear ordering tasks like the following three-term series problem:

George is older than Peter.
George is younger than Donald.
What about Peter and Donald?

Do you place George, Peter and Donald on a mental scale and then derive your judgment? The example shows some of the difficulties connected with imagery codes. While the task does only require a linear order, the visual code does more than that, it must introduce a spacing (for a discussion see Evans, 1982, 49 ff.). But all we can say from experiments is that the order information is easily used serially and nothing is known about spacing.

Often imagery updating is simple : "... if we add a new city to a map, we need only put it in the right place." (Barr and Feigenbaum, 1981, 201) But Barr and Feigenbaum also give an interesting counter-example in which the imagery code does not allow the representation of incomplete knowledge: "... analogical representations become unwieldy for certain kinds of incomplete information. That is, if a new city is added to a map, its distance from other cities is obtained easily. But suppose that its location is known only indirectly, for example, that it is equidistant from cities Y and Z. Then the distance to other cities must be represented as equations, and the power of the analogue has been lost." (206)

## 5. PROCEDURAL REPRESENTATION

### 5.1 Production Systems

I am sure you know how to bind your shoe-laces, but try to give a verbal description of it! For some parts of our knowledge it is much easier to demonstrate it by doing than to give a description in a declarative form. Here, knowledge is represented in a process-code. "In a procedural representation, knowledge about the world is contained in procedures - small programs that know how to do specific things, how to proceed in well-specified situations." (Barr and Feigenbaum, 1981, 155)

In cognitive psychology most research work focuses on the perceptual input and the internal symbol processing. Much less research work has been done on the behavior and action generating processes. One exception, of course, is decision theory, which tries to explain choice behavior. Another exception are production systems. "Productions provide the connection between declarative knowledge and behavior." (Anderson, 1983, 215) A production system consists of a set of condition-action pairs of the form

IF <condition> THEN <action>.

If the left-hand-side condition is fulfilled in the working memory the right-hand-side action is triggered. The interaction between rules is minimized, one rule does not call forth another one; the rules are autonomous, modular, tiny pieces of data-action pairs.

The main architectural components of a production system are its working memory, its production memory and its control structure. A classical production system does not contain a separate declarative data base (long term memory) containing facts or another type of knowledge. All permanent information is contained in the productions. A "psychological" exception is Anderson's ACT or ACT* model; it contains not only the working memory and the production memory but also an additional declarative memory (Anderson, 1976, 1983). The action part of a production adds, deletes, or modifies data elements in the working memory. The control structure determines the methods of <u>conflict resolution</u> and of <u>matching</u>.

A conflict results if the left hand side of more than one production is instantiated by the data in the working memory. Conflict resolution methods define the selection of just one production (in a serial system) out of the conflict set. McDermott and Forgy (1978) distinguish five major categories of different conflict resolution strategies:

<u>1. Order rules</u> use a pre-established priority ordering (total or partial) on the productions;

<u>2. Specificity rules</u> prefer those productions which are more specific; there are three subtypes: (i) priority is given to the more specific left hand side of productions, (ii) priority is given to the more specific data in the working memory and (iii) priority is given to both the more specific data and the more specific left hand side of the production taking also negated condition elements into account.

<u>3. Recency rules</u> use the amount of time that elements have been in working

memory; the time is measured by the number of actions that have been performed or by interpreter cycles.
4. Distinctiveness rules select productions on the basis of their similarity or dissimilarity to previously fired productions; they avoid repetition and prohibit doing things twice.
5. Random selection just uses randomization in the cases of ties.

We usually keep track on the conflict resolution strategies we apply. Thus, we have knowledge about the size of the conflict set and an intelligent assessor of betting ratios may just use this size to adjust his betting ratios. We know how complex the conflict resolution strategies in a task were, how much time and effort we needed to resolve the conflict, or how distinct or unique was the information we processed. This indicates that we make extensive use our auto-control knowledge in the quantification of incomplete knowledge.

### 5.2 Partial programs

When you buy a new computer you get a description of its processor, its memory, its instruction set and so on. This is a description of the possible states and the actions of the machine. Call, for a moment, a function from the possibles states to the legal actions, an abstract machine. The abstract machine contains all the programs which may be writte for it. This list of all the possible programs is, in fact, a maximally incomplete program. You may start now reducing the incompleteness by writing down commands in an arbitrary language. Each command introduces a number of constraints upon the previous states and actions. The more commands you write down the more constraints are introduced. Reducing incompleteness works by the method of subtraction or suppression.

This idea was elaborated by Genesereth (1984). It is fascinating to see how closely it resembles de Finetti's first stage of induction described at the beginning of this paper. The psychological knowledge about partial programs in human information processing is not only incomplete but completely missing.

## 6. CONCLUSIONS

1) There are many local and highly specialized principles by which incomplete knowledge is handled by both man and machine. There is not just one uni-dimensional global "feeling of uncertainty" that is used in the processing of probabilities; different effects upon the decoding, enconding and internal combination of probabilities are predicted for different representational formats.

2) Subjective probabilities are not subjective in the sense of being opaque and scientifically unaccessible. They are subjective only in the sense of belonging to one knowledge system only. As knowledge structures become more and more describable they also become more and more "objective". This is completely in the spirit of de Finetti. Why shouldn't machines have subjective probabilities?

3) Subjective probability assessments are a summarizing auto-description of a knowledge state. They need an appreciable amount of de-coding of internal information and - most important - of meta-knowledge.

4) The probability assessor may be trapped by fallacies through superficial or misleading heuristics. Beside hunting for fallacies it is also interesting to look for the fundamental cognitive principles of knowledge processing in some detail.

5) Once partial knowledge states are de-coded and summarized by betting ratios, probability distributions etc. they are ready for communication and for entering a scientific community or a group of decision makers. The probabilities now take on the role of an inter-lingua and are public. The process of understanding communicated probabilities, which is the inverse of the assessment process, may lead to the activation of already stored mental models and prototypes. Only with a considerable amount of training will the probabilities, through abstraction processes, become more and more autonomous. An explicit declarative representation of uncertainty is only possible when this level has been reached (compare Humphreys and Berkeley, 1983; Hogarth, 1975).

## REFERENCES


Anderson, J.R. (1983), The Architecture of Cognition, Cambridge, MA: Harvard University Press.

Anderson, J.R: (1976), Language, Memory, and Thought, Hillsdale, NJ: Erlbaum.

Anderson, R.J. & Bower, G.H. (1973), Human Associative Memory, Washington: Winston.

Armstrong, S., Gleitman, L.G., & Gleitman, H. (1983), What some concepts might not be, Cognition, 13, 263-308.

Barr, A. & Feigenbaum, E. (Eds.) (1981), The Handbook of Artificial Intelligence, Vol. I. Reading, MA: Addison-Wesley.

Barsalou, L.W. & Bower, G.H. (1984), Discrimination nets as psychological models, Cognitive Science, 8, 1-26.

Bobrow, D.G. & Winograd, T. (1977), An overview of KRL, a kowledge representation language, Cognitive Science, 1, 3-46.

Bobrow, D.G. & Winograd, T. (1979), KRL, another perspective, Cognitive Science, 3, 29-42.

Braine, M.D. (1978), On the relation between the natural logic of reasoning and standard logic, Psychological Review, 85, 1-21.

Clark, K.L. & McCabe (1979), The control facilities of IC-PROLOG, in: D. Michie (Ed.), Expert Systems in the Micro Electronic Age, Edinburgh: Edinburgh University Press, pp.122-149.

Collins, A.M. & Loftus, E.F. (1975), A spreading-activation theory of semantic processing, Psychological Review, 82, 407-428.

Collins, A., Warnock, E.H., Aiello, N., & Miller, M.L. (1975), Reasoning from incomplete knowledge, in: D.G.Bobrow & A.Collins (Eds.), Representation and Understanding, New York: Academic Press, pp.383-415.

De Finetti, B. (1974), Theory of Probability, Vol. I. London: Wiley (Original work published 1970).

Duda, R., Gaschnig, J., & Hart, P. (1979), Model design in the prospector consultant system for mineral exploration, in: D.Michie (Ed.), Expert Systems in the Micro Electonic Age, Edinburgh: Edinburgh University Press, pp.153-167.

Evans, J.S.B.T. (1982), The Psychology of Deductive Reasoning, London: Routledge.

Finke, R.A. (1986), Bildhaftes Vorstellen und visuelle Wahrnehmung, Spektrum der Wissenschaft,(Mai (5)), 78-86.

Futo, I., Darvas, F., & Szeredi, P. (1978), The application of prolog to the development of QA and DBM systems, in: H.Gallaire & J.Minker (Eds.), Logic and Data Bases, New York: Plenum, pp.347-376.

Genesereth, M.R. (1984), Partial programs, Stanford University, Computer Science Department Stanford, California 94305 Heuristic Programming Project Report HPP-84-1.

Genesereth, M.R. & Nilsson, N.J. (1987), Logical Foundations of Artificial Intelligence, Preliminary Draft, Stanford University, Department of Computer Science.

Gentner, D. & Collins, A. (1981), Studies of inference from lack of knowledge, Memory & Cognition, 9, 434-443.
Ginsberg, M.L. (1984), Analyzing incomplete information, Department of Computer Science, Stanford University Standford HPP 84-17.
Glucksberg, S. & McCloskey, M. (1981), Decisions about ignorance: knowing that you don't know, Journal of Experimental Psychology: Human Learning and Memory, 7, 311-324.
Glymour, C. (1985), Independence assumptions and Bayesian updating, Artificial Intelligence, 25, 95-99.
Graesser, A.C. & Clark, L.F. (1985), Structures and Procedures of Implicit Knowledge, Norwood, NJ: Ablex.
Harvey, L.D., Jr. (1986), in: F.Klix & H.Hagendorf (Eds.), Human Memory and Cognitive Capabilities Amsterdam: North-Holland.
Hogarth, R.G. (1975), Cognitive processes and the assessment of subjective probability distributions, Journal of the American Statistical Association, 70, 271-289.
Huber, O. (1983), The information presented and actually processed in a decision task, in: P.Humphreys, O.Svenson, & A.Vari (Eds.), Analysing and Aiding Decision Processes, Amsterdam: North-Holland, pp.441-454.
Humphreys, P. & Berkeley, D. (1983), Problem structuring calculi and levels of knowledge representation in decision making, in: R.W.Scholz (Ed.), Decision Making under Uncertainty, Amsterdam: North-Holland, 121-157.
Kadesch, R.R. (1986), Subjective inference with multiple evidence, Artificial Intelligence, 28, 333-341.
Kahneman, D. & Tversky, A. (1972), Subjective probability: a judgment of representativeness, Cognitive Psychology, 3, 430-454.
Kahneman, D., Slovic, P., & Tversky, A. (Eds.) (1982), Judgment under uncertainty: Heuristics and biases, New York: Cambirdge Univ.Press.
Kleiter, G.D. (1981), Bayes Statistik, Berlin: de Gruyter.
Kosslyn, S.M. (1980), Image and Mind, Harvard: Harvard University Press.
Kosslyn, S.M. (1984), Mental representation, in: J.R.Anderson & S.M.Kosslyn, Tutorials in Learning and Memory, San Francisco: Freeman, pp.91-117.
Kuipers, B.J. (1975), A frame for frames: representing knowledge for recognition, in: D.G.Bobrow & A.Collins (Eds.), Representation and Understanding, New York: Academic Press, pp.151-184.
Lehnert, W. & Wilks, Y. (1979), A critical perspective on KRL, Cognitive Science, 3, 1-28.
Lifschitz, V. (1985), Closed-world databases and circumscription, Artificial Intelligence, 27, 229-235.
Marr, D. (1982), Vision, San Francisco: Freeman.
McCarthy, J. (1980), Circumscription- a form of non-monotonic reasoning, Artificial Intelligence, 13, 27-39.
McCarthy, J. (1980), Addendum: Circumscription and other non-monotonic formalisms, Artificial Intelligence, 13, 171-172.
McCarthy, J. (1986), Applications of circumscription to formalizing commonsense knowledge, Artificial Intelligence, 28, 89-116.
McDermott, D. & Doyle, J. (1980), Non-monotonic logic I, Artificial Intelligence, 13, 41-72.
McDermott, J. & Forgy, C. (1978), Production system conflict resolution strategies, in: D.A.Waterman & F.Hayes-Roth (Eds.), Pattern Directed Inference Systems, New York: Academic Press, pp.177-199.
Medin, D.L. & Smith, E.E. (1984), Concepts and concept formation, Annual Review of Psychology, 35, 113-138.
Minsky, M. (1975), A framework for representing knowledge, in: P.H.Winston (Ed.), The Psychology of Computer Vision New York: McGraw-Hill.
Moore, R.C. (1985), Semantical considerations on nonmonotonic logic, Artificial Intelligence, 25, 75-94.
Nilsson (1986), Probabilistic Logic, Artificial Intelligence, 28, 21-87.
Norman, D.A. (1973), Memory, knowledge, and the answering of questions, in: R.L.Solso (Ed.), Contemporary Issues in Cognitive Psychology: The Loyola Symposium, Washington: Winston, pp.135-165.

Osherson, D.N. & Smith, E.E. (1981), On the adequacy of prototype theory as a theory of concepts, Cognition, 9, 35-58.

Osherson, D.N. & Smith, E.E. (1982), Gradedness and conceptual combination, Cognition, 12, 299-318.

Pearl, J. (1986), On evidential reasoning in a hierarchy of hypotheses, Artificial Intelligence, 28, 9-15.

Pednault, E., Zucker, S., & Muresan, I. (1981), On the independence assumption underlying subjective Bayesian updating, Artificial Intelligence, 16, 213-222.

Quillian, M.R. (1968), Semantic memory, in: M.Minsky (Ed.), Semantic Information Processing Cambridge, MA: MIT Press.

Reiter, R. (1978), On closed world data bases, in: H.Gallaire & J.Minker (Eds.), Logic and Data Bases, New York: Plenum Press, pp.55-76.

Reiter, R. (1980), A logic for default reasoning, Artificial Intelligence, 13, 81-132.

Robinson, E.J. & Robinson, W.P. (1982), Knowing when you don't know enough: children's judgments about ambiguous information, Cognition, 12, 267-280.

Robinson, J.A. (1979), Fundamentals of machine-oriented deductive logic, in: D.Michie (Ed.), Introductory Readings in Expert Systems, New York: Gordon and Breach, pp.81-92.

Robinson, J.A. (1979), The logical basis of programming by assertion and query, in: D.Michie (Ed.), Expert Systems in the Micro Electronic Age, Edinburgh: Edinburgh University Press, pp.105-121.

Rosch, E. (1975), Cognitive representations of semantic categories, Journal of Experimental Psychology, General, 104, 192-233.

Rumelhart, D.E., Lindsay, P.H., & Norman, D.A. (1972), A process model for long-term memory, in: E.Tulving & W.Donaldson (Eds.), Organization of memory, New York: Academic Press, pp.198-246.

Schank, R.C. (1982), Dynamic Memory, Cambridge: Cambridge University Press.

Schank, R.C. (1982), Depth of knowledge, in: B.de Gelder, Knowledge and Representation, London: Routledge & Kegan Paul, pp.170-193.

Schank, R.C. & Abelson, R.P. (1977), Scripts plans goals and understanding, Hillsdale, NJ: Erlbaum.

Shepard, R.N. & Cooper, L.A. (1982), Mental Images and their transformations, Cambridge, MA: MIT Press.

Shepard, R.N., Kilpatric, D.W., & Cunningham, J.P. (1975), The internal representation of numbers, Cognitive Psychology, 7, 82-138.

Shortliffe, E.H. (1976), Computer-Based Medical Consultations: MYCIN, New York: Elsevier.

Slovic, P. & MacPhillamy, D. (1974), Dimensional commensurability and cue utilization in comparative judgment, Journal of Organizational Behavior and Human Performance, 11, 172-194.

Smith, E.E. & Medin, D.L. (1981), Categories and Concepts, Cambridge, MA: Harvard University Press.

Tversky, A. & Kahneman, D. (1983), Extensional versus intuitive reasoning: the conjunction fallacy in probability judgment, Psychological Review, 90, 293-315.

Warren, D. (1979), PROLOG on the DECsystem-10, in: Michie (Ed.), Expert Systems in the Micro Electronic Age, Edinburgh: Edinburgh University Press, pp.112-121.

Zadeh, L.A. (1982), A note on prototype theory and fuzzy sets, Cognition, 12, 291-297.

COMPARISON OF SOME STATISTICAL METHODS

FOR COUNTING PROCESS OBSERVATIONS

Giorgio Koch

Dept. Mathematics "Guido Castelnuovo"
University of Roma - La Sapienza
Rome, Italy

ABSTRACT

In reliability theory and survival analysis, the problem often arises of estimating unknown parameters affecting the failure rate,.or equivalently the intensity process for the observed counting process.

In the infinite dimensional parameter case, classical methods in statistics lead to maximum likelihood estimators (MLE), or to the heuristic but powerful Aalen estimators.

Bayesian methods are also quite effective and take advantage from the semimartingale theory and the filtering theory for counting process observations.

In the paper the three estimators are compared both on theoretical ground and application to specific examples. Conditions are provided for the coincidence of Aalen estimators and MLE. Then they are compared to the output of bayesian estimators (filters) with a convenient choice of the a priori distribution.

INTRODUCTION

The estimation problem. In reliability and survival analysis we often consider nonnegative random variables $T_1, T_2, \ldots, T_n, \ldots$ with the meaning of failure (birth, death,...) times.

Denoting by $T_1, T_2 \ldots$ the ordered statistics (to keep notation simple enough!), we also assume that:

$$\bar{F}_{n+1}(t) = 1 - F_{n+1}(t) = P(T_{n+1} > t | T_1, T_2, \ldots T_n) \tag{1}$$

is absolutely continuous with respect to Lebesgue measure, n=0,1,2... and is given by

$$\begin{aligned} \bar{F}_{n+1}(t) &= \exp\left(-\int_{T_n}^{t} \lambda_s ds\right), \quad t \geq T_n \\ &= 1 \quad\quad\quad\quad\quad\quad\quad\;\;, \quad t < T_n \end{aligned} \tag{2}$$

Eq. (2) can be reversed:

$$\lambda_t = \frac{f_{n+1}(t)}{\bar{F}_{n+1}(t)} \quad , T_n \leq t < T_{n+1} \tag{3}$$

where $f_{n+1}$ is the density of $F_{n+1}$. The quantity $\lambda_t$ is the failure (birth, death,...) rate, and of course $\lambda_t \equiv \lambda$ corresponds to the happy case of $\{T_i - T_{i-1}\}$ i.i.d. with exponential $\lambda$ distribution.

Quite often, however, we have to face the situation where:

$$\lambda_t = \lambda(X_t) \tag{4}$$

in which $\lambda$ is a known function and $X_t$ an unknown quantity (for instance a stochastic process). This happens if we have an estimation problem for the failure rate, or if the "system" affects the failure rate by its "state" $X_t$. The latter may play the role of an underlying disturbance parameter, or be attached a physically meaningful value on its own. Of course, the case $\lambda_t = X_t$ is also possible. Then (1) deserves a new definition:

$$\begin{aligned} \bar{F}_{n+1}(t) &= P(T_{n+1} > t \mid T_1, T_2, \ldots T_n;\ X_s,\ s \leq t) = \\ &= \exp\left(-\int_{T_n}^{t} \lambda(X_s)\,ds\right), \quad t \geq T_n \\ &= 1 \qquad\qquad t < T_n \end{aligned} \tag{5}$$

The problem now arises of estimating $X_t$ (or $\lambda(X_t)$) given $T_1, T_2, \ldots T_n \leq t$ (and $T_{n+1} > t$).

Here we prefer to discuss the above problem by looking at it from a dynamical point of view (Brémaud, 1981; Koch, 1985). That is, we substitute sequences $\{T_i\}$ of nonnegative random variables by trajectories of a counting process $Y_t$ defined as:

$$Y_t = \Sigma_i\, I_{(T_i \leq t)} \tag{6}$$

Given now a probability space $(\Omega, \mathcal{F}, P)$ supporting X and Y, and the flow of $\sigma$-algebras (filtration) $\{\mathcal{F}_t\}$, $\mathcal{F}_t \subset \mathcal{F}$:

$$\mathcal{F}_t = \sigma(Y_s, X_s,\ s \leq t)$$

we know that (6) enjoys the decomposition (Liptser and Shiryaev, 1978; Brémaud, 1981; Jacobsen, 1982; Del Grosso et al. 1986):

$$Y_t = \int_0^t \lambda(X_s)\,ds + M_t \tag{7}$$

In (7), $M_t$ is a $\{\mathcal{F}_t\}$ martingale, and therefore is such that

$$E(M_t \mid \mathcal{F}_s) = M_s \quad , \quad t \geq s \tag{8}$$

Thus the compensator $A_t$

$$A_t = \int_0^t \lambda(X_s)\,ds \tag{9}$$

fully accounts for the mean increment of $Y_t$ conditioned upon the past.

It follows that if we consider the smaller flow of $\sigma$-algebras $\{F_t^y\}$:

$$F_t^y = \sigma(Y_s, s \leq t)$$

then (Liptser and Shiryaev, 1978; Brémaud, 1981):

$$Y_t = E[Y_t | F_t^y] = \int_0^t E(\lambda(X_s) | F_s^y) ds + M'_t \tag{10}$$

where $M'_t$ is a $\{F_t^y\}$ martingale. Also:

$$E(Y_t) = E(A_t) = \int_0^t E(\lambda(X_s)) ds \tag{11}$$

The fact that the compensator of a counting process $Y_t$ is a random process with absolute continuous (with probability one) trajectories, as in (9), is equivalent to the existence of a density for $F_{n+1}$, n=0,1,2,... And its derivative $\lambda(X_t)$ is then by definition the intensity (process) of $Y_t$ (with respect to the flow $\{F_t\}$. From (10) it follows that $E[\lambda(X_t) | F_t^y]$ is the intensity of $Y_t$ with respect to $\{F_t\}$.

Again, by a theorem due to Watenabe, $\lambda(X_t) \equiv \lambda$ corresponds to the case of $Y_t$ Poisson process with mean value $\int_0^t \lambda ds = \lambda t$ (Brémaud, 1981). This dynamical point of view is clearly equivalent to the more traditional one, which uses sequences of nonnegative random variables. But it offers the advantage of direct and natural exploitation of results in the theory of martingales and stochastic processes, to the fullest convenient extent.

Taking this point, in this paper we discuss

- a heuristic estimator (Aalen)
- an instance of maximum likelihood estimator
- a bayesian estimator (filter)

All of them enjoy a recursive structure and (of course) offer interesting features. But, as it will be evident, the sounder the statistical meaning of the estimator is, the heavier the effort is requested in the modeling of the problem and computing of the solution. The nice thing is that even in the bayesian estimator this effort still appears to be quite feasible. In some cases the estimators coincide, giving some clues for a more comprehensive analysis to be carried out.

The product model. For sake of simplicity, we take the following product model (Aalen, 1978; Jacobsen, 1982) as a reference model for all estimators (but most results do hold in wider generality):

$X_t$ is one-dimensional, $\forall t$

$Y_t = (Y_t^{(1)}, Y_t^{(2)}, \ldots, Y_t^{(r)})$ is an r-dimensional counting process, with independent components.

For each component $Y_t^{(i)}$, i = 1,2...r the intensity has the structure

$$\lambda_t^{(i)} = X_t \; Z_t^{(i)} \tag{12}$$

where $Z^{(i)}$ is measurable with respect to $F_t^{y^{(i)}} = \sigma\{Y_s^{(i)}, s \leq t\}$. A possible interpretation is that we run r independent "experiments"; in each of them the intensity depends on the past of the experiment itself, via a common proportionality quantity ("unitary" rate) $X_t$.

Let us introduce the notations:

$$\tilde{Y}_t = \sum_1^r {}_i Y_t^{(i)} \quad , \quad \tilde{Z}_t = \sum_1^r {}_i Z_t^{(i)} \tag{13}$$

Then $\lambda(X_t) = X_t \cdot \tilde{Z}_t$ is the intensity for $\tilde{Y}_t$ (Jacobsen, 1982). By $R_t$ we denote the <u>risk set</u> at time t:

$$R_t \subset \{1,2\ldots r\} \quad : j \in R_t \iff Z_{t-}^{(j)} > 0$$

Let $T_i$ be the i-th jump time for $Y_t$.

We shall denote by $R_i$ the risk set at $T_i$ and by $|R_i|$ its cardinality. Finally by $a^+$ we denote the generalized inverse:

$$a^+ = a^{-1} I_{(a>0)} \quad \begin{array}{ll} = a^{-1} & , a \neq 0 \\ = 0 & , a = 0 \end{array} \tag{14}$$

The following technical requirements are introduced on Z (Jacobsen, 1982):

$$Z_t^{(i)} \leq a + b\, Y_t^{(i)} \quad , \forall\, t,\ i = 1,2,\ldots r \quad \text{for some nonnegative constants } a,b \tag{15}$$

(this prevents $Y_t$ from explosion in a finite time)

$$\sup_{0 \leq t \leq T} |Z_{t-}^{(i)}|^+ \leq K_T^{(i)}, \ \forall\, T > 0,\ i = 1,2,\ldots \text{ for some nonnegative constant } K_T \tag{16}$$

Let us mention <u>some</u> (classical) <u>examples</u> to support the relevance of the model.

(i) $T^{(1)}$ is a failure time with distribution

$$F(t) = P(T^{(1)} \leq t) = 1 - \exp(-\textstyle\int_0^t X_s\, ds) \tag{17}$$

$Y_t^{(1)}$ jumps from 0 to 1 at $T^{(1)}$, being constant otherwise. Then

$$Y_t^{(1)} = I_{(T^{(1)} \leq t)} \tag{18}$$

and (12) holds with r = 1 and

$$Z^{(1)} = 1 - Y_t^{(1)} \tag{19}$$

(ii) Take $T^{(1)}, T^{(2)}, \ldots T^{(r)}$ independent failure times identically distributed according to (17). Repeat (i) for 1,2,...r. Then again (12) holds for all components of $Y_t$ and (13) yields:

$$\tilde{Y}_t = \sum_1^r {}_i I_{(T^{(i)} \leq t)} \quad ; \quad \tilde{Z}_t = r - \tilde{Y}_t \tag{20}$$

(iii) Take $U_1, U_2, \ldots U_r$ (known) censoring times and consider again the example (ii), with each i-th component censored at $U_i$. Then we easily check that:

$$\tilde{Y}_t = \sum_1^r {}_i I_{(T^{(i)} \leq t \wedge U_i)} \quad ; \quad \tilde{Z}_t = \sum_1^r {}_i I_{(T^{(i)} \wedge U_i > t)} \tag{21}$$

<u>The definition</u>. In a classical statistical framework, X is taken as an unknown deterministic function with very weak constraints described by the assumption:

$$X \in H,\ H = \{f : [0,\infty) \to (0,\infty),\ f \text{ right continuous with left limits},\ \int_0^t X_s ds < \infty,\ \forall t \geq 0\} \qquad (22)$$

Thus we assume we virtually know nothing about X.

Due to the same reasons that suggest us to switch from the estimation problem of a probability density to the estimation problem of its distribution function (the estimate will eventually be a discontinuous distribution function), we look for an estimate of $\int_0^t X_s ds$. Going one step further, the following quantity is taken as a more sensible goal for our estimation problem (Jacobsen, 1982):

$$\xi_t = \int_0^t X_s I_{(\tilde{Z}_s > 0)} ds \qquad (23)$$

Indeed, being the constraints on X so weak, no information can be obtained on $X_t$ if $\tilde{Z}_t = 0$ so that all components of $Y_t$ have zero intensity anyway. The following estimate of (23) is suggested on heuristic bases:

$$\hat{\xi}_t = \int_0^t \frac{1}{\tilde{Z}_{s-}} I_{(\tilde{Z}_{s-} > 0)} d\tilde{Y}_s = \sum_{T_i \leq t} \tilde{Z}^{+}_{T_i -} \qquad (24)$$

Notice how in (24) we stay "neutral" in between jump times, while the information carried by jumps is immediately used in the estimate, to update its actual (and therefore future) value. But we never "look backward" (no smoothing).

<u>Properties of the estimator</u>. Now it can be proved from martingale theory (Aalen, 1978; Jacobsen, 1982) that $\forall X$, $\hat{\xi}_t - \xi_t$ is a $P_x$ martingale where $P_x$ is the measure induced by $Y_t$, whose components have intensities (12), on the space of r-dimensional counting process trajectories. In particular ($E_x$ being the mean value with respect to $P_x$):

$$E_x(\xi_t - \hat{\xi}_t) = 0 \qquad (25)$$

Also, we can obtain estimates for the variance:

$$E_x((\xi_t - \hat{\xi}_t)^2) \simeq \int_0^t (\tilde{Z}^2_{s-})^{+} d\tilde{Y}_s = \sum_{T_i \leq t} \tilde{Z}^{2+}_{T_i -} \qquad (26)$$

and consequently build an asymptotic theory and hypothesis testing.

As we said, such an estimator of $\xi_t$ is <u>quite heuristic</u>. Indeed, rather than $X_t$ we estimate:

$$E_x(\xi_t) = E_x(\int_0^t X_s I_{(\tilde{Z}_s > 0)} ds) \qquad (27)$$

In the case of example (ii) above, it is not difficult to show that (27) leads to:

for $r = 1$ : $E_x(\xi_t) = 1 - \exp(-\int_0^t X_s ds) = F_1(t)$

for $r > 1$ : $E_x(\xi_t) = \sum_1^r k \frac{[F_1(t)]^k}{k}$

But clearly $\hat{\xi}_t$ has the advantage of a very easy computation with a recursive feature (notice how $\tilde{Z}_s$, $\tilde{Y}_s$, $s \leq t$ are sufficient statistics):

$$\hat{\xi}_t \text{ is constant for } T_{i-1} \leq t < T_i$$

and updated at $T_i$ by:

$$\Delta\hat{\xi}_{T_i} = \tilde{Z}^+_{T_i^-} \tag{28}$$

In the above examples, (28) reduces to $|R_i|^{-1}$.

Notice however that we estimate a continuous process $\xi_t$ by a (piecewise constant) discontinuous quantity $\hat{\xi}_t$. Indeed, the original model, defined by

$$A^{(i)} = \int_0^t X_s Z^{(i)} ds \quad ; \quad X_s \in H \tag{29}$$

has been implicitely extended to:

$$A_t^{(i)} = \int_0^t Z_{s-}^{(i)} d\bar{X}_s \quad ; \quad \bar{X}_s \in \bar{H} \tag{30}$$

where:

$$\bar{H} = \{f : [0,\infty) \to [0,\infty),\ f(0) = 0, \text{ right continuous, nondecreasing bounded variation}\}$$

Then $\hat{\xi}_t$ is an estimate of $\bar{X}_t$ over the set $\{t: \tilde{Z}_{t-} > 0\}$.

The data obviously follow a model as in (29), which corresponds to an absolutely continuous compensator, and therefore to inaccessible jump times for $Y_t$. On the contrary, the extended model allows for discontinuous compensators, and therefore for not inaccessible jump times. Moreover differently from (29) the model (30) allows for nonzero probability for more than one component of Y jumping at the same time.

## A MAXIMUM LIKELIHOOD ESTIMATOR

The estimator. For each $X \in H$, and $t > 0$, let $P_{x,t}$ be the restriction of $P_x$ to $F_t^y$; and let $P_t$ be the similar restriction of the measure induced by an r-dimensional standard Poisson process.

Since $A_t$ is absolutely continuous, $P_{x,t}$ is dominated by $P_t$ and we can define the likelihood functional $L_t = dP_{x,t}/dP_t$, which, for each Y, takes the value:

$$\begin{aligned} L_t(Y) = \frac{dP_{x,t}}{dP_t}(Y) &= \prod_{i=1}^r \exp\{\int_0^t \ln \lambda_{s-}^{(i)} dY_s^{(i)} - \int_0^t (\lambda_s^{(i)} - 1) ds\} = \\ &= c \cdot \exp\{\int_0^t \ln X_{s-} d\tilde{Y}_s - \int_0^t X_s \tilde{Z}_s ds\} \end{aligned} \tag{31}$$

where c is a constant which does not depend on X.

Expression (31) evidentiates the sufficiency of $\tilde{Z}_s$, $\tilde{Y}_s$, $s \leq t$. However, if we attempt to maximize the value of $L_t$, at the observed trajectory Y, over H to achieve a maximum likelihood estimate of X, we would find an X which vanishes outside the set $\{T_i\}$ of jump times for Y, while $X_{T_i-}$ diverges to $+\infty$, i = 1,2...

Thus, again, we better switch to the model (30). As a matter of fact, (30) does not define compensators correctly, since we have to prevent compensators to have jump of size larger than 1 (Liptser and Shiryaev, 1978). Moreover, we have to enlarge the trajectory space for Y, and include trajectories with more than one component jumping at the same time.

We adopt the model:

$$A_t^{(i)} = \int_o^t Z_{s-}^{(i)} \, I_{(Z_{s-}^{(i)} \Delta \bar{X}_s \leq 1)} \, d\bar{X}_s \tag{32}$$

which now for each $\bar{X} \in \bar{H}$ properly defines a (unique) probability measure $P_{\bar{X}}$ on the enlarged trajectory space. Let $P_{\bar{X},t}$ be the restriction of $P_{\bar{X}}$ up to time t. For a discontinuous $\bar{X}$, $P_{\bar{X}}$ (Y has more than one component jumping at the same time) > 0; thus $P_{\bar{X},t}$ is not dominated by Poisson measure any more and a new definition of maximum likelihood estimate is called for (Jacobsen, 1984).

Let us introduce the notation:

$$C_i \subset \{1,2,\ldots r\} : j \in C_i \iff \Delta Y_{T_i}^{(j)} = +1 \tag{33}$$

($C_i$ is the index set of those components which jump at $T_i$)

Also, $\bar{X}^c$ denotes the continuous part of $\bar{X}$, and we use the shorthand notation:

$$Z_{s-}^{(i)}(\bar{X}) = Z_{s-}^{(i)} \, I_{Z_{s-}^{(i)} \Delta \bar{X}_s \leq 1)} \tag{34}$$

Then we obtain from (32):

$$\begin{aligned} P_{\bar{X},t}(Y) = {} & \Pi_j [\{\exp(-\int_{T_m}^t Z_{s-}^{(j)}(\bar{X}) d\bar{X}_s^c)\} \prod_{T_m < s \leq t} (1 - Z_{s-}^{(j)}(\bar{X}) \Delta \bar{X}_s)] \cdot \\ & \cdot \prod_{j \in C_m} (Z_{T_m}^{(j)}(\bar{X}) \Delta \bar{X}_{T_m}) \prod_{j \in R_m - C_m} (1 - Z_{T_m-}^{(j)}(\bar{X}) \Delta \bar{X}_{T_m}) \cdot \\ & \cdot \Pi_j [\{\exp(-\int_{T_{m-1}}^{T_m} Z_{s-}^{(j)}(\bar{X}) d\bar{X}_s^c)\} \prod_{T_{m-1} < s < T_m} (1 - Z_{s-}^{(j)}(\bar{X}) \Delta \bar{X}_s) \ldots \\ & \cdot \Pi_j [\{\exp(-\int_o^{T_1} Z_{s-}^{(j)}(\bar{X}) d\bar{X}_s^c)\} \prod_{0 < s < T_1} (1 - Z^{(j)}(\bar{X}) \Delta \bar{X}_s), \quad T_m \leq t < T_{m+1} \end{aligned} \tag{36}$$

From (36) it appears that for any trajectory Y in the enlarged trajectory space (except for trajectories such that $Z_{T_i-}^{(j)} = 0$, for some i and some $j \in C_i$) and for any t, there is an $\bar{X} \in \bar{H}$ such that $P_{\bar{X},t}(Y) > 0$. As a matter of fact it is enough to take $\bar{X}$ such that:

$$\Delta \bar{X}_{T_i} \neq (Z_{T_i-}^{(j)})^{-1}, \ \forall \ j \in R_i - C_i \ ; \ \Delta \bar{X}_s \neq (Z_{s-}^{(j)})^{-1}, \ j \in R_{T_i}, \ T_i < s < T_{i+1}$$

$$0 < \Delta \bar{X}_{T_i} \leq (\max_{j \in C_i} Z_{T_i-}^{(j)})^{-1} \tag{37}$$

We then define the maximum likelihood estimate of $\bar{X}$ as an $\hat{\bar{X}} \in \bar{H}$ for which at the observed Y:

$$P_{\hat{\bar{X}},t}(Y) \geq P_{\bar{X},t}(Y) \ , \quad \forall\ \bar{X} \in \bar{H} \tag{38}$$

Again, in order to get uniqueness, we consider:

$$\bar{\xi}_t = \int_0^t I_{(\tilde{Z}_{s-} > 0)}\ d\bar{X}_s \tag{39}$$

and we notice that, $\forall\ \bar{X}$, the model (32) generates the same measure $P_{\bar{X}}$ as (and therefore is indistinguishable from) the <u>model</u>:

$$A_t^{(i)} = \int_0^t Z_s^{(i)}\ I_{(Z_{s-}^{(i)}\ \Delta\bar{\xi}_s \leq 1)}\ d\bar{\xi}_s \tag{40}$$

We then substitute $\bar{\xi}_t$ for $\bar{X}_t$ in (35) and look for the estimate $\bar{\xi}$ such that

$$P_{\hat{\bar{\xi}},t}(Y) \geq P_{\bar{\xi},t}(Y) \ , \quad \forall\ \bar{\xi} \text{ of the type (39)} \tag{41}$$

<u>The maximization</u>. Looking for the maximum value of $P_{\bar{\xi},t}(Y)$ immediately leads to:

$$\hat{\bar{\xi}}_s = \text{constant} \quad T_{i-1} < s < T_i \tag{42}$$

We are then left, for each $T_i$, with the maximization problem for:

$$\prod_{j \in C_i} (Z_{T_i^-}^{(j)}(\bar{\xi})\ \Delta\bar{\xi}_{T_i}) \prod_{j \in R_i - C_i} (1 - Z_{T_i^-}^{(j)}(\bar{\xi})\ \Delta\bar{\xi}_{T_i}) \tag{43}$$

Due to (37), (43) is positive only for $\Delta\xi_{T_i}$ belonging to one of the intervals:

$$(0, x_1),\ (x_1, x_2), \ldots\ (x_k, x_{k+1}]$$

where $x_{k+1} = (\max_{j \in C_i} Z_{T_i^-}^{(j)})^{-1}$ and $x_1, x_2, \ldots x_k$ are those (ordered) values $(Z_{T_i^-}^{(j)})^{-1}$, $j \in R_i - C_i$ which occur to fall in $(0, x_{k+1}]$.

Now, if $R_i - C_i = \emptyset$, the maximum of (43) is achieved at:

$$\hat{\Delta\xi}_{T_i} = (\max_{j \in R_i} Z_{T_i^-}^{(j)})^{-1} \tag{44}$$

Otherwise, in the first interval $(0, x_1)$, the factor (43) becomes:

$$\prod_{j \in C_i} (Z_{T_i^-}^{(j)}\ \Delta\bar{\xi}_{T_i}) \prod_{j \in R_i - C_i} (1 - Z_{T_i^-}^{(j)}\ \Delta\bar{\xi}_{T_i}) \tag{45}$$

and its maximum is achieved (Del Grosso et al., 1986) at the unique solution of:

$$|R_i| = \sum_{h = R_i - C_i} [1 - Z_{T_i^-}^{(h)}\ \hat{\Delta\xi}_{T_i}]^{-1} \tag{46}$$

In each of the additional intervals $(x_\ell, x_{\ell+1})$, $\ell = 1,2 \ldots k$, the factor (43) would appear to be:

$$\prod_{j \in C_i} (Z_{T_i^-}^{(j)}\ \Delta\bar{\xi}_{T_i}) \prod_{j \in R^{(\ell)} - C_i} (1 - Z_{T_i^-}^{(j)}\ \Delta\bar{\xi}_{T_i}) \tag{47}$$

where $R^{(\ell)}$ denotes the index set of all those components for which $Z^{(j)}_{T_i-}(\bar{\xi}) > 0, \Delta\bar{\xi}_{T_i} \in (x_\ell, x_{\ell+1})$. And the supremum of (47) is achieved at the solution of:

$$|R_i^{(\ell)}| = \sum_{h \in R_i^{(\ell)} - C_i} [1 - Z^{(h)}_{T_i-} \Delta\hat{\xi}_{T_i}|^{-1} \tag{48}$$

or coincides with the value of (47) at $x_\ell$ if (48) does not have a solution. But we shall not consider the maxima of (43) outside $(0,x_1)$, since it is only in $(0,x_1)$ that the risk set of the original product model (30) coincides with that one of the model (32) which is not a product model. Additional maxima of (43) outside $(0,x_1)$ are then to be taken as artifacts introduced by (32).

Summing up, the maximum likelihood estimate of $\bar{\xi}$ may be represented as:

$$\hat{\bar{\xi}}_t = \int_0^t I_{(\tilde{Z}_{s-} > 0)} f_{s-} d\tilde{Y}_s \tag{49}$$

where f is a real nonnegative process, $\{F_t^y\}$-measurable, such that:

$$f_{T_i-} = \frac{\Delta\hat{\bar{\xi}}_{T_i}}{|C_i|}, \tag{50}$$

and $\Delta\hat{\xi}_{T_i}$ is the solution of (46) in $(0,(\max_{j \in R_i} Z^{(j)}_{T_i-})^{-1})$. (We refer to Jacobsen, 1984 for another approach).

The estimate (49) is a little less easy to be computed than (24)(at each $T_i$, we must solve a nonlinear equation). Still, it enjoys a recursive feature (now the sufficient statistics are $Y_s$, $Z_s$, $s \leq t$) and the only difference from (24) is in the jump size.

However, it is clear from (46) that anytime we have:

$$Z^{(j)}_{T_i-} = z_i \quad , \forall\, j \in R_i - C_i \tag{51}$$

(that is any time the Y-dependent factor in the intensity is the same for all components at risk which do not jump) it follows:

$$\Delta\hat{\xi}_{T_i} = \frac{|C_i|}{\tilde{Z}_{T_i-}} \tag{52}$$

that is the maximum likelihood estimate $\hat{\xi}_t$ and the Aalen estimate $\hat{\xi}_t$ coincide. This trivially occurs, for instance, whenever Y is one dimensional. (Also notice that for a trajectory following the original model (29), $|C_i| = 1$).

Some examples. Let us go back to example (ii) of the previous section, with $\tilde{Y}$ and $\tilde{Z}$ given by (20). Assume we only observe $\tilde{Y}$ (one dimensional). Then both estimators yield:

$$\hat{\xi}_t = \hat{\bar{\xi}}_t = \sum_{T^{(i)} \leq t} \frac{1}{\tilde{Z}_{T^{(i)}-}} = \sum_{i=1}^{\tilde{Y}_t \wedge r} \frac{1}{r-(i-1)} \tag{53}$$

that is the popular Nelson estimator (Jacobsen, 1982).

The same result is achieved if we observe the whole r-dimensional process Y, since (51) holds with $z_i = 1$.

But suppose now that the r components of Y are put in two groups, and we are only able to observe $(r_1 + r_2 = r)$:

$$\tilde{Y}_t^{(i)} = \sum_1^{r_1} Y_t^{(i)} \quad , \quad \tilde{Y}_t^{(2)} = \sum_{r_1+1}^{r_1+r_2} Y_t^{(i)} \tag{54}$$

Then (51) is not guaranteed any more, since the intensities for the two processes (54) are $\tilde{\lambda}_t^{(1)} = X_t \tilde{Z}_t^{(1)}$, $\tilde{\lambda}_t^{(2)} = X_t \tilde{Z}_t^{(2)}$, with:

$$\tilde{Z}_t^{(1)} = r_1 - \tilde{Y}_t^{(1)} \quad , \quad \tilde{Z}_t^{(2)} = r_2 - \tilde{Y}_t^{(2)} \tag{55}$$

The Aalen estimator again yields:

$$\hat{\xi}_t = \sum_{T^{(i)} \leq t} \frac{1}{\tilde{Z}_{T^{(i)}-}} = \sum_1^{\tilde{Y}_t \wedge r} \frac{1}{r_1 - \tilde{Y}^{(1)}_{T_i-} + r_2 - \tilde{Y}^{(2)}_{T_i-}} \tag{56}$$

while (46) is solved by:

$$\Delta\hat{\hat{\xi}}_{T_i} = \frac{1}{2(r_h - \tilde{Y}^{(h)}_{T_i-})} \quad , \quad \text{for } h \in R_i - C_i, \tag{57}$$

if $|R_i - C_i| = 1$, and $\Delta\hat{\hat{\xi}}_{T_i} = \Delta\hat{\xi}_{T_i}$ if $R_i - C_i = \emptyset$. Similarly, if we observe $(r_1+r_2+r_3 = r)$:

$$\tilde{Y}_t^{(1)} = \sum_1^{r_1} Y_t^{(i)}, \quad \tilde{Y}_t^{(2)} = \sum_{r_1+1}^{r_1+r_2} Y_t^{(i)} \quad , \quad \tilde{Y}_t^{(3)} = \sum_{r_1+r_2+1}^{r_1+r_2+r_3} Y_t^{(i)} \tag{58}$$

we get:

$$\hat{\xi}_t = \sum_{T^{(i)} \leq t} \frac{1}{\tilde{Z}_{T^{(i)}-}} = \sum_1^{\tilde{Y}_t \wedge r} \frac{1}{r_1 - \tilde{Y}^{(1)}_{T_i-} + r_2 - \tilde{Y}^{(2)}_{T_i-} + r_3 - \tilde{Y}^{(3)}_{T_i-}} \tag{59}$$

while (46) is solved by:

$$\Delta\hat{\hat{\xi}}_{T_i} = \frac{(r_{h_1} - \tilde{Y}^{(h_1)}_{T_i-}) + (r_{h_2} - \tilde{Y}^{(h_2)}_{T_i-}) - \sqrt{(r_{h_1} - \tilde{Y}^{(h_1)}_{T_i-})^2 + (r_{h_2} - \tilde{Y}^{(h_2)}_{T_i-})^2 - (r_{h_1} - \tilde{Y}^{(h_1)}_{T_i-})(r_{h_2} - \tilde{Y}^{(h_2)}_{T_i-})}}{3(r_{h_1} - \tilde{Y}^{(h_1)}_{T_i-})(r_{h_2} - \tilde{Y}^{(h_2)}_{T_i-})} \tag{60}$$

for $h_1, h_2 \in R_i - C_i$

if $|R_i - C_i| = 2$, and $\Delta\hat{\hat{\xi}}_{T_i}$ as in the previous examples if $|R_i - C_i| = 1$ or 0.

The censored case may be dealt with in a similar way.

## A BAYES ESTIMATE (FILTER)

The model for X. To set up a Bayes estimator, we now need to supplement the model with a distribution on the trajectory space for Y.

In the usual formulation of non parametric Bayes problem (Ferguson, 1973, Leonard, 1978; Thorburn, 1986), one puts a distribution on a space of probability measures and then finds (possibly analytically) the posterior distribution of the measure $\mu$ given the data and assumed the latter ones to be i.i.d. with distribution $\mu$.

In the filtering set up, we circumvent the problem of assigning a distribution or an infinite dimensional space, such as the space of $\mu$. Rather, we model a finite dimensional parameter X (the state) by assuming it to be (for instance) a Markov process, a semi-martingale, a solution for a stochastic differential equation. Then we link the distribution of $Y_t$ to $X_s$, $s \leq t$ by a suitable model for Y and we look for an estimate of $X_t$ itself rather than of the distribution of Y (the state value is interesting on its own).

Possible dynamical (Markov) models for X are obtained by assuming it to be the solution of ($t \geq 0$):

a) $X_t = X_o$ , that is $X_t$ = const, with a given distribution $\pi_o$ for $X_o$

b) $X_t = \int_0^t b(X_s)ds + \int_0^t c(X_s)dw_s$, that is a diffusion ($w_t$ Wiener process)

c) $X_t = \int_0^t \int_U K(X_{s-},u)N(ds,du)$, that is a jump process ($N_t$ Poisson random measure)

d) a linear combination of a), b), c).

This model accomodates for any Markov additive semimartingale (Cinlar et al., 1980).

General conditions for existence and uniqueness of solutions are found in Gikhman and Skorokhod, 1979; Liptser and Shiryaev, 1978; Jacod, 1979; Athreya, Kliemann and Koch, 1986.

Of course, in our case, provisions are required on the equation for X or on its boundary conditions, in order to guarantee $X_t \geq 0$, $\forall\ t > 0$.

<u>The filtering problem</u>. Taking for Y the model (7) suffices to define the distribution of its jump times and therefore the probability measure on the space of its trajectories (given X).

Let us introduce the notation $\pi_t$ for the conditional distribution of $X_t$ given $F_t^Y$. From now on by $\widehat{f(X_t)}$ we shall denote the conditional mean with respect to $\pi_t$:

$$\widehat{f(X_t)} = \int f(x)\, d\pi_t(x) \tag{61}$$

The following remarks are worthwhile:

(i) the solution of the filtering problem is $\pi_t$

(ii) since now X is given a dynamical model, $\pi_t$ does not stay constant in between jump times of Y (we don't stay "neutral")

(iii) since now X is given a dynamical model, its estimate turns out to be smooth: there is no need here of using integrals of X, nor of artificially extend its trajectory space.

(iv) $\pi_t$ provides us with estimates of $\lambda(X_t)$, such as $\widehat{\lambda(X_{t^+})}$, and therefore with the failure rate of $Y_t$ given its own past as it follows from (10) and (5):

$$P(T_{n+1} \geq t | T_1, T_2, \ldots T_n) = \exp\left(-\int_{T_n}^t \widehat{\lambda(X_s)}ds\right), \quad T_n \leq t < T_{n+1} \tag{62}$$

(v) again, the estimate does not look backward (no smoothing).

The solution. Assume $\pi_t$ has a density $p_t$ (otherwise we deal with weak solutions). Then $p_t$ enjoys the following representation (Liptser and Shiryaev, 1978, Brémaud, 1981):

In between jump times, $T_{i-1} \leq t < T_i$:

$$p_t(x) = p_{T_{i-1}}(x) + \int_{T_{i-1}}^{t} L^* p_s(x)ds - \int_{T_{i-1}}^{t} R^* p_s(x)ds - \\ - \int_{T_{i-1}}^{t} x \tilde{Z}_s p_s(x)ds + \int_{T_{i-1}}^{t} \hat{X}_s \tilde{Z}_s p_s(x)ds \tag{63}$$

$p_o(x)$ = density of $\pi_o$

At jump times:

$$p_{T_i}(x) = (\hat{X}_{T_{i-}} Z^{(j)}_{T_{i-}})^{-1} (x Z^{(j)}_{T_{i-}} p_{T_{i-}}(x) + R^* p_{T_{i-}}(x)), \ j \in C_i \tag{64}$$

In (63) L denotes the generator of X, and L* its adjoint; R* is the (adjoint of the) quadratic covariance operator between martingale parts of X and Y, which is zero if X, Y do not jump at the same time (with probability 1). In (64) j is the only index in $C_i$, since admissible trajectories for the model (7) do jump once at a time.

Eq. (63) is in general a nonlinear (because of the term in $\hat{X}$) deterministic integro-differential equation; (64) provides a nonlinear instantaneous updating.

In Kliemann et al., 1986, eq. (63),(64) are proved to have a unique solution. In addition it is shown that this solution $p_t$ can be obtained via a linearization-normalization procedure:

- Solve for the linear problem:

$$q_t(x) = q_{T_{i-1}}(x) + \int_{T_{i-1}}^{t} L^* q_s(x)ds - \int_{T_{i-1}}^{t} R^* q_s(x)ds - \\ - \int_{T_{i-1}}^{t} x \tilde{Z}_s q_s(x)ds \tag{65}$$

$$q_o(x) = p_o(x)$$

$$q_{T_i}(x) = x Z^{(j)}_{T_{i-}} q_{T_{i-}}(x) + R^* q_{T_{i-}}(x) \tag{66}$$

- Normalize:

$$p_t(x) = q_t(x) / \int q_t(x)dx \tag{67}$$

It clearly follows from the above that:
- the filter estimator provides a recursive estimate;
- besides updating at jump times, the estimate evolves according to (65). The solution of (65) may well require a significant computational burden, (especially as compared to the no computation case of previous estimates) but its deterministic and linear character do keep it at a feasible level;
- again, $Y_s$, $Z_s$, $s \leq t$ provide a sufficient statistic.

A finite dimensional example. In some cases, (65), (66) admit a finite dimensional solution. Let us consider the following problem (r = 1):

$$X_t = X_o \tag{68}$$

$$Y_t = \int_o^t X_s(n - Y_s)ds + M_t \qquad (69)$$

and take for the initial distribution $\pi_o$ of $X_o$ a gamma distribution $\Gamma(\alpha,\beta)$ with parameter $\alpha$ and $\beta$ (for a motivation of this example see Koch and Spreij,1983). Then $L^* = R^* = 0$ and (65), (66), (67) easily lead to the solution:

$$\pi_t = \Gamma(\alpha + Y_t, \beta + (n-Y_t)t + \Sigma_1^{Y_t} T_i) \qquad (70)$$

Thus the two parameters in the gamma distribution (70) are enough to describe the evolution of the whole $\pi_t$. Furthermore, the mean value $\hat{X}_t$ of (70) can be checked to be the solution of:

$$d\hat{X}_t = (\beta + (n-Y_t)t + \Sigma_1^{Y_t} T_i)^{-1} (dY_t - \hat{X}_t(n-Y_t)dt) \qquad (71)$$

$$\hat{X}_o = \alpha/\beta$$

Therefore, in this case we get an equation involving just the conditioned mean value. And for $t > 0$, the solution of (71) with $\alpha = \beta = 0$ coincides with the maximum likelihood estimate of $X_o$.

## CONCLUSIONS

We were able to find a general connection between the Aalen estimator and the maximum likelihood estimator, which hinges on the comparison between (24) and (49).

It would be interesting to further explore connections between maximum likelihood and filter estimators, in various distances of models for X and prior distributions fo $X_o$. This would permit achieving a general framework for estimators with counting process observations.

## REFERENCES

Koch, G., 1985, A dynamical approach to reliability theory, in: "Theory of Reliability", Nuovo Cimento.

Liptser, R.S., Shiryaev, A.N., 1978, "Statistics of random processes", v. I and II, Springer Verlag, New York.

Brémaud, P., 1981, "Point processes and queues. Martingale dynamics", Springer Verlag, New York.

Del Grosso, G., Gerardi, A., Koch, G., Nappo, G., Spizzichino, F., 1986, "Statistica matematica per processi di punto", Quaderni di Calcolo delle Probabilità, n.2, Dept. Mathematics, Univ. of Rome "La Sapienza".

Jacobsen, M., 1982, "Statistical analysis of counting processes",Springer Verlag, New York.

Aalen, O.O., 1978, Non parameter influence for a family of counting processes, Ann. Stat., 6, 701-726.

Jacobsen, M., 1984, Maximum likelihood estimation in the multiplicative intensity model: a survey, Int. Stat. Rev., 2, 193-207.

Ferguson, T.S., 1973, A Bayesian analysis of some nonparametric problems, Ann. Stat., 1, 209-230.

Leonard, T., 1978, Bensity estimation, stochastic processes and prior information (with discussion), J. R. Stat. Soc., B40, 113-46.

Thorburn,D., 1986, A Bayesian approach to density estimation, Biometrika, 73, 65-75.

Cinlar, Jacod, J., Protter, P., Sharpe, M.J., 1980, Semimartingales and Markov processes, Z. Wahrsch. ver. G., 54,161-219.

Jacod, J., 1979, "Calcul stochastique et problemes de martingales", Lecture Notes in Mathematics, n. 714, Springer Verlag, New York.

Athreya, K.B., Klieman, W., Koch, G., 1986, On sequential construction of solutions of stochastic differential equations with jump terms, to appear.

Gikhman, I.I., Skorokhod, A.V., 1979, "The theory of stochastic processes", v. III, Springer Verlag, New York.

Kliemann, W., Koch, G., Marchetti, F., 1986, On the unnormalized solution of the filtering problem for counting process observations, to appear.

Koch, G., Spreij, P.J.C., 1983, Software reliability as an application of martingale and filtering theory; EEE Trans. Reliab., R - 32.

# BAYES INFERENCE IN LIFE TESTS WHEN SAMPLES SIZES ARE FIXED OR RANDOM

G.S. Lingappaiah

Department of Mathematics
Sir George Williams Campus
Concordia University
Montreal, Canada

SUMMARY

This paper predicts the order statistics in the future sample in terms of order statistics in the earlier sample in a life test based on the exponential model. The sizes of the samples are either fixed or random variables, having the Poisson distribution. For the prediction purpose, predictive distributions are obtained. Three situations are considered such as (a) both the sample sizes are random (b) size of one sample is fixed while that of the other is a random variable (c) both sizes are fixed. For each of these three cases, Bayesian Prediction Regions (BPR) are obtained, and also, the variance of these predictive distributions for all these three situations, is put in closed forms.

## 1. INTRODUCTION

This paper deals with the problem of prediction in life tests based on exponential model. This problem is to predict the order statistics in future samples in terms of order statistics in the earlier samples, when a series of independent samples are drawn from a life test which has exponential distribution as the model. This problem has received much attention in recent years. Lawless [6], Lingappaiah [7, 8], and Kaminsky & Nelson [5] deal with this problem from the classical point of view, while Dunsmore [3], Lingappaiah [9, 10, 11, 12, 13] and Padgett [14] approach the same problem from Bayesian point of view. Dunsmore [3] and Padgett [14] are for two sample case only while Lingappaiah [9-13] extends to more than two samples. In all these works, sample size at all stages (of all samples) is fixed. This paper attempts this prediction problem when the sample sizes are fixed or random. This new aspect of randomness of the sample size has been dealt with by Burnham [1], Raghunandan and Patil [15], Consul [2], and Gupta & Gupta [4]. In all these works, distribution of order statistics, when the sample size is a random variable, is given when the sample size has various distributions like Poisson, generalised Poisson, or generalised negative binomial. What is being done in this paper is to combine these two concepts such as prediction and randomness of the sample size. For this purpose three different situations are considered with three samples. They are, (a) predicting $y_2$ ($k_2$-th order statistics in sample 2 of size $n_2$) in terms the $y_1$ ($k_1$-th order statistics in the sample 1 of size $n_1$) and the total-test-time $\hat{\theta}$ from sample 0 of size $n_0$, when the sample sizes in samples 1 and 2 are both random variables (b) predicting $y_2$ in terms $y_1$ and $\hat{C}$ when the size of sample 1 is fixed while that of sample 2 is random (c) predicting $y_2$ when the sample sizes of both samples 1 and 2 are fixed. The case (c) is already treated in Lingappaiah [10]. Concentration is on cases (a) and (b). For this prediction purpose, predictive distribution of $y_2$ is obtained for cases (a) and (b) and the Bayesian Prediction Regions (BPR), $P(y_2 \geq z) = \beta$ are evaluated for all three cases so that the comparison can be made. Also, the variance of the predictive distributions in all the three cases are put in closed forms.

## 2. PREDICTIVE DISTRIBUTIONS

Let three independent samples, 0, 1, 2 of sized $n_o$, $n_1$, $n_2$ respectively, be drawn from a life test based on the exponential model

$$f(x) = \theta \exp(-\theta x), \ \theta > 0, \ x > 0 \tag{1}$$

Then from sample 0, total-test-time to the r-th failure be represented by

$$\hat{\theta} = \sum_{i=1}^{r} x_{(i)o} + (n_o - r)x_{(r)o} \tag{2}$$

where $x_{(i)j}$ represents the i-th order statistic in the j-th sample, $j = 0, 1, 2$.

Now it is well known that $\hat{\theta}$ has the pdf

$$f(\hat{\theta},\theta) = e^{-\theta\hat{\theta}} \, (\theta\hat{\theta})^{r-1} \, \theta/\Gamma(r) \tag{3}$$

If the prior for $\theta$ is

$$g(\theta) = e^{-\theta h}(\theta h)^{g-1} \, h/\,\Gamma(g) \tag{4}$$

Then from (3) and (4), we have

$$f(\theta \mid \hat{\theta}) = e^{-\theta H} \, H(\theta H)^{G-1} \, /\Gamma(G) \tag{5}$$

where $H = h + \hat{\theta}$ , $G = g + r$ .

<u>2a: Both the sample sizes are random</u>

Now, let the size of the sample 1 be a random variable having its pdf as the Poisson distribution, given by

$$f(x) = e^{-\lambda} \lambda^{x}/\, x! \ , \ x = 0, 1, 2, \ldots \tag{6}$$

$$\lambda > 0 \, .$$

The distribution of the $k_1$-th order statistics $y_1 = x_{(k_1)1}$, when the sample size is a random variable is (Gupta and Gupta)

$$f(y_1 \mid \theta) = \frac{1}{P(n_1 \geq k_1)} \sum_{k=k_1}^{\infty} f(y_1 \mid k) P(n_1 = k) \tag{7}$$

where $P(n_1 = k) = e^{-\lambda} \lambda^{k} / k!$ and $P(n_1 \geq k_1) = \sum_{x=k_1}^{\infty} e^{-\lambda} \lambda^{x} / x! = \phi(k_1)$

$$f(y_1 \mid k) = \frac{k!}{(k-k_1)!\,(k_1-1)!} \, F^{k_1-1} \, (1-F)^{k-k_1} \, dF \tag{7a}$$

where F is the distribution function and $F = (1 - e^{-\theta x})$ for the exponential case (1).

Using (7a), (7) reduces to

$$f(y_1 \mid \theta) = C_1 \sum_{u_1=0}^{\infty} \sum_{j_1=0}^{k_1-1} \frac{\Omega(j_1)\lambda^{u_1}\theta}{u_1!} \, e^{-\theta y_1(1+u_1+j_1)} \tag{8}$$

where $C_1 = e^{-\lambda} \lambda^{k_1} \big/ \phi(k_1)(k_1-1)!$ ,

$$\Omega(j_1) = \binom{k_1-1}{j_1} (-1)^{j_1}$$

Now, from (5) and (8), we get

$$f(y_1 \mid \theta)f(\theta \mid \hat{\theta}) = C_1 \sum_{u_1} \sum_{j_1} \Omega(j_1) \frac{\lambda^{u_1} \theta^G e^{-\theta[H + y_1 a_1]} H^G}{u_1!\,\Gamma(G)} \tag{9}$$

with $a_1 = u_1 + j_1 + 1$ and from (9), we get

$$f(y_1, \hat{\theta}) = C_1 \sum_{u_1} \sum_{j_1} \frac{\Omega(j_1)\lambda^{u_1}\,\Gamma(G+1)H^G}{u_1!\,\Gamma(G)(H + a_1 y_1)^{G+1}} \tag{10}$$

Now suppose the size of the second sample is also a random variable following the <u>same</u> Poisson distribution (6), we have from (8),

$$f(y_2 \mid \theta) = C_2 \sum_{u_2} \sum_{j_2} \frac{\Omega(j_2)\lambda^{u_2}}{u_2!} \left\{ \theta e^{-\theta y_2 a_2} \right\} \tag{11}$$

where $y_2 = x_{(k_2)2}$, $k_2$-th order statistics in the sample 2. $a_2 = 1 + u_2 + j_2$, and $C_2$ similar to $C_1$ and $C_2 = e^{-\lambda}\lambda^{k_2} / \phi(k_2)(k_2 - 1)!$ From (9) and (10) we have

$$f(\theta \mid y_1, \hat{\theta}) = \frac{\sum\limits_{u_1} \sum\limits_{j_1} \frac{\Omega(j_1)\lambda^{u_1}}{u_1!}\, \theta^G e^{-\theta(H + a_1 y_1)}}{\sum\limits_{u_1} \sum\limits_{j_1} \frac{\Omega(j_1)\lambda^{u_1}}{u_1!} \left\{ \frac{\Gamma(G+1)}{(H + a_1 y_1)^{G+1}} \right\}} \tag{11a}$$

Now using (11a) and (11), we get

$$f(y_2 \mid \theta)f(\theta \mid y_1, \hat{\theta}) = C_2 \left[ \prod_{i=1}^{2} \sum_{u_i} \sum_{j_i} \frac{\Omega(j_i)\lambda^{u_i}}{u_i!} \right]$$

$$\cdot \left[ \frac{e^{-\theta[H + a_1 y_1 + a_2 y_2]}\, \theta^{G+1}}{\sum\limits_{u_1} \sum\limits_{j_1} \frac{\Omega(j_1)\lambda^{u_1}\,\Gamma(G+1)}{u_1!(H + a_1 y_1)G+1}} \right] \tag{12}$$

From (12), the predictive distribution of the $k_2$-th order statistics in sample 2 is

$$f(y_2 \mid y_1, \hat{\theta}) = \int f(y_2 \mid \theta)f(\theta \mid y_1, \hat{\theta})d\theta \tag{13}$$

and from (13), we get

$$f(y_2 \mid y_1, \hat{\theta}) = \frac{C_2 \left\{ \prod_{i=1}^{2} \sum_{u_i} \sum_{j_i} \frac{\Omega(j_i)\lambda^{u_i}}{u_i!} \right\} \left\{ \frac{\Gamma(G+2)}{(H+a_1y_1+a_2y_2)^{G+2}} \right\}}{\sum_{u_1} \sum_{j_1} \Omega(j_1) \frac{\lambda^{u_1}}{u_1!} \left\{ \frac{\Gamma(G+1)}{(H+a_1y_k)^{G+1}} \right\}} \tag{14}$$

It is easy to see from (14)

$$\int_0^\infty f(y_2 \mid y_1, \hat{\theta})dy_2 = C_2 \sum_{u_2} \sum_{j_2} \Omega(j_2) \frac{\lambda^{u_2}}{u_2!a_2}$$

$$= C_2 \sum_{u_2} \frac{\lambda^{u_2}(k_2-1)!}{(u_2+k_2)!} \tag{15}$$

using $\sum_{i=0}^{n} \binom{n}{i} (-1)^i \frac{1}{(i+1)} \cdot \frac{n!}{\prod_{i=0}^{n} (i+1)}$ (15a)

and (15) is equal to 1.

From (14), we get $P(y_2 > z) = \beta$ as

$$\beta = \frac{C_2 \left\{ \prod_{i=1} \sum_{u_i} \sum_{j_i} \Omega(j_i) \frac{\lambda^{u_i}}{u_i!} \right\} \cdot \frac{1}{a_2(H+a_1y_1+a_2z)^{G+1}}}{\sum_{u_1} \sum_{j_1} \Omega(j_1) \frac{\lambda^{u_1}}{u_1!} \left\{ \frac{1}{(H+a_1y_1)^{G+1}} \right\}} \tag{16}$$

From (16), one can calculate $\beta$ for given $\lambda$, $k_1$, $k_2$ and $\hat{\theta}$ .

2b: First sample size is fixed, second sample size is a variable

Now, suppose the size of sample 1, n, is fixed, the distribution of the $k_1$-th order statistics $y_1 = x_{(k_1)1}$ is

$$f(y_1 \mid \theta) = C_1^o \left(1 - e^{-\theta y_1}\right)^{k_1 - 1} \cdot \left(e^{-\theta y_1}\right)^{n_1 - k_1} \theta e^{-\theta y_1} \tag{17}$$

where $C_1^o = n_1!/(k_1-1)!(n_1-k_1)!$ and (17) is

$$f(y_1 \mid \theta) = C_1^o \sum_{j_1} \Omega(j_1)\theta e^{-\theta y_1(n_1-k_1+j_1+1)} \tag{18}$$

Now from (5) and (18) one gets

$$f(y_1, \theta, \hat{\theta}) = C_1^o \sum_{j_1} \Omega(j_1)\theta^G H^G e^{-\theta[H+b_1y_1]} [\Gamma^{-1}(G)] \tag{19}$$

where $b_1 = n_1 - k_1 + j_1 + 1$ and (19) gives

$$f(\theta \mid y_1, \hat{\theta}) = \frac{\sum_{j_1} \Omega(j_1)\theta^G e^{-\theta[H+b_1y_1]}}{\sum_{j_1} \Omega(j_1) \left\{ \frac{\Gamma(G+1)}{(H+b_1y_1)^{G+1}} \right\}} \tag{20}$$

Now suppose the size of second sample is a random variable following the Poisson distribution (6), then the distribution of the $k_2$-th order statistics in this sample 2 is given by (11). Now from (20) and (11), we get

$$f(y_2 \mid \theta) f(\theta \mid y_1, \hat{\theta}) = C_2 \sum_{u_2} \sum_{j_1} \sum_{j_2} \Omega(j_1)\Omega(j_2) \left\{ \lambda^{u_2} / u_2! \right\}$$

$$\cdot \frac{\theta^{G+1} e^{-(\theta H + b_1y_1 + a_2y_2)}}{\sum_{j_1} \Omega(j_1) \left\{ \frac{\Gamma(G+1)}{(H+b_1y_1)^{G+1}} \right\}} \tag{21}$$

where $a_2 = u_2 + j_2 + 1$, $b_1$, $C_2$ as before. From (21) predictive distribution of $k_2$-th order statistics follows as

$$f(y_2 \mid \hat{\theta}, y_1) = \frac{C_2 \sum_{u_2} \sum_{j_1} \sum_{j_2} \Omega(j_1)\Omega(j_2) \left\{ \frac{\lambda^{u_2}\, \Gamma(G+2)}{u_2!\,(H+b_1y_1+a_2y_2)^{G+2}} \right\}}{\sum_{j_1} \Omega(j_1) \left\{ \frac{\Gamma(G+1)}{(H+b_1y_1)^{G+1}} \right\}} \tag{22}$$

and from (22), $P(y_2 > z) = \beta$ is

$$\beta = \frac{C_2 \sum_{u_2} \sum_{j_1} \sum_{j_2} \Omega(j_1)\Omega(j_2) \left\{ \frac{\lambda^{u_2}}{a_2 u_2!\,(H+b_1y_1+a_2z)^{G+1}} \right\}}{\sum_{j_1} \Omega(j_1) \left\{ \frac{1}{(H+b_1y_1)^{G+1}} \right\}} \tag{23}$$

From (23), it is easy to see that if $z = 0$, (23) reduces to

$$C_2 \sum_{u_2} \sum_{j_2} \Omega(j_2) \frac{\lambda^{u_2}}{u_2! a_2} \quad (23a)$$

using (15a)

$$= C_2 \sum_{u_2=0}^{\infty} \frac{\lambda^{u_2+k_2}}{(u_2+k_2)!} \quad (23b)$$

and (23b) is equal to 1.

2c: Both the sample sizes are fixed

Suppose both $n_1$ and $n_2$ are fixed, the predictive distribution of $y_2 = x_{(k_2)2}$, $k_2$-th order statistics in the sample 2 is given in Lingappaiah [9] as

$$f(y_2 \mid y_1, \hat{\theta}) = \frac{C_2^o \sum_{j_1} \sum_{j_2} \Omega(j_1)\Omega(j_2) \left\{ \frac{\Gamma(G+2)}{(H+b_1y_1+b_2y_2)^{G+2}} \right\}}{\sum_{j_1} \Omega(j_1) \left\{ \frac{\Gamma(G+1)}{(H+b_1y_1)^{G+1}} \right\}} \quad (24)$$

where $C_2^o = n_2!/(n_2-k_2)!(k_2-1)!$, $b_2 = n_2 - k_2 + j_2 + 1$ and from (24), $P(y_2 > z) = \beta$ is

$$\beta = \frac{C_2^o \left\{ \prod_{i=1}^{2} \sum_{j_i} \Omega(j_i) \right\} \left\{ \frac{1}{b_2(H+b_1y_1+b_2z)^{G+1}} \right\}}{\sum_{j_1} \Omega(j_1) \left\{ \frac{1}{(H+b_1y_1)^{G+1}} \right\}} \quad (25)$$

Obviously if $z = 0$ in (25), then (25) reduces to

$C_2^o \sum_{j_2} \Omega(j_2)(1/b_2)$, using (15a),

$$= C_2^o[(k_2-1)!(n_2-k_2)!/n_2!] \quad (25a)$$

and (25a) is equal to 1.

3. SPECIAL CASES: (Prediction of Minimum)

Suppose we set $k_1 = k_2 = 1$, then we are predicting $y_2^o = x_{(1)2}$, the minimum in the sample 2 in terms of minimum in the sample 1, $y_1^o = x_{(1)1}$ and $\hat{\theta}$. Now the corresponding distributions of $y_2^o$ from (14), (22) and (24) are

$$f(y_2^o \mid y_1^o, \hat{\theta}) = \frac{\bar{C}_2 \sum_{u_1} \sum_{u_2} \frac{\lambda^{u_1+u_2}}{u_1!\, u_2!} \left\{ \frac{\Gamma(G+2)}{(H + a_1^o y_1^o + a_2^o y_2^o)^{G+2}} \right\}}{\sum_{u_1} \frac{\lambda^{u_1}}{u_1!} \left\{ \frac{\Gamma(G+1)}{(H + a_1^o y_1^o)^{G+1}} \right\}} \tag{26}$$

where $\bar{C}_2 = C_2$ when $k_2 = 1$, that is $\bar{C}_2 = e^{-\lambda}\lambda/(1-e^{-\lambda})$ and $a_i^o = (1+u_i)$, $i = 1, 2$ ($a_i^o$ is $a_i$ when $k_i = 1$) $i = 1, 2$. Now (22) reduces to

$$f(y_2^o \mid y_1^o, \hat{\theta}) = \frac{\bar{C}_2 \sum_{u_2} \frac{\lambda^{u_2}}{u_2!} \left\{ \frac{\Gamma(G+2)}{(H + b_1^o y_1^o + a_2^o y_2^o)^{G+2}} \right\}}{\left\{ \frac{\Gamma(G+1)}{(H + b_1^o y_1^o)^{G+1}} \right\}} \tag{27}$$

where $b_i^o = n_i$ ($b_i^o = b_i$ when $k_i = 1$, $i = 1, 2$) and similarly (26) reduces to

$$f(y_2^o \mid y_1^o, \hat{\theta}) = \frac{(G+1)n_2 \left\{ 1 \Big/ (H + b_1^o y_1^o + b_2^o y_2^o)^{G+2} \right\}}{\left\{ 1 \Big/ (H + b_1^o y_1^o)^{G+1} \right\}} \tag{28}$$

From (26), (27) and (28), $\beta_o = P(y_2^o > z_o)$ can be evaluated. Now (26) gives

$$\beta_o = \frac{\bar{C}_2 \sum_{u_1} \sum_{u_2} \frac{\lambda^{u_1+u_2}}{u_1!(u_2+1)!} \left[ \frac{1}{(H + a_1^o y_1^o + a_2^o z_o)^{G+1}} \right]}{\sum_{u_1} \frac{\lambda^{u_1}}{u_1!} \left[ \frac{1}{(H + a_1^o y_1^o)^{G+1}} \right]} \tag{29}$$

and (27) gives

$$\beta_o = \bar{C}_2 \sum_{u_2} \frac{\lambda^{u_2}}{(u_2+1)!} \left[ \frac{H + b_1^o y_1^o}{H + b_1^o y_1^o + a_2^o z_o} \right]^{G+1} \tag{30}$$

and (28) gives

$$\beta_o = \left( \frac{H + b_1^o y_1^o}{H + b_1^o y_1^o + b_2^o z_o} \right)^{G+1} \tag{31}$$

## 4. VARIANCES

From (26), we get for random-random case

$$E(y_2^o)^{r-1} = \bar{C}_2 \sum_{u_1} \sum_{u_2} \frac{\lambda^{u_1+u_2}}{u_1!u_2!} \cdot \frac{B(r, G+2-r)(G+1)}{(a_2^o)^r (Q)^{G+2-r} \left\{ \phi(H, y_1^o) \right\}} \tag{32}$$

where $Q = (H + a_1^o y_1^o)$, $\phi(H, y_1^o) = \sum_{u_1} \frac{\lambda^{u_1}}{u_1!} \frac{1}{Q^{G+1}}$.

Similarly (27) gives for fixed-random case,

$$E(y_2^o)^{r-1} = \bar{C}_2 \sum_{u_2} \frac{\lambda^{u_2}}{u_2!} \cdot \frac{(G+1)B(r, G+2-r)}{(a_2^o)^r \ (Q_o)^{G+2-r} \ \phi_o(H, y_1^o)} \tag{33}$$

where $Q_o = (H + b_1^o y_1^o)$, $\phi_o(H, y_1^o) = \left\{ 1 \Big/ (H + b_1^o y_1^o)^{G+1} \right\}$.

In the same way (28) gives for fixed-fixed case

$$E(y_2^o)^{r-1} = b_2^o \frac{B(r, G+2-r)(Q_o)^{G+1}(G+1)}{(b_2^o)^r \ (Q_o)^{G+2-r}} \tag{34}$$

It is easy to see that if $r = 1$, then (32), (33) and (34) reduce to 1. From these three equations, Var $y_2^o$ can be evaluated. From (30), we get

$$\text{Var}(y_2^o) = \left\{ \frac{2\ \bar{C}_2}{G(G-1)} \sum_{u_1} \sum_{u_2} \frac{\lambda^{u_1+u_2}}{u_1!u_2!} \frac{1}{(u_2+1)^3 Q^{G-1} \ \phi(H, y_1^o)} \right\}$$
$$- \left( \frac{\bar{C}_2}{G} \sum_{u_1} \sum_{u_2} \frac{\lambda^{u_1+u_2}}{u!u_2!(u_2+1)^2} \frac{1}{Q^G \phi(H, y_1^o)} \right)^2 \tag{35}$$

Similarly for the case where the size of the first sample fixed and that of second, a variable, is from (33)

$$\text{Var}(y_2^o) = \frac{2\ \bar{C}_2}{G(G-1)} \left( \sum_{u_2} \frac{\lambda^{u_2}}{u_2!(u_2+1)^3} \ \frac{1}{(Q_o)^{G-1} \phi_o(H, y_1^o)} \right)$$
$$- \left( \frac{\bar{C}_2}{G} \sum_{u_2} \frac{\lambda^{u_2}}{u_2!(u_2+1)^2} \ \frac{1}{(Q_o)^G} \ \frac{1}{\phi_o(H, y_1^o)} \right)^2 \tag{36}$$

and finally for both the sample sizes fixed case, we get from (34),

$$\text{Var } y_2^o = \frac{2}{G(G-1)} \left\{\frac{Q_o}{n_2}\right\}^2 - \left\{\frac{Q_o}{Gn_2}\right\}^2 \tag{37}$$

$$= (Q_o/n_2)^2 \left\{\frac{G+1}{G^2(G-1)}\right\} \tag{37a}$$

Comments: 1. Obviously, the simplest case of prediction is for $k_1 = k_2 = 1$, that is, predicting minimum in the second sample in terms of the minimum in the first sample. In this case, second set of sums on $j_1, j_2$ vanish. 2. In the case of predicting any other statistics $k_2 > 1$, then two sets of sums, one on $u_i$'s and another on $j_i$'s have to be taken care of. This means more computation. For large $\lambda$, sums may run longer. 3. Though the workload in the case of $k_1 = k_2 = 1$ is simpler, the prediction procedure remains the same for the general case $k_1, k_2 > 1$. 4. $\lambda$ in $f(n_1)$ and $f(n_2)$ need not be the same. It could just as well be $\lambda_1$ and $\lambda_2$ in which case, sums on $u_1, u_2$ have to be run on $\lambda_1$ and $\lambda_2$ respectively. This separate $\lambda_1, \lambda_2$ does not affect the analysis in any other way except more tables for $\beta$ and variances for different values of $\lambda_1$ and $\lambda_2$. 5. The values of h and g affect the results, expecially large g. But h may not have too much effect unless it is very large since $H = h + \hat{\theta}$. 6. If need be, much complex f(x) other than Poisson can be chosen such as Generalised Poisson or generalised negative binomial and the like. In this case, only computation will be more and analysis procedure remains the same. The complexity of f(x) in (6) may depend on the nature of randomness of the sample sizes. Again, these complex f(x) add more parameters to analysis and hence more tables of $\beta$ and variances, for different values of these new parameters introduced. 7. As can be seen easily, the case of both sample sizes fixed is the simplest while the case when both sample sizes are variable is the hardest as far as computation is concerned. In this case two sets of sums have to be evaluated one set on $u_i$'s and another on $j_i$'s. The case of first sample size fixed and the second random needs medium size computation only on $u_2$ and $j_1, j_2$. 8. The gamma prior for $\theta$ in sample 0 is taken for simplicity sake. It can be replaced by a complex prior if need arises. Again, the procedure of analysis remains the same. 9. In Section 2b, it could as well, the first sample size is variable while that of second sample is fixed. Then in (22), it will be $a_1$ and $b_2$ instead of $a_2$ and $b_1$ and instead of $C_2$ we have $C_1$. The analysis remains virtually the same. In the denominator of (22), there will be two sums, one on $u_1$ and another on

$j_1$. 10. In the case 2b in section 2, variable $u_2$ could have been just a dummy variable u. But it is taken as $u_2$ just because it is for sample 2. No special meaning for subscript 2 in $u_2$.

REFERENCES


Buhrman, J.M. (1973) On order statistics when sample size has a binomial distribution. Statistica Neerlandica, Vol. 27, pp. 125-126.

Consul, P.C. (1984) On the distribution of order statistics for a random sample size. Statistica Neerlandica, Vol. 38, no. 4., pp. 249-256.

Dunsmore, I.R. (1974) The Bayesian predictive distribution in life testing models. Technometrics, Vol. 16, pp. 455-460.

Gupta, D. and Gupta, R.C. (1984) On the distribution of order statistics for a random sample size. Statistica Neerlandica, Vol. 38, pp. 13-19.

Kaminsky, Kenneth S. and Nelson, Paul I. (1975) Best linear unbiased prediction of order statistics in location and scale families. J. American Statistical Association, Vol. 70, pp. 145-150.

Lawless, J.F. (1971) A prediction problem concerning samples from the exponential distribution with application to life testing. Technometrics, Vol. 13, pp. 725-270.

Lingappaiah, G.S. (1973) Prediction in exponential life testing. Canadian J. of Statistics, Vol. 1, pp. 113-117.

Lingappaiah, G.S. (1974) Prediction in samples from the gamma distribution as applied to life testing. The Australian J. of Statistics, Vol. 1, pp. 113-117.

Lingappaiah, G.S. (1978) Bayesian approach to the prediction problem in the exponential population. IEEE Trans. Reliability, Vol. R-27, pp. 222-225.

Lingappaiah, G.S. (1979) Bayesian approach to the prediction problem in complete and censored samples from the gamma and exponential populations. Communications in Statistics, Vol. A8, pp. 1403-1424.

Lingappaiah, G.S. (1981) Sequential life testing with spacings. Exponential model. IEEE Trans. Reliability, Vol. R-30, pp. 370-374.

Lingappaiah, G.S. (1984) Bayesian prediction regions for the extreme order statistics. Biometrische Zeitschrift (Biometrical Journal), Vol. 26, pp. 49-56.

Lingappaiah, G.S. (1985) A study of shifting models in life tests via Bayesian approach using semi or used priors. (SOUPS) Annals of Institute of Statistical Mathematics, Vol. 37, pp. 151-163.

Padgett, W.J. (1982) An approximate prediction interval for the mean of the future observations from the inverse gaussian distribution. J. of Statistical Computation and Simulation, Vol. 14, pp. 191-199.

Raghunandan, K. and Patil, S.A. (1972) On order statistics from random sample size. Statistica Neerlandica, Vol. 26, pp. 121-126.



Key words: Prediction; order statistics; random sample size; Bayesian Prediction Regions; Poisson distributions

AMS Classification: 62 E 15, 62 F 15.

ON COX'S CONFIDENCE DISTRIBUTION

Jochen Mau

Statistics Project
Institute of Medical Biometry
University of Tübingen
Hausserstr. 11, D-7400 Tübingen, F.R.G.

SUMMARY

A confidence distribution function is a graphical tool for flexible statistical analyses. It provides one- and two-sided tests of simple and interval hypotheses for any size, central and symmetrical confidence intervals of any level. Given an interval of equivalent values, it quantifies the strength of evidence for "no material difference" between two populations in a set of data, but is independent of the particular choice of such an interval.

## 1. INTRODUCTION

To describe the <u>context of observation</u>, let

(i) $(\Omega, \mathcal{A}, \mathbb{P})$ be the underlying probability space in the usual triple notation,

(ii) $(\mathcal{X}, \mathfrak{B}_{\mathcal{X}})$ be the sample space, where $\mathcal{X}$ is a Polish space and $\mathfrak{B}_{\mathcal{X}}$ its $\sigma$-algebra of Borel sets, and

(iii) $X : \Omega \to \mathcal{X}$ a measurable mapping which is observable. Thus X represents the data.

The <u>statistical problem</u> is introduced via the distribution $X(\mathbb{P})$ of X. Assume that it depends on a vector of unknown parameters $(\Theta, \xi) \in \Xi \times \mathbb{R}^k$, $\Xi \subset \mathbb{R}$ an interval, $\Theta \in \Xi$, and consider the family of one-sided test problems for $\Theta$,

$$H_\gamma : \Theta = \gamma \text{ versus } K_\gamma : \Theta > \gamma \ (\gamma \in \Xi). \quad (1.1)$$

For inference about $\Theta$, one will define a <u>random distribution</u> on the measure space $(\Xi, \mathfrak{B}_\Xi)$, using a test statistic $\mathcal{T}: \Xi \times \mathcal{X} \to \mathbb{R}$ for (1.1). $\xi$ represents nuisance parameters. The choice of $\mathcal{T}$ may be based either on convention (e.g. from a particular application) or on some optimality criterion (e.g. uniformly most powerful unbiased tests in a context of multiparameter exponential families; cf. Lehmann, 1959, Sect.4.4).

The confidence distribution - a term coined by Cox (1958) - permits

(i) the statement of an <u>observed confidence level</u> as a measure of the strength of evidence for a hypothesis

$$\Theta \in [\gamma_1, \gamma_2] \subset \Xi, \qquad (1.2)$$

with some given $\gamma_1 < \gamma_2$, for any data $x = X(\omega)$ and a selected test statistic (Section 2), and

(ii) the simple construction of observed <u>central</u> or <u>symmetrical confidence intervals</u> for $\Theta$ at any level $1-\alpha$, $\mathbb{P}\{[\underline{\Theta}, \overline{\Theta}] \ni \Theta\} \geq 1-\alpha$, aided by a plot of the observed confidence distribution function (Section 3).

(iii) In the light of further data, the confidence distribution from a previous data set may be "updated" to provide a measure of the strength of evidence from the combined data (Section 4).

## 2. CONSTRUCTION OF THE CONFIDENCE DISTRIBUTION

One first needs some notation and assumptions. Let $\mathcal{T}_\Theta$ and $\mathcal{T}_x$ denote the $\Theta$- and the x-section of $\mathcal{T}$ for any $\Theta \in \Xi$ and $x \in \mathcal{X}$, respectively.

<u>ASSUMPTIONS 2.1</u>.

(i) For any $\gamma \in \Xi$, $\mathcal{T}_\gamma : \mathcal{X} \to \mathbb{R}$ is continuous,

(ii) For any $x \in \mathcal{X}$, $\mathcal{T}_x : \Xi \to \mathbb{R}$ is nonincreasing and left-continuous with right-hand limits,

(iii) If $\Theta = \gamma$, i.e. $\gamma$ is the true value of $\Theta$, the law of $\mathcal{T}_\gamma(X)$ is independent of the particular values of $(\Theta, \xi)$. If $G(t)$, $t \in \mathbb{R}$, denotes its distribution function, assume $G(\mathcal{T}_\gamma(x)) \to 1$ as $\gamma \to \inf \Xi$, for any $x \in \mathcal{X}$ and whatever $(\Theta, \xi)$.

<u>REMARK 2.1</u>. By (i) and (ii), $\mathcal{T}$ is $\mathcal{B}_\Xi \otimes \mathcal{B}_\mathcal{X} - \mathcal{B}_\mathbb{R}$ - measurable (cf. Rudin, 1970, Chap. 7, Ex. 8).

<u>THEOREM 2.1</u>. There exists a random measure $\mathcal{C}$ on $\mathcal{B}_\Xi$, such that for any $x \in \mathcal{X}$,

$$\mathcal{C}(x)\, I_\gamma = \mathcal{T}_\Theta[X(\mathbb{P})_{\Theta,\xi}\,]\,[\mathcal{T}_\gamma(x), +\infty[$$

$$= 1 - G(\mathcal{T}_\gamma(x)), \ \gamma \in \Xi \qquad (2.1)$$

gives a distribution on $\Xi$, $I_\gamma = \,]\inf \Xi, \gamma[$.
$\mathcal{C}(X)$ is called a confidence distribution on $\mathcal{B}_\Xi$.

PROOF: Denote the righthand side of (2.1) by $P(\gamma;x)$. Then, $P(\gamma;\,.) : \mathcal{X} \to [0, 1]$ are measurable, $P(.;\,x) : \Xi \to [0, 1]$ are measurable, nondecreasing, and leftcontinuous with $P(\gamma;x) \to 0$ as $\gamma \to \inf \Xi$, for any $\gamma \in \Xi$ and $x \in \mathcal{X}$, respectively. By Rudin (1970, Theo. 8.14) there exists a unique Borel measure $\mathcal{C}(x)$ on $\Xi$ for any $x \in \mathcal{X}$ such that (2.1) holds.
Let $\mathcal{M}$ be the set of all Borel measures $\mu$ on $\Xi$, $\mu\,\Xi \leq 1$, and the $\sigma$-algebra $\mathcal{A}$ generated by the mappings $\mu \to \mu B$, $\mu \in \mathcal{M}$, for any $B \in \mathcal{B}_\Xi$. Hence, $\mathcal{C}(x) \in \mathcal{M}$ for any $x \in \mathcal{X}$. It remains to show that $\mathcal{C}$ is $\mathcal{B}_\mathcal{X}$-$\mathcal{A}$-measurable (cf. Kallenberg, 1976).
The projections $\pi_B$, $\pi_B\mu = B$, $\mu \in \mathcal{M}$, are $\mathcal{A} - \mathcal{B}_{[0,1]}$-measurable for any $B \in \mathcal{B}_\Xi$. Since $\pi_B\,\mathcal{C}(x) = P(\gamma_2;\,x) - P(\gamma_1;\,x)$ for any $\gamma_1, \gamma_2 \in \Xi$, $\gamma_1 < \gamma_2$, $B = \,]\gamma_1, \gamma_2[$, and $x \in \mathcal{X}$, it follows that $x \to \pi_B\,\mathcal{C}(x)$ is $\mathcal{B}_\mathcal{X} - \mathcal{B}_{[0,1]}$- measurable for any $B \in \mathcal{B}_\Xi$. The

desired property now follows from a standard result (cf. Bauer, 1968, Theo. 7.4).∎

The construction of $\mathfrak{C}$ as a random measure implies the consideration of concepts like the distribution of $\mathfrak{C}$, its intensity in the sense of Kallenberg (1976) and its Laplace transform. In particular, the distribution of $\mathfrak{C}$ on $\mathcal{M}$ is

$$\mathfrak{C}(X(\mathbb{P})_{\Theta,\xi})\, M = \mathbb{P}_{\Theta,\xi}\{\, \mathfrak{C}(X) \in M\},\ M \in \mathcal{M},$$

and its intensity, again a measure on $\mathcal{B}_{\Xi}$, turns out to be the expectation of the confidence distribution:

$$\int_{\mathcal{X}} \mathfrak{C}B \; d\, X(\mathbb{P})_{\Theta\xi} = \int_{\Omega} \int dP(\gamma;\, X)\; d\, \mathbb{P}_{\Theta\xi}$$

$$= \mathbb{E}_{\Theta\xi}\, \mathfrak{C}(X)\; B,\ B \in \mathcal{B}_{\Xi}.$$

With $B = I_{\Theta}$,

$$\mathbb{E}_{\Theta\xi}\mathfrak{C}(X)\; I_{\Theta} = \int_{\Omega}[1 - G(\mathcal{T}_{\Theta}(X))]\; d\, \mathbb{P}_{\Theta\xi}\ .$$

But this is, by Assumption 2.1 (iii), the expectation of a random variable with a uniform distribution on [0, 1]. This proves the main theoretical justification for the construction of the confidence distribution: the true parameter $\Theta$ is the median of the expectation of the confidence distribution.

<u>COROLLARY 2.1</u>. $\mathbb{E}_{\Theta\xi}\,\mathfrak{C}(X)\; I_{\Theta} = \frac{1}{2}$.

The following result represents the principal justification for the use of the confidence distribution in practice and suggests its interpretation as an objective measure of the strength of evidence for an interval hypothesis (1.2). For any fixed $\gamma_1, \gamma_2 \in \Xi$, $\gamma_1 < \gamma_2$, and $0 < \alpha < 1/2$ consider the problem of testing the null hypothesis, H, against the alternative hypothesis, K,

$$H : \Theta \notin [\gamma_1, \gamma_2], \qquad (2.2)$$

$$K : \Theta \in [\gamma_1, \gamma_2], \qquad (2.3)$$

and apply the following decision rule in terms of the confidence distribution of $\Theta$ :

"reject H, if

$$\mathfrak{C}(x)\; I_{\gamma_1} < \alpha/2 \text{ and } \mathfrak{C}(x)\; I_{\gamma_2} > 1 - \alpha/2, \qquad (2.4)$$

and accept H, otherwise."

<u>LEMMA 2.1</u>. The level of the test of H against K based on (2.4), does not exceed $\alpha/2$.

PROOF: In Section 3, central confidence intervals of $\Theta$ for a confidence level of $1-\alpha$ will be introduced. By (3.1), decision rule (2.4) is equivalent to an »inclusion rule«: reject H, if the central $(1-\alpha)$-confidence interval is completely contained in $[\gamma_1, \gamma_2]$, and accept H, otherwise. This confidence-interval test obviously has a level of at most $\alpha/2$.∎

By this rule, $\mathfrak{C}(x)\, [\gamma_1, \gamma_2] \geq 1 - \alpha$ is a necessary condition for rejection of H. The smaller the value of $\alpha$ can be without accepting H for a given data set, the stronger is the evidence for K in terms of the observed confidence distribution, and vice versa (with the qualification that the confidence mass outside of the interval is roughly equal on both sides). This interpretation has much in common with the well-established interpretation of an observed significance level in usual hypothesis testing: the latter quantifies the strength of evidence against a null hypothesis in the light of given data, cf. Cox (1977). In either case, the quantification depends on the chosen test statistics.

REMARK 2.2. If $\gamma_2 > 0$, $\gamma_1 = -\gamma_2$ in (2.2) and (2.3), then one can replace (2.4) by

$$\mathfrak{C}(x)\,[\gamma_1, \gamma_2] > 1 - \alpha,$$

which is equivalent to a confidence-interval test based on an inclusion rule with a symmetrical $(1-\alpha)$-confidence interval. This test has a level of at most $\alpha$.

## 3. CENTRAL AND SYMMETRICAL CONFIDENCE INTERVALS

We call those confidence intervals central which have equal probability outside either endpoint. Confidence intervals which are symmetrical around zero, play a role in some applications where zero is contained in $\Xi$ (cf. Mandallaz and Mau, 1981). The observed confidence intervals of either kind are easily read from a plot of the observed confidence distribution function.

ASSUMPTION 3.1. $\mathbb{P}\{\mathfrak{C}(X)\,\Xi = 1\} = 1$ and the distribution function G is continuous.

PROPOSITION 3.1. Let $0 < \alpha < 1/2$, then the equations

$$\left.\begin{aligned} &\mathfrak{C}(X)\,I_{\gamma'} = \alpha/2,\ \gamma' \in \Xi, \\ &\mathfrak{C}(X)\,I_{\gamma''} = 1-\alpha/2,\ \gamma'' \in \Xi, \end{aligned}\right\} \qquad (3.1)$$

$\gamma' < \gamma''$, have unique (random) solutions $\Theta_u < \Theta_o$ in $\Xi$, $\mathbb{P}$-almost-surely. Further,

$$\mathbb{P}_{\Theta\xi}\ \{[\Theta_u, \Theta_o] \ni \Theta\} \geq 1 - \alpha \qquad (3.2)$$

whatever the values of $(\Theta, \xi) \in \Xi \times \mathbb{R}^k$.

PROOF: Existence follows from Assumptions 2.1 (iii) and 3.1, since $\Xi$ is an interval. By the measurability of the $\gamma$-section of P (cf. proof of Theorem 2.1), the sets

$$\begin{aligned} A(\gamma) &= \{x \in \mathfrak{X} : \Theta_u \leq \gamma \leq \Theta_o\} \\ &= \{x \in \mathfrak{X} : \mathfrak{C}(x)\, I_{\Theta_u} \leq \mathfrak{C}(x)\, I \quad \leq \mathfrak{C}(x)\, I_{\Theta_o}\} \end{aligned}$$

are in $\mathfrak{B}_{\mathfrak{X}}$ for any $\gamma \in \Xi$. Then, $X(\mathbb{P})_{\Theta\xi}\, A(\Theta) \geq 1-\alpha$. ■

REMARK 3.1. Continuity of G is not necessary to prove (3.2), if one defines the endpoints of the confidence interval by supremum and infimum of $\mathfrak{C}(X)\, I_{\gamma}$ below $\alpha/2$ and above $1 - \alpha/2$, respectively.

ASSUMPTION 3.2. $\Xi$ is a symmetrical interval around zero, and G is continuous.

PROPOSITION 3.2. Let $0 < \alpha < 1/2$. Then, the equation

$$\mathfrak{C}(X)\,[-\gamma, \gamma] = 1 - \alpha,\ \gamma \in \Xi,\ \gamma > 0, \tag{3.3}$$

has a unique (random) solution $\Theta_s$ in $\Xi$, $\mathbb{P}$-almost surely. Further,

$$\mathbb{P}_{\Theta\xi}\{[-\Theta_s, \Theta_s] \ni \Theta\} \geq 1 - \alpha \tag{3.4}$$

whatever the values of $(\Theta, \xi) \in \Xi \times \mathbb{R}^k$.

PROOF: By the definition of $\Theta_s$, one has for $\Theta > 0$ $\{-\Theta_s \leq \Theta \leq \Theta_s\} = \{\mathfrak{C}(X)[-\Theta, \Theta] \leq 1 - \alpha\} \supset \{1 - \mathfrak{C}(X)I_\Theta \geq \alpha\}$ except on an $\omega$-set of P-measure zero. The result follows from Assumption 2.1. (iii). Treat $\Theta < 0$ analogously.■

REMARK 3.2. Again, continuity of G is not necessary to prove (3.4) when $\Theta_s$ is defined as $\inf\{\mathfrak{C}(X)[-\gamma, \gamma] \geq 1 - \alpha\}$.

EXAMPLE 3.1. To demonstrate a typical application, assume that we conduct an experiment twice, each repetition comprising two series, A and B, of Bernoulli experiments of equal size: Firstly, data x, $n_A = n_B = 150$ replications, estimated probabilities of a success in A and B of $\hat{\pi}_A = 0.50$ and $\hat{\pi}_B = 0.55$, secondly, data y, $n_A = n_B = 150$, $\hat{\pi}_A = 0.50$ and $\hat{\pi}_B = 0.70$.

The commonly used test statistic for (1.1) with $\hat{\Theta} = \hat{\pi}_A - \hat{\pi}_B$ is $\mathcal{T}_\gamma(x) = (\hat{\Theta} - \gamma)/\hat{\sigma}$, where $\hat{\sigma}^2 = \hat{\pi}_A(1-\hat{\pi}_A)/n_A + \hat{\pi}_B(1-\hat{\pi}_B)/n_B$, which is approximately standard normally distributed under $H_\gamma$. The observed confidence distribution functions, $P(\gamma; x)$ or $P(\gamma; y) = 1 - \Phi_{0,1}[(\hat{\Theta} - \gamma)/\hat{\sigma}]$, $\gamma \in \Xi$, are shown in Fig. 1, their densities are plotted in Fig. 2. (Here, $\Phi_{0,1}$ denotes the standard normal distribution function.)

Let the interval [-0.15, 0.15] represent practically irrelevant values of $\Theta$. Its observed confidence levels are,

$$\mathfrak{C}(x)\,[-0.15, 0.15] = 0.958$$
$$\mathfrak{C}(y)\,[-0.15, 0.15] = 0.183,$$

which is visualized by the areas under the respective densities in Fig. 2. The observed 95%-confidence intervals and their overlap with [-0.15, 0.15] are shown in Fig. 1. This situation is typical of many clinical trials which compare successful treatment with two competitive drugs, say, in samples of patients. An interval of clinical equivalence can often be stated, though not unanimously. However, the plot of the observed confidence distribution function is independent of the choice of such an interval.

## 4. ANALYSIS OF ACCUMULATING DATA

To investigate the potential of the confidence distribution concept for the analysis of data which accumulates in batches, e.g. sequential clinical trials with interim analyses, assume that one observed a data set $x = X(\omega_1)$ and is completing observation of a further data set $y = X(\omega_2)$. In Assumption 2.1 (iii), the law of $\mathcal{T}_\gamma(X)$ was only given for $\Theta = \gamma$.

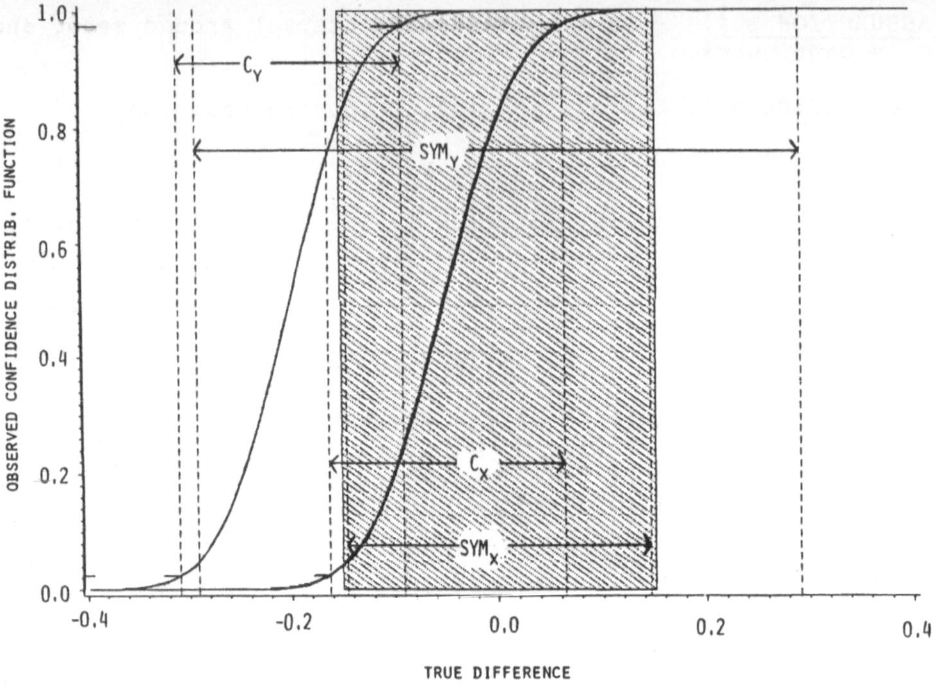


Fig. 1. Observed confidence distribution functions, central (Cx, Cy) and symmetrical (SYMx, SYMy) 0.95-confidence intervals for the true difference, based on data x (slim line) and y (bold line). The interval of practical equivalence is [-0.15, +0.15] (shaded area).

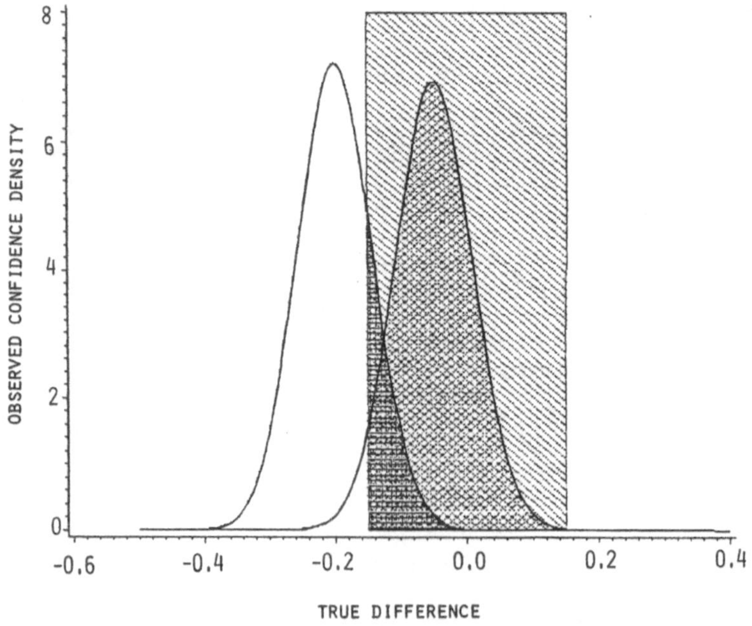


Fig. 2. Observed confidence densities for the true difference, based on data x (right) and y (left), with an interval of practical equivalence as in Fig. 1.

Upon observation of y, one has already derived $\mathfrak{C}(x)$ as a measure on $(\Xi, \mathcal{B}_\Xi)$. Thus,

$$\begin{array}{l} \omega_2 \in (\Omega, \mathcal{A}, \mathbb{P}) \\ \big\downarrow X \\ y \\ \cap \\ (\Xi \times \mathcal{X}, \mathcal{B}_\Xi \otimes \mathcal{B}_\mathcal{X}, \mathfrak{C}(x) \otimes X(\mathbb{P})) \rightarrow (\mathbb{R}, \mathcal{B}_\mathbb{R}). \end{array}$$

By Remark 2.1. $\mathcal{T}$ introduces a measure $\mathcal{T}[\mathfrak{C}(x) \otimes X(\mathbb{P})]$ on $\mathcal{B}_\mathbb{R}$. In view of (2.1), one agrees on

DEFINITION 4.1. If $x = X(\omega_1)$ and $y = X(\omega_2)$ are observed values of X, the observed confidence distribution on $\Xi$ from y given x is

$$\mathfrak{C}(y|x)\, I_\gamma = \mathcal{T}[\mathfrak{C}(x) \otimes X(\mathbb{P})][\mathcal{T}_\gamma(y), \infty[$$

for any $\gamma \in \Xi$.

Denote the observed confidence distribution functions of $\mathfrak{C}(y|x)$, $\mathfrak{C}(x)$, and $\mathfrak{C}(y)$ by $P(.; y|x)$, $P(.; x)$, and $P(.; y)$, respectively.

PROPOSITION 4.1. With the assumptions of Definition 4.1.,

$$P(\gamma; y|x) = P(\gamma; y), \ \gamma \in \Xi.$$

PROOF: Recall first that $\mathfrak{C}(x)$ represents a distribution of the true $\Theta$ given x. Hence, for any $B \in \mathcal{B}_\mathbb{R}$,

$$\mathcal{T}[\mathfrak{C}(x) \otimes X(\mathbb{P})_{\Theta\xi}\,](B) = \int_\Xi \int_{\mathcal{T}_\Theta^{-1}[B]} dX(\mathbb{P})_{\Theta\xi}\, \mathfrak{C}(x)(d\Theta)$$

[by Assumption 2.1 (iii)] $\qquad = \int_\Xi dP(\Theta; x) \int_B d\,G(t)$

$$= \int_B dG(t)$$

In particular, $B = [\mathcal{T}_\gamma(y), \infty[$ implies by (2.1)

$$P(\gamma; y|x) = [1 - G(\mathcal{T}_\gamma(y))]$$

$$= P(\gamma; y)$$

for whatever value of $\gamma$.■

REMARK 4.1. Note that independent observations are not required. The result is essentially due to Assumption 2.1. (iii).

4.2. As a consequence of the proposition, the joint confidence distribution of $\Xi^2$ based on x and y, $\mathfrak{C}(x, y)$, can be represented as a product of the marginals, $\mathfrak{C}(x)$ and $\mathfrak{C}(y)$. Considering $\mathfrak{C}(x, y)$ only on the diagonal, then gives $P(\gamma; x)\, P(\gamma; y)$, $\gamma \in \Xi$, as a possible choice for a confidence distribution function with density

$$p(\gamma,x)\, P(\gamma;y)+p(\gamma;y)\, P(\gamma;x), \ \gamma \in \Xi, \qquad (4.1)$$

where $p(.; x)$ and $p(.; y)$ denote the densities (w. r. t. Lebesgue measure) of $\mathfrak{C}(x)$ and $\mathfrak{C}(y)$, respectively.

4.3. Considering the bivariate observed confidence density

function associated with $\mathcal{C}(x, y)$ also on the diagonal, gives rise to yet another distribution via a density

$$p(\gamma;\, x,\, y) = \frac{p(\gamma;x)\; p(\gamma;\, y)}{\int_{\Xi} p(\gamma';\, x)\; p(\gamma';\, y)\; d\gamma'},\quad \gamma \in \Xi, \qquad (4.2)$$

with distribution function

$$P(\gamma;\, x,\, y) = \frac{\int_{I_\gamma} p(\gamma';\, x)\; p(\gamma';\, y)\; d\gamma'}{\int_{\Xi} p(\gamma';\, x)\; p(\gamma';\, y)\; d\gamma'},\quad \gamma \in \Xi.$$

4.4. A third way to combine the evidence from independent experiments was mentioned by Mandallaz and Mau (1981). It is based on a chi-square criterion and is seen to give similar results as the above methods in an example considered in Mau (1986).

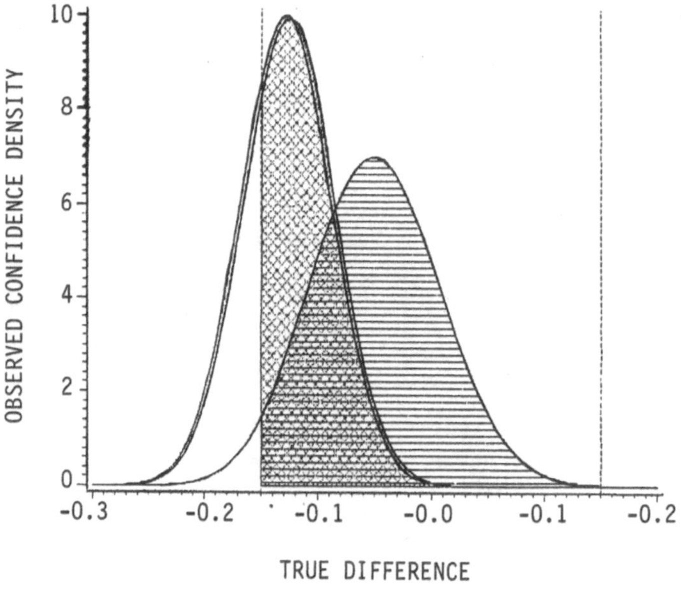


Fig. 3. Observed confidence densities, for the true difference, based on - from left to right - the pooled data z and the updating formulae (4.2) and (4.1).

EXAMPLE 4.1. For the data x and y from Example 3.1, the observed confidence densities according to (4.1) and (4.2) are plotted in Fig. 3. For a comparison, the distribution function and density of $\mathcal{C}(z)$ are also shown, where z is obtained from pooling the data contained in x and y, data z: $n_A = n_B = 300$, $\hat{\pi}_A = 0.50$, $\hat{\pi}_B = 0.625$.
The observed confidences for [-0.15, 0.15], the interval of practically irrelevant differences $\Theta$, are

with (4.1): 0.966,
with (4.2): 0.710,
with $\mathcal{C}(z)$: 0.733.

This underlines the nature of the "updating" formula (4.1): it is not an averaging, as it is obtained by the analysis of the pooled data z and by (4.2).

REMARK 4.5. Note that (4.1) is not a useful way of combining evidence for an interval of practical equivalence, $[\gamma_1, \gamma_2]$. Obviously, this formula will always give more weight to the distribution located right most. As an extreme situation, consider two very steep distribution functions based on x and y, one located at zero ($\gamma_1 < 0 < \gamma_2$), the other far to the right from $\gamma_2$. The combined distribution function would be practically identical to the latter!

## 5. DISCUSSION

Though the confidence distribution was derived from a pure frequentist's viewpoint, its implications are close to those of Bayesian analysis except for the explicit use of subjective prior distributions. In fact, Mandallaz and Mau (1981) obtained it as a Bayesian posterior distribution for an improper vague prior and Mau (1983) derived it as a fiducial distribution along the outline of Pedersen (1978), with standard two-sample tests for multiplicative and additive effects in a normal theory linear model, respectively.

If the likelihood of $\Theta$, given observation of X = y, is proportional to the density of $\mathcal{C}(y)$,

$$p(\gamma; y) = -g\ (\mathcal{T}_\gamma(y))\ d\ \mathcal{T}_\gamma(y)/d\gamma,\ \gamma \in \Xi,$$

then (4.2) is essentially Bayes' formula.

The wide-spread applicability of the confidence distribution concept rests upon its computational simplicity with at least approximately normally distributed test statistics for shift alternatives and a symmetrical interval hypothesis as in (1.2). This situation is frequently met in comparative clinical trials, where up to now it has mostly been exploited for sample size determinations (cf., e.g., Spiegelhalter and Freedman, 1986).

The confidence distribution is specifically important for a quantitative assessment of clinically equivalent efficacy after non-significant tests with two differently treated groups of patients (cf. Mau, 1986, for a more detailed consideration of this application).

Plots of the confidence distribution functions provide a useful summary of the main features of the data in reports on experimental results, since any reader can try his own beliefs about the proper size of an important difference, captured analytically by $\gamma_1$ and $\gamma_2$. This aspect might be appealing to regulatory authorities and review bodies.

## ACKNOWLEDGEMENTS

This research was done while the author visited the Department of Theoretical Statistics, University of Aarhus/Denmark, and the Department of Mathematics, University of Nottingham/England with the support of the Deutsche Forschungsgemeinschaft, SFB 175 Implantology.

Stimulating discussions with O. Barndorff-Nielsen, and J.G. Pedersen are gratefully acknowledged.

## REFERENCES

Bauer, H., 1968, "Wahrscheinlichkeitsrechnung und Grundzüge der Maßtheorie," de Gruyter, Berlin.

Cox, D. R., 1958, Some problems connected with statistical inference, Ann. Math. Statist., 29:357.

Cox, D. R., 1977, The role of significance tests (with discussion), Scand. J. Statist., 4:49.

Kallenberg, O., 1976, "Random Measures," Akademie-Verlag, Berlin.

Lehmann, E. L., 1959, "Testing Statistical Hypothesis," Wiley, New York.

Mandallaz, D. and Mau, J., 1981, Comparison of different approaches to the assessment of bioequivalence, Biometrics, 37:213.

Mau, J., 1983, Die Verwendung von Fiduzialwahrscheinlichkeiten zur Beurteilung der Bioäquivalenz, in: "Biometrie in der chemisch-pharmazeutischen Industrie 1," Vollmar, J., ed., Gustav Fischer, Stuttgart.

Mau, J., 1986, A statistical assessment of clinical equivalence, Res. Rep. 4/86, Statistics Proj. SFB 175, Univ. of Tübingen, Tübingen.

Pedersen, J. G., 1978, Fiducial inference, Internat. Statist. Rev., 46:147.

Rudin, W., 1970, "Real and Complex Analysis," McGraw - Hill, London.

Spiegelhalter, D. J. and Freedman, L. S., 1986, A predictive approach to selecting the size of a clinical trial, based on subjective clinical opinion, Statistics in Medicine, 5:1.

# A BAYESIAN ANALYSIS OF A GENERALIZED SLOPE RATIO BIOASSAY

M. Mendoza

Departamento de Matemáticas,UNAM, Fac. Ciencias, México/Departamento de Estadística, Univ.de Valencia, Spain

SUMMARY

The reference posterior distribution for the parameter of interest in a widely used class of biological assays is obtained. This class contains as a particular instance the very well known *slope ratio* assay. The results obtained avoid the classical difficulties associated to the frequentist estimation of ratios and generalize some previous analysis.

## 1. INTRODUCTION

In a previous paper (Mendoza 1986) a Bayesian analysis was provided for the simplest version of the slope ratio bioassay. There, an experiment was considered where p doses $(X_{11},\ldots,X_{1p})$ of a first stimulus and q doses $(X_{21},\ldots,X_{2q})$ of a second stimulus were assayed to obtain a set $\{Y_{1jk};\ j=1,..,p;\ k=1,..,n;\ Y_{2jk};\ j=1,\ldots,q;\ k=1,\ldots,n\}$ of $n(p+q)$ conditionally independent Normal observations with common variance $\sigma^2$ and such that

$$\begin{aligned} E(Y_{1jk}) &= \alpha + \beta X_{1j} \qquad k=1,\ldots,n\ ;\ j=1,\ldots,p \\ E(Y_{2jk}) &= \alpha + \rho\beta X_{2j} \qquad k=1,\ldots,n\ ;\ j=1,\ldots,q \end{aligned} \tag{1}$$

In order to assess the relative *potency* of these stimuli (the main objective of a comparative assay) it suffices to produce inferences about the slope ratio $\rho = \rho\beta/\beta$ since, under the assumptions stated, this parameter describes the ratio of the first to the second stimulus for every pair of doses leading to the same expected response (equivalent doses). In Mendoza (1986) a reference posterior distribution was obtained for $\rho$ and some of its properties were discussed.

In this paper, a natural, well known extension of (1) (see Finney 1978, chap.7) is considered. Suppose that an experiment is performed where p doses ( $X_{11},\ldots,X_{1p}$) of a first stimulus and q doses $(X_{21},\ldots,X_{2q})$ of a second stimulus are assayed to obtain a set $\{Y_{1jk};\ j=1,..,p;\ k=1,..,n;\ Y_{2jk};\ j=1,..,q;\ k=1,..,n\}$ of $n(p+q)$ conditonally independent Normal observations with common variance

$\sigma^2$ and such that

$$E(Y_{1jk}) = \alpha + \beta(X_{1j})^{\lambda} \qquad k=1,\ldots,n\ ;\ j=1,\ldots,p$$
$$E(Y_{2jk}) = \alpha + \rho\beta(X_{2j})^{\lambda} \qquad k=1,\ldots,n\ ;\ j=1,\ldots,q \qquad (2)$$

This structure is the same as (1) except for the inclusion of the parameter $\lambda$ which is assumed to have a positive value and contains the previous model as a particular instance ($\lambda=1$). As usually, the doses are assumed to be non-negative ($X_{ij} \geq 0$).

This extension allows the experimenter to deal with a variety of curves for the dose-response relationship with respect to the relative potency it can be easily shown that if $X_1$ is a dose of the first stimulus and $X_2$ is a dose of the second stimulus, such that the associated expected responses are equal, then $(X_1/X_2)^{\lambda} = \rho$, so that the relative potency of these two stimuli may be defined as $\phi = \rho^{1/\lambda} = (\rho\beta/\beta)^{1/\lambda}$ and then, the assay still is of the *slope ratio* type. As an initial approximation, in what follows the analysis of the assay will be developed conditional on $\lambda$ (i.e. $\lambda$ will be assumed to have a *known* value). In the absence of additional assumptions the existency of $\phi$, at least for some cases, may be disputed. It will be argued, however, that a very natural restriction on $\rho$ can be imposed to avoid such a problem.

For the sake of simplicity and in order to make the parallelism with the results obtained in Mendoza (1986) more evident, let us define $W_{ij} = (X_{ij})^{\lambda}$. Therefore, for a given set of data D, the likelihood function of $(\rho,\alpha,\beta,\sigma)$ is given by

$$p(D \mid \rho,\alpha,\beta,\sigma) \propto \sigma^{-n(p+q)} \exp\{-[\Sigma_j\Sigma_k(Y_{1jk}-\alpha-\beta W_{1j})^2$$
$$+\Sigma_j\Sigma_k(Y_{2jk}-\alpha-\rho\beta W_{2j})^2]/(2\sigma^2)\}.$$

## 2. THE REFERENCE POSTERIOR DISTRIBUTION

In order to produce inferences about the parameter of interest, the information provided by the experiment must be combined, via Bayes' theorem, with the available initial information described by means of an appropiate prior distribution. However, in many cases the experimenter has only relatively vague initial information or thinks that he must let the experimental information 'speak by itself'. Whatever the cause may be, in such situation the use of a 'non-informative' or reference prior distribution may be the solution.

Bernardo's (1979) procedure already considered in Mendoza (1986) can be used here to obtain a reference posterior distribution for the relative potency. However, two points must be made initially: firstly, the model described is parametrized by the vector $\boldsymbol{\theta} = (\rho,\alpha,\beta,\sigma)^t$ but the parameter of interest is $\phi = \rho^{1/\lambda}$; secondly, $\phi$ can *always* be considered as a well defined transformation (one-to-one) of $\rho$ since, without loss of generality, $\rho$ can be assumed to have a positive value.

The rationale for the second point is as follows. A negative value for $\rho$ implies that the slopes have an opposite sign. Hence, and since $W = X^{\lambda}$ is an increasing funtion of the (non-negative) dose X for every possible value of $\lambda$, it follows that the stimuli have an opposite effect in the response i.e. a dose increase of one stimulus causes an increase on the expected response whereas a dose increase of the other stimulus causes a decrease of the expected response. It is clear that a comparison, based on the idea of relative potency, of such a pair of stimuli has no sense.

We may therefore, assume that the experimenter is assaying two stimuli with the same qualitative effect in the response and that, consequently, $\rho$ is positive. This restriction for the values of $\rho$ guarantees that $\phi$ exists and may be considered a one-to-one transformation of $\rho$. Under such circumstances, the reference posterior distribution for $\phi$ can be obtained. It must be recalled that the procedure proposed by Bernardo(1979) is invariant under one-to-one transformations of the parameter of interest, in the sense that the desired distribution is simply derived by the appropiate change of variable, if *a priori* the nuisance parameters and the parameter of interest are independent. Moreover, the effect of a truncation of the parameter space can be accomplished by imposing the truncation on the original reference distribution. Hence, the reference posterior distribution for $\phi$ can be derived from that for $\rho$ obtained in Mendoza(1986) replacing $W_{ij}$ by $X_{ij}$, imposing the restriction $\rho > 0$ and then applying the appropriate change of variable from $\rho$ to $\phi$.

As in section 4 of Bernardo 1979, we have that if $\pi(\rho)$ and $\pi(\alpha,\beta,\sigma|\rho)$ are defined as the operational priors which respectively maximize the missing information about $\rho$ and the missing residual information about $(\alpha,\beta,\sigma)$ given $\rho$, associated to the experiment described in the previous section, then

$$\pi(\alpha,\beta,\sigma|\rho) \propto \exp\{-\int p(D|\alpha,\beta,\sigma,\rho)H[p^*(\alpha,\beta,\sigma|\rho,D)]dD\},$$

$$\pi(\rho) \propto \exp\{-\int p(D|\rho)H[p^*(\rho|D)]dD\},$$

where H[.] is the well known entropy operator whereas $p^*(\rho|D)$ and $p^*(\alpha,\beta,\sigma|\rho,D)$ respectively represent the asymptotic posterior distribution of $\rho$ and the asymptotic posterior distribution of $\alpha$, $\beta$ and $\sigma$ given $\rho$. The asymptotic Normality of the joint posterior distribution can be verified so that after some calculus we have that

$$\begin{aligned}\pi(\rho,\alpha,\beta,\sigma) &= \pi(\alpha,\beta,\sigma|\rho)\pi(\rho)\\ &= \{\sigma^{-3}\}\{Q(\rho)\}^{-1/2}\end{aligned} \tag{4}$$

where,

$$Q(\rho) = c_2\rho^2+c_1\rho+c_0 ,$$

$$c_2 = (p+q)\Sigma_j(W_{2j})^2 - (W_{2}.)^2 ,$$

$$c_1 = -2W_1.W_2. ,$$

$$c_0 = (p+q)\Sigma_j(W_{1j})^2 - (W_{1}.)^2 ,$$

$$W_i. = \Sigma_j W_{ij} \quad ; \ i=1,2.$$

Combination of this prior distribution with the likelihood function (3) leads to the posterior reference distribution

$$\pi(\rho,\alpha,\beta,\sigma|D) \propto \{Q(\rho)\}^{-1/2}\ \sigma^{-M} \exp\{-[\Sigma_j\Sigma_k(Y_{1jk}-\alpha-\beta W_{1j})^2 + \Sigma_i\Sigma_j(Y_{2jk}-\alpha-\rho\beta W_{2j})^2]/(2\sigma^2)\} \qquad (5)$$

for every $\rho \in \mathbf{R}$, $\alpha \in \mathbf{R}$, $\beta \in \mathbf{R}$, $\sigma > 0$; $(M = n(p+q)+3)$.

The marginal density of $\rho$ is obtained integrating out from $\pi(\rho,\alpha,\beta,\sigma|D)$ the nuisance parameters $\alpha$, $\beta$ and $\sigma$ so that,

$$\pi(\rho|D) = \iiint \pi(\rho,\alpha,\beta,\sigma|D)\ d\alpha\, d\beta\, d\sigma$$

$$\propto \{Q(\rho)\}^{(m-1)} \ /\ \{Q(\rho)S_Y^2 - v\ [S_{WY1} + \rho S_{WY2}]^2\}^m$$

for every $\rho \in \mathbf{R}$ and where,

$$m = n(p+q)/2\ ; \qquad v = (p+q)/n\ ;$$

$$S_Y^2 = \Sigma_i\Sigma_j\Sigma_k\ (Y_{ijk})^2 - (\Sigma_i\Sigma_j\Sigma_k\ Y_{ijk})^2\ /(2m)\ ;$$

$$S_{WY1} = \Sigma_j\Sigma_k\ Y_{1jk}W_{1j} - (\Sigma_i\Sigma_j\Sigma_k\ Y_{ijk})(\Sigma_j W_{1j})\ /\ (p+q)\ ;$$

$$S_{WY2} = \Sigma_j\Sigma_k\ Y_{2jk}W_{2j} - (\Sigma_i\Sigma_j\Sigma_k\ Y_{ijk})(\Sigma_j W_{2j})\ /\ (p+q)\ .$$

As discussed in Mendoza (1986), $\pi(\rho|D)$ is a proper distribution whenever $p+q \geq 3$ and $n \geq 2$, and may have one or two modes. If the positiveness restriction on $\rho$ is imposed we have that

$$\pi(\rho|D) = \begin{cases} C\{Q(\rho)\}^{(m-1)} \ /\ \{Q(\rho)S_Y^2 - v[S_{WY1} + \rho S_{WY2}]^2\}^m\ ; & \rho > 0 \\ 0 & \text{elsewhere.} \end{cases}$$

where C is an adequate constant such that $\int \pi(\rho|D)d\rho = 1$. Now, since $\phi = \rho^{1/\lambda}$, we have $\rho = \phi^\lambda$ and hence, the derivative of $\rho$ with respect to $\phi$ is given by $\rho' = \lambda\phi^{(\lambda-1)}$ so that the reference posterior distribution for the relative potency $\phi$ can finally be written as

$$\pi(\phi|D) = \begin{cases} C^*\{\phi^{(\lambda-1)}[Q(\phi^\lambda)]^{(m-1)}\}/\{Q(\phi^\lambda)S_Y^2 - v[S_{WY1}+\phi^\lambda S_{WY2}]^2\}^m; & \phi > 0 \\ 0 & \text{elsewhere,} \end{cases}$$

where, $C^*$ is an adequate constant such that $\int \pi(\phi|D)d\phi = 1$.

The most important result is that obviously, $\pi(\phi|D)$ is also a proper distribution whenever $p+q \geq 3$ and $n \geq 2$ so that inferences about the parameter of interest may be obtained without any difficulty for any sensible design. It is worthwhile to recall that this is *not* the situation with the *frequentist* approach where the procedures applied to produce the so-called 'confidence intervals' have proved to be rather controversial (Fieller,1954).

Some other characteristics of $\pi(\phi|D)$ may depend upon the specific value of $\lambda$; the next section includes some examples which may provide some insight on the behaviour of $\pi(\phi|D)$.

## 3. NUMERICAL EXAMPLES

As has been stated in the previous section, the reference

posterior $\pi(\phi|D)$ is a proper distribution for any sensible experimental design. However, the constant of proportionality cannot be determinated analytically so that implementation of the procedure described requires the use of computer routines for numerical integration. In this section, two simulated examples are considered in order to contribute to the understanding of the general behaviour of $\pi(\phi|D)$. For the first example a set of parameter values ($\alpha=1, \beta=5, \rho=0.75, \sigma=1, \lambda=0.5$) have been selected to simulate a convex dose-response relationship. Figure 1 shows the curves associated to the expected responses.

Two independent samples ($D_1$ and $D_2$) with the same experimental design were generated using these parametric values. The resulting data is shown in Table 1. The respective reference posterior distributions, $\pi(\phi|D_1)$ and $\pi(\phi|D_2)$, are shown in Figure 2.

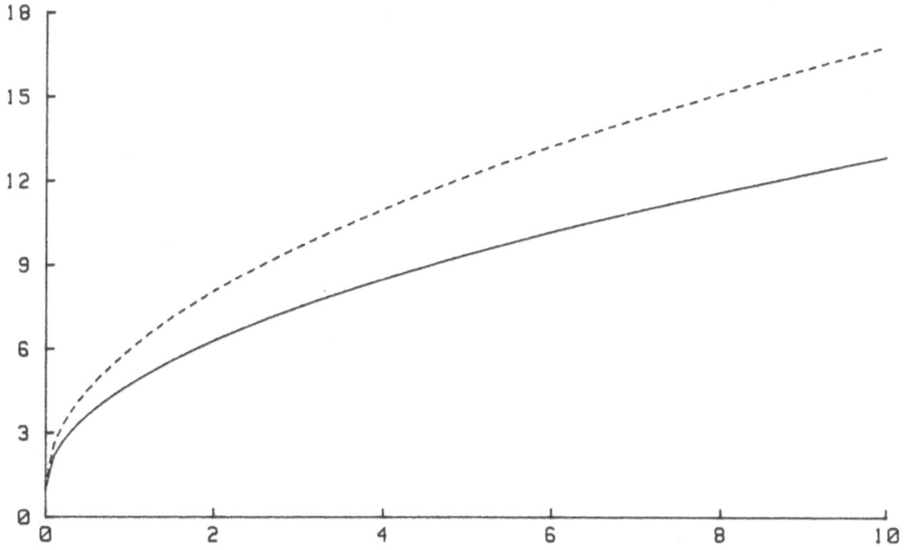


Figure 1. Expected responses first example (--:first stimulus, —:second stimulus).

Table 1. Simulated data, example 1 (n=3, p=q=4 )

| Stimulus | 1 | | | | 2 | | | |
|---|---|---|---|---|---|---|---|---|
| Doses | 1 | 4 | 7 | 10 | 2 | 4 | 6 | 8 |
| $D_1$ | 6.588 | 8.593 | 13.279 | 16.395 | 5.122 | 9.253 | 11.606 | 12.765 |
| | 6.889 | 11.473 | 14.465 | 15.974 | 7.273 | 8.559 | 11.634 | 11.050 |
| | 5.664 | 10.692 | 15.815 | 17.888 | 3.614 | 9.596 | 10.469 | 12.059 |
| $D_2$ | 4.959 | 10.980 | 14.721 | 17.954 | 5.678 | 9.062 | 9.327 | 11.679 |
| | 6.070 | 11.176 | 13.416 | 17.642 | 5.170 | 7.958 | 13.197 | 11.104 |
| | 5.084 | 10.215 | 15.459 | 17.686 | 5.545 | 9.178 | 8.692 | 11.553 |

Recalling that the true value of the parameter of interest is given by $\phi = \rho^2 = 0.5625$, we have that both unimodal distributions concentrate the mass of probability near the true value of $\phi$ even though the variation among samples seems to be appreciable for this experimental design.

The data for the second example was generated using a set of parameter values such that, as opposed to the first example, the obtained dose-response relationship is described by a concave curve ($\alpha=10, \beta=0.5, \rho=2, \sigma=1, \lambda=2$). Figure 3 shows the expected response curves for both stimuli. Again, two independent samples were generated according to this model. The information ($D_3$ and $D_4$) is displayed in Table 2.

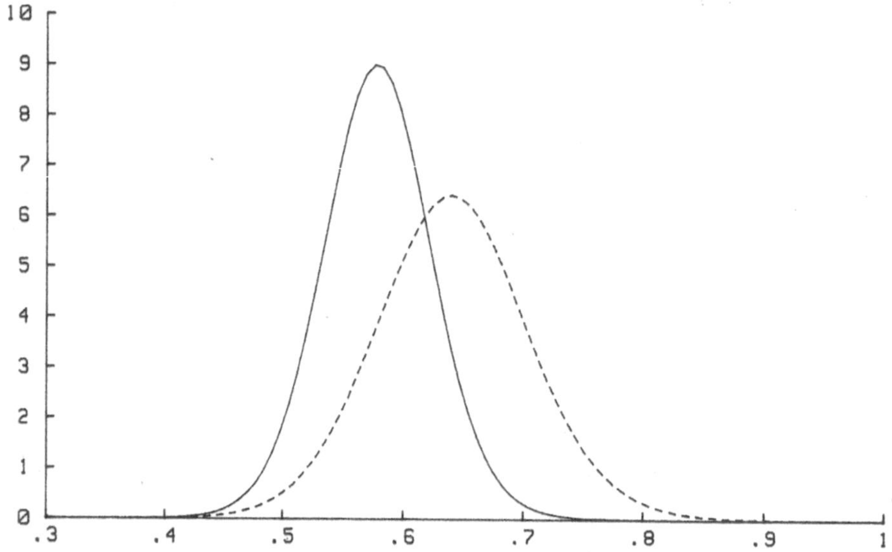


Figure 2. Reference posterior distributions (--: $\pi(\phi|D_1)$, —: $\pi(\phi|D_2)$).

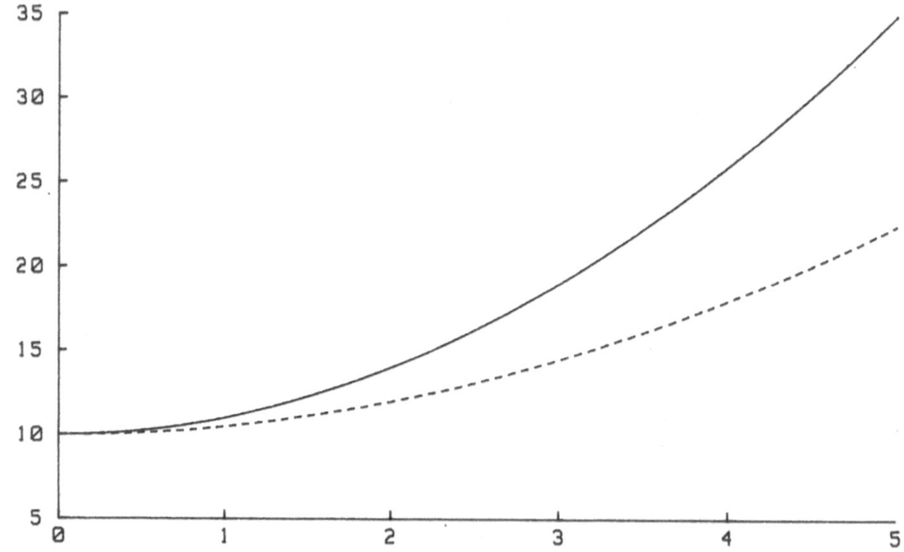


Figure 3. Expected responses second example (--: first stimulus, —: second stimulus).

Table 2. Simulated data, example 2 (n=3, p=q=4)

| Stimulus | 1 | | | | 2 | | | |
|---|---|---|---|---|---|---|---|---|
| Doses | 1 | 3 | 4 | 5 | 1 | 3 | 4 | 5 |
| | 12.135 | 16.993 | 17.512 | 22.275 | 9.702 | 19.460 | 28.564 | 35.786 |
| $D_3$ | 11.035 | 14.381 | 18.381 | 22.348 | 11.792 | 19.818 | 25.746 | 35.835 |
| | 11.697 | 13.531 | 17.220 | 21.627 | 11.830 | 20.039 | 24.939 | 36.826 |
| | 11.015 | 13.849 | 17.872 | 23.090 | 8.825 | 18.957 | 26.865 | 34.946 |
| $D_4$ | 10.380 | 13.644 | 17.434 | 22.717 | 10.145 | 20.851 | 25.028 | 35.324 |
| | 9.384 | 12.789 | 17.533 | 21.706 | 10.314 | 19.391 | 25.198 | 35.496 |

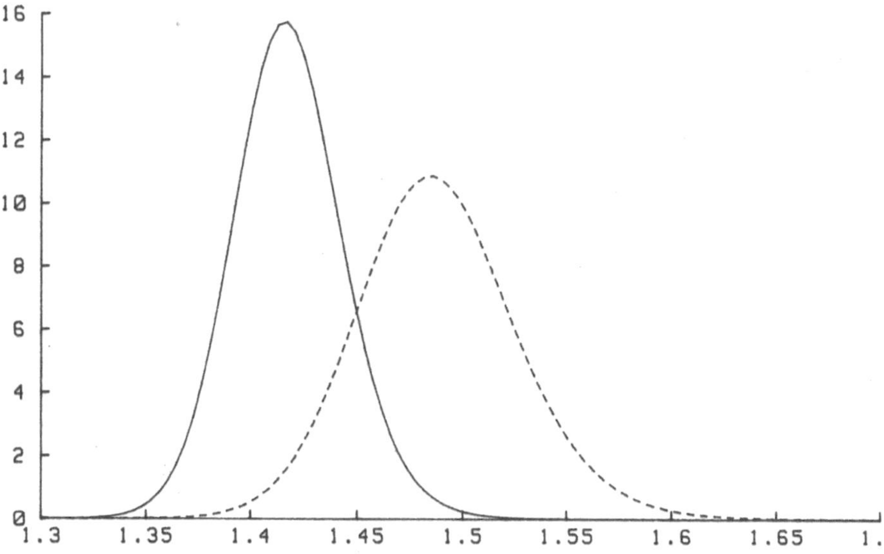


Figure 4.Reference posterior distributions (--:$\pi(\phi|D_3)$,—:$\pi(\phi|D_4)$)

Since $\phi = \rho^{1/2} = \sqrt{2}$, it follows from Figure 4 that similar conclusions to those obtained in example 1 can be produced.Both posterior distributions are unimodal and concentrate the mass of probability on a neighborhood of the true value of $\phi$. Another noticeable similarity is that for the selected design the variation among samples may be again considered appreciable.

## 4. CONCLUDING REMARKS

The procedure described in this paper can be used to deal with a large class of biological assays of the slope ratio type. However, it requires the value of the parameter $\lambda$ to be known. An additional effort may be necessary in order to investigate the situation where $\lambda$ is unknown and, hence, has to be considered as another nuisance parameter. An alternative approach which is already available using only the results contained in this paper may be based on the idea of a *sensitivity analysis* of the posterior distribution $\pi(\phi|D)$ for a range of values of $\lambda$. Indeed, it often happens that the experimenter's prior information about $\lambda$ reduces to the specification of a lower bound and an upper bound for this parameter. Under such circumstances, particularly if the interval defined is rather narrow, it may suffice to produce a (usually small) number of conditional analysis for a given set of $\lambda$-values within the interval to obtain a useful idea of the behaviour of $\pi(\phi|D)$.

## ACKNOWLEDGEMENT

This work was partly supported by the Sistema Nacional de Investigadores (SNI) of México.

## REFERENCES

Bernardo, J.M., 1979, Reference posterior distributions for Bayesian inference(with discussion), J.R.Statist.Soc., B, 41:113-147.

Fieller, E.C., 1954, Some problems in interval estimation, J.R. Statist. Soc., B, 16:175-185.

Finney, D.J., 1978, "Statistical Method in Biological Assay", Griffin, London.

Mendoza, M., 1986, A Bayesian analysis of the slope ratio assay, Submitted for publication.

# ON ABSOLUTE CONTINUITY OF MEASURES DUE TO GAUSSIAN LOCALLY STATIONARY PROCESSES

Jiří Michálek

Institute of Information Theory
and Automation
Prague, Czechoslovakia

The notion of a locally stationary process was introduced and first studied by Silvermann[1]. His results were generalized by Michálek[2] where a spectral decomposition of a locally stationary harmonizable process is investigated. The notions of a harmonizable covariance function and of a harmonizable process were introduced by Loève; a short note on a spectral theory of harmonizable processes is given e. g. in Loève[3].

Let $x(t)$, $t \in T \subset R_1$ be a locally stationary harmonizable process with a spectral density function $h(\lambda,\mu)$. The property of local stationarity enables to express

$$h(\lambda,\mu) = h_1\ \left(\frac{\lambda+\mu}{2}\right)\ h_2(\lambda-\mu)$$

where $h(\lambda,\mu)$ is a locally stationary covariance again, i. e. $h_1 \geq 0$ and $h_2$ is a stationary covariance. Such a process, see Michálek[2], can be expressed in the form of a stochastic integral understood in the quadratic mean sense

$$x(t) = \int_{-\infty}^{+\infty} e^{it\lambda} z(\lambda)d\lambda \qquad (1)$$

where $z(\lambda)$ is a locally stationary process again having $h(\lambda,\mu)$ as its covariance function. Further, let us suppose that the process $x(t)$, $t \in T$ is Gaussian; with respect to a measure $P_0$ $E_0\{x(t)\} = 0$ for every $t \in T$ and with respect to another measure $P_1$ $E_1\{x(t)\} = a(t)$ on $T$ under assumption

$$E_1\{x(s)\overline{x(t)} - a(s)\overline{a(t)}\} = E_0\{x(s)\overline{x(t)}\}$$

holding for every pair $(s,t) \in T \times T$. At once the following question arises: under which conditions put on the function $a(t)$, $t \in T$ the measures $P_0$, $P_1$ shall be equivalent?

Let us consider a linear set $U = \{u: u = \sum_{k=1}^{n} c_k x(t_k)$, $c_k$ complex, $t_k \in T$, $k = 1,2,\dots,n\}$ of all linear combinations defined by means of values $x(t)$, $t \in T$. Thanks to the spectral decomposition (1) one can immediately write

$$u = \int_{-\infty}^{+\infty} \sum_{k=1}^{n} c_k e^{it_k\lambda} z(\lambda)d\lambda$$

and hence there exists a one-to-one-mapping among the elements of $U$ and the functions from $S = \{\sum_{k=1}^{n} c_k e^{it_k\lambda}\}$ if we identify such elements $u_1$, $u_2 \in U$ for which

$$E_0\{|u_1-u_2|^2\} = 0, \text{ i. e.}$$

$$\iint_{-\infty}^{+\infty} \Big(\sum_{k=1}^{n} c_k e^{it_k\lambda} - \sum_{\ell=1}^{m} d\, e^{is_\ell\lambda}\Big)\Big(\sum_{k=1}^{n} \overline{c}_k\, e^{-it_k\mu} - \sum_{\ell=1}^{m} \overline{d}_\ell e^{-is_\ell\mu}\Big) \times$$

$$\times\, h_1\Big(\frac{\lambda+\mu}{2}\Big)\, h_2(\lambda-\mu)d\lambda d\mu = 0$$

when $u_1 = \sum_{k=1}^{n} c_k x(t_k)$, $u_2 = \sum_{\ell=1}^{m} d_\ell x(s_\ell)$.

In this way we can introduce a scalar product $\langle u_1,u_2\rangle$ on $U$, namely

$$\langle u_1,u_2\rangle = E_0\{u_1\overline{u}_2\} = \iint_{-\infty}^{+\infty} f_1(\lambda)\overline{f_2(\mu)}h_1\Big(\frac{\lambda+\mu}{2}\Big)h_2(\lambda-\mu)d\lambda d\mu$$

where $f_1(\lambda) = \sum_{k=1}^{n} c_k e^{it_k\lambda}$, $f_2(\lambda) = \sum_{\ell=1}^{m} d\, e^{is_\ell\lambda}$.

Let $\overline{U}$ be the closure of $U$ with respect to the norm $||u|| = \langle u,u\rangle^{\frac{1}{2}}$. Then $\overline{U}$ is a Hilbert space because $\overline{U}$ is a subspace in $L_2(\Omega,\Sigma,P_0)$ where the process $x(t)$, $t \in T$ is defined. By means of the one-to-one mapping between $U$ and $S$ we can construct a closure $\overline{S}$ of $S$ under the norm

$$||f|| = \Big(\iint_{-\infty}^{+\infty} f(\lambda)\overline{f}(\mu)h_1\Big(\frac{\lambda+\mu}{2}\Big)h_2(\lambda-\mu)d\lambda d\mu\Big)^{\frac{1}{2}}.$$

when $f, g \in \overline{S}$ then the scalar product induced by that one-to-one mapping into $\overline{S}$ has the form

$$(f,g) = \iint_{-\infty}^{+\infty} f(\lambda)\overline{g(\mu)}h_1\left(\frac{\lambda+\mu}{2}\right)h_2(\lambda-\mu)\,d\lambda d\mu.$$

Thus, for every random variable $\eta \in \overline{U}$ there exists an element $f_\eta(\lambda) \in \overline{S}$ such that

$$\eta = \int_{-\infty}^{+\infty} f_\eta(\lambda)z(\lambda)\,d\lambda$$

and

$$E_0\{|\eta|^2\} = \iint_{-\infty}^{+\infty} f_\eta(\lambda)\overline{f_\eta(\mu)}h_1\left(\frac{\lambda+\mu}{2}\right)h_2(\lambda-\mu)\,d\lambda d\mu.$$

It follows from the construction of the closure $\overline{S}$ that $\overline{S}$ is a subspace in the space $L_2(h_1,h_2)$ of all complex functions of a real variable for which

$$\iint_{-\infty}^{+\infty} f(\lambda)\overline{f(\mu)}h_1\left(\frac{\lambda+\mu}{2}\right)h_2(\lambda-\mu)\,d\lambda d\mu$$

exists. A general theory of absolute continuity of Gaussian measures presented in Rozanov[4] gives then a necessary condition for absolute continuity of $P_0$, $P_1$.

*Theorem.* A necessary condition for absolute continuity of measures $P_0$, $P_1$ corresponding to Gaussian locally stationary harmonizable processes distinguishing in expected values only is a possibility to express the difference $a(t)$, $t \in T$ of these expected values in the following form

$$a(t) = \iint_{-\infty}^{+\infty} f(\lambda)e^{-it\mu}h_1\left(\frac{\lambda+\mu}{2}\right)h_2(\lambda-\mu)\,d\lambda d\mu.$$

If $f(\lambda) \in \overline{S}$ then this condition is sufficient too. It means when the closure $\overline{U}$ is izometric to the whole space $L_2(h_1,h_2)$ then this condition will be necessary and sufficient. This situation occurs, e. g. if $T = (-\infty,+\infty)$ because then an inversion formula

$$z(\lambda) = \frac{1}{2\pi}\int_{-\infty}^{+\infty} e^{-it\lambda}x(t)\,dt$$

expressing $z(\lambda)$ by means of $x(t)$, $t \in T$ holds and hence the both processes $x(t)$, $t \in T$ and $z(\lambda)$, $\lambda \in (-\infty,+\infty)$ have the same range of values.

REFERENCES

1. R. A. Silvermann, Locally Stationary Processes. IRE Transactions of Information Theory, Vol IT-3, 183:187 (1957).
2. J. Michálek, Spectral Decomposition of Locally Stationary Random Processes, Kybernetika, No 3, 244:255 (1986).
3. M. Loève, "Probability Theory", D. van Nostrand Company, Toronto - New York - London (1960).
4. Y. A. Rozanov, "Gaussian Infinite-dimensional Probability Distributions", Nauka, Moscow (1968) (in Russian).

# BAYESIAN ADAPTIVE DECISION THEORY VERSUS DYNAMIC GAMES AS MODELS FOR ECONOMIC PLANNING AND POLICY-MAKING UNDER UNCERTAINTY

Reinhard Neck

University of Economics, Vienna
Vienna, Austria

## INTRODUCTION

One of the main problems in theoretical and applied studies of quantitative economic policy and planning is concerned with the potential achievements of stabilization policies aimed at controlling a dynamic economic system and guided by an intertemporal objective function, which exhibits trade-offs between different target variables. Optimization methods, particularly those of optimal control theory and dynamic programming including adaptive control theory, have been applied to many theoretical and empirical models in order to obtain insights into this question. During the last years, however, this research has come under increasing attack from several authors. One of the main arguments against these optimization studies is the assertion that optimizing stabilization policies cannot achieve their aims because of the high degree of uncertainty inherent in socio-economic (as opposed to physical) systems. But if the basic decision-theoretic framework of the theory of economic policy is accepted, this claim is largely lacking a theoretical foundation. In particular, it can be shown by methods of adaptive (dual) optimal control theory that a combination of cautious active policy-making and learning about the system response in general can improve the performance to be achieved, even under substantial uncertainties of several kinds (see, e. g., Kendrick, 1981).

One possible justification of the critical attitude of some opponents to active stabilization policies may be found in the idea that in economic systems there is not only uncertainty in a stochastic sense, but also uncertainty arising from strategic reactions of other (public and private sectors') decision-makers upon economic policies planned by the government. Although this position by itself need not substantiate a verdict against stabilization policies, it may lead to the methodological consequence of abandoning decision theory, including Bayesian adaptive control theory, in favor of dynamic game models, resulting in a different theoretical framework for analyzing stabilization policies. In this paper we will investigate whether subjectivist decision theory becomes obsolete in a game-theoretic framework and whether stochastic and adaptive decision and control theory or dynamic game theory are more adequate as models for economic planning and policy-making under uncertainty. These issues will be discussed first on a general methodological level and then illustrated with the help of a simple analytical economic model.

From the perspective of methodology, the main question concerns the relative advantages and disadvantages of subjectivist decision theory and game theory, especially in a dynamic framework. This is a more general problem not confined to the theory of economic planning and policy, which has been discussed recently in a series of papers following Kadane and Larkey (1982; see also their exchange with Harsanyi; Kahan, 1983; Roth and Schoumaker, 1983; Rothkopf, 1983; Kadane and Larkey, 1983, and the comments by Shubik). Although so far no consensus has been reached about appropriate research strategies, the main arguments of this discussion concerning the purpose of the investigation at hand, the presumption of rationality, and the necessity of specifying strategic interactions and particular solution concepts can be applied to the framework of economic policy as well. Some of these consequences will be discussed in the present section.

## Purpose of the Study

The first issue to be clarified by a researcher is the purpose of the investigation he (she) is carrying out. In particular, a fundamental distinction can be made between positive studies, concerned with "is"-statements, and normative ones, concerned with "ought"-statements. Positive theories may be descriptive (e. g., describing institutions carrying out stabilization policies), explanatory (specifying possible causal relations; e. g. influences of economic policy variables on targets), or predictive (e. g., giving forecasts about future developments of target variables contingent upon specific policy actions), although usually more than one of these purposes will be pursued in a positive investigation. Normative theories, on the other hand, may be speculative (specifying only a criterion, e. g., an objective function for an economic policy-maker) or prescriptive (providing also some procedure to obtain the "optimal" or at least a "better" value of the criterion; e. g., specific policy measures to be adopted); furthermore, normative theories may serve as an advice for an actual decision-maker (for economic planning in this case it should contain also informations on how to implement the policy recommendations) or merely for comparing different institutional arrangements or outcomes of actual and hypothetical policies, including those for previous periods.

In general, normative and positive statements will be different unless one believes that the way decisions are made is always the best one possible. Both kinds of theories are useful, but they serve different research purposes. Kadane and Larkey (1983), however, maintain that they are often confused in developing and applying theories of decision-making in the social sciences, especially in game theory. Although it is true that one must carefully distinguish between normative game theory, which may serve to advise particular players or to get qualitative insights into results of recommended behavior, and positive theories of actual behavior in game situations, this applies mutatis mutandis to subjectivist decision theory, which is also primarily prescriptive; whether either of the two approaches can be used as a good predictive theory is primarily an empirical question. This might be one of the reasons why so far attempts to develop a positive theory of economic policy and planning on the basis of either decision or game theory had only limited success, and most studies in this area show (at least implicitly) a normative orientation: Empirically reliable positive theories of actual policy-makers' behavior are largely lacking. This is unfortunate, because even if the advice-perspective of a normative theory of economic policy is accepted, at least some elements of such a positive theory are required to assess the feasibility (with respect to implementation) of policy recommendations.

A particular problem arises, however, in a game theoretic approach, namely that of the behavior of the opponents to a specific decision-maker.

Here we have to distinguish between the perspective of an outside observer of the game (the scientist) and the perspective of a single player or his (her) adviser. For the purposes of a normative theory usually only the latter will be relevant, in the context of economic policy in most cases as an advice to the government or a specific politician. However, both for positive and for prescriptive purposes positive theories of the behavior of the other players and of the interaction between the players are required. An adviser needs not only a prescriptive theory to guide the government's decisions, but also a predictive theory about other decision-makers, which may be different from the recommended decision procedure of the government. Since both predictions and prescriptions of government behavior must be conditional on the behavior of the others, behavioral differences must be recognizable in a model of game theory for economic policy. Whether the assumptions about the other players include their rationality is primarily an empirical question; Bayesian decision theory could also allow for revisions of those assumptions during the course of accumulation of information about the game. However, as Shubik has correctly pointed out, game theory cannot be blamed for neglecting this, because different assumptions about the rules of the game and about the solution concept also allow for a variety of possible behaviors of the other players. Investigating the consequences of different solution concepts for a given problem of economic planning with more decision-makers can provide useful insights, both for positive and for normative purposes; whether this procedure results in a loss of information about possible behavior of the opponents depends on the scope of the solution concepts considered and to some extent also on the specific problem at hand.

## Rationality

As in other fields of the social sciences, the task of modeling human behavior creates tremendous problems for a positive theory of economic policy: Cognitive processes are highly complex, experiments are usually not feasible, and decision processes must be recovered from data about behavior in nonrepetitive situations, which may depend on the context in an essential way. Psychological experiments of decisions in laboratory situations have shown that human beings often are not able to conform with the coherence postulates of Bayesian prescriptions, and the same is true for game-theoretic concepts such as the minimax strategy. On the other hand, there exist a few successful applications of game theory to experimental negotiations. One of the reasons for the mixed evidence is the impossibility of deriving unique models of decision processes from data about behavior, even under "ideal" experimental conditions. Already Simon (1956) has shown that we must distinguish between subjective rationality (given the goals and perceptions of the decision-maker) and objective rationality (judged from the experimenter or the observer): There may be subjective without objective rationality if decisions are rational given perceptions which themselves may be irrational. For a predictive theory this results in a loss of forecasting capability, since a model of the decision-maker's view of the alternatives and their consequences would be required. The difficulty of objectifying subjective probabilities and utilities may be responsible for the seemingly "irrational" behavior of decision-makers as seen by an outside observer.

Although this gives some support to the approach of subjectivist decision theory, it is not necessarily an argument against game theory, because for both theories the same basic model of "rational man" is fundamental, and not every game-theoretic analysis needs the assumption of well-defined utility functions about uncertain alternatives with objective probabilities. Whereas for a normative analysis the concept of rationality (both in a game-theoretic and a decision-theoretic context, depending upon whether strategic interactions are essential or not) seems unquestionable, at least for the decision-maker to be advised, for a positive theory different methodological views are possible. Harsanyi's position, which claims that behavior must be interpreted either as rational or as psychologically understandable deviation from rationality, may be too rigid for problems of economic policy, but some positive theory of behavior is certainly required, and often subjectively rational optimization will be more appropriate than an incompletely specified alternative theory. Furthermore, rational theories of decision-making may help the decision-maker in better understanding the problem and the situation, both his (her) own problem and that of his (her) opponents. This didactic value of both subjectivist decision theory and game theory can be particularly important in economic planning, where repeated normative uses of concepts of rationality may contribute to more widespread rationality of actual decisions and may hence improve their usefulness for a positive theory. Research on elicitation of prior probabilities and preferences, for instance, and training of planners in these abilities may help closing the gap between normative rationality and actual decision-making.

Nevertheless theories of "bounded rationality" or "semirational" behavior should not be discarded as alternatives to both decision and game theory. There may be situations where costs of obtaining and processing informations enforce using a simpler decision procedure than optimization. But even then it would be desirable to formulate this kind of behavior in a theoretically satisfactory way, in particular to deduce it from more general assumptions and principles (including that of using the "optimal" decision procedure in the presence of informational costs), before it should be applied to a particular problem of modeling economic policies. "Applied modeling is an art" (Rothkopf, 1983: 1345) is not sufficient if one wants to avoid "methodological anarchism". Modeling should be directed towards showing that empirically observed behavior can be explained systematically by a general theory, which often (though not always) will contain some elements of (at least subjective) rationality. This does not preclude a careful consideration of details for concrete modeling, but points to the necessity of having an organizing principle for analysing these details. Economic policy studies based on ad-hoc schemes of bounded rationality (e. g. Mosley, 1976) could be critizised for lacking such a principle.

Strategic Interactions

The main difference between Bayesian decision theory and game theory is the explicit recognition of strategic interactions between different decision-makers (players) by the latter. Decisions by other players are regarded not as results of random processes but of conscious deliberations; the decision variables of the other players are given to each decision-maker, but endogenous with respect to the game model. Every player decides on the basis of his (her) expectations about the expectations of the other rational players. In situations where there is such an essential interaction which is understood by all participants, game theory can provide the adequate model. On the other hand, the interactive structure may be inessential if there are too many other players, or if the other players do not react upon the decisions of one player, or if there is substantial uncertainty about the "rules of the game". In these cases the situation may be modeled as a one-person game against "nature", i. e. against the aggregate of the passive or unknown other players, and a Bayesian approach may be appropriate (cf. Kahan, 1983). For economic policy problems this may be true if the government is confronted with a private sector composed of a great number of households and firms who do not react strategically on government policies. When "big" institutions (firms, unions, associations) or other policy-makers on a national (e. g., the central bank) or an international level (governments of other countries, especially if they are "big" in the sense of theoretical international economics) are involved, then usually a game-theoretic model will be required.

Subjectivist one-person decision theory may also be applied if, for some interaction structure, it is known that the other players do not act rationally, provided there is a theory (supported by empirical or other evidence) about probable strategies of these irrational players (or about the probable "errors" in their strategies). Bayesian objections against game theory thus have some relevance for asymmetrical situations, where the decision-maker does not expect his (her) opponents to react rationally. But even then, not any subjective probability should be acceptable as prior; some (psychological or other) theory is required, and game-theoretic results may be used as reference standards. It may be dangerous for a policy-maker to underestimate the rationality of the other players; some of the policy ineffectiveness and time inconsistency results of economic theory are due to government's neglect of strategic reactions of the private sector. This means that there is a theoretical shortcut; the recursion of "I think that he (she) thinks that I think..." is cut off by arbitrarily assuming the other players not to be fully rational. Unless there are convincing reasons for this lack of rationality, subjectivist decision theory can serve at most as a substitute for a game-theoretic analysis when the latter cannot (yet) be used due to mathematical intractability (see also Lindley, 1982: 217).

## Solution Concepts

Even if there are rational, strategically interacting decision-makers, the problem remains whether the structure of the interaction and hence the solution concept of the game is unambiguous from the outset. The theory of games has developed a great number of solution concepts, both for non-cooperative and for cooperative games. Although game theorists universally agree that their most important theoretical task is to develop exact definitions of rational behavior in situations of strategic interactions, there is much less agreement about the requirement of a single solution concept valid for all such situations. Whereas some game theorists (notably Harsanyi) aim at developing a general solution concept leading to a unique and predictable outcome for all situations of strategic interactions, most game theorists confine their task to restricting possible outcomes to a certain range and to developing several solution concepts. For instance, in a two-person zero-sum (or constant-sum) game the minimax solution is normatively compelling if the other player also acts rationally and hence plays also minimax. However, if the other player is irrational, or if the game is not two-person constant-sum, or if it is a game against nature, then game theorists will readily accept the Bayesian argument of minimax being inconsistent for such situations (though not necessarily the conclusion that only subjectivist decision theory provides the adequate analytical tool).

Since there are rarely situations in economic policy where a two-person constant-sum game could be regarded as a correct model, the lack of a universally accepted solution concept for other games is relevant for assessing game versus decision theory models of economic policies. Game theorists respond to this challenge by emphasizing the importance of modeling the assumptions and rules of the game prior to specifying the solution concept. First it has to be determined whether the game is assumed to be cooperative or noncooperative. The former will be better suited for interactions between a few players who can easily communicate and agree on "fair" commitments. If there are more players and communication possibilities are low, the choice of the solution concept becomes more complicated. But unless all but one player can be aggregated into a fictitious "passive" player, one-person games against nature or subjectivist decision theory will not solve the problem, especially because of the infinite recursion problem. Moreover, for games against nature with uncertainty concerning the payoffs and the strategies there exists no universally accepted behavioral theory so far. Thus in general there is no easy way to bypass the requirement of carefully modeling the interactions between the players and examining several solution concepts for each particular problem.

One of the complications behind the multitude of solution concepts lies in the formation of expectations by decision-makers, in particular about the other players' actions and expectations. If the solution concept is not obvious to all participants in the strategic interaction, it must rely on their expectations about possible actions of the others. Assuming rationality for the other players, however, restricts the class of prior probability distributions for each player in a Bayesian approach. Neglecting rational actions of other players and their reciprocal expectations of rational actions amounts to throwing away essential informations, as Harsanyi has correctly pointed out. Therefore for rational players it is necessary to eliminate systematically those priors which contradict certain classes of solution concepts which seem relevant for the problem under consideration. The extremely subjectivist point of view, which insists on Bayesian methods also for the elicitation of prior beliefs, is not adequate for situations where one or several solution concepts seem potentially applicable. But the extremely deterministic point of view requiring a unique solution to every game is not adequate either, unless there are strong arguments for a particular solution concept to be applied. Experiments (Roth and Schoumaker, 1983) have shown that expectations and hence subjective probabilities become relevant for determining the outcomes of many games, where further assumptions about the behavior and the expectations of the players are required to define rational behavior. Thus informations from the solution concept(s) and (empirical or theoretical) informations about expectations of other players should be combined in forming priors in a subjectivist decision theory approach; conversely, in a game-theoretic analysis more than one solution concept could be tried, the outcomes could be compared and assessed with the help of additional problem-specific informations.

Accepting a Bayesian approach to the analysis of a situation of strategic interaction amounts to assuming that the policy-maker has subjective probabilities about the actions of his (her) opponents and their consequences, including their expectations about his (her) own actions, and so on. Particular solution concepts in this framework generate special prior distributions and vice versa. From a theoretical point of view it is interesting to know how a rational player would choose his (her) prior probability distribution of the opponents' strategies if he (she) expects that they will act rationally and that they also expect every player to act rationally. In particular, which restrictions upon the prior beliefs of the players and on their behavior given prior beliefs are implied by the assumptions or axioms of different solution concepts? This problem has been studied recently by Tan and Werlang (1986), who transform a simultaneous game in normal form into a Bayesian decision problem, assume for each player a space of uncertainty over the strategic choices of the other players with priors over the sets of strategies of the other players, over their priors and so on for the infinite recursion of beliefs, and derive the noncooperative solution concepts of iterative elimination of strictly dominated strategies, of rationalizable equilibrium, of Nash equilibrium, and of correlated equilibrium from explicit assumptions about behavior, information and beliefs of the players. The results show that iterative elimination of strictly dominated strategies is the solution concept for Bayesian players if it is only assumed that rationality is "common knowledge", that is, if every player knows the structure of the model, the rules of the game and the rationality of all players, and every player knows that every player knows the above, and so on. Rationalizable strategies are obtained when, in addition, common knowledge of independent actions of all players is assumed, whereas for Nash and correlated equilibrium additional assumptions about the players' priors are required, which can be interpreted as demanding some coordination between them. In a much simpler approach Kadane (1985) has shown that in a two-person zero-sum game both players act in accordance with Bayesian decision theory if their utilities express opposite interests (have constant sum) and their prior probabilities coincide. These results show that widely used solution concepts, such as the Nash equilibrium, impose nontrivial restrictions on the priors of a Bayesian player, which should be taken into account for applications of these solution concepts.

For problems of economic policy and planning an additional complication arises from the dynamics usually inherent in them. In a Bayesian approach this means that not only probabilities have to be coherent over time according to Bayes' theorem and the likelihood principle of Bayesian statistics, but also the principle of optimality of dynamic programming must hold for the sequence of decisions. Dynamic one-person decision problems under uncertainty can be formulated as stochastic control problems; when learning about the economic system is explicitly allowed, adaptive control methods are required, which again may use Bayesian estimates of unknown states, parameters, etc. Unfortunately, a full analytical solution of an adaptive control problem for an optimal closed-loop policy cannot be obtained so far, even for extremely simple models. Only approximations are available, obtained for example by simplifying the information structure or by restricting the admissible control law, sometimes supported by a sensitivity analysis with respect to approximation and model errors. On the other hand, the theory of dynamic games has developed several feasible solution methods, but a full characterization of equilibria for closed-loop information structures with memory strategies is also not available so far. Furthermore, the problem of multiple equilibrium solutions is aggravated in a dynamic game model by "informational nonuniqueness", implying the existence of a continuum of Nash equilibria for closed-loop information structures, for example. However, there are some solution concepts, notably the memoryless feedback Nash equilibrium, which have desirable properties (subgame perfectness) and can be obtained analytically or numerically, at least for simple (e. g., linear-quadratic) models. Thus from the point of mathematical tractability, in a dynamic context both Bayesian decision (control) theory and game theory at the present state of the art put severe limitations on the complexity of an economic policy problem that can be solved; the previous considerations about the appropriateness of either Bayesian decision theory or game theory apply, nevertheless, also for the dynamic case.

## AN ECONOMIC POLICY EXAMPLE

The methodological discussion, although providing some hints to situations where a subjectivist decision-theoretic analysis could be more useful than a game-theoretic one and vice versa, does not give a clear-cut conclusion for deciding between the two approaches for problems of economic policy-making and planning. But it suggests that for problems where strategic interactions are essential but the solution concept cannot be determined unambiguously a priori, a combination of Bayesian and game-theoretic insights might be most helpful. Many economic policy problems are of this type, especially when their dynamic nature and hence the ambiguity with respect to the information structure are taken into account. This may be illustrated with the help of a simple example from the theory of stabilization policy. For reasons of lack of space, it is only sketched here; a more extensive discussion of its different solutions is given in Neck (1986).

We consider the following analytical model of the trade-off between unemployment and inflation in a closed economy:

$$p(t) = \lambda h(t) + p^*(t), \ \lambda > 0, \tag{1}$$

$$\hat{u}(t) = u(t) - u_N = -\delta h(t), \ \delta > 0, \tag{2}$$

$$h(t) = \beta[m(t) - p(t)] + \gamma g(t), \ \beta > 0, \ \gamma > 0, \tag{3}$$

$$\dot{p}^*(t) = \eta[p(t) - p^*(t)], \ \eta > 0, \ p^*(0) = p^*_o > 0. \tag{4}$$

Here $p(t)$ is the actual rate of inflation, $p^*(t)$ the expected rate of inflation, $h(t)$ aggregate excess demand, $u(t)$ the rate of unemployment, $u_N$ the constant natural rate of unemployment, $m(t)$ the growth rate of money supply, and $g(t)$ the growth rate of real public expenditures for goods and services. $g(t)$ is a policy variable of the government (player 1 in a game, or "the decision-maker" in a Bayesian setting), and $m(t)$ is the policy variable of the central bank. The government aims at minimizing

$$J_1 = (1/2) \int_0^\infty \exp(-rt)[a_1\hat{u}^2(t) + b_1p^2(t) + c_1g^2(t)]dt, \tag{5}$$

and the central bank wants to minimize

$$J_2 = (1/2) \int_0^\infty \exp(-rt)[a_2\hat{u}^2(t) + b_2p^2(t) + d_2m^2(t)]dt. \tag{6}$$

Extensions to models with stochastic disturbances are given in Neck (1986).

The outcomes of several solution concepts for the above dynamic game have been determined analytically; here we consider only a cooperative Pareto-optimal solution and the feedback Nash equilibrium without memory. Pareto-optimal policies can be obtained by minimizing $J = \alpha J_1 + (1-\alpha)J_2$ for some $\alpha \in (0,1)$, yielding linear feedback policy rules:

$$g^P(t) = (g_1^P + g_2^PV^P)\, p^*(t), \tag{7}$$

$$m^P(t) = (m_1^P + m_2^PV^P)\, p^*(t), \tag{8}$$

where the $g_i^P$, $m_i^P$, $i = 1,2$, and $V^P$ are constants to be calculated from the parameters of the model and the objective functions. For the feedback Nash equilibrium solution, we get similar feedback rules:

$$g^N(t) = [g_1^N + g_2^NV_1^N + g_3^NV_2^N]p^*(t), \tag{9}$$

$$m^N(t) = [m_1^N + m_2^NV_1^N + m_3^NV_2^N]p^*(t), \tag{10}$$

but with different coefficients $g_i^N$, $m_i^N$, $i = 1,2,3$, $V_j^N$, $j = 1,2$. Also the optimal values of the objective functions can be calculated in each case, giving $J_i^{P*}$ and $J_i^{N*}$, $i = 1,2$, as functions of the parameters.

If we assume that the government does not know exactly whether its interaction with the central bank is cooperative or noncooperative, but only considers a (particular) Pareto-optimum and the feedback Nash equilibrium as possibilities, it may have a subjective probability of p for the cooperative solution being played (or for the central bank to cooperate) and probability of 1-p for the noncooperative equilibrium. Then without learning the government could minimize, over $g(t)$, $pJ_1^P + (1-p)J_1^N$ subject to a system

$$\dot{p}^*(t) = e_1[pm^P(t) + (1-p)m^N(t)] + e_2g(t) - e_3p^*(t), \tag{11}$$

where $e_i$, $i=1,2,3$, are the coefficients of the reduced form of the model (1) - (4), and $J_1^P$ and $J_1^N$ are functions of $\{g(t),\ t \in [0,\infty)\}$ obtained from (5) by inserting $m(t) = m^P(t)$ and $m(t) = m^N(t)$, respectively. A more realistic assumption would be that the government reoptimized periodically after some time interval, simultaneously revising its prior probability p in the light of the results achieved so far, according to the Bayes formula. Although there may be analytical problems in obtaining revised estimates of p, this approach is more flexible than relying on one particular solution concept, since it allows for the possibilities of both cooperative and noncooperative behavior of the opponent (the central bank). Extensions to the introduction of subjective probabilities about the parameters of the model (including $\alpha$) and of $J_2$ as well as to allowing for other solution concepts (like Stackelberg equilibrium) and other information structures (like open-loop, memory, etc.) could also be conceived of. The point to be made by this example is that even in an economic policy problem with clearly strategic interactions, Bayesian adaptive decision theory can be used in addition to dynamic game methods when the solution concept is not beyond dispute. For more sophisticated models analytical calculations will be prohibitively difficult, but we hope to have shown that developing feasible procedures to solve such problems, which combine methods of Bayesian decision theory and game theory, can be an intersting task for further research.

REFERENCES


Kadane, J. B., 1985, Opposition of interest in subjective Bayesian theory, Man. Science, 31: 1586 - 1588.

Kadane, J. B., and Larkey, P. D., Subjective probability and the theory of games, Man. Science, 28: 113 - 125 (with comment by J. C. Harsanyi, reply and rejoinder).

Kadane, J. B., and Larkey, P. D., The confusion of is and ought in game theoretic contexts, Man. Science, 29: 1365 - 1383 (with comments by M. Shubik).

Kahan, J. P., On choosing between Rev. Bayes and Prof. von Neumann, Man. Science, 29: 1334 - 1336.

Kendrick, D., 1981, "Stochastic Control for Economic Models", McGraw-Hill, New York et al.

Lindley, D. V., 1982, The subjectivist view of decision-making, Europ. J. of Op. Res., 9: 213 - 222.

Mosley, P., 1976, Towards a "satisficing" theory of economic policy, Econ. J., 86: 59 - 72.

Neck, R., 1986, Kann Stabilisierungspolitik unter Unsicherheit und Risiko "optimal" sein? Working Paper, University of Economics, Vienna.

Roth, A. E., and Shoumaker, F., 1983, Subjective probability and the theory of games: Some further comments, Man. Science, 29: 1337 - 1340.

Rothkopf, M. H., 1983, Modeling semirational competitive behavior, Man. Science, 29: 1341 - 1345.

Simon, H. A., 1956, A comparison of game theory and learning theory, Psychometrika, 21: 267 - 272.

Tan, T.C.-C., and Werlang, S.R.d.C., 1986, The Bayesian foundations of solution concepts of games, Working Paper, Graduate School of Business, University of Chicago.

# REMARKS ON FOUNDATIONS OF BAYESIAN STATISTICS AND ECONOMETRICS

Wolfgang Polasek

University of Vienna
Institute for Statistics
Universitaetsstr. 4, A-1010 Vienna, Austria

ABSTRACT

Econometrics can be viewed as applying statistics in economics. Applied statistics is a toolbox of coherent methods to deal with empirical uncertainties. Therefore a reasonable conjecture is that any Bayesian statistical method is applicable to some economic problem. Statistical methods in econometrics are breaking new grounds in two areas with very specific problems: a) the non-experimental nature of almost all economic data, and b) simultaneous equation systems.

For this reason we concentrate on foundational statistical issues in the first part of the paper and switch to a brief survey on new econometric developments in the second part. The review follows the "search approach" to econometrics, proposed for non-experimental data by Leamer (1978). This includes robust Bayesian methods, or in more fashionable term the extreme bound analysis (EBA), hierarchical models, smoothness priors for multivariate time series models, and Bayesian regression diagnostics. Furthermore, we review recent developments of numerical integration techniques (importance functions) in Bayesian simultaneous equation systems. Finally we discuss the acceptance of Bayesian methods in econometrics and possible future developments.

## 1. INTRODUCTION

D. Hume (1739, p45) introduces his philosophy in the following way: "All perceptions of the human mind resolve themselves into two distinct kinds, which I shall call impressions and ideas. The difference betwixt these consist in the degrees of force and liveliness, with which they strike upon the mind, and make their way into our thought or consciousness. Those perceptions which enter with most force and violence, we may name impressions; ... By ideas I mean the faint images of these in thinking and reasoning. ... Every one of himself will readily perceive the difference betwixt feeling and thinking."

From a statistical point of view D. Hume introduction into his philosophical work is actual and modern as 250 years ago. About the same time as Th. Bayes designed his solution to the inference problem we find a similar approach by D. Hume. Despite many technical progresses since that time, some basic issues remain the same: Shall we trust more our impressions (data) or our ideas (prior knowledge). Also the distinction between data (likelihood) and the prior is sometimes very fuzzy, especially in hierarchical (prior) structures, and often we have the problem: Shall we believe in our data-impressions or shall we stick more to our prior-ideas?

## 2. CHALLENGES

As a young science, econometrics was founded in the 30's and 40's, and the basic groundwork was layed up to 1960. The first booming decade was the 60's, followed by a critical wake-up period, which does not seem to have ended. The early beliefs were, that because econometrics has found its genuine research subject "simultaneous equation systems", it could be decoupeled from other empirical sciences. Almost all recent developments in statistics can be found in modern econometric textbooks (like e.g. Amemiya 1985 or Judge et al. 1985): time series analysis, asymptotics, qualitative choice models, etc. What are the challenges in econometrics, what makes the subject so difficult?

Up to now econometrics might be characterized as a semi-empirical science. Usually empirical sciences are data-driven, but for largely historical reasons econometrics is theory-driven. Econometrics means measuring economics by theoretical concepts which are entirely embedded in economics. This attitude would seem to be a fruitful nourishing field for Bayesian econometrics. But this is not the case, since there are substantial doubts for the measurement process. One finds the curious attitude that whole models are simply ignored or dismissed, because variables are 'wrong' or wrongly measured. Only those methods and economic variables are celebrated which fit to prevailing theories. Is it an immature or even lack of scientific attitude? A similar desperate search for confirmations of theories by data can be found in medical sciences.

Given this background it is not surprising that crises in econometrics are more severe than in ordinary empirical sciences. 'Ordinary' means: compiling data evidence in endless measurements, cautiously creating hypotheses until new theories are formed. Econometricians like the character of a brilliant genius, who comes up with a completely new idea and smashes the audiences around the world by having found a data set which matches his theory perfectly. While there is still the desire for such heroic ideals, people have found out some major sources of practical inconvenience:

(1) Forecasting performances of many new models are generally poor, especially "out of sample" as one has to add nowadays. Comparison is made via "thresholds" of naive models, like no-change or constant change models.

(2) Causality gap in multivariate time series models. While it is generally acknowledged that economics is a difficult science, because everything is interrelated with each other, large scaled economic models don't pay off in better explanatory powers, like substantial improvements in causality or forecasting.

(3) Simple use of time series models face problems of too many parameters and non-stationarities. Classical models assume constant variances and parameters, Kalman filter models are few. From a Bayesian view these problems are ideally solvable by the information updating mechanism of Bayes theorem (West, Harrison and Migon 1985).

The paper deals in the first part with those foundational arguments in Bayesian statistics, which are frequently found in discussions about foundational issues in econometrics. My claim is that the desire for true models and objectivity is paradoxically more pronounced in econometrics, because of the historical and nonexperimental nature of data in economics. Hardly any econometric textbook addresses this issue, but in general we find a strong tendency to more technical and complicated models.

The only book which draws consequences for the modeling process is Leamer's (1978) concept of 'specification searches'. Therefore recent developments in econometrics will be viewed from the "search perspective" and is discussed in section 6. Section 3 summarizes the discussion on the objectivity versus subjectivity controversy. Section 4 deals with probabilty interpretation problem and its consequences on statistical methods. Section 5 gives a critical review for the scope of Bayesian methods in econometrics. A final section summarizes the present and speculates abou the future state of the art in Bayesian econometrics.

## 3. THE SUBJECTIVE-OBJECTIVE CONTROVERSY

How can subjective methods ever outdate the objective statistical procedures? Objectivity is the goal of science: true knowledge has to be found. Honesty is the silent moral codex of all science. Subjectivity is the laymans domain, objectivity can only be obtained by accumulating knowledge. Why don't Bayesians stick to all these ideas? Why can they be proud too be subjective, personalistic, etc. Only renegades of the true science, sects of the pure religion of wisdom can adhere to this fanatism!

All Bayesians stick to this ideal as well! How is this possible? By the nature of the (Bayesian) learning process every prior information can be overruled by data. Only dogmatic (the word stemming obviously from many religious examples in the history of science, where religion forces you to ignore data) or orthodox priors (degenerate priors with variance 0) cannot be changed by data at all.

### 3.1 Probabilistic Reductionism

Stegmueller (1972) calls a major difference between the classical and the subjective interpretation of probabilities the "probabilistic reductionism" argument. This implies the claim that the notion of statistical probabilities have to be explained by already known terms. The main advocats of this position can be seen in B. deFinetti or the predictive school which want to express everything in terms of observables.

Stegmueller (1972, p.224): "Only if the probabilistic reductionism argument is valid, the subjectivists are right. But is this reductionism valid?". Stegmueller argues that the notion of chance is not definable by observables, but is partially interpretable as a theoretical term. This means that there is no contradiction between objectivism and subjectivism but a controversy between reductionist and non-reductionist. "Subjectivists are only victorious if the pendulum swings in favour to the first alternative (i.e. reductionism)". Further: "The times for a general belief in reductionistic programs have passed. In discussions of foundations of sciences the position prevails that theoretical terms play an important role in those disciplines." But this doesn't imply that reductionism doesn't work in statistics and therefore the notion of chance has to be a theoretical term. Partial intepretable theoretical terms are not concepts which are welcomed but have to be considered as a necessary evil. "In one decisive aspect the personalist finds himself in a better position than his reductionistic collegues in other faculties: the question of proof. ... It has to be shown that those terms cannot to be introduced into the scientific language as fully understandable notions." The subjective school claims that they have a theory available to introduce probabilities as a non-theoretical quantities.

Stegmueller points out that there are many variations of the reductionist arguments, a famous one being the mathematical branch of constructivism. A predictive type of reductionism has been found recently in the Bayesian discussion, called the 'Greek hinterland controversy'. Lindley (p6) in his reply to Efron (1986) notes: "Sampling-theory statistics takes place in a Greek hinterland (see below) that diminishes this connection with reality." The implied reductionist version in this statement is: Reduce the Bayesian controversy to observables and forget about the concept of theoretical and unobservable parameters (written generally in Greek alphabet).

### 3.2 The High Ground Of Objectivity

A widespread common belief is: Objectivity guarantees that one can solve empirical problems once and forever. But Bayesians have scruples: If everything depends on your subjective knowledge, maybe you have forgotten something to specify, maybe you should have done something else? Are the observations really independent and is the likelihood correct?

Using classical methods one sleeps well: If you have used the recent most powerful objective procedures there are no sleepless nights. You have achieved something, what hasn't been around before and therefore it must be something real and good. At least as long as the methods are modern this result will hold, what could be better!

Efron (1986) notes in the summary of his article 'Why not everyone is a Bayesian' that "Objectivity: the high ground of scientific objectivity has been seized by the frequentists." This view was heavily attacked by Lindley in his comment: "It is not true that 'strict objectivity is one of the crucial factors seperating scientific thinking from wishful thinking'. The objective element is the data: interpretation of the data is subjective, as anyone who has interacted with scientists knows. Furthermore the Bayesian view accepts the data, whereas the sampling-theory has to make a subjective embedding of them in a sample space."

Objectivity is defined differently by Bayesians. It rests on the notion of empirical learning. By accumulating data people with different prior opinions will converge finally to the same value. Objectivity can be obtained if prior information is dominated enough by the data. Subjectivity is just a lack of appropriate or enough data information.

## 4. THE PROBABILITY INTERPRETATION CONTROVERSY

Despite the technical progress which statistical methods have experienced in the last 50 years, there is no common accepted definition of probability. The Kolmogorov axioms are a nice device for how to use probabilities, but do not say how to measure or assign them. Bayesians and philosophers of science agree that all concepts so far proposed are circular, and therefore not valid (classicals need equiprobable atomic events, frequentists the notion of a probabilty limit). The Bayesian solution is that everything depends on You and therefore Your attitude has simply to be measured directly, by bets, lotteries, etc. Orthodox Bayesians think that You have certainly one definite attitude towards uncertainty in a particular real situation, which can be elicited to any degree of accuracy if necessary. This measurement process can take very long, but one would argue that if You need a result in a certain situation or experiment, then it is very well worthwile to find out about your prior information.

Again it pays to adopt a very tolerant position. Statistical clients are quite different in attitudes and utilities toward numerical and statistical results. While some want to know it very precisely, some like just a quick overview and don't want to be asked about prior opinions at all.

It is fascinating to speculate if the present definitions of probabilities and the associated interpretation of statistical inference results are really the ultima ratio of our profession. Will different fields of empirical sciences develop their own measurement devices or will a more appropriate new interpretation concept emerge? I think a similar situation and development has been going on for almost 250 years (since D. Hume) about the notion of causality. I don't think we have made too much progress for a general accepted causality definition, will probability definitions have a similar fate?

Frequentist and therefore the asymptotic interpretation seem to be very appealing, at least in experimental sciences. What can be wrong if one has only enough data? It is simply the forecasting problem: Why should something which is correct in the limit, happen to be correct in the next instance?" (Stegmueller 1973, p.246).

On the other side all these problems have been raised many times, but people don't seem to worry. They are happy with the results, else one cannot explain the heavy demand for statistical methods and advice. The theoretical unsolved problem how to justify the transition from a sequence of results to a single event is solved in practice daily. And it works, or at least seems to. Why?

Maybe our theoretical tools are still inappropriate. Our state of the art is a "proxy-solution" and we are still away from a right statistical language. E.g. 100 years ago differential calculus was in a similar position until the notion of limiting sequences has been made precise, which puzzeled mathematicians for decades. Therefore the only explanation for me how to justify asymptotic results is that practical results seem to justify the procedure, despite theoretical insufficiencies. If the results were useless, other methods will replace them pretty soon.

Nevertheless by pointing out these insufficiencies we should encourage researchers to replace our present knowledge by better methods.

4.1 Statistics: Normative (Prescriptive) or Descriptive?

Is statistics a science which has for every data problem one and only one solution or are several ways possible? The general tendency to objectivity would certainly prefer the unique statistical data -> method mapping. I think this is also the underlying but never questioned assumption of todays developments of expert systems. They follow a classical principle: If we only think hard enough we will find a special question-answering system which lead us to a unique answer to a data problem. And of course, since it is the computer who gives the answer based on expert knowledge, this will be objective as well.

This is another challenge to Bayesian statistics: Since the first working expert systems will be non-Bayesian, classical statistics will be perpetuated even more and longer. The demand for subjective judgement will be found obscure, since now 'the computer' tells all these results.

But lets turn to the other problem. Suppose we don't agree that there is only one solution to a data problem. How different can solution be? I think that we are more often confronted with the latter case and we have to face the problem that if theory offers you a range of methods, people can choose their methods according to a utility function. But we should make clear that we can communicate results. This is also the rational behind the demand of Leamer et al. (1983) that the "reporting style" of empirical studies matters. If we cannot agree to a common statistical approach to solutions, we should at least try to propagate a common reporting style.

But even inside a Bayesian framework we are often confronted with the problem: Shall we act normatively (= prescriptively) or descriptively? Is there only one likelihood function for a problem and only one prior, or do we have a choice between several likelihood functions and more priors? This leads to the problems of classical and Bayesian robustness.

Robustness is another area of common interests in classical and Bayesian statistics, but hardly noticed. Bayesian robustness is a descriptive device. Given the knowledge that prior opinions can differ in a certain range, what can be said about the resulting class of posterior distributions? It has two advantages: it solves some aspects of the communication problem and it narrows the gap between too much subjectivity and heroic objectivity. Summarizing I think that a descriptive or robust Bayesian approach is a plus for the Bayesian position. It allows to explain why different persons can end up with different conclusions for the same data set, but allow you at the same time to point out where your position relative to others is.

## 5. THE SCOPE OF BAYESIAN STATISTICS

There exists a frequent misunderstanding about the scope of Bayesian statistics. In- and outsiders usually overemphasize the inference aspect. E.g. Lindley (1986, p6) notes: "It is not true that 'Bayesian theory concentrates on inference.' If it concentrates on anything, it is decision analysis and has often been unjustly accused of marketplace philosophy. It embraces all of the topics mentioned, including randomisation and experimental design. It is a way of 'thinking about things'; it is relevant to everyone (Lindley 1985)."

Unfortunately in econometrics it seems to be true that all Bayesian results concentrates on inference. While design of experiments was not a subject up to now, economic decision theory, which has an enjoyable subjective character, is a subject only in management science or economic utility theory.

### 5.1 Personal or Main Frame Statistics?

Personal statistics allows You to solve Your personal problems. You are the decison maker, You are faced with unknowns, You can update you information, nobody else is involved except some family member or good friends.

Main frame statistics is concerned with your analysis and the rest of the world. The hostile environment forces you to make your analysis look good even in the worst circumstances. Therefore you share a main frame methodology with the rest of the world, which reduces communication to normed interfaces like 1% or 5% significances, accepted acronyms like BLUE, MMSE, UMPIU, BAN, etc. Personal judgments are considered as weakness in a frustrated scientific society looking for sublimed heroes. Objectivity can only be obtained if your personal involvement is minimized, but general acceptance in the storage of mainframe libraries is maximized.

Consultancies may of course involve personal judgements, but these are personal rewards obtained by the licence of previous objective masterpieces. Therefore as an offspring of your objective knowledge it is of course feasable to get your advice in one single circumstance. But personal consulting doesn't mean that one has to elicit the clients needs and utitlities. They are of course happy to follow the latest state of the art in the subject, where you are the celebrated (and objective) representative.

Can Bayesian methods be accepted in the memory of mainframe statistics? It turns out yes, to a certain degree. Recent progress of this kind has been made with econometric forecasts by Litterman (1986a) using smoothness priors. Smoothness priors are a special case of hierarchical priors, where many first stage parameters are "smoothed" into fewer hyperparameters which are assigned diffuse priors. So they are some kind of semi-objective priors, recently also called "of-the-shelf priors".

Akaike (1986) also developed the notion of Bayesian Information Criteria (BIC or ABIC) which can viewed as an approach to model selection by some semi-objective priors: BIC can be obtained as limiting expression of posterior odds ratios for regression models (see Leamer 1978 or Zellner 1978). ABIC is an selection criterion based on type II likelihoods: Hyperparameters are estimated in such a way that the resulting prior has a minimum effect on the data (or likelihood).

A further example is the smoothness approach adopted for seasonal adjustement in BAYSEA (Akaike 1980). To be most flexible, there are in the first stage more parameters than observations, which are restricted by estimated hyperparameters.

### 5.2 Prior Information Yes, But Not About Parameters

It is interesting that the existence of prior information is usually accepted by objectivists, but they don't think that it is on the same level as the likelihood information. Classical prior information is of the 0/1 type, like the inclusion/exclusion restriction problems in the identification problem of simultaneous equation system. Parameters are the holy secrets of statistical inference, their holy nature cannot reveiled by mortals. Therefore they are not on the same level as error measurements.

Up to now Bayesian statistics was concerned only with probabilistic prior information, should there also be a deterministic version? Following the principle "thinking about things" would partially imply this. Is Bayesian statistics a general language for empirical reasoning or just a probabilistic one? Research in this area of "metastatistics" has been scarce so far.

## 6. THE LEAMER CLASSIFICATION

Leamer (1978) has given a constructive review of the current state of econometrics by observing the discrepancy between deeds and talks of econometricians, even by the high priests in this discipline. He proposes a system of "metastatistics", called specification searches, based on Bayesian theory, and suggested a new search-type classification for econometric problems. This classification has not been challenged so far until recently, only a special application of the sensitivity analysis for variable selection problem in encompassing models has been disputed recently by McAleer et al.(1985). The econometric modeling process for nonexperimental data is divided into 6 searches:

1) Hypothesis testing searches: How to choose a true model.

2) Interpretative searches: Interprete multidimensional (regression) evidence.

3) Simplification searches: How to construct a "fruitful model".

4) Proxy searches: Find adequate substitutes for variables, which cannot be observed directly.

5) Data-selection search: Select an appropriate data set.

6) Postdata model construction: Improve an existing model

In the following we try to fit recent advances in Bayesian econometrics into this framework. To start with the easy part: there had been no new suggestions for modeling nonexperimental data. Most people work in their educated schools, deserters are rare.

Also there had been no new approaches to data instigated models and hypothesis searches. Posterior odds ratios are pushed heavily by Zellner (1984), but applications in econometrics seem to be rare.

Leamer (1978) and Zellner (1978) have shown that BIC can be obtained as limiting case for posterior odds in nested regression models, but except for time series it has not found any applications. AIC is more pursued by the "Japanese school of statistics" (Sakamoto et al. 1986), and is accepted rather by applied statisticians than by econometricians.

Most progress has been made in interpretative searches. Bayesian techniques for simultaneous equation systems have been elaborated by Dreze and Richard (1983), Richard (1984). Computational limits are given by the analytical intractability of posterior distributions, and Monte Carlo integration is seen as only alternative (Kloek and VanDijk 1985, Stewart 1985, Smith 1986). But there are also Bayesian claims that 'Monte Carlo integration is fundamentally unsound' (O'Hagan 1986), mainly because on principal grounds that sampling theory is used without taking into account the available prior information. Also there is Lindley's argument that doing something analytically gives more insights, than simple reporting of numerical results.

### 6.1 Importance Functions

An increasingly important numerical integration technique, developed by Hammersley and Handscomb (1964), is the so-called "importance sampling method" which was first applied in econometrics by Kloek and VanDijk (1978). An importance function is a density function with the following two properties:

1) It must be a good approximation for (the kernel) of a posterior density, and
2) it must be possible to generate random numbers for that density easily.

If such a function can be found, then the posterior moments can be calculated as a quotient of two sums. In a recent book Bauwens (1984, pp26) has compared 3 types of importance functions, called poly-t fixed condition (PTFC), poly-t drawn condition (PTDC), and poly-t student ( PTST). Comparison of importance functions with respect to different econometric models led Bauwens to the following conclusion: "No importance function is uniformly more efficient; conversely PTDC is always least efficient" (p. 65). If initial checks are not satisfactory, Bauwens recommends to start with a student importance function, and then to switch to PTFC or to PTST; this assures "robust" bahaviour for unfavorable alternatives.

As Stewart has pointed out, importance sampling is also preferable by comparing computing costs per analysis:

Importance sampling: 2$ Sample from prior: 10-250$

### 6.2 Bayesian Robustness

A further area of rapid development is Bayesian sensitivity analysis and Bayesian robustness. An excellent review is given by Berger (1984) in a special volume covering that topic edited by Kadane (1984). This is different from the classical robustness concept, which relies on ad hoc assumptions of influence functions. Smith (1983) showed that these can be derived by derivatives of proper posterior densities. Since Leamer's (1983) provocative article this version of Bayesian robustness analysis has become known as extreme bound analysis (EBA). An EBA analysis reports the set of all possible posterior means, if one specifys the conjugate normal prior distribution in a regression framework only partially. Given a fixed prior mean, but any positive definite prior covariance matrix, then the posterior mean is constrained to lie in an ellipsoid. The projections of the ellipsoid onto the coordinate axis are the extreme bounds for every coefficient.

If one can restrict the class of prior covariance matrices by an upper or an lower bound, then it can be shown that the ellipsoids shrink in size within the original feasible ellipsoid. This property of prior covariance restrictions arises almost naturally in a hierarchical (prior) framework. Polasek (1984,1986) has formulated this hierarchical approach for multivariate regression systems. Models which can be easily analysed by a hierarchical extreme bound analysis (HEBA) are seasonal distributed lag models (Polasek 1985) and multivariate regression systems with exchangeability properties.

Such hierarchical or exchangeabilty assumptions can be often made for so-called 2-dimensional models. In Leamer's framework this problem is of the "data-selection search"-type, but the methods of inferences are related strongly to "interpretative" methods. Hierarchical models lead to 'shrinkage estimates', mainly in form of a matrix weighted average between prior and ML-location. Such set-ups can always be analysed by EBA or HEBA. Recently Poetzelberger (1986) developed a method which describes the set of posterior distributions by so-called high fiduciary (HiFi-) contours, or envelope curves of the union of HPD regions of size $\alpha$ .

## 6.3 Local Sensitivity Analysis

The local sensitivity concept has two applications in Bayesian statistics. First, it is a tool to find out if posterior results react sensitive to (prior) input parameters. Second, it can be used for Bayesian regression diagnostics, as e.g. in Polasek (1985) or Zellner and Moulton (1985). If a model specification is changed because of diagnostic warnings, then we are in the framework of data instigated models (Leamer 1978). A satisfactory treatment of the inference problem when samples are re-used still not available.

Further methods belonging to the data-selection class are outlier identification (Petit and Smith 1984) and time varying parameters, like discount Bayesian modeling (West et al. 1985). All these areas are currently highly active research fields.

## 6.4 Bounding the Influence of Proxies

Research in this area has seen contributions from different fields in recent years. The results are interesting, but a homogeneous and practical useful general approach is not in sight. The underlying econometric problem, the error-in-variables problem, is "very deep and nontrivial" (Kalman 1982a) and challenges the present use of regression models. While Kalman (1982a) thinks that the whole statistical paradigm breaks down in the general case and has therefore to be replaced by a less 'prejudiced' method, like system-realisation theory, I think the robust Bayesian framework is flexible enough, to adjust also to this challenge: Even if the set of estimates and linear relations can be described easily, one has to make at some point a decison what model has to be used in a certain situation. Klepper and Leamer (1983) have shown how to attack the problem by Bayesian techniques. Krasker and Pratt (1986) and Kroch (1985) are working in similar areas. Causality analysis will get a different flavour if these methods will become practicable. But it will not solve the non-stationarity problem. Present pilot studies (by Kalman) show, that the set of functional relations react sensitive to inclusion of different data points.

## 7. CONCLUSIONS

Stegmueller (1972, p76) notes: "In at least two respects statistics resembles philosophy. Firstly, numerous directions exist fighting against each other and declaring other ones as nonsense. Secondly, there is a strong tendency in both fields for thinking in schemes. In statistics as in philosophy this is realised in such a way, that questions which extends into different dimensions are treated as questions of one and only one type."

Statistics is full of unsolved foundational issues, and econometrics shares a similar fate. If data are not informative enough (to convince strong prior view) then philosophical aspects matter. So I disagree with D. Lindley by claiming that Bayesian issues provoke philosphical issues. Econometric conferences are often a marketplace for philosophies, and I see the Stegmueller statement confirmed year by year. Since prior views by econometricians seems to be particularly strong, convergence and learning speed is slow.

In theoretical terms, Bayesian econometrics is developing in about the same speed as the remaining statistical methods. But on the practical side I would like to see more convincing applications. Except for recent progresses in Bayesian forecasting, like West et al. (1985) or Litterman (1986) other successful applications are few. Simulations, asymptotics, test principles, time series and qualitative models dominate econometrics in a classical style.

Another strong influencing force would be Bayesian econometric program packages. Except Leamer's SEARCH, Litterman's RATS, and Zellner's BRAP program, no 'canned' Bayesian econometric programs are on the market; big omnibus packages (SAS, SPSS, S) contain no Bayesian methods at all. While econometric estimation could be done by most other (Bayesian) packages as well, there is a psychological barrier. Econometricians have a tendency to a personal type of application, and therefore personal econometrics would help. In summary we need: More convincing examples and more Bayesian software!

In general one has to admit, that Bayesian methods are more difficult to understand, need more tedious input requirements, like the elicitation of your prior, and takes often longer to compute. In business and economics, where 'time is money', this is a serious constraint. Most econometricians are only part time statisticians and are pressed to get often a quick answer. For such purposes classical statistics is more handy than Bayesian statistics. This side constraints will challenge the Bayesian 'thinkers about things' quite a long time, which will make a 21th century of Bayesian econometrics a long way to go.

## REFERENCES


AKAIKE H. (1980) Seasonal Adjustment by Bayesian Modeling, J. of Time Series Analysis 1, 1-13.

AKAIKE H. (1986) The Selection of Smoothness Priors for Distributed Lag Estimation, in: P.K. GOEL and A. ZELLNER (eds.), Bayesian Inference and Decison Techniques, North-Holland, 109-493.

AMEMIYA T. (1985) Advanced Econometrics, Harvard Univ. Press, Cambridge MA.

BAUWENS L. (1984) Bayesian Full Information Analysis of Simultaneous Equation Models Using Integration by Monte Carlo, Springer Verlag, Berlin.

BERGER J.O. (1985) The Robust Bayesian Viewpoint, in: J. KADANE (ed.) Robustness in Bayesian Statistics, North-Holland, 63-144.
BERGER J.O. (1985) Statistical Decision Theory and Bayesian Analysis, 2nd ed., Springer, NY.
BERGER J. (1986) Bayesian Salesmanship, in: GOEL P. and A. ZELLNER eds. (1986) Bayesian Inference and Decison Techniques, North-Holland, 473-488.
BERGER J. and WOLPERT R. (1984) The Likelihood Principle: A Review and Generalisations, Monograph of the IMS.
BERNARDO J.M. (1979) Reference Posterior Distribution for Bayesian Inference, JRSS B 41, 113-142, with discussion.
BERNARDO J.M., DEGROOT M.H., LINDLEY D.V. and SMITH A.F.M. eds. (1985) Bayesian Statistics 2: Proc. of the 2nd Valencia Int. Meeting 1983, North-Holland and Val. Univ. Press.
BOOS D.D. and J.F. MONAHAN (1983) The Bootstrap for Robust Bayesian Analysis: An Adventure in Computing, in: K.W. HEINER et al. (eds.) Computer Science and Statistics: Proc. of the 14th Symp. on the Interface, Springer, 101-107.
BOX G.E.P. and G.C. TIAO (1973) Bayesian Analysis in Statistical Analysis, Addison-Wesley, Reading, MA.
BROEMELING L.D. (1985) Bayesian Analysis of Linear Models, Marcel Dekker, NY.
COOK P. (1985) Bayesian Autoregressive Spectral Analysis, Comm. in Stat. 14, 1001-1018.
DEMPSTER A. (1967) Upper and Lower Bound Analysis Induced by Multivalued Mapping, Ann. Math. Stat. 28, 325-39.
DICKEY J.M. (1974) Bayesian Alternative to the F-Test and Least Squares Estimates in the Normal Linear Model, in: S.E. Fienberg and A. Zellner (eds.) Studies in Bayesian Econometrics and Statistics, North-Holland.
DICKEY J.M., D.V. LINDLEY, and S.J. PRESS (1985) Bayesian Estimation of the Dispersion Matrix of a Multivariate Normal Distribution, Comm. in Stat. 14, 1019-1034.
DOAN T., LITTERMAN R. and C. SIMS (1984) Forecasting and Conditional Projection Using Realistic Prior Distributions, Econometric Reviews 3, 1-144, with discussions.
DREZE J.Q. and J.F. RICHARD (1983) Bayesian Analysis of Simultaneous Equation Systems, in: Z. Griliches and M.D. Intriligator (eds.), Handbook of Econometrics I, North-Holland, 517-598.
EFRON B. (1986) Why Isn't Everyone a Bayesian?, Am. Stat. 40, 1-11, with discussion.
FIENBERG S.E. and ZELLNER A. eds. (1975) Studies in Bayesian Econometrics and Statistics in Honor of Leonard J. Savage, North-Holland.
GARCIA-FERRER A., R.A. HIGHFIELD, F. PALM and A. ZELLNER (1986) Macroeconomic Forecasting using Pooled International Data, to appear in J. of Bus.&Ec. Statistics.
C.Z. GILSTEIN and E.E. LEAMER (1983) The Set of Weighted Regression Estimates, JASA 78, 942-948.
GOEL P. and A. ZELLNER eds. (1986) Bayesian Inference and Decison Techniques, North-Holland.
GEWEKE J.F. (1982) Measurement of Linear Dependence and Feedback Between Multiple Time Series, JASA 77, No. 378, with discussion, 304-324.
GOOD I.J. (1965) The Estimation of Probabilities: An Essay on Modern Bayesian Methods, MIT Press: Cambridge.
GRANGER C.W.J. (1969) Investigating Causal Relations by Econometric Models and Cross-Spectral Methods. Econometrica 37, 424-438.
HAMMERSLEY J.M. and D.C. HANDSCOMB (1964) Monte Carlo Methods, Menthuen, London (also Chapman and Hall 1979).

HENDRY D.F. and K.F. WALLIS eds. (1984) Econometrics and Quantitative Economics, Blackwell, London.
HUME D. (1739) A Treatise of Human Nature I, reprinted by Fontana/ Collins 1962.
JEFFREYS H. (1961) Theory of Probabilty (3rd ed) Oxford, Clarendon Press.
JUDGE G.G., HILL R.C., GRIFFITHS W.E., LUETKEPOHL H. and LEE T.C.(1982) Introduction to the Theory and Practice of Econometrics, Wiley NY.
JUDGE G.G., GRIFFITHS W.E., HILL R.C., LUETKEPOHL H. and LEE T.C.(1985) The Theory and Practice of Econometrics, Wiley NY.
KALMAN R. (1981) Identifiabilty and Problems of Model Selection in Econometrics, in: the Proc. of the 4th World congress of the econometric society, North-Holland.
KALMAN R. (1982a) Identification from Real Data, in: M. HAZEWINKEL and A.H.G. RINNOOY KAN (eds.) Current Developments in the Interface: Economics, Econometrics, Mathematics, Reidel, 161-196.
KALMAN R. (1982b) System Identification from Dynamic Systems II, in: A.R. BEDNAREK and L. CESARI (eds.), Dynamical Systems: Proceedings of a University of Florida International Symposium, Academic Press, 135-164.
KLEPPER S. and E.E. LEAMER (1984) Consistent Sets of Estimates for Regressions with Error in All Variables, Econometrica 52, 163-83.
KLOEK T. and VANDIJK H.K. (1978) Bayesian Estimates of Equation System Parameters: An Application of Integration by Monte Carlo, Econometrica 46, 1-19.
VANDIJK H.K. and KLOEK T. (1985) Experiments with some Alternatives for simple Importance Sampling in Monte Carlo Intergration, in: J.M. BERNARDO et al. (eds.) Bayesian Statistics 2, North-Holland, 511-530.
KRASKER W.S. and PRATT J.W. (1986 )Bounding the Effects of Proxy Variables on Regression Coefficients, Econometrica 54, 641-656.
KROCH E. (1985) Searching for Econometric Specification with Data Proxies, Univ. of Philadelphia, Ec. Dep.
LEAMER E.E. (1972) A Class of Informative Priors and Distributed Lag Analysis, Econometrica 40, 1059-81.
LEAMER E.E. (1978) Specification Searches, Wiley.
LEAMER E.E. (1981) Sets of Estimates of Locations, Econometrica 49, 193 -204.
LEAMER E.E. (1982) Sets of Posterior Means with Bounded Variance Priors, Econometrica 50, 725-36.
LEAMER E.E (1983a) Let's Take the Con out of Econometrics, Am. Ec. Review 73, 31-45.
LEAMER E.E. (1983b) Model Choice and Specification Analysis, in: Z. Griliches and M.D. Intriligator (eds.), Handbook of Econometrics I, North-Holland, 285-330.
LEAMER E.E. (1984) Sources of International Comparative Advantage, Theory and Evidence, MIT Press, Cambridge MA.
LEAMER E.E. (1985) Vector Autoregression for Causal Inference?, in: K. Brunner and A.H. Meltzler (eds.) Carnegie-Rochester Conference Series 22, 255-304
LEAMER E.E. and LEONARD H. (1983) Reporting the Fragilty of Regression Estimates, Rev. Ec. & Stat. 65, 306-17.
LINDLEY D.V. (1971) Bayesian Statistics, A Review, SIAM, Philadelphia.
LINDLEY D.V. and A.F.M. SMITH (1972) Bayes Estimates for the General Linear Model, JRSS B, 1-41.
LINDLEY D.V. (1985) Making Decisions, John Wiley, London.
LITTERMAN R.B. (1986a) A Statistical Approach to Economic Forecasting, J. of Bus.&Ec. Stat. 4, 1-4.
LITTERMAN R.B. (1986b) Forecasting With Bayesian Vector Autoregressions - Five Years of Experience, J. of Bus.&Ec. Stat. 4, 25-38.
LITTERMAN R.B. (1986c) Specifying Vector Autoregressions for Macroeconomic Forecasting, 79-94.

MCALEER M., PAGAN A.R. and P.A. VOLKER (1985) What Will Take the Con Out of Econometrics, Am. Ec. Rev., 293-307.
MIZON G.E. and RICHARD J.J. (1986) The Encompassing Principle and Its Application to Testing Nonnested Hypothesis, Econometrica 54, 657-678.
MORALES J.A. (1971) Bayesian Full Information Structural Analysis, Berlin, Springer Verlag.
O'HAGAN A. (1986) Bayesian Quadrature, Dep. of Statistik, No. 82, U. of Warwick
PAGAN A. (1986) On the Role of Simulation in the Statistical Evaluation of Econometric Models, Disc. Paper #6, U. of Warwick.
PETIT L.I. and SMITH A.F.M. (1984) Bayesian Model Comparisons in the Presence of Outliers, Bull. 44th ISI Meeting, Madrid, 292-306.
PHILLIPS L.D. (1973) Bayesian Statistics for Social Sciences, Nelson, London.
PIERCE D.A. and L.D. HAUGH (1977) Causality in Temporal Systems, Characterisation and a Survey, J. of Econometrics 5, 265-293.
POETZELBERGER K. (1986) HPD-Regions for the Linear Model, mimeo, TU-Wien.
POLASEK W. (1982) Two Kinds of Pooling Information in Cross Sectional Regression Models, in: W. Grossmann et al. (eds.). Probability and Statistical Inference, Reidel, 297-306.
POLASEK W. (1983) Hierarchical Load Curve Models, The Statistician 32, No. 1+2, 188-194.
POLASEK W. (1984a) Extreme Bound Analysis for Residential Load Curve Models: Empirical Evidence for a 16-household Example, Empirical Economics 9/3, 165-181.
POLASEK W. (1984b) An Application of the Hierarchical Extreme Bound Analysis: The Grunfeld Data, Preprint No. 44, to appear in a J. of Econometrics supplement.
POLASEK W. (1984c) Multivariate Regression Systems: Estimation and Sensitivity Analysis for Two-Dimensional Data, in: J. KADANE (ed.) Robustness in Bayesian Statistics, North-Holland, 229-309.
POLASEK W. (1985a) A Dual Approach for Matrix Derivatives, Metrika 32, 275-292.
POLASEK W. (1985b) Hierarchical Models for Time Series Data, in: J.M. BERNARDO et al. (eds.) Bayesian Statistics 2, North-Holland, 723-732.
POLASEK W. (1986a) Local Sensitivity Analysis and Bayesian Regression Diagnostics, in: P. GOEL and A. ZELLNER (eds.) Bayesian Inference and Decison Techniques, North-Holland, 375-387.
POLASEK W. (1986b) Bounds for Rounding Errors in Linear Regression, Research Rep. BSSY, Warwick Univ., to appear in The Statistician.
RAIFFA H. and R. SCHLAIFER (1961) Applied Statistical Decision Theory, Harvard U. Press, Cambridge MA.
REIERSOL O. (1950) Identifiabilty of a Linear Relation between Variables which are Subject to Error, Econometrica 18, 375-389.
RICHARD J.F. (1973) Posterior and Predictive Densities for Simultaneous Equation Models, Springer Verlag, Berlin.
RICHARD J.F. (1984) Classical and Bayesian Inference in Incomplete Simultaneous Equation Models, in: HENDRY D.F. and K.F. WALLIS (eds.) Econometrics and Quantitative Economics, Blackwell, 61-102.
SHILLER R.J. (1973) A Distributed Lag Estimator Derived from Smoothness Priors, Econometrica 41, 755-888.
SIMS C.A. (1972) Money, Income, and Causality, Am. Econ. Rev. 62, 540-552.
SIMS C.A. (1980) Macroeconomics and Reality, Econometrica 48, 1-48.
SIMS C.A. (1981) An Autoregressive Index Model for the US 1948-75, in: J. KMENTA and B. RAMSEY eds., Large Scale Macro Econometric Models, North-Holland.
SMITH A.F.M. (1973) A General Bayesian Linear Model, JRSS B, 67-75.
SMITH A.F.M. (1983) Bayesian Approaches to Outliers and Robustness, in: Specifying Statistical Models (Florens et. al. eds.), Springer Verlag New York, 13-35.

SMITH A.F.M. (1986) Numerical Integration Methods for Bayesian Inference, forthcoming The Statistician.
STEGMUELLER W. (1973) 'Jenseits von Popper und Carnap': Die logischen Grundlagen des statistischen Schliessens, Studienausgabe Teil D, Springer Verlag.
STEWART L. (1985) Multiparameter Bayesian Inference Using Monte Carlo Integration: Some Techniques for Bivariate Analysis, in: J.M. BERNARDO et al. (eds.) Bayesian Statistics 2, North-Holland, 495-510.
WEST M., HARRISON J. and MIGON H.S. (1985) Dynamic Generalized Linear Models and Bayesian Forecasting, JASA 80, 73-97, with discussion.
ZELLNER A. (1971) An Introduction into Bayesian Inference in Econometrics, Wiley, NY.
ZELLNER A. (1978) Jeffreys-Bayes Posterior Odds Ratio and the Akaike Information Criterion for Discriminating between Models, Economic Letters 1, 337-42 (reprinted in Zellner 1984).
ZELLNER A. (1984) Basic Issues in Econometrics, U. of Chicago Press.
ZELLNER A. and B.R. MOULTON (1985) Bayesian Regression Diagnostics with Applications to International Consumption and Income Data, J. of Econometrics 29, 187-211.
ZELLNER A. (1986) A Tale of Forecasting 1001 Series: The Bayesian Knight Strikes Again, to appear in J. of Forecasting.
ZELLNER A. and W. VANDAELE (1975) Bayes-Stein Estimators for k-Means, Regression and Simultaneous Equation Models, in: S.E. Fienberg and A. Zellner eds,. Studies in Bayesian Econometrics and Statistics in Honor of L.J. Savage, 627-53, North Holland.

# HPD-REGIONS FOR THE LINEAR REGRESSION MODEL

Klaus Pötzelberger

Technische Universität Wien
Wien, Austria

## INTRODUCTION

In this paper we are concerned with the influence of the a-priori precision of $\theta$ on the HPD-regions for the linear regression model $y = X\theta + \varepsilon$ . For fixed a-priori mean the set of a-posteriori means of $\theta$ has been examined in detail by Leamer and Chamberlain (1976) and Polasek (1984). It has been shown that this set forms an ellipsoid, the so-called feasible ellipsoid.

HPD-regions give more information about the parameter $\theta$ than the a-posteriori mean. We shall examine the union of all HPD-regions of fixed level $\alpha$. This set may be viewed as a robust alternative for HPD-regions (if H stands for high, not highest). We shall show that this union is most often not much larger than the HPD-region with noninformative prior. We shall compute its Lebesgue measure and give a representation of its boundary. Surprisingly, it may have edges.

First, some notation. The linear regression model is given by

$$y = X\theta + \varepsilon, \tag{1}$$

where $y,\varepsilon \in \mathbb{R}^r, \theta \in \mathbb{R}^n$ and X is a rxn matrix of full rank. A-priori, $\varepsilon$ and $\theta$ are normally distributed, $\varepsilon$ with mean 0 and precision matrix P and $\theta$ with mean $b_o$ and precision matrix $\Sigma$, so that the posterior distribution of $\theta$ is $N(b(\Sigma), \Sigma + X'PX)$, where $\Sigma + X'PX$ is the posterior precision and $b(\Sigma)$ is the posterior mean, given by

$$b(\Sigma) = (\Sigma + X'PX)^{-1}(X'Py + \Sigma b_\circ). \tag{2}$$

Throughout the paper we shall assume $b_\circ = 0$, so that (2) reduces to

$$b(\Sigma) = (\Sigma + X'PX)^{-1}X'Py. \tag{3}$$

The closure of the set of posterior means is denoted by F,

$$F = \text{closure } \{ b(\Sigma) \mid \Sigma \text{ positive definite \& symmetric}\}. \tag{4}$$

This set is the feasible ellipsoid mentioned above. For $\alpha \in (0,1)$ $\Omega_\alpha(\Sigma)$, the HPD-region of level $\alpha$ is characterized by $P_\Sigma(\theta \in \Omega_\alpha(\Sigma)) = \alpha$ and $p_\Sigma(\theta_1) \geq p_\Sigma(\theta_2)$ for $\theta_1 \in \Omega_\alpha(\Sigma)$ and $\theta_2 \notin \Omega_\alpha(\Sigma)$, where $p_\Sigma(\theta)$ denotes the posterior density of $\theta$. A $\beta \geq 0$ exists, such that

$$\Omega_\alpha(\Sigma) = \{ \theta \mid (\theta - b(\Sigma))'(\Sigma + X'PX)(\theta - b(\Sigma)) \leq \beta^2\} . \tag{5}$$

This $\beta$ is independent of $\Sigma$ and given by $\beta^2 = \chi^2_{n,\alpha}$, where $\chi^2_{n,\alpha}$ denotes the $\alpha$-fractile of the $\chi^2$-distribution with n degrees of freedom. The closure of the union of all $\Omega_\alpha(\Sigma)$ is denoted by $H_\alpha$,

$$H_\alpha = \text{closure} \bigcup_\Sigma \Omega_\alpha(\Sigma). \tag{6}$$

The dependence of F, $H_\alpha$, $\Omega_\alpha(\Sigma)$ on X'PX and X'Py is indicated by subscripts, if necessary. The boundary of one of these sets is denoted by $\partial$, such as $\partial F$. The Lebesgue measure of any Borel set $A \subseteq \mathbb{R}^n$ is denoted by $\lambda(A)$ and the determinant of a matrix M by det(M).

REDUCTION OF THE PROBLEM

To simplify the computations for $H_\alpha$ and $\lambda(H_\alpha)$, we transform the set H linearly. Let $y_\circ = (1,0,\ldots,0)'$, $t = \|(X'PX)^{-1/2}X'Py\|^{-1}$, U an orthogonal nxn matrix with $U^{-1}y_\circ = t(X'PX)^{-1/2}X'Py$ and $V = U(X'PX)^{1/2}$. Then

$$\begin{aligned} b(\Sigma) &= (\Sigma + X'PX)^{-1}X'Py = (\Sigma + X'PX)^{-1}(X'PX)^{1/2}t^{-1}U^{-1}y_\circ \\ &= (tV)^{-1}(V'^{-1}\Sigma V^{-1} + V'^{-1}X'PXV^{-1})^{-1}y_\circ = (tV)^{-1}(\tilde{\Sigma} + I)^{-1}y_\circ \\ &=: (tV)^{-1}\tilde{b}(\tilde{\Sigma}), \end{aligned}$$

where $\tilde{\Sigma} = V'^{-1}\Sigma V^{-1}$. Additionally, we get

$$z \in \Omega_{\alpha,X'PX,X'Py}(\Sigma) \text{ iff } (z - b(\Sigma))'(\Sigma + X'PX)(z - b(\Sigma)) \leq \beta^2$$
$$\text{iff } (tVz - \tilde{b}(\tilde{\Sigma}))'(\tilde{\Sigma} + I)(tVz - \tilde{b}(\tilde{\Sigma})) \leq (\beta t)^2$$
$$\text{iff } tVz \in \Omega_{\tilde{\alpha},y_\circ,I}(\tilde{\Sigma}),$$

where $(\beta t)^2 = \chi^2_{n,\tilde{\alpha}}$. We conclude

$$tVF_{X'PX,X'Py} = F_{I,y_\circ}, \qquad (7)$$

$$tVH_{\alpha,X'PX,X'Py} = H_{\tilde{\alpha},I,y_\circ} \qquad (8)$$

and $$\lambda(H_{\alpha,X'PX,X'Py}) = \lambda(H_{\tilde{\alpha},I,y_\circ})\det(tV)^{-1}. \qquad (9)$$

Robustness is a conditional feature, depending on the data. Note that $H_{\tilde{\alpha},I,y_\circ}$ depends on X,P and y only through $\tilde{\alpha}$.

GEOMETRY OF $H_\alpha$

Prior distributions can hardly ever be quantified exactly. Therefore it is important to know how sensitive the posterior distribution is about changes of the prior. If no prior information is available, the use of noninformative priors is indicated. There is no agreement what prior is noninformative. Beside this fact noninformative priors are informative, they are chosen with the intention to give maximum weight to the data and minimum weight to the prior, for instance. If $\dim\theta = 1$, it may happen that $H_\alpha = \Omega_\alpha(0)$ ($\Omega_\alpha(0)$ is the HPD-region for the noninformative prior). Thus $\Omega_\alpha(0)$ is the largest HPD-region, not only in terms of its Lebesgue measure. It contains all others. If a fixed vector z does not belong to the HPD-region, when no prior information is available, it does not, if prior information is at hand. If $\dim\theta \geq 2$, this will never happen. A certain vector z may belong to some HPD-region (with informative prior), but not to that with noninformative prior and not to that with highly informative prior.

We shall now describe the geometrical features of H . Let for $n \geq 2$

$$W = \{1\}\times\mathbb{R}^{n-1}. \qquad (10)$$

<u>Lemma 1.</u> Let $X'PX = I$, $X'Py = y_\circ$, $w \in W$ and define a set $S(w)$ to be the closure of $\{s \in \mathbb{R} \mid \exists\Sigma\colon b(\Sigma) = sw\}$. Then

$$S(w) = [0,1/\|w\|^2]. \qquad (11)$$

Proof: It is easy to see that $S(w)$ is an interval with $0 \in S(w)$. If $s \in S(w)$, then $b(\Sigma) = (\Sigma + I)^{-1}y_\circ = sw$ implies $y_\circ = (\Sigma + I)sw$, so that $1 = w'y_\circ = sw'(\Sigma + I)w \geq s\|w\|^2$, showing $S(w) \subseteq [0, 1/\|w\|^2]$. To prove the lemma, we have to show that precision matrices $\Sigma_k$ exist with $b(\Sigma_k) \to (1/\|w\|^2)w$. Let $\Sigma_k = (k + 1/k)I - (k/\|w\|^2)ww'$. Then $b(\Sigma_k) = k/(k^2+k+1)y_\circ + (k/(k+1) - k/(k^2+k+1))/\|w\|^2 w$ and $b(\Sigma_k) \to (1/\|w\|^2)w$.

Corollary. If $X'PX = I$, $X'Py = y_\circ$, then $\partial F$, the boundary of $F$, is given by

$$\begin{aligned} \partial F &= \{(1/\|w\|^2)w \mid w \in W\} \cup \{0\} \\ &= \{z \mid \|z - 0.5y_\circ\| = 0.5\}. \end{aligned} \tag{12}$$

Lemma 2. Let $X'PX = I$, $X'Py = y_\circ$. If $z \in \partial H_\alpha$, then a sequence of precision matrices $(\Sigma_k)$, a $b \in F$ and a $\phi \in \mathbb{R}$ exist, such that

$$b(\Sigma_k) \to b, \tag{13}$$

$$(z - b(\Sigma_k))'(\Sigma_k + I)(z - b(\Sigma_k)) \to \beta^2 \tag{14}$$

and

$$b = \phi z. \tag{15}$$

Proof: If $z \in \partial H_\alpha$, then for all precision matrices $\Sigma$, $(z - b(\Sigma))'(\Sigma + I)(z - b(\Sigma)) \geq \beta^2$ holds. A sequence $(\Sigma_k)$ exists, such that (14) holds. $(b(\Sigma_k))$ is a sequence of vectors in $F$, a compact subset of $\mathbb{R}^n$. By passing to a subsequence, if necessary, we can assume that (13) holds for a $b \in F$. We shall show that (15) holds for this $b$ and a $\phi \in R$.

Suppose, this is not true. A $u \in \mathbb{R}^n$ exists, such that $\|u\| = 1$, $u'z = 0$ and $u'b \neq 0$. Let $0 < h < 1$ and define $\tilde{\Sigma}_k$ by $\tilde{\Sigma}_k = \Sigma_k + huu'$. We have $u'b(\tilde{\Sigma}_k) = u'b(\Sigma_k) - hu'(\Sigma_k + I)uu'b(\Sigma_k)$ and $\beta^2 \leq (z - b(\tilde{\Sigma}_k))'(\tilde{\Sigma}_k + I)(z - b(\tilde{\Sigma}_k)) = -hu'b(\tilde{\Sigma}_k)u'b(\Sigma_k)$
$+ (z - b(\Sigma_k))'(\Sigma_k + I)(z - b(\Sigma_k)) =$
$= -h(u'b(\Sigma_k))^2(1 - hu'(\Sigma_k + I)^{-1}u) + (z - b(\Sigma_k))'(\Sigma_k + I)(z-b(\Sigma_k))$
Then (14) implies $0 \leq \lim (-h(u'b(\Sigma_k))^2(1 - h\,u'(\Sigma_k + I)^{-1}u))$
$\leq (u'b)^2(-1 + hu'u) < 0$, a contradiction.

Lemma 3. Let $X'PX = I$, $X'Py = y_\circ$ and $z \in \partial H_\alpha$. If a sequence $(\Sigma_k)$ of precision matrices satisfies (14) and if $b(\Sigma_k) \to 0$, then $y_\circ$ and $z$ are orthogonal ($y_\circ'z = 0$).

Proof. Orthogonal matrices $U_k$ and positive diagonal matrices $D_k = \mathrm{diag}(d_1^{(k)}, \ldots, d_n^{(k)})$ exist, such that $\Sigma_k = U_k'D_kU_k$. We may assume that $U_k \to V$ for an orthogonal matrix $V$. Let $Vy_\circ = v$ and $Vz = u$ ($v = (v_1, \ldots, v_n)'$, $u = (u_1, \ldots, u_n)'$). We have to show that

$u'v = 0$. $b(\Sigma_k) \to 0$ implies $\|U_k'(D_k + I)^{-1}U_k y_\circ\| = \|(D_k + I)^{-1}U_k y_\circ\| \to 0$, so that $v_i \neq 0$ implies $d_i^{(k)} \to \infty$ (for $k \to \infty$). (14) implies $z'(\Sigma_k + I)z - 2y_\circ'z + b(\Sigma_k)'y_\circ \to \beta^2$, so that a constant C exists with $z'(\Sigma_k + I)z \leq C$. $(U_k z)'(D_k + I)(U_k z) \leq C$ means $\lim_k \sup\, (d_i^{(k)} + 1)u_i^2 < \infty$ for $1 \leq i \leq n$, which gives $d_i^{(k)} \not\to \infty$ if $u_i \neq 0$, so that $u_i \neq 0$ implies $d_i^{(k)} \not\to \infty$, which again implies $v_i = 0$. Thus $u'v = 0$.

Theorem. Let $X'PX = I$, $X'Py = y_\circ$, $\beta = (\chi^2_{n,\alpha})^{1/2} \geq 1$. Then $\partial H_\alpha$ = closure of $\partial H_\alpha^+ \cup \partial H_\alpha^-$, where

$$\partial H_\alpha^+ = \{\, \gamma^+(w)w \mid w \in W \,\}, \tag{16}$$

$$\partial H_\alpha^- = \{\, \gamma^-(w)w \mid w \in W \,\}, \tag{17}$$

$$\gamma^+(w) = 1/\|w\|^2 + \beta/\|w\| \tag{18}$$

and

$$\gamma^-(w) = \begin{cases} 1/\|w\|^2 - \beta/\|w\| & \text{if } 2 \leq \beta \text{ or } 2/\beta \leq \|w\| \\ -\beta^2/4 & \text{else.} \end{cases} \tag{19}$$

Proof: Let $z \in \partial H_\alpha$, $w \in W$, $r \in \mathbb{R}$ with $z = rw$. We shall show that $r = \gamma^+(w)$ or $r = \gamma^-(w)$ holds. If z and $y_\circ$ are not orthogonal, then such r and w exist. Furthermore, we can find a sequence $(\Sigma_k)$, a vector $b \in F$ and a $\phi \in \mathbb{R}$, such that in addition to (13), (14) and (15) $\phi \neq 0$ holds. Let $\psi = 1/\phi$, so that $z = \psi b$. (14) implies $(\psi - 1)^2 = \lim_k \beta^2/b(\Sigma_k)'(\Sigma_k + I)b(\Sigma_k) = \lim_k \beta^2/y_\circ'(\Sigma_k+I)^{-1}y_\circ$, which gives $1 \pm \beta(y_\circ'(\Sigma_k + I)^{-1}y_\circ)^{-1/2} \to \psi$, so that $z = \lim_k (1 \pm \beta(y_\circ'(\Sigma_k + I)^{-1}y_\circ)^{-1/2})b(\Sigma_k)$. We have $r = ry_\circ'w = y_\circ'z$ $= \lim_k (1 \pm \beta(y_\circ'(\Sigma_k + I)^{-1}y_\circ)^{-1/2})y_\circ'(\Sigma_k + I)^{-1}y_\circ =$ $= \lim_k (p_k \pm \beta p_k^{1/2})$ with $p_k = y_\circ'(\Sigma_k + I)^{-1}y_\circ$. Note that $p_k = y_\circ'b(\Sigma_k)$. If $z \in \partial H_\alpha$, then

$$r = \sup\{\, p + \beta p^{1/2} \mid \exists s^+ \in \mathbb{R}, \Sigma: p = y_\circ'b(\Sigma) \text{ and } z = s^+b(\Sigma)\} \tag{20}$$

or

$$r = \inf\{\, p - \beta p^{1/2} \mid \exists s^- \in \mathbb{R}, \Sigma: p = y_\circ'b(\Sigma) \text{ and } z = s^-b(\Sigma)\} \tag{21}$$

which means that $r = \sup\{\, p + \beta p^{1/2} \mid p \in S(w)\,\}$ or $r = \inf\{\, p - \beta p^{1/2} \mid p \in S(w)\}$. Lemma 1 tells us that $\sup\{\, p + \beta p^{1/2} \mid p \in S(w)\,\} = 1/\|w\|^2 + \beta/\|w\|$, which is $\gamma^+(w)$. To compute (21), we note that $\inf\{\, p - \beta p^{1/2} \mid p \in S(w)\} =$ $= -\beta^2/4$, if $\beta^2/4 \in S(w)$, and $1/\|w\|^2 - \beta/\|w\|$, if $\beta^2/4 \notin S(w)$.

$\beta^2/4 \in S(w)$ is equivalent to $\|w\| \leq 2/\beta$. This and $\|w\| \geq 1$ implies that in the case of (21), $r = \gamma^-(w)$.

Remark. (a) If $1 \leq \beta < 2$, the set $\{ \gamma^-(w)w \mid w \in W, \|w\| = 2/\beta \} = \{ -(\beta^2/4)w \mid w \in W, \|w\| = 2/\beta \}$ is an "edge".

(b) For $z = (z_1,\ldots,z_n)' \in \mathbb{R}^n$ let

$$p_F(z) = z_1^2 + \ldots + z_n^2 - z_1, \tag{22}$$

$$p_\Omega(z) = z_1^2 + \ldots + z_n^2 + 1 - 2z_1 - \beta^2, \tag{23}$$

$$p_H(z) = z_1^2 + (z_1^2 + \ldots + z_n^2)(z_1^2 + \ldots + z_n^2 - 2z_1 - \beta^2). \tag{24}$$

If $\beta \geq 2$, then

$$\partial F = \{ z \mid p_F(z) = 0 \}, \tag{25}$$

$$\partial\Omega_\alpha(0) = \{ z \mid p_\Omega(z) = 0 \} \tag{26}$$

and

$$\partial H_\alpha = \{ z \mid p_H(z) = 0 \}. \tag{27}$$

SIZE OF $H_\alpha$

In this section we shall compute $\lambda(H_\alpha)$, the Lebesgue measure of $H_\alpha$, for the case $\beta \geq 2$. Let $X'PX = I$, $X'Py = y_\circ$, $\beta = (\chi^2_{n,\alpha})^{1/2} \geq 2$ and $f_\alpha = \lambda(H_\alpha)$. Then

$$f_\alpha = \int |\omega_n \gamma^+(w)^{n-1} \frac{\partial\gamma^+}{\partial\omega}(w)| \, d\omega_2 \ldots d\omega_n$$

$$+ \int |\omega_n \gamma^-(w)^{n-1} \frac{\partial\gamma^-}{\partial\omega_n}(w)| \, d\omega_2 \ldots d\omega_n. \tag{28}$$

Partial integration gives

$$f_\alpha = \frac{1}{n}\int (|\gamma^+(w)|^n + |\gamma^-(w)|^n) \, d\omega_2 \ldots d\omega_n. \tag{29}$$

Lemma 4. For $n \geq 2$ we define $I(s,n)$ by $I(s,n) = \int \|w\|^{-2s} \, d\omega_2 \ldots d\omega_n$. If $I(s) = I(s,2)$, then $I(1) = \pi$, $I(3/2) = 2$, $I(s+1) = (1 - 1/2s)I(s)$ and for $n \geq 3$

$$I(s,n) = \prod_{i=0}^{n-2} I(s-i/2). \tag{30}$$

Useful expressions for $f_\alpha$ are:

$$f_\alpha = \frac{2}{n} \sum_{\substack{k=0 \\ k \in 2\mathbb{N}_\circ}}^{n} \binom{n}{k} \beta^{n-k} \, I((n+k)/2,n), \tag{31}$$

$$f_\alpha = \frac{2}{n} \sum_{\substack{k=0 \\ k \in 2\mathbb{N}_0}}^{n} \binom{n}{k} \beta^{n-k} \prod_{i=0}^{n-2} I((n+k+i)/2). \tag{32}$$

To compare $f_\alpha$ and $\lambda(\Omega_\alpha(0))$, the Lebesgue measure of the HPD-region in the noninformative case, we have computed $f_\alpha$ and $\lambda(\Omega_\alpha(0))$ for n = 2, 3, 4. For large $\beta$, $f_\alpha$ and $\lambda(\Omega_\alpha(0)$ do not differ much.

Table 1. $f_\alpha$ and $\lambda(\Omega_\alpha(0))$

| n | $f_\alpha$ | $\lambda(\Omega_\alpha(0))$ |
|---|---|---|
| 2 | $\pi(\beta^2 + \frac{1}{2})$ | $\pi\beta^2$ |
| 3 | $\frac{4\pi}{3}(\beta^3 + \beta)$ | $\frac{4\pi\beta^3}{3}$ |
| 4 | $\frac{\pi^2}{2}(\beta^4 + \frac{3\beta^2}{2} + \frac{1}{8})$ | $\frac{\pi^2\beta^4}{2}$ |

Table 2. $f_\alpha$ and $\lambda(\Omega_\alpha(0))$ for $\alpha = 0.95$ and for $\beta = 2$

| n | $\alpha$ | $\beta$ | $f_\alpha$ | $\lambda(\Omega_\alpha(0))$ |
|---|---|---|---|---|
| 2 | 0.87 | 2.00 | 14.14 | 12.57 |
| | 0.95 | 2.45 | 20.43 | 18.86 |
| 3 | 0.74 | 2.00 | 41.89 | 33.51 |
| | 0.95 | 2.80 | 103.68 | 91.95 |
| 4 | 0.20 | 2.00 | 109.18 | 78.96 |
| | 0.95 | 3.08 | 514.93 | 444.09 |

## REFERENCES


Chamberlain, G. and Leamer, E. E., 1976, Matrix weighted averages and posterior bounds, Journal Roy. Statist. Soc. Ser. B 38, 73-84.

Polasek, W., 1984, Multivariate regression systems - estimation and sensitivity analysis of two-dimensional data, in: Robustness of Bayesian analyses, J. B. Kadane, ed., North-Holland, Amsterdam.

# A VERY GENERAL DE FINETTI-TYPE THEOREM

Paul Ressel

Math.-Geogr. Fakultät
Katholische Universität Eichstätt
8078 Eichstätt, W. Germany

## INTRODUCTION

A few years ago it turned out that De Finetti's famous theorem concerning exchangeable 0 - 1 valued random variables can also be proved by harmonic analysis means, applied to the special semigroup $\{(k,n) \in \mathbb{N}_0^2 | k \leq n\}$. This is no pure coincidence; a careful inspection of the new proof revealed that many other De Finetti-type theorems, old and new ones, could be shown the same way, among them Schoenberg's representation of spherically symmetric random sequences, Hewitt and Savage's far-reaching generalisation of De Finetti's original result, and numerous characterisations of mixtures of i.i.d.-sequences with concrete prescribed distributions.

So far we only considered countable infinite random sequences. More recently, some interesting results concerning mixtures of stochastic processes could be proved, among them characterizations of mixtures of Brownian motions, Brownian bridges and Poisson processes. It was tempting to look for one general De Finetti-type theorem from which all above mentioned results would follow straightforwardly. Finally this goal is achieved now, and De Finetti's invention turns out to be the prototype of a powerful integral representation theorem in commutative harmonic analysis.

Starting with the classical result we will extend it in three steps, providing detailed proofs that each time a proper generalization is obtained.

## DE FINETTI'S THEOREM

In 1931 De Finetti proved the following result:

*Let* $X = (X_1, X_2, \ldots)$ *be an infinite sequence of* $\{0,1\}$-*valued random variables which is exchangeable in the sense that*

$$(*) \quad P(X_1 = x_1,\dots,X_n = x_n) = P(X_1 = x_{\pi(1)},\dots X_n = x_{\pi(n)})$$

*holds for all* $n \in \mathbb{N}$, $x_i \in \{0,1\}$ *and all permutations* $\pi$ *of* $\{1,\dots,n\}$ . *Then there is a unique probability measure* $\mu$ *on* $[0,1]$ *such that*

$$P(X_1 = x_1,\dots,X_n = x_n) = \int_0^1 p^{\Sigma x_i}(1-p)^{n-\Sigma x_i}\, d\mu(p) ,$$

*or, equivalently,* $\mathcal{D}(X) = \int_0^1 B(1,p)^{\infty}\, d\mu(p)$; *in other words:* $X$ *is a unique mixture of coin tossing processes.*

Condition (*) may be reformulated this way:

$$P(X_1 = x_1,\dots,X_n = x_n) = \varphi_n(\sum_{i=1}^{n} x_i) = \varphi(\sum_{i=1}^{n} x_i,n)$$

$$= \varphi(\sum_{i=1}^{n} (x_i,1)) = \varphi(\sum_{i=1}^{n} v(x_i))$$

with $v(x) := (x,1)$, $\varphi$ being defined on $S := \{(k,n)|\in \mathbb{N}_o^2\ k \leq n\}$, the subsemigroup of $\mathbb{N}_o^2$ generated by $v(\{0,1\}) = \{(0,1),(1,1)\}$, and De Finetti's result shows $\varphi$ to have the form

$$\varphi(k,n) = \int_0^1 p^k(1-p)^{n-k}\, d\mu(p),$$

i.e. $\varphi$ is a mixture of the functions $\rho_p(k,n) := p^k(1-p)^{n-k}$, and these $\rho_p$'s are *characters* on $S$, that is they are multiplicative.

An arbitrary character $\rho$ on $S$ has the form $\rho(k,n) = u^k v^{n-k}$, $u,v \in \mathbb{R}$, so that $S^*$, the set of all characters, can be identified with $\mathbb{R}^2$; the non-negative characters $S_+^*$ then correspond to $\mathbb{R}_+^2$, and the bounded characters $\hat{S}$ may be identified with $[-1,1]^2$.

Note that only the "small" part $W := \{(u,v) \in \hat{S}|u + v = 1\}$ enters in the above representation of $\varphi$ .

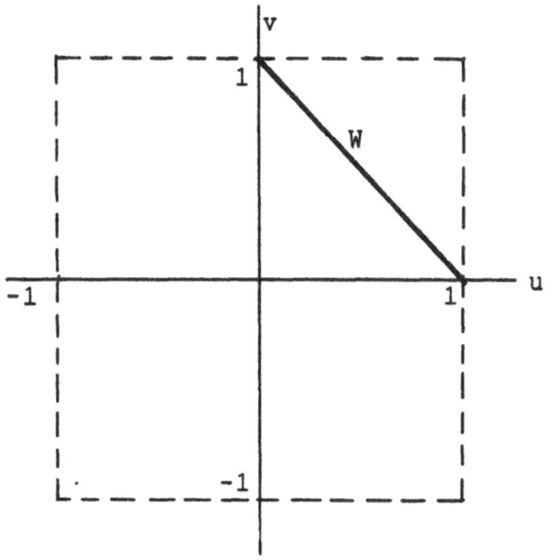


$[-1,1]^2 \simeq \hat{S}$ where $S = \{(k,n)\} \in \mathbb{N}_0^2 | k \leq n\}$

A FIRST EXTENSION

Let $X$ be an at most countable set, $S$ an abelian semigroup, $v : X \longrightarrow S$ a mapping such that $v(X)$ generates $S$, abbreviated $S = \$(v(X))$.

Theorem 1. *A probability measure* $P \in M_+^1(X^\infty)$ *has the property*

$$P(x_1,\ldots,x_n) = \varphi(\sum_{j=1}^{n} v(x_j))$$

*for all* $n \in \mathbb{N}$, $x_1,\ldots,x_n \in X$ *if and only if the function* $\varphi$ *has the (unique) representation*

$$\varphi(s) = \int_W \rho(s)\, d\mu(\rho)$$

*where* $\mu$ *is a probability measure on* $W := \{\rho \in \hat{S}_+ | \sum_{x \in X} \rho(v(x)) = 1\}$.
*Equivalently:* $P = \int_W \kappa_\rho^\infty\, d\mu(\rho)$ *with* $\kappa_\rho(\{x\}) := \rho(v(x))$.

Example 1. De Finetti's theorem.

Example 2. $X = \mathbb{N}_o,\ P(x_1,\dots,x_n) = \varphi_n(\sum_{i=1}^{n} x_i) = \varphi(\sum_{i=1}^{n} (x_i,1)),$

$S = \$(v(X)) = \$(\mathbb{N}_o \times \{1\}) = (\mathbb{N}_o \times \mathbb{N}) \cup \{(0,0)\}.$

$\rho \in S^* \Leftrightarrow \exists\, u,v \in \mathbb{R} : \rho(k,n) = u^k v^n$

$\rho \in W \Leftrightarrow u,v \geq 0 \quad \text{and} \quad \Sigma_{k=0}^{\infty} u^k v = 1$

$\qquad \Leftrightarrow u \in [0,1[ \quad \text{and} \quad v = 1-u\ .$

Hence geometrically distributed i.i.d.-sequences are the extreme points in the integral representation of $P$ .

Example 3. $X = \mathbb{N},\ P(x_1,\dots,x_n) = \varphi_n(\max_{1\leq i\leq n} x_i) = \varphi(\text{¢}(x_i,1))$

[with $\text{¢}(x_i,x_i) := (\max x_i, \Sigma y_i)$]

$S = \mathbb{N}^2 \cup \{(1,0)\} \subseteq (\mathbb{N},\vee) \times (\mathbb{N}_o,+)$

$\rho \in S^* \Leftrightarrow \exists\, j \in \overline{\mathbb{N}},\ v \in \mathbb{R} : \rho(k,n) = 1_{\{1,\dots,j\}}(k)v^n$

$\rho \in W \Leftrightarrow v \geq 0 \quad \text{and} \quad \Sigma_{k=1}^{\infty} \rho(k,1) = 1$

$\qquad \Leftrightarrow j < \infty \quad \text{and} \quad v = 1/j.$

Hence i.i.d.-sequences with a uniform distribution on one of the discrete intervals $\{1,\dots,j\}$, $j \in \mathbb{N}$, are the extreme solutions.

## A SECOND EXTENSION

$(S,+,*)$ is a *$*$-semigroup* if $(S,+)$ is a semigroup and $* : S \longrightarrow S$ fulfills $(s+t)^* = s^* + t^*$, $s^{**} = s$. We always assume that $S$ is abelian and contains a neutral element $0$ .

A *character* $\rho$ on $S$ is then by definition a function $\rho : S \longrightarrow \mathbb{C}$ such that

(1) $\rho(s+t) = \rho(s)\,\rho(t)$

(2) $\rho(s^*) = \overline{\rho(s)}$

(3) $\rho(0) = 1$ .

If $* = \text{id}$, then all characters are real-valued. Any abelian group is f. ex. a $*$-semigroup if we put $s^* := -s$. Another example is $(S,+,*) := (\mathbb{C},\cdot,^{-})$.

A function $\varphi : S \longrightarrow \mathbb{C}$ is *positive definite* iff

$$\sum_{j,k=1}^{n} c_j \bar{c}_k \varphi(s_j + s_k^*) \geq 0$$

$\forall\, n \in \mathbb{N}$, $c_j \in \mathbb{C}$, $s_j \in S$ . Let $P(S)$ denote the set of all positive definite functions on $S$; then $S^* \subseteq P(S)$ because $\Sigma\, c_j \bar{c}_k\, \rho(s_j + s_k^*) = |\Sigma\, c_j\, \rho(s_j)|^2 \geq 0$ for $\rho \in S^*$ .

We note the following fundamental result:

*If* $\varphi \in P(S)$ *is bounded then* $\exists!\ \mu \in M_+(\hat{S})$ *with* $\varphi(s) = \int \rho(s)\, d\mu(\rho)$ , $s \in S$ . (Lindahl and Maserick, 1971)

Let us list a few examples:

(i) $S = \mathbb{N}_0$, $S^* = \{n \longmapsto x^n | x \in \mathbb{R}\} \simeq \mathbb{R}$ , $\hat{S} \simeq [-1,1]$. Bounded positive definite functions on $\mathbb{N}_0$ are moment functions on $[-1,1]$ .

(ii) $S = \mathbb{R}_+$, $S^*$ "not known", $\hat{S} = \{s \longmapsto e^{-\lambda s} | 0 \leq \lambda \leq \infty\}$, where $e^{-\infty \cdot s} := 1_{\{0\}}(s)$. Bounded positive definite functions on $\mathbb{R}_+$ are the "usual Laplace transforms".

(iii) If $S$ is a (discrete) abelian group, then $S^* = \hat{S}$ is the usual dual group ; the above characterization reduces to Bochner's theorem.

Let $(X,\mathcal{B})$ denote a measureable space and let $F$ be a set of complex-valued measurable functions on $X$ , bounded by 1, stable under multiplication as well as conjugation, and containing the constant 1 (in other words, $F$ is itself a *-semigroup w.r. to pointwise multiplication and conjugation).

Let $(\Omega,A,P)$ be a probability space and $X_1, X_2, \ldots : \Omega \longrightarrow X$ a sequence of random variables. Let $S$ be another *-semigroup and $v : F \longrightarrow S$ a mapping such that $v(1) = 0$, $v(\bar{f}) = (v(f))^*\ \forall\, f \in F$ and $S = \$(v(F))$.

<u>Theorem 2</u>. *If* $E(\Pi_{j=1}^n f_j \circ X_j)$ *only depends on* $\Sigma_{j=1}^n v(f_j)$, *i.e.* $E(\Pi_{j=1}^n f_j \circ X_j) = \varphi(\Sigma_{j=1}^n v(f_j))$ *where* $\varphi : S \longrightarrow \mathbb{C}$ *is some function, then* $\varphi$ *is positive definite (and automatically bounded), and in fact a mixture of those* $\rho \in \hat{S}$ *for which* $\rho \circ v \in P(F)$.

Both Theorem 1 and (a slightly different version of) Theorem 2 have been proved in Ressel (1985). Before giving some examples we shall show how to get Theorem 1 from Theorem 2: put $\mathcal{F} := \{1_{\{x\}} | x \in X\} \cup \{0,1\}$, enlarge the semigroup $S$ from Theorem 1 by an *absorbing element* $\zeta$, i.e. consider $T := S \cup \{\zeta\}$ together with the rules $s + \zeta = \zeta + s = \zeta + \zeta = \zeta$ for all $s \in S$; extend $v : X \longrightarrow S$ to $w : \mathcal{F} \longrightarrow T$ by $w(1_{\{x\}}) := v(x)$, $w(0) := \zeta$ and $w(1) := 0$, and finally extend $\varphi$ to $T$ by $\varphi(\zeta) := 0$. The assumption $P(x_1,\ldots,x_n) = \varphi(\Sigma\, v(x_i))$ then translates into $E(\Pi_{j=1}^n f_j \circ X_j) = \varphi(\Sigma\, w(f_j))$. By Theorem 2 the extended $\varphi$ is a mixture of those $\tau \in \hat{T}$ for which $\tau \circ w \in P(\mathcal{F})$. It is easy to see that a function $g : \mathcal{F} \longrightarrow \mathbb{R}$ is positive definite iff $g(f) \geq g(0) \geq 0$ for all $f \in \mathcal{F}$. If $\mu$ is the measure on $\hat{T}$ representing $\varphi$ then $0 = \varphi(\zeta) = \int \tau(\zeta)\, d\mu(\tau)$, so $\tau(\zeta) = 0$ for all $\tau \in \mathrm{supp}(\mu)$. Therefore $\mu$ concentrates on those $\tau$ for which $\tau \circ w \geq \tau(w(0)) = \tau(\zeta) = 0$; such $\tau$ may be identified with its restriction $\rho$ to $S$, hence $\mu$ may be considered as a measure on $\hat{S}_+$. Since

$$1 = \sum_{x_1,\ldots,x_n \in X} P(x_1,\ldots,x_n) = \int [\sum_{x \in X} \rho(v(x))]^n \, d\mu(\rho)$$

for $n = 1,2,\ldots$, we see that indeed $\mu$ is carried by those $\rho \in \hat{S}_+$ for which $\Sigma_{x \in X}\, \rho(v(x)) = 1$.

Example 4. Let $X = \mathbb{R}$, $\mathcal{F} = \{e^{it\cdot} | t \in \mathbb{R}\}$ the group of complex exponentials, $v(e^{it\cdot}) := t^2$, $S = \mathbb{R}_+$. Then $E(\Pi_{j=1}^n e^{it_j \cdot X_j}) = E(e^{i\langle t,X\rangle}) = \varphi(\Sigma_{j=1}^n t_j^2) = \varphi(\|t\|^2)$ for some $\varphi$ on $\mathbb{R}_+$ means that $X = (X_1,X_2,\ldots)$ is spherically symmetric. From Theorem 3 we get $\varphi(x) = \int_0^\infty e^{-\lambda s}\, d\mu(\lambda)$ for some probability measure $\mu$ on $\mathbb{R}_+$, translating into $E(e^{i\langle t,X\rangle}) = \int e^{-\lambda\|t\|^2}\, d\mu(\lambda)$ for all $t = (t_1,t_2,\ldots)$ with only finitely many $t_j \neq 0$, and thus $\mathcal{D}(X) = \int_0^\infty N(0,2\lambda)^\infty\, d\mu(\lambda)$, i.e. $X$ is a variance mixture of centred normal i.i.d.-sequences. In some disguise this result goes back to Schoenberg (1938).

We may replace the function $t^2$ here by $|t|^p$, $0 < p < \infty$. Then if $p \leq 2$ we get symmetric stable distributions instead of centred normal ones, whereas for $p > 2$ only the trivial solution $X \equiv 0$ exists, since for no $\lambda > 0$ the function $\exp(-\lambda|t|^p)$ is positive definite on $\mathbb{R}$.

Example 5. Let $\mathcal{X}$ and $\mathcal{F}$ be as before but let now $v(t) = (t,t^2)$. If $X = (X_1,X_2,\dots)$ is a sequence of real-valued random variables such that $E(e^{i\langle t,X\rangle}) = \varphi(\Sigma\, t_j, \Sigma\, t_j^2)$ then there is a unique $\mu \in M_+^1(\mathbb{R} \times \mathbb{R}_+)$ with $\varphi(u,v) = \int \exp(i\, u\, a - \lambda v)\, d\mu(a,\lambda)$, i.e. $\mathcal{D}(X) = \int N(a,2\lambda)^\infty\, d\mu(a,\lambda)$. See Ressel (1985), pp. 916-918 for the technical details. The semigroup $S$ is in this case $(\mathbb{R} \times ]0,\infty[) \cup \{(0,0)\}$ with the involution $(x,y)^* = (-x,y)$.

Example 6. Let $\mathcal{X} = \mathbb{R}_+$, $\mathcal{F} = \{1_{[a,\infty[} | a \ge 0\}$, $v(1_{[a,\infty[}) = a$, $S = \mathbb{R}_+$. If $X = (X_1,X_2,\dots)$ is a sequence of non-negative random variables such that $E(\Pi_{j=1}^n 1_{[a_j,\infty[} \circ X_j) = P(X_j \ge a_j,\ j = 1,\dots,n) = \varphi(\Sigma_{j=1}^n a_j)$ then $\varphi(s) = \int_{[0,\infty]} e^{-\lambda s}\, d\mu(\lambda)$ for some probability measure $\mu$ on $[0,\infty]$, or $\mathcal{D}(X) = \int_{[0,\infty]} \varepsilon_\lambda^\infty\, d\mu(\lambda)$ where $\varepsilon_\lambda$ denotes the exponential distribution with parameter $\lambda$ ($\varepsilon_\infty$ is the Dirac measure in 0).

Example 7. Let $\mathcal{X}$ be any compact Hausdorff space and $\mathcal{F}$ the semigroup of all $[0,1]$-valued continuous functions on $\mathcal{X}$. Let further $\Omega = \mathcal{X}^\infty$ and denote by $X_1,X_2,\dots$ the canonical projections on $\Omega$. If $P \in M_+^1(\Omega)$ is exchangeable the expectation $E(\Pi_{j=1}^n f_j \circ X_j)$ can be written as $\varphi(\Sigma_{j=1}^n \delta_{f_j})$ with $\delta_f$ the Dirac measure in $f$, i.e. $\delta_f \in S := \mathbb{N}_0^{(\mathcal{F})}$, the free abelian semigroup over $\mathcal{F}$. With a little extra work (Ressel (1985), pp. 904 / 905) we get the theorem of Hewitt and Savage (1955), namely the unique representation $P = \int \kappa^\infty\, d\mu(\kappa)$ of $P$ as a mixture of product measures, where $\mu \in M_+^1(M_+^1(\mathcal{X}))$.

## THE THIRD EXTENSION

Let $A$ be a nonempty set, $S$ a $*$-semigroup. An $S$-valued kernel $\psi : A \times A \longrightarrow S$ will be called *almost additive* iff given $\{s_1,\dots,s_n\} \subseteq S$, $\{x_1,\dots,x_m\} \subseteq A$ and $N \in \mathbb{N}$ there exist $\{x_{jp\sigma} | j \le n, p \le m, \sigma \le N\} \subseteq A$ such that

$$\psi(x_{jp\sigma},x_{kq\tau}) = s_j + s_k^* + \psi(x_p,x_q) \quad \text{for } (j,p,\sigma) \neq (k,q,\tau)\ .$$

Theorem 3. *Let $\psi : A \times A \longrightarrow S$ be almost additive and $\varphi : S \longrightarrow \mathbb{C}$ be bounded. Then $\varphi \circ \psi$ is a positive definite kernel iff $\varphi$ is a positive definite function and in fact a mixture of those $\rho \in \hat{S}$ for which $\rho \circ \psi$ is positive definite.*

Before looking to some applications we'll show how to derive Theorem 2 from Theorem 3 (the proof of the latter is given in Ressel, 1986). Put $A := \mathcal{F}^{(\infty)} = \{\underline{f} = (f_1, f_2, \ldots) | f_i \in \mathcal{F},\ f_i = 1$ for all but finitely many $i\}$ and define $\psi : A \times A \longrightarrow S$ by $\psi(\underline{f}, \underline{g}) := \Sigma\, v(f_i \bar{g}_i)$. Then

$$\sum_{j,k=1}^{n} c_j \bar{c}_k\, \varphi(\psi(\underline{f}^{(j)}, \underline{f}^{(k)})) = E\{|\sum_{j=1}^{n} c_j \prod_i f_i^{(j)} \circ X_i|^2\} \geq 0 ,$$

i.e. $\varphi \circ \psi$ is a positive definite kernel. We'll see that $\psi$ is almost additive: let $s_1, \ldots, s_n \in S$, $\underline{f}^{(1)}, \ldots, \underline{f}^{(m)} \in A$, then $f_i^{(p)} = 1$ for all $i > I$ and all $p \leq m$, if $I \in \mathbb{N}$ is suitably chosen. We have $s_j = \Sigma_{\ell=1}^{L} v(f_{j,\ell})$ with $f_{j,\ell} \in \mathcal{F}$ by assumption. Given $N \in \mathbb{N}$ define $\underline{f}^{(jp\sigma)} \in A$ by

$$f_i^{(jp\sigma)} := \begin{cases} f_i^{(p)} & , i = 1, \ldots, I \\ f_{j,1}, \ldots, f_{j,L} & , i \text{ running through } N_{jp\sigma} \\ 1 \text{ else} & \end{cases}$$

with pairwise disjoint subsets $N_{jp\sigma} \subseteq \mathbb{N} \setminus \{1, \ldots, I\}$ of cardinality $|N_{jp\sigma}| = L (j \leq n, p \leq m, \sigma \leq N)$. For $(j,p,\sigma) \neq (k,q,\tau)$ we get $\psi(\underline{f}^{(jp\sigma)}, \underline{f}^{(kq\tau)}) = \psi(\underline{f}^{(p)}, \underline{f}^{(q)}) + s_j + s_k^*$, i.e. $\psi$ is indeed almost additive. By Theorem 3 $\varphi$ is a mixture of those $\rho \in \hat{S}$ for which $\rho \circ \psi$ is positive definite. Since $\rho(\psi(\underline{f}, \underline{g})) = \rho(\Sigma\, v(f_i \bar{g}_i)) = \Pi\, \rho(v(f_i \bar{g}_i))$, using Schur's lemma, this property is equivalent with positive definiteness of $\rho \circ v$.

We shall now describe some applications of Theorem 3 to stochastic processes. Let $T$ be an infinite (time-)set and consider a stochastic process $X = (X_t, t \in t)$ indexed by $T$. The distribution $P = \mathcal{D}(X)$ is then a (Baire-) probability measure on $\mathbb{R}^T$, and is determined by all its finite dimensional marginals. The *characteristic functional* $\hat{P}$ of $P$ may be defined on Mol(T), the vector space of all real-valued "molecular" measures on $T$ (i.e. measures with finite support) by

$$\hat{P}(\nu) := E\{\exp[i \int X_t \, d\nu(t)]\} \quad , \nu \in \text{Mol}(T)$$

and evidently determines P , i.e. $\hat{P} = \hat{Q}$ iff $P = Q$ .

If f.ex. X is a centred gaussian process with covariance $K(s,t) = E(X_s X_t)$, then $\int X_t \, d\nu(t)$ is a normal r.v. with mean 0 and variance $\int \int K(s,t) \, d\nu(s) \, d\nu(t)$, whence

$$\hat{P}(\nu) = \exp(-\frac{1}{2} \int K \, d\nu \otimes \nu) \quad , \nu \in \mathrm{Mol}(T).$$

If $P_\lambda$ denotes the law of the centred gaussian process with covariance $\lambda \cdot K$, $\lambda \geq 0$, then a mixture

$$P = \int_0^\infty P_\lambda \, d\mu(\lambda) \qquad , \text{where } \mu \in M_+^1(\mathbb{R}_+)$$

will have the characteristic functional

$$\hat{P}(\nu) = \int_0^\infty \hat{P}_\lambda(\nu) \, d\mu(\lambda) = \int_0^\infty \exp(-\frac{\lambda}{2} \int K \, d\nu \otimes \nu) \, d\mu(\lambda)$$

depending only via the "quadratic form" $\int K \, d\nu \otimes \nu$ on $\nu$ .

It turns out that this is a characteristic property of such mixtures if we impose on K the natural (and evidently necessary) condition to be "non-degenerate" in the sense that for any $n \in \mathbb{N}$ there exist $t_1, \ldots, t_n \in T$ such that the matrix $(K(t_j, t_k))_{j,k \leq n}$ is non-singular. (Equivalently: the RKHS of K is infinite dimensional.)

Theorem 4. *Let* $P \in M_+^1(\mathbb{R}^T)$ *be a probability measure whose characteristic functional only depends on* $\int K \, d\nu \otimes \nu$, K *denoting a non-degenerate positive definite kernel on* $T \times T (T \neq \emptyset)$. *Then* P *is a "scale mixture" of centred gaussian processes with covariances* $\lambda \cdot K$, $\lambda \geq 0$ .

Example 8. $T = \mathbb{R}_+$, $K(s,t) = s \wedge t$

If $X = (X_t, t \geq 0)$ is a process whose characteristic functional depends on $\int s \wedge t \, d\nu(s) \, d\nu(t)$ then X is a mixture of centred Brownian motions.

Example 9. $T = [0,1]$, $K(s,t) := s \wedge t - s \cdot t$

If $X = (X_t, 0 \leq t \leq 1)$ is a process whose characteristic functional depends on $\int (s \wedge t - st) \, d\nu(s) \, d\nu(t)$ then X is a mixture of centred Brownian bridges.

With some extra effort the following can be shown:

Examples 8',9'. Let $X = (X_t, t \geq 0)$, resp. $(X_t, 0 \leq t \leq 1)$ be a process whose characteristic functional depends on $\int t\, d\nu(t)$ and $\int s \wedge t\, d\nu(s)\, d\nu(t)$, resp. $\int t\, d\nu(t)$ and $\int (s \wedge t - s \cdot t)\, d\nu(s)\, d\nu(t)$. Then $X$ is a unique "scale/drift"-mixture of Brownian motions resp. bridges, or:

$$(X_t) \overset{\mathcal{D}}{=} (Y \cdot t + V \cdot Z_t)$$

where $Y$ is a real, $V$ a non-negative r.v., $Z$ is standard Brownian motion resp. bridge, and $(Y,V)$ is independent of $Z$ .

Out last application concerns Poisson processes and related random measures. Let $\mathcal{X}$ be locally compact (second countable) with a fixed reference measure $\nu \in M_+(\mathcal{X})$ of infinite mass, f.ex. $\mathcal{X} = \mathbb{R}^n$ and $\nu$ the Lebesgue measure. Let $b : \mathbb{R}_+ \longrightarrow \mathbb{R}_+$ be any non-trivial Bernstein function (i.e. $b$ is continuous, $b(0) = 0$, $(-1)^n\, b^{(n+1)}(x) \geq 0$ for $n = 0,1,2,\ldots, x > 0$). An important example is given by $b(s) = 1 - e^{-s}$.

A random measure on $\mathcal{X}$ is by definition a probability measure $P$ on $M_+(\mathcal{X})$, and its *Laplace functional* $\hat{P}$ is defined on $C_+^c(\mathcal{X})$ by

$$\hat{P}(f) := \int \exp(-\int f\, d\kappa)\, dP(\kappa) \ .$$

For example the Poisson process with intensity measure $\nu$ has the Laplace functional $\exp[-\int(1-e^{-f})d\nu]$. Likewise for any Bernstein function $b$ there is a corresponding Laplace functional $\exp(-\int b \circ f\, d\nu)$ whose underlying random measure $P_b$ is characterized in the following way: denoting $\xi_B(\kappa) := \kappa(B)$ the random mass given to a bounded Borel subset $B \subseteq \mathcal{X}$ , and letting $(\tau_\lambda)_{\lambda \geq 0} \subseteq M_+^1(\mathbb{R}_+)$ be the convolution semigroup determined by $b$ — i.e. $\exp(-\lambda b)$ is the Laplace transform of $\tau_\lambda$ for all $\lambda \geq 0$ — we have

i) $\mathcal{D}(\xi_B) = \tau_{\nu(B)}$

ii) $\xi_{B_1}, \ldots, \xi_{B_n}$ are independent for pairwise disjoint $B_1, \ldots, B_n$ .

Theorem 5. *Let* P *be a random measure whose Laplace functional depends only on* $\int b \circ f \, d\nu$. *Then* P *is a unique scale mixture of the underlying* $P_b$, *i.e.* $P = \int_0^\infty P_{\lambda b} \, d\mu(\lambda)$ *for some unique probability* $\mu$ *on* $\mathbb{R}_+$ .

For the proofs of Theorems 4 and 5 we again refer to Ressel (1986).

REFERENCES


De Finetti, B., 1931, Funzione caratteristica di un fenomeno allatorio, Atti della R. Accademia Nazionale dei Lincii Ser. 6, Memorie, Classe di Scienze, Fisiche, Matematiche e Naturali, 4 : 251.

Hewitt, E., Savage, L.J., 1955, Symmetric measures on Cartesian products, Trans. Amer. Math. Soc., 80 : 470.

Lindahl, R.J., Maserick, P.H., 1971, Positive-definite functions on involution semigroups, Duke Math. J., 38 : 771.

Ressel, P., 1985, De Finetti-type theorems: an analytical approach, Ann. Prob., 13 : 898.

Ressel, P., 1986, Integral representations of symmetric stochastic processes, to appear in Prob. Theory Rel. Fields.

Schoenberg, I.J., 1938, Metric spaces and positive definite functions. Trans. Amer. Math. Soc., 44 : 522.

## A BAYESIAN APPROACH TO ESTIMATING THE PARAMETERS OF A HYDROLOGICAL FORECASTING SYSTEM

Sylvia Schnatter, Dieter Gutknecht and Robert Kirnbauer

Institut für Hydraulik, Gewässerkunde und Wasserwirtschaft
Technische Universität Wien
Vienna, Austria

### SUMMARY

In this paper certain aspects of estimating model parameters in hydrological short term forecasting are dealt with. In estimating parameters of hydrological flood routing models difficulties arise when the input-output relationship of the model is affected by violations of the mass condition. In this paper an estimation procedure is presented which can handle this problem. The procedure is based upon a Bayesian algorithm for recursive estimation of the parameters of a dynamic linear model. The unknown volume increase is dealt with a volume correction coefficient which is estimated using the Kalman Filter. Finally, an application of the model to a real world example is given.

### HYDROLOGICAL SHORT TERM FORECASTING

Hydrological short term forecasting is applied to predict the future runoff in a river system some hours or days ahead. Forecasting models are frequently based upon rainfall-runoff models and flood routing models.

#### Flood Routing Models

Flood routing models describe how an inflow is transformed into an outflow by flowing through a river reach. The relationship between inflow $q^Z(t)$ and outflow $q^A(t)$ is based upon a time independent and linear (in the system theoretical sense) impulse response function $h(\tau)$:

$$q^A(t) = \int_0^\infty h(\tau) \cdot q^Z(t-\tau)d\tau \tag{1}$$

Nash (1958) was the first to introduce a gamma probability density function for $h(\tau)$:

$$h(\tau) = h(\tau,\alpha,\beta) = \frac{1}{\Gamma(\alpha)\beta^\alpha} \tau^{\alpha-1} e^{-\tau/\beta} \tag{2}$$

whose parameters $\alpha$ and $\beta$ have an interesting hydrological meaning: the parameters' product $t_L = \alpha.\beta$ is the mean travel time of the flood wave between inflow and outflow section of the river reach. As $t_L$ can be associated with hydraulic properties of the river bed, prior information about the range of $t_L$ is available.

Model (1) is based upon two main assumptions:

- Firstly, the assumption of constant mass, meaning that inflow and outflow volumes have the same amount. This assumption is often violated especially if there are unknown inflows from small tributaries to the main stretch.
- Secondly, the assumption, that the transformation does not depend upon time (time invariance) and upon inflow (linearity). In case of inundation this assumption might be invalid. In order to compensate for this effect a second transfer function model of type (1) is introduced (see formula(16)).

## A DYNAMIC LINEAR MODEL FOR HYDROLOGICAL SHORT TERM FORECASTING

As the assumption of constant mass normally is violated, the forecasting model has to estimate the actual increase of volume. In the model proposed this is done by updating a coefficient $c_k$ of volume increase in a one-dimensional state space formulation:

$$\begin{aligned} &\text{system equation:} && c_k = c_{k-1} + w_{k-1} && w_{k-1} \sim N(0,Q) \\ &\text{observation equation:} && q_k^A = H_k \cdot c_k + v_k && v_k \sim N(0,R_k), \end{aligned} \tag{3a}$$

where $q_k^A$ is the runoff observed at the end of the reach at time $t_k$ and $H_k$ is the runoff at this point assuming that the condition of constant mass is valid:

$$H_k = \int_0^\infty h(\tau,\alpha,\beta) \;.\; q^Z(t_k-\tau)d\tau \tag{3b}$$

For hydrological reasons $R_k$ will depend upon the current runoff rate:

$$R_k = (d\cdot q_{k-1}^A)^2 \tag{4}$$

Therefore system (3a) has to be modified to a conditionally gaussian sequence:

$$\begin{aligned} &\text{system equation:} && c_k = c_{k-1} + w_{k-1} && w_{k-1} \sim N(0,Q) \\ &\text{observation equation:} && q_k^A = H_k\cdot c_k + d\cdot q_{k-1}^A\cdot v_k^* && v_k^* \sim N(0,1) \end{aligned} \tag{5}$$

It can be shown (Liptser and Shirayayev(1977)) that the conditional distribution $p(c_k|q^{A,k})$ is Gaussian:

$$p(c_k|q^{A,k}) \sim N(\hat{c}_k,\hat{P}_k)$$

Troughout the paper $q^{A,k}$ will denote the entire sequence of values from 1 up to k:

$$q^{A,k} = (q_1^A,q_2^A,\ldots,q_k^A) = (q^{A,k-1},q_k^A)$$

Liptser and Shirayayev(1977) give a closed system of recursive equations for $\hat{c}_k$ and $\hat{P}_k$:

$$\begin{aligned} &c_k \sim N(\hat{c}_k \;,\; \hat{P}_k) \\ &\hat{c}_k = \hat{c}_{k-1} + K_k \cdot (q_k^A - H_k \cdot \hat{c}_{k-1}) \end{aligned} \tag{6}$$

$$\hat{P}_k = (\hat{P}_{k-1} + Q)\ (1 - H_k \cdot K_k)$$

$$K_k = \frac{(\hat{P}_{k-1} + Q) \cdot H_k}{(\hat{P}_{k-1} + Q) \cdot (H_k)^2 + (d \cdot q_{k-1}^A)^2}$$

The application of (5) and (6) to real world problems requires
- knowledge of the initial distribution of $c_k$
- knowledge of the statistical properties of the error terms, i.e. the values of Q and d
- knowledge of the values of the parameters α and β in (3b)

The initial value $\hat{c}_0$ was choosen to be 1 assuming no volume increase at starting time. The variance $\hat{P}_0$ can be set to any plausible number since it turns out that the final result is insensitive to the choice of $\hat{P}_0$.

Reasonable values of Q and d were derived by simulation studies. Choosing certain quantities d and Q runoff hydrographs were generated by ( 5). By judging the hydrological plausibility of the computed outflow hydrographs values of d about 0.01 and of Q about $(0.01)^2$ or $(0.02)^2$ were found to suite quite well.

Knowledge of α and β is rather fuzzy: as mentioned above information concerning the range of $t_L = \alpha . \beta$ could be found by hydraulic considerations, but usually it is not possible to determine the parameters precisely. To handle this uncertainty the dynamic linear model (5) was combined with a Bayesian algorithm for estimating the current posterior density of the parameters. In the following investigations we will estimate $(t_L, \beta)$ instead of $(\alpha, \beta)$.

## A BAYESIAN ALGORITHM FOR ESTIMATING $\underline{\theta}$ AND $c_k$

As $\underline{\theta} = (t_L, \beta)$ is assumed to be a random variable also the term $H_k$ in system (5) will be random. Any estimation of $c_k$ has to take this uncertainty into account. If the measurement noise were not depend upon the observation $q_{k-1}^A$ one could use the algorithm for adaptive Kalman filtering given by Magill(1965), Lainiotis(1971), Harrison and Stevens(1976) or Szöllösi-Nagy and Wood(1976). Similar results can be found for a conditionally Gaussian sequence.

We start with a probability density function (p.d.f) $p(c_{k-1}, \underline{\theta} | q^{A,k-1})$ of $c_{k-1}$ and $\underline{\theta}$ which can be expressed in the following form:

$$p(c_{k-1}, \underline{\theta} | q^{A,k-1}) = p(c_{k-1} | \underline{\theta}, q^{A,k-1}) \cdot p(\underline{\theta} | q^{A,k-1}) \qquad (9)$$

with $p(c_{k-1} | \underline{\theta}, q^{A,k-1})$ Gaussian with mean $\hat{c}_{k-1}(\underline{\theta})$ and variance $\hat{P}_{k-1}(\underline{\theta})$ and an arbitrary p.d.f. for $p(\underline{\theta} | q^{A,k-1})$. Using the system equation in (5) it can be easily shown that the prior p.d.f. $p(c_k | \underline{\theta}, q^{A,k-1})$ of $c_k$ given $\underline{\theta}$ and observations until k-1 is also Gaussian with mean $\hat{c}_{k-1}(\underline{\theta})$ and variance $\hat{P}_{k-1}(\underline{\theta})+Q$:

$$p(c_k | \underline{\theta}, q^{A,k-1}) \sim N(\hat{c}_{k-1}(\underline{\theta}), \hat{P}_{k-1}(\underline{\theta}) + Q) \qquad (10)$$

The posterior p.d.f. $p(c_k, \underline{\theta} | q^{A,k})$ of $c_k$ and $\underline{\theta}$ given observations until k can be calculated by means of the Bayesian theorem:

$$p(c_k, \underline{\theta} | q^{A,k}) = p(c_k, \underline{\theta} | q_k^A, q^{A,k-1}) \propto p(q_k^A | c_k, \underline{\theta}, q^{A,k-1}) \cdot p(c_k, \underline{\theta} | q^{A,k-1})$$

$$\propto p(q_k^A | c_k, \underline{\theta}, q^{A,k-1}) \cdot p(c_k | \underline{\theta}, q^{A,k-1}) \cdot p(\underline{\theta} | q^{A,k-1}) \qquad (11)$$

The likelihood function $p(q_k^A | c_k, \underline{\theta}, q^{A,k-1})$ is given by the observation equation in (5):

$$p(q_k^A | c_k, \underline{\theta}, q^{A,k-1}) = \frac{1}{(d \cdot q_{k-1}^A)} \cdot \exp\{-\frac{1}{2} \frac{(q_k^A - H_k \cdot c_k)^2}{(d \cdot q_{k-1}^A)^2}\} \qquad (12)$$

Using the prior p.d.f. (10) and the likelihood function (12) the posterior p.d.f. (11) can be expressed by the following form:

$$p(c_k, \underline{\theta} | q^{A,k}) \propto \frac{1}{d\, q_{k-1}^A \, (\hat{P}_{k-1}(\underline{\theta}) + Q)^{1/2}} \cdot \exp\{-\frac{1}{2}(\frac{(q_k^A - H_k \cdot c_k)^2}{(d \cdot q_{k-1}^A)^2} + \frac{(c_k - \hat{c}_{k-1}(\underline{\theta}))^2}{\hat{P}_{k-1}(\underline{\theta}) + Q})\}$$

By completing the squares of $c_k$ in the exponent we obtain:

$$p(c_k, \underline{\theta} | q^{A,k}) \propto \frac{1}{\hat{P}_k(\underline{\theta})^{1/2}} \exp\{-\frac{1}{2} \cdot \frac{(c_k - \hat{c}_k(\underline{\theta}))^2}{\hat{P}_k(\underline{\theta})}\} \qquad (13)$$
$$\frac{1}{N_k(\underline{\theta})^{1/2}} \exp\{-\frac{1}{2} \cdot \frac{(q_k^A - H_k \cdot \hat{c}_{k-1}(\underline{\theta}))^2}{N_k(\underline{\theta})}\}$$

with

$$N_k(\underline{\theta}) = H_k(\underline{\theta})^2 \cdot (\hat{P}_{k-1}(\underline{\theta}) + Q) + (d\, q_{k-1}^A)^2$$
$$\hat{c}_k(\underline{\theta}) = \hat{c}_{k-1}(\underline{\theta}) + K_k(\underline{\theta}) \cdot (q_k^A - H_k(\underline{\theta}) \cdot \hat{c}_{K-1}(\underline{\theta}))$$
$$\hat{P}_k(\underline{\theta}) = (\hat{P}_{k-1}(\underline{\theta}) + Q)\,(1 - H_k(\underline{\theta}) \cdot K_k(\underline{\theta}))$$
$$K_k(\underline{\theta}) = \frac{(\hat{P}_{k-1}(\underline{\theta}) + Q) \cdot H_k(\underline{\theta})}{(\hat{P}_{k-1}(\underline{\theta}) + Q) \cdot (H_k(\underline{\theta}))^2 + (d \cdot q_{k-1}^A)^2}$$

$p(c_k, \underline{\theta} | q^{A,k})$ can therefore be split up in the same way as (9) as a product of the $\underline{\theta}$-conditional posterior p.d.f of $c_k$ which is Gaussian and a posterior p.d.f of $\underline{\theta}$ independ of $c_k$:

$$p(c_k, \underline{\theta} | q^{A,k}) = p(c_k | \underline{\theta}, q^{A,k}) \cdot p(\underline{\theta} | q^{A,k}) \qquad (14)$$

The marginal p.d.f. of $c_k$ is no more Gaussian:

$$p(c_k | q^{A,k}) = \int_{\Theta} p(c_k | \underline{\theta}, q^{A,k}) \cdot p(\underline{\theta} | q^{A,k})\, d\underline{\theta}$$

An optimal adaptive estimate of $c_k$ with respect to a quadratic loss function is given by the weighted $\underline{\theta}$-contional means:

$$\hat{c}_k = \int_{\Theta} \hat{c}_k(\underline{\theta}) \cdot p(\underline{\theta} | q^{A,k})\, d\underline{\theta} \qquad (15)$$

## Model Modification: The Inundation Case

If inundation occurs the parameters $t_L$ and $\beta$ will change suddenly. In this case the linear transformation (1) is no longer appropriate. We overcome this difficulty by using two impulse-response functions:

$$H_k = \int_0^\infty \min(q^Z(t_k-\tau), q^Z_{AUS}) \cdot h^1(\tau, t^1_L, \beta^1)d\tau + \int_0^\infty \max(q^Z(t_k-\tau) - q^Z_{AUS}, 0) \cdot h^2(\tau, t^2_L, \beta^2)d\tau \quad (16)$$

$q^Z_{AUS}$ is the threshold value of runoff above which inundation starts. It can be computed approximately from the shape and size of the riverbed. The unknown parameters $\underline{\theta} = (t_L{}^1, \beta^1, t_L{}^2, \beta^2)$ are estimated by means of the Bayesian algorithm presented above. It should be noted that the Bayesian approach yields posterior distribution of $t_L{}^2$ and $\beta^2$, although prior information about these parameters is very poor and thus is particularly well suited to this problem.

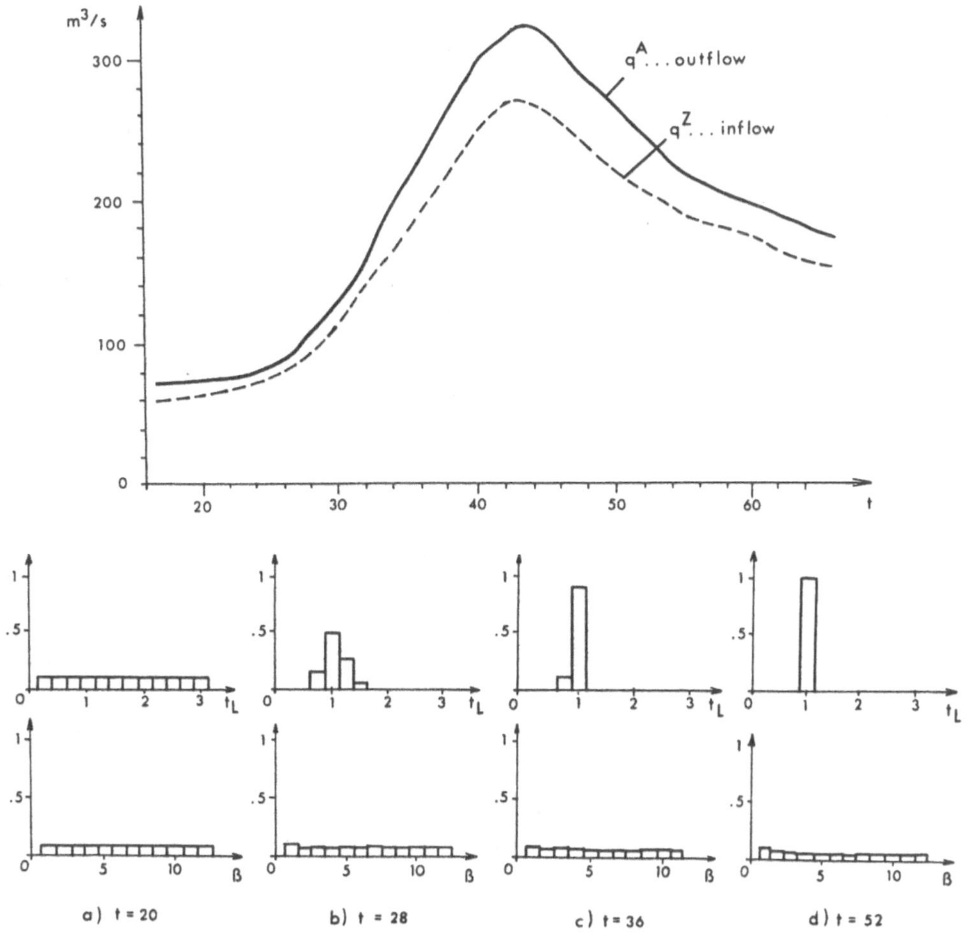


Figure 1. Flood event of May 1978

A REAL WORLD APPLICATION

The model presented above was applied to real world data. Short term forecasting based on flood routing was performed for certain rivers in Carinthia (Austria). Starting with a diffuse prior of travel time $t_L$ and $\beta$ within a reasonable range the posterior probability density function was calculated for each time step k using (6) and (13). As $H_k(\underline{\theta})$ is a nonlinear function of $\underline{\theta}$ (see (2) and (3b)) no closed form for the posterior p.d.f. could be found. Therefore the parameter space was discretized. The posterior value (13) was calculated for each parameter.

As an example fig. 1 shows the observed inflow and the observed outflow of the reach between the gauging stations Oberdrauburg and Greifenburg at the river Drau for the flood event of may 1978. Starting with a diffuse prior at t = 20, the marginal posteriors of $t_L$ and $\beta$ are shown at t = 28, t = 36 and t = 56. The marginal posterior of $t_L$ converges rather quickly while the marginal posterior of $\beta$ hardly changes.

For the event of July 1981 inundation is observed. Prior information concerning the parameters ${t_L}^2$ and $\beta^2$ of the second impulse-response function (see (1 )) is rather poor. Therefore the prior was chosen within a wide range (see fig. 2). The updating algorithm starts at t = 48 when inundation sets in. While the marginal posterior p.d.f. of ${t_L}^2$ converges within a few hours, the marginal posterior p.d.f. of $\beta^2$ changes more slowly (see fig.2)).

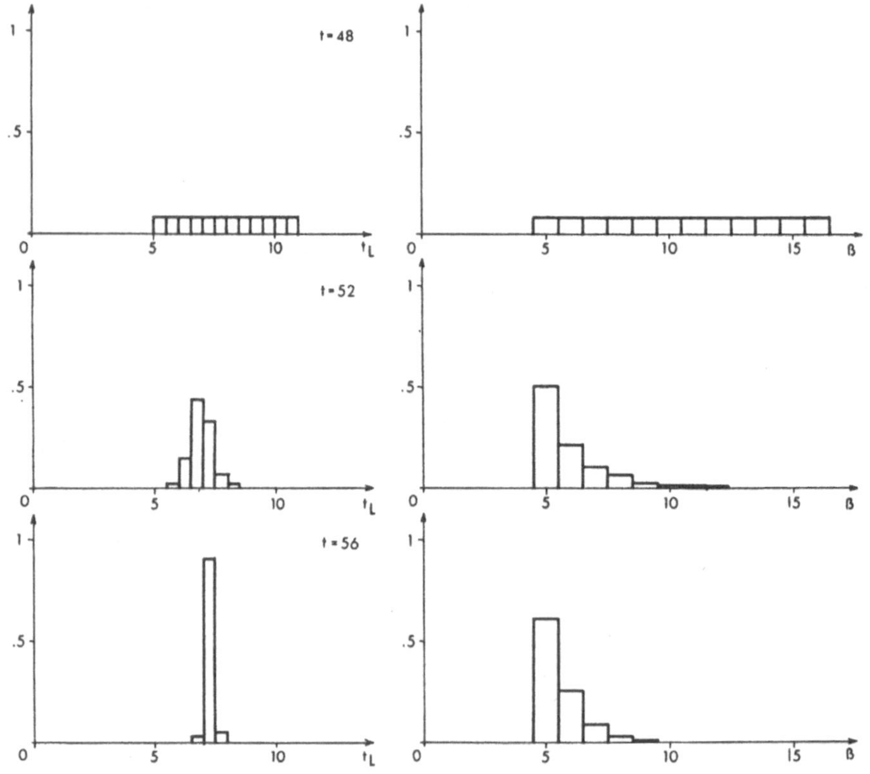


Figure 2. Flood event of July 1981

Evaluating the result obtained it may be concluded that the method is well suited for being incorporated in an operational forecasting model for the river Drau.

REFERENCES

Lainiotis,D.G., 1971, Optimal adaptive parameter estimation, IEEE Trans. Autom. Control, 16:260-270.

Liptser,R.S. and Shiryayev,A.N., 1977, Statistics of random processes, Springer, New York/Heidelberg/Berlin.

Magill,D.T., 1965, Optimal adaptive estimation of sampled stochastic processes, IEEE Trans. Autom. Control, 10:434-439.

Harrison,P.J. and Stevens,C.F., 1976, Bayesian forecasting, Journal of the Royal Statistical Society, Ser.B, 38:205-247.

Nash,J.E., 1958, The form of the instantaneous unit hydrograph, in: A.I.H.S. Publication, 45:114-119.

Szöllösi-Nagy,A. and Wood,E.F., 1976, Bayesian strategies for controlling dynamic water resources, in:Real-time forecasting/control of water recource systems, IIASA Proceeding Series, 8:203-214, Pergamon Press, Oxford/New York.

# THE EXTENDED BAYES-POSTULATE, ITS POTENTIAL EFFECT ON STATISTICAL METHODS AND SOME HISTORICAL ASPECTS

Friedrich Schreiber

Aachen University of Technology

## Summary

The principal problem of statistics is considered where the value of a single parameter or of a parameter vector is a priori unknown. In this case the extended Bayes-postulate requiring the statement of two prior uniform distributions provides a unique parameter representation (leaving no freedom for nonlinear parameter transformations) and unique posterior statements which are useful for small as well as for large sample sizes. A short survey is given of recent work in this field which has been named the "Bayes-Laplace-statistics" and of its historical background.

## 1. Bayes-Postulate and Fisher's Transformation Argument

1.1 For introductory purposes we consider the principal task of statistics in its most elementary form: a given stationary random process is described by a known convolutional type prediction p.f. $P(y \mid \mu)$, where the sum-r.v.

$$y = \sum_{i=1}^{n} x_i \; ; \quad E\{y/n\} = \mu \tag{1}$$

will be gained from the random vector $(x_1, x_2, \cdots, x_n)$ to be measured in n independent trials and where the expectation $\mu$ is the only parameter whose value is *unknown* and shall be determined from the value of $y$. Due to the *Bayes-postulate* the prior situation "value of $\mu$ unknown" means in accordance with logical insight that all values of $\mu$ must be equally possible. Therefore the prior knowledge about $\mu$ has to be described within the given limits $\mu_I \leq \mu \leq \mu_{II}$ by the uniform distribution with density

$$f(\mu) = C_\mu = 1/(\mu_{II} - \mu_I) \,. \tag{2a}$$

After $y$ has been measured and since $f(\mu)$ is a constant $C_\mu$ the inversion law for conditional probabilities (Bayes-theorem) yields the well defined posterior density

$$f(\mu \mid y) = P(y \mid \mu) / \int_\mu P(y \mid \mu) d\mu \,. \tag{3}$$

This posterior knowledge statement with respect to the value of the expectation $\mu$ whose formal simplicity corresponds to its fundamental importance has been given by Laplace [11] following Bayes [1].

1.2 About 65 years ago in his address to the Royal Society [5] Fisher argued that any nonlinear parameter transformation $\varsigma(\mu)$ leads to an equivalent description of the r.v. $y$ by the p.f. $P(y \mid \varsigma)$ instead of $P(y \mid \mu)$; that the parameter $\varsigma$ could be considered to be unknown as well as $\mu$ and could therefore also be described by a prior uniform distribution with constant density $f(\varsigma) = C_\varsigma$

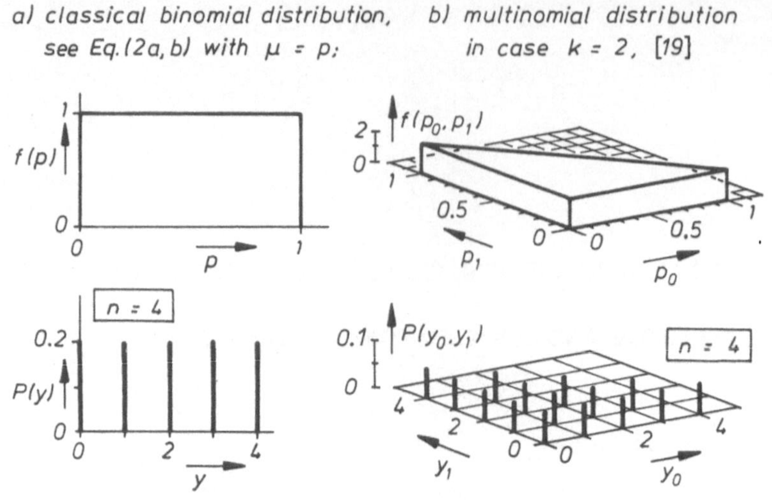


Fig. 1 Examples of the extended Bayes-postulate with two prior uniform distributions.

and that the resulting posterior density $f(\varsigma \mid y) = P(y \mid \varsigma)/\int_{\varsigma} P(y \mid \varsigma)d\varsigma$ would be obviously inconsistent with eq.(3). Fisher concluded that any result obtained on the basis of the Bayes-postulate suffers from total ambiguity and is therefore worthless.

This transformation argument against the Bayes-postulate has been commonly accepted and seems to be the cause for the emergence of different statistical schools and is the main root of subjective Bayesianism: if the parameter representation could be chosen arbitrarily then indeed we also seem to be free to choose the description of the prior knowledge about a parameter according to our personal judgement. Moreover it can be shown that by acceptance of the transformation argument the application range of any statistical inference method is reduced to the large sample case.

1.3 In [17] the author has pointed out that the prior situation is characterized by *two* unknown quantities each to be described by a uniform distribution: the parameter $\mu$ by $f(\mu) = C_\mu$ eq.(2a) and the future value of the sum-r.v. $y$ eq.(1) by the prior prediction p.f.

$$P(y) = C_\mu \int_\mu P(y \mid \mu)d\mu = C_y = 1/(1 + y_{II} - y_I)\,, \qquad (2b)$$

assuming $y$ to be discrete in the range $y = y_I,\ y_I + 1, \cdots,\ y_{II}$. The requirement for the simultaneous existence of eq.(2a) and eq.(2b) being called the *extended Bayes-postulate* is only fulfilled if the parameter is uniquely chosen to be the *expectation* $\mu$ (see Fig.1a) and any nonlinear parameter transformation $\varsigma(\mu)$ is ruled out. Therefore the expectation $\mu$ might be called the "natural parameter".

The extended Bayes-postulate is applicable only to convolutional type p.f.'s $P(y \mid \mu)$ resp. densities $f(y \mid \mu)$ i.e. to the processing of a measured $x$-sequence by the sum operation eq.(1). It is unalteredly valid for those important distributions where the parameter $\mu$ is defined on the half or total real axis: the uniform distributions for $\mu$ and $y$ are then described by constants $C_\mu \to 0$ resp. $C_y \to 0$ ( see Rényi [14] ) and also in this limit case $C_\mu$ will cancel in the inversion formula as usual. In [19] it has been shown that the extended Bayes-postulate is also applicable to *k-dimensional* cases like the multinomial-distribution where in the prior situation the expectation parameter vector $(p_0, p_1, \cdots, p_k)$ and the sum random vector $(y_0, y_1, \cdots, y_k)$ have to be described by k-dimensional uniform distributions, see Fig.1b.

The extended Bayes-postulate has been found to be true "in all cases investigated so far" and it represents presumably a general law whose proof will follow sooner or later.

1.4 Following the above insights it is also advisable to make eventual *nonuniform* prior statements $f(\mu) \neq C_\mu$ with respect to the expectation parameter $\mu$ only; this would ease the comparison of different prior assumptions.

In this case we should always check whether the associated prior prediction statement about $y$ corresponding to eq.(2b)

$$P(y) = \int_\mu P(y \mid \mu) \cdot f(\mu) d\mu \neq C_y \tag{4}$$

is really in accordance with our prior intuition, see the examples for "prejudices" in [17]. Obviously $P(y)$ is not merely a formal normalizing expression in the denominator of the inversion formula which needs no further attention but is equally important as $f(\mu)$ and it is the pair of prior functions $f(\mu), P(y)$ which must be justified. These considerations indicate the problematic nature of intuitive prior assumptions (see also section 2.2) and that it is preferable - wherever possibly - to ignore intuitive prior information about $\mu$ resp. y and rather retreat to the objective prior pair $C_\mu, C_y$ eq.(2a,b).

## 2. The Bayes-Laplace-Statistics[1]

This name can be attributed to all statistical methods based on the extended Bayes-postulate. The BL-statistics provide for *any sample* size unique posterior statements of the type eq.(3) for all important distributions with discrete or continuous sum-r.v. $y$ including finite population cases like the hypergeometric distribution and also for multi-parameter cases including Markov chains. Within the immense variety of different posterior formulae which might be offered due to different subjective prior assumptions the posterior formulae of the BL-statistics are distinguished by the fact that they constitute the objective principal case where the natural parameter representation $\mu$ is mandatory and where any prior information about $\mu$ and $y$ does not exist resp. has been ignored. A few guide lines shall be given for the typical use and the future development of this statistical concept.

2.1 Usually simple formulae for the $j$-th order moment $M_j\{\mu \mid y\}$ can be derived from the posterior density of the type $f(\mu \mid y)$ eq.(3). Mainly we need the *mean parameter value* $\overline{\mu}$, the *absolute error* (mean quadratic deviation from $\overline{\mu}$) $\sigma_\mu$ and the *relative error* (coefficient of variation) $d_\mu = \sigma_\mu/\overline{\mu}$ :

$$\overline{\mu} = M_1 \; ; \; \sigma_\mu = \left(M_2 - M_1^2\right)^{1/2} ; \; d_\mu = \left(M_2/M_1^2 - 1\right)^{1/2} . \tag{5a, b, c}$$

It is recommended to use the error measure $\sigma_\mu$ resp. $d_\mu$ instead of confidence interval statements, which contain subjective elements [16].

2.2 Sometimes the *small sample case* cannot be avoided due to the cost of an experiment. Then the formulae eq.(3) and eq.(5) represent the unbiased, conservative posterior statements not affected by subjective prior assumptions. This shall be explained by the following example of considerable practical importance: the sum-r.v. $y$ eq.(1) is described by the normal density $f(y \mid \mu, v)$ with the expectation $\mu$ and the variance $v = \sigma^2$ being both a priori unknown. Then as shown in [15] the posterior knowledge of $\mu$ is described by the *objective t-distribution* with density

$$\left.\begin{aligned} f(t) &= \frac{\Gamma[(n-2)/2]}{[\pi(n-1)]^{1/2}\,\Gamma[(n-3)/2]}[1+t^2/(n-1)]^{-(n-2)/2} \; ; \; n \geq 4 \; ; \\ t &= (\mu - y/n)/[\frac{\eta}{n(n-1)}]^{1/2} \; ; \; \eta = \sum_{i=1}^{n} x_i^2 - y^2/n \end{aligned}\right\} \tag{6}$$

and not by the commonly used Student's t-distribution [7], [9] whose density we denote by $f^*(t)$.

[1] Abbreviation: BL-statistics. The term "objective Bayes- statistics" being used in some former publications has been abandoned. Nevertheless all statements concerning a principal problem in statistics which are based on the extended Bayes-postulate and which can be verified by a proper computer random experiment (see section 3) might be qualified to be "objective" because they do not depend on subjective prior assumptions.

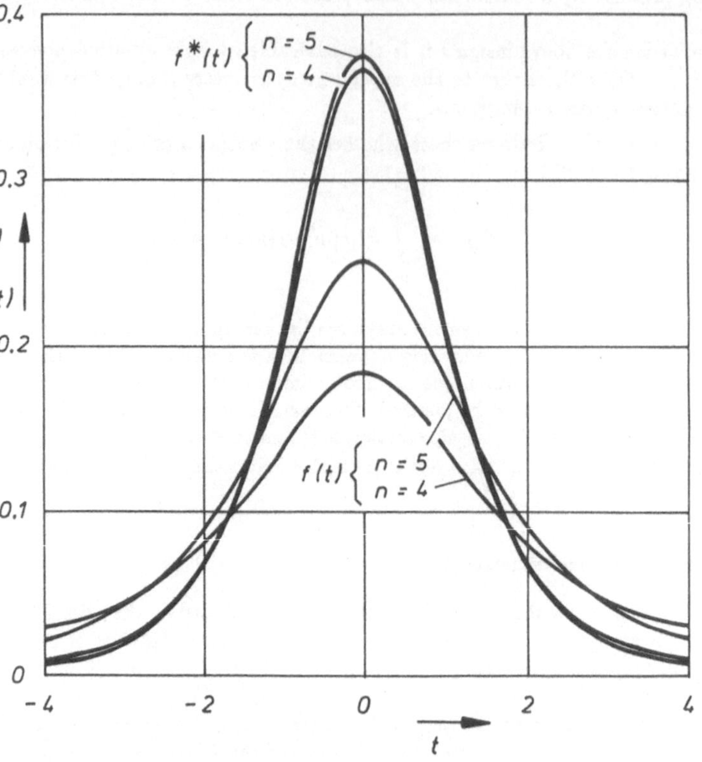


Fig. 2 Small sample case: comparison of Student's $t$-distribution density $f^*(t)$ with the density $f(t)$ Eq.(6) of the objective $t$-distribution [15].

The comparison of both densities in Fig.2 reveals that for small samples [2] $f(t)$ is much broader than $f^*(t)$ i.e. that any error measure derived from $f^*(t)$ would represent an optimistic delusion. It is interesting that Box and Tiao [2] derive student's density $f^*(t)$ on basis of the prior assumption $f(\sigma) \sim \sigma^{-1}$ which they believe to be "noninformative" and all this shows exemplarily how questionable the results of subjective assumptions might become for small samples.[3]

2.3 In the *large sample case* we usually have $\overline{\mu} \approx y/n$ independently from the prior assumption. Then the relative error formula eq.(5c) becomes our main interest in order to control the evaluation of a large data volume. The LRE-algorithm [4] for obtaining the empirical d.f. $F_n(x)$ of *independent* $x$-sequences [18] gives a useful example for this. A more elaborate version of this algorithm for *correlated* $x$-sequences can be derived from the statistical analysis of Markov chain models.

2.4 The investigation of multidimensional problems with correlation phenomena etc. might lead to troublesome mathematical barriers. Then - irrespective of a personal philosophy concerning prior assumptions - the relative simplicity of all formulae derived from posterior densities of the fundamental type eq.(3) can be of great help.

---

2 In typical applications like the "batch means"-method [6] the sample size $n$ is often small, e.g. $n$=10.

3 It should be mentioned that under the rules of the BL-statistics the proper prior statement cannot be made with respect to the standard deviation $\sigma$ but only with respect to the variance $v=\sigma^2 = E\{\eta/(n-1)\}$ being the natural parameter of the $\chi^2$-distribution with prediction density $f(\eta|v)$, see [15].

4 LRE = Limited Relative Error.

| Statement | math. Description |
|---|---|

a) "μ unknown"

$$f(\mu) = C_\mu = 1 / (\mu_{II} - \mu_I)$$

"Uniform Distribution"

b) "μ relatively known"

$f(\mu)$

"Nonuniform Distribution"

c) "μ known"

$$P(\mu) = \begin{cases} 1, & \mu = \mu_I \\ 0, & else \end{cases}$$

"Deterministic Distribution"

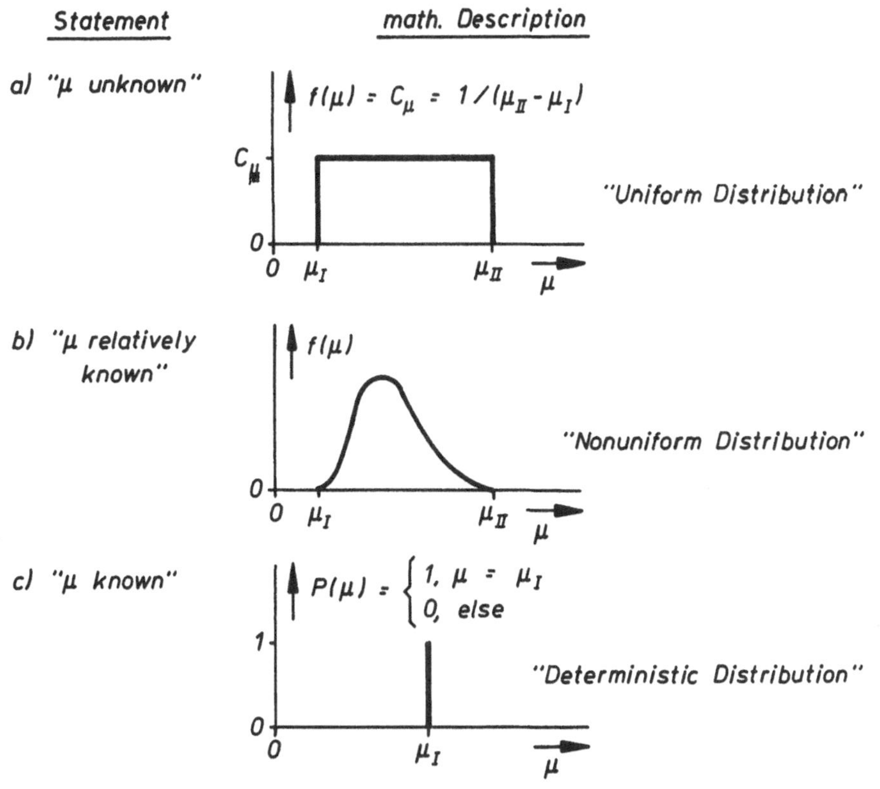

Fig. 3    Terminology "knowledge probability" with respect to the value of a parameter $\mu$.

2.5 After the value $y = y_n$ eq.(1) has been measured in $n$ trials, the BL-statistics allow objective prediction statements for the outcome $y_m$ of $m$ future trials of the same random process

$$P(y_m \mid y_n) = \int_\mu P(y_m \mid \mu) \cdot f(\mu \mid y_n) d\mu . \qquad (7)$$

A posterior statement of this type has been given for the first time by Laplace in his early period [10] and has been extended to the multinomial distribution by Lubbock [13], see also [19].

## 3. Computer Random Experiments

In queueing theory, computer performance analysis, physics and in many other fields statistical simulation techniques on large computers ("Monte Carlo methods") are used to find the random behaviour of a complex system [6]. But beyond this practical aspect it is of great interest that such computer random experiments may provide a universal experimental background for the introduction and verification of statistical concepts and may thus help to reduce subjective influences on our reasoning and to overcome certain historical controversies on fundamental principles in statistics. This is possible due to the following properties of computer random experiments [5] [17].

---

[5] The randomness of such experiments may rely on pseudo random generators [6] or better on a quasi-ideal table generator which contains a large volume of stored random bits gained from a physical source, preferably a radioactive material [8].

3.1 *Flexibility* : they are universally applicable for the modelling and implementation of any type of random process resp. random network;

3.2 *High speed* : they allow a true and effective verification of statements in probability and statistics;

3.3 *Independent intelligence* : they are autonomously executed without intervention and human influences from outside;

3.4 *Reproduction* : they can be repeated any time at different locations by different observers.

Due to property 3.3 the computer is able to carry out a complete statistical experiment including the prior generation of the value of a parameter $\mu$ according to a given prior density $f(\mu)$ and it will not disclose this value of $\mu$ to the external world. Thus we have a perfect simulation of the prior situation where the observer must describe his knowledge about a physically existing parameter $\mu$ by means of the density $f(\mu)$ being known to him and the whole domain of "knowledge probability" is no longer a fictitious idea but is firmely related to a random experiment (and can therefore be verified by frequency investigations if desired) in much the same way as the common domain of "prediction probability" [17].

Also as shown in Fig.3 we see that the statement "$\mu$ unknown" can be described mathematically and experimentally only by the uniform distribution, and any nonuniform distribution must be associated with the statement "$\mu$ relatively known" resp. "$\mu$ known". The terminology used here might be replaced by another one; important is only that we do have three categories of knowledge probability statements and that the uniform distribution is uniquely distinguished from all other distributions.

## 4. Some Historical Aspects (Fig.4)

Quite independently from [17] Stigler has recently pointed out [20] that possibly Bayes himself has expressed in [1] the need for the two prior uniform distributions [6] as expressed by eq.(2a,b). Obviously only eq.(2a) concerning the parameter has been understood by the posterity and it is the neglect of eq.(2b) that has caused so much misunderstanding and dispute. Nevertheless following Laplace the Bayes-postulate has been correctly applied by Gauss (1809), Lubbock (1830) [13], Poisson (1837), De Morgan (1845) and others. The criticism of the postulate began with Boole (1854) and Venn (1866). Chrystal (1895) condemned the principle of inverse probability totally; Fisher (1922) did the same with great effect [5] taking no notice of the strong opposition against Chrystal by E.T. Whittaker, Lidstone (1920) and by other members of the Edinburgh faculty of actuaries, see literature in [19].

Three decades before Fisher the parameter transformation problem has been stated very clearly by Edgeworth (1885) [4] who in a way seems to be the first entering the path to the subjective Bayesianism of our time which later on has been elaborated by De Finetti, Savage and others [12], [3].

## 5. Final Remark

Many participants of the Innsbruck symposium will remember the song "Bayesians in the Night" which was introduced at the end of the conference dinner by the gentlemen B. Natvig and M.H. DeGroot alias "Frank Sinatra". This song might be interpreted as a subtle, charming parody of the relationship between subjective and objective bayesians. Perhaps some day a further discussion of the issues involved will take place.

*Acknowledgement*

The author is indebted to Prof. Dr.rer.nat. O. Krafft, Aachen who has previewed and commented all publication scripts on the BL-statistics during the past decade.

---

[6] Stigler believes that the principle with two prior uniform distributions introduced by Bayes is restricted to the binomial case; but - as we have seen - under the conditions "sum operation eq.(1); expectation parameter $\mu$" this principle is generally applicable to all distributions with prediction p.f. $P(y|\mu)$ resp. density $f(y|\mu)$.

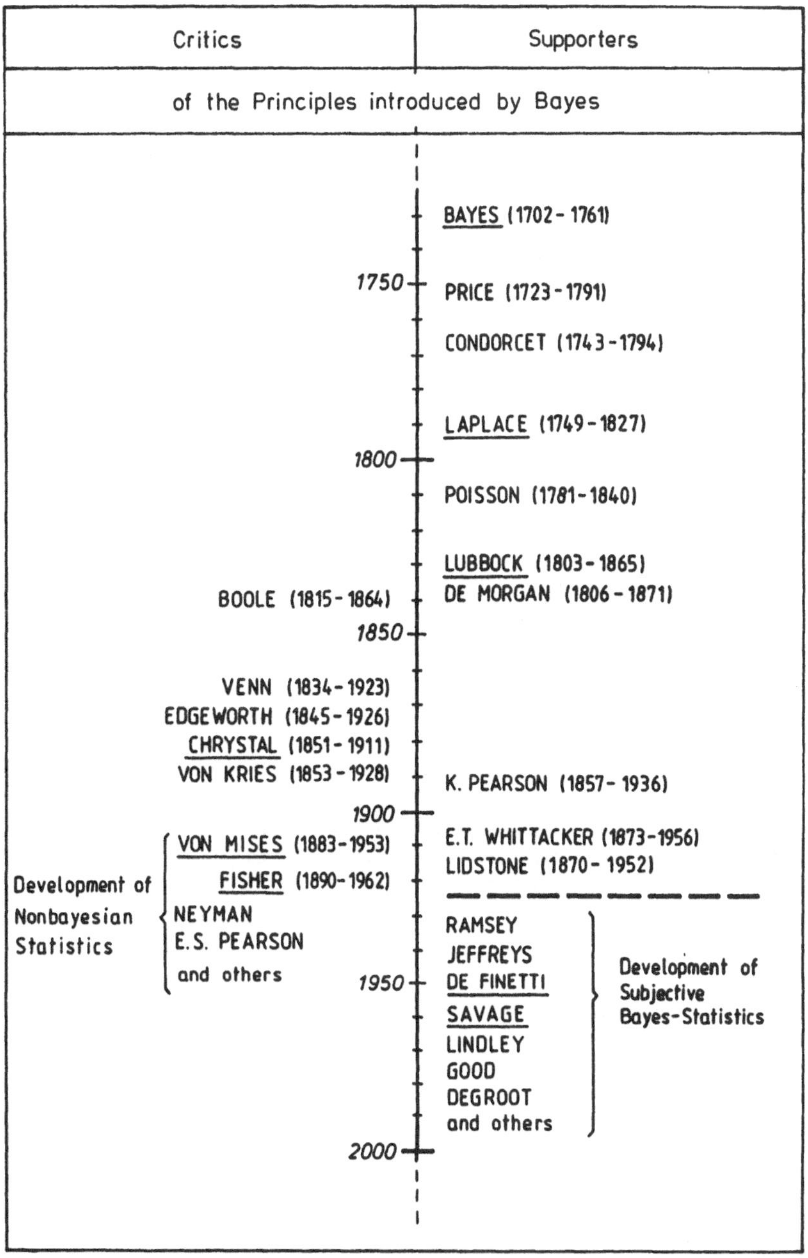


Fig. 4 Chronology of names indicating the changing appreciation of Thomas Bayes' principles for statistical inference.

## References


[1] Bayes, T., An essay towards solving a problem in the doctrine of chances. Phil. Transact. Roy. Soc. 53 (1763), 370-418.

[2] Box, G. E. P. and Tiao, G. C., Bayesian inference in statistical analysis. Addison-Wesley, Reading, Mass., 1973.

[3] Cox, D.R. and Hinkley, D. V., Theoretical statistics. Chapman and Hall, London 1974.

[4] Edgeworth, F.Y., Observations and statistics. An essay on the theory of errors of observation and the first principle of statistics. Transact. Cambridge Phil. Soc. 14 (1885), 138-169.

[5] Fisher, R.A., On the mathematical foundation of theoretical statistics. Phil. Transact. Roy. Soc. A, 222 (1922), 309-368.

[6] Fishman, G.S., Concepts and methods in discrete event digital simulation. J. Wiley & Sons, New York 1973.

[7] Gosset, W.S., ("Student"), The probable error of a mean. Biometrika 6 (1908), 1-25.

[8] Gude, M., Concept for a high performance random number generator based on physical random phenomena. Frequenz 39 (1985), 187-190.

[9] Kendall, M.G. and Stuart, A., The advanced theory of statistics; Vol.1. C. Griffin & Co., London 1969.

[10] Laplace, P.S., Mémoire sur la probabilité des causes par les événements. Mémoires de math. et de phys. présentes a l'Academie par divers savans 6 (1774), 621-656.

[11] Laplace, P.S., Théorie analytique des probabilités. Oeuvre complétes de Laplace, Vol.7. Gauthier-Villars, Paris 1886.

[12] Lindley, D.V., Bayesian statistics, a review. S.I.A.M., Philadelphia 1972.

[13] Lubbock, J.W., On the calculation of annuities and on some questions in the theory of chances. Transact. Cambridge Phil. Soc. 3 (1830), 141-154.

[14] Rényi, A., On a new axiomatic theory of probability. Acta Math. Acad. Sci. Hung. 6 (1955), 285-335.

[15] Schreiber, F., Anwendung der objektiven Bayes-Statistik auf die Normal- und $\chi^2$-Verteilung. AEÜ 32 (1978), 363-368.

[16] Schreiber, F., Improved simulation by application of the objective Bayes-statistics. AEÜ 34 (1980), 243-249.

[17] Schreiber, F., The extended Bayes-postulate, computer random experiments and the unique solution of an elementary problem in statistics. AEÜ 35 (1981), 34-43.

[18] Schreiber, F., Time efficient simulation: the LRE- algorithm for producing empirical distribution functions with limited relative error. AEÜ 38 (1984), 93-98.

[19] Schreiber, F., The Bayes-Laplace-statistic of the multinomial distribution. AEÜ 39 (1985), 293-298.

[20] Stigler, S.M., Thomas Bayes's bayesian inference. J. Roy. Statist. Soc. 145 (1982), 250-258.


---

AEÜ: Archiv für Elektronik und Übertragungstechnik (Electronics and Communication)

# THE ANALYSIS OF WEIBULL LIFETIME DATA INCORPORATING EXPERT OPINION

Nozer D. Singpurwalla
The George Washington University
Washington, DC 20052

Mao Shi Song
East China Normal University
Shanghai, PRC

Abstract

In this paper, we present a new approach for the analysis of lifetime data that are assumed to be described by a two-parameter Weibull distribution. The novel feature of our approach pertains to the incorporation of expert opinion into the analysis. Provision is also made for incorporating our own opinions on the expertise of the experts and also on the lifetimes of items. Our approach involves the use of Laplace's approximation and this results in formulae which are easy to compute.

## 1. INTRODUCTION AND OVERVIEW

The use of expert opinion in several practical applications of reliability and risk analysis is on the increase. One can look at some recent articles in the engineering and scientific journals to get an appreciation for the above - see for example, Okrent (1975), Apostolakis and Mosleh (1979), Martz and Bryson (1984), and Mosleh and Apostolakis (1986), to name a few. Unfortunately little has been written on this important topic in the statistical journals which addresses reliability problems; an exception is Lindley and Singpurwalla (1986). Here, we develop a procedure for undertaking the reliability assessment of biological and engineering items whose life lengths can be described by the Weibull distribution. The key features of our approach are:

1. The elicitation, codification and modulation of expert opinion in a formal manner;

2. The use of historical data, on identical copies of the item, in conjunction with 1 above; and

3. The use of approximations which greatly facilitate our ability to undertake the necessary computations and make our approach attractive to a user.

The methodology here is based on a theme described by Lindley (1983).

## 2. PRELIMINARIES

Let T denote the time to failure of a fresh unit. We wish to make statements of uncertainty about T conditional on the information that we have. If $H$, the background information, is all that we have then we are able to specify $R(t|H) \overset{\text{def}}{=} P(T \geq t|H)$, the <u>reliability</u> of the unit for a mission of duration t. To facilitate a specification of the above, we extend the conversation to a scale (shape) parameter $\theta(\beta)$ and introduce the Weibull as a chance distribution for T. That is

$$R(t|\underset{\sim}{\omega}) = \int_{\theta}\int_{\beta} P(T \geq t|\theta,\beta,\underset{\sim}{\omega})\ \pi(\theta,\beta|\underset{\sim}{\omega})d\beta d\theta, \tag{2.1}$$

where $\pi(\theta,\beta|\omega)$ describes our uncertainty about $\theta$ and $\beta$ conditional on a vector of specified parameters $\underset{\sim}{\omega}$, and $P(T \geq t|\theta,\beta,\underset{\sim}{\omega}) = P(T \geq t|\theta,\beta) = \exp(-a\ (t/\theta)^{\beta})$, with $a = \ell n2$. In writing the above we are of course making the assertion that given the *median* $\theta$ and $\beta$, T is independent of $\underset{\sim}{\omega}$.

The usual way of parameterizing the Weibull distribution is via a scale (shape) parameter $\alpha(\beta)$, so that $P(T \geq t|\alpha,\beta) = \exp(-(t/\alpha)^{\beta})$. The reason for our parameterization in terms of $\theta$ is that it is easier to elicit expert opinion about the median life than about an arbitrary scale parameter $\alpha$.

The focus of our paper is the elicitation and the codification of expert opinion to enable us specify $\pi(\theta,\beta|\omega)$ and the development of approximations which facilitates its use. If we do not have any lifetime data on units which can be judged exchangeable with the unit in question, then all that we have to go by is $\pi(\theta,\beta|\omega)$ and the resulting $R(t|\omega)$, given by (2.1). Note that $R(t|\omega) = E[R(\tilde{t}|\theta,\beta)]$, where E(x) denotes expectation of X. If a statement of uncertainty about $R(t|\theta,\beta)$ is also desired, then this is provided by

$$P\{R(t|\theta,\beta) \leq c\} = \int_{\theta^*}\int_{\beta^*} \pi(\theta^*,\beta^*|\underset{\sim}{\omega})d\beta^* d\theta^*,$$

where (2.2)

$$\{(\theta^*,\beta^*):\ R(t|\theta^*,\beta^*) \leq c\}\ ,\quad \text{for } 0 \leq c \leq 1.$$

If and when we obtain $\underset{\sim}{d}$, failure data on items which are exchangeable with the item in question, then expert opinion as exemplified by $\pi(\theta,\beta|\underset{\sim}{\omega})$ must be updated to obtain $\pi(\theta,\beta|\underset{\sim}{\omega},\underset{\sim}{d})$, where

$$\pi(\theta,\beta|\underset{\sim}{d},\underset{\sim}{\omega}) \propto L(\underset{\sim}{d}|\theta,\beta)\ \pi(\theta,\beta|\underset{\sim}{\omega}) \tag{2.3}$$

and $L(d|\theta,\beta)$ is the likelihood of $\theta$ and $\beta$ given $\underset{\sim}{d}$. If $\underset{\sim}{d} = (t_1,\ldots,t_n)$, where the $t_i$'s denote the observed times to failure, then

$$L(\underset{\sim}{d}|\theta,\beta) = \frac{\beta^n}{\theta^{n\beta}}\left(\prod_{i=1}^{n} t_i^{\beta-1}\right) e^{-a\sum_{i=1}^{n}(t_i/\theta)^{\beta}}\ . \tag{2.4}$$

Once we obtain $\pi(\theta,\beta|\underset{\sim}{d},\underset{\sim}{\omega})$ we update our assessment about the reliability from $R(t|\underset{\sim}{\omega})$ to

$$R(t|d,\underset{\sim}{\omega}) = \int_{\theta}\int_{\beta} \exp(-a(t|\theta)^{\beta})\ \pi(\theta,\beta|\underset{\sim}{d},\underset{\sim}{\omega})d\beta d\theta,$$

where (2.5)

$$R(t|\underset{\sim}{d},\underset{\sim}{\omega}) \overset{\text{def}}{=} P(T \geq t|\underset{\sim}{d},\underset{\sim}{\omega}) = E[R(t|\underset{\sim}{\theta},\underset{\sim}{\beta},d)].$$

If an updated statement of uncertainty about $R(t|\theta,\beta)$ is also desired, then this is provided by

$$P\{R(t|\theta,\beta,\underset{\sim}{d}) \leq c\} = \int_{\theta^*}\int_{\beta^*} \pi(\theta^*,\beta^*|\underset{\sim}{d},\underset{\sim}{\omega})d\beta^* d\theta^*, \qquad (2.6)$$

where $\{(\theta^*,\beta^*) : R(t|\theta^*,\beta^*) \leq c\}$, for $0 \leq c \leq 1$.

A comparison of (2.1) with (2.5) and (2.2) with (2.6) indicates the effect of the data on our assessment of reliability based on expert opinion alone.

Note that once the expert opinion is solicited and codified, the assessments $R(t|\omega)$ and $P\{R(t|\alpha,\beta) \leq c\}$ remain the same over all time, for any mission of duration t. However, since $\underset{\sim}{d}$ changes over time, the quantities $R(t|\underset{\sim}{d},\underset{\sim}{\omega})$ and $P\{R(t|\theta,\beta,\underset{\sim}{d}) \leq c\}$ also change with time giving their plots (as a function of t), a dynamic feature.

The foregoing material describes the overall strategy underlying our procedure. The principle though simple and straightforward, poses difficulties with respect to computation. These difficulties have been overcome using a recently proposed approximation.

## 3. THE ELICITATION, MODULATION, AND CODIFICATION OF EXPERT OPINION

In the published literature on a Bayesian analysis of the Weibull distribution, the priors used have been chosen for their analytical convenience rather than their ability to meaningfully represent expert opinion: see for example Soland (1969), Tsokos (1972), Barlow and Proschan (1981) and Erto (1982). A departure from the above is Singpurwalla (1986); however the latter approach emphasizes computer graphics, and yields results which are not in closed form. A possible reason for choosing priors based on analytical convenience stems from the fact that a Weibull distribution is typically parameterized in terms of a general scale parameter $\alpha$ which conceptually difficult to interpret. Expert opinion about measures of central tendency, such as the median $\theta$, is easier to come by [cf. Martz, Bryson and Waller (1984)] and this is what we will do. The shape parameter $\beta$ characterizes ageing, and opinion on it from engineers and scientists is easy to elicit.

### 3.1 Elicitation of Expert Opinion on Median Life

Suppose that an expert $\mathcal{E}$ conceptualizes his/her uncertainty about the unknown $\theta$ via some distribution with mean m and standard deviation s. This means that in $\mathcal{E}$'s view, 50 percent of similar units if observed until failure are most likely to fail by m. The quantity s is a measure of the expert's uncertainty in specifying m.

Suppose that $\mathcal{E}$ declares to an analyst $\mathcal{A}$, two numbers m and s which describe $\mathcal{E}$'s uncertainty about $\theta$. It is not essential that $\mathcal{A}$ be cognizant of $\mathcal{E}$'s conceptualized distribution; $\mathcal{A}$ only needs to be told that m is a measure of location and s a measure of scale. Given m and s, $\mathcal{A}$'s judgment about $\theta$ can be described, using Bayes law, in terms of the probability distribution

$$p(\theta|m,s) \propto p(m,s|\theta)\ p(\theta), \qquad (3.1)$$

where $p(\theta)$ describes $\mathcal{A}$'s view of $\theta$ before the receipt of $\mathcal{E}$'s advice, and $p(m,s|\theta)$ is the likelihood of $\theta$. It is to be emphasized that (3.1) is specified by $\mathcal{A}$ not $\mathcal{E}$, and for convenience $\mathcal{H}$, $\mathcal{A}$'s background information, has been suppressed. The likelihood $p(m,s|\theta)$ will also be specified by $\mathcal{A}$, and is written to describe $A$'s judgment of the *expertise of the expert*. For convenience, we write the likelihood as

$$p(m,s|\theta) = p(m|s,\theta)\ p(s|\theta), \tag{3.2}$$

and consider some possible assumptions that $A$ can make about the right hand side of (3.2).

Assumption 1 (A1). $p(s|\theta)$ does not depend on $\theta$.

This says that in $A$'s view, s on its own gives no information about $\theta$.

Assumption 2 (A2). For some $c \neq 0$ and $\nu > 0$

$$\frac{c^2 \nu \theta^2}{sm^2} \sim \chi^2(\nu/s)$$

where the notation "$X \sim \chi^2(n)$" denotes the fact that X has a chi-square distribution with n degrees of freedom. The constants c and $\nu$ are specified by $A$, and they reflect $A$'s view of $E$'s biases and precision in dealing with m and s. The assumption A2 implies that

$$p(m|s,\theta,\nu,c) \propto \theta^{\nu/s}\ m^{-((\nu/s)+1)} e^{-(c^2\nu\theta^2/2sm^2)}$$

and a way to conceptualize the above is to say that were $\theta$ to be known to $E$, then having specified s, $E$ (in the opinion of $A$) would specify m in such a manner that with probability of .95, log m $\in$[logc+log$\theta$) $\pm\sqrt{2s/\nu}$], or that m $\in[w^{-1}\ c\theta,\ wc\theta]$ where $\log w = \sqrt{2s/\nu}$. The term log c denotes a bias in $E$'s specification of log m. Specifically, c=1 implies that there is no bias, whereas $c < 1$ (>1) implies that $E$ underestimates (overestimates) $\theta$ in specifying m. The parameter $\nu$ allows $A$ to express opinion on how precise $E$ is. With $\nu <$ (>) 1/2, $A$ thinks $E$ tends to exaggerate (is overcautious about) the precision of his/her assessment.

If $A$ has full faith in the expertise of $E$ or if $A$ does not wish to modulate $E$'s inputs, then $A$ will set c=1 and $\nu$=1/2.

Assumption 3 (A3). $p(\theta)$ is effectively constant.

This says that $A$'s initial knowledge of $\theta$, before receiving expert testimony, is weak.

The assumptions A1 and A3 say that $p(s|\theta)$ and $p(\theta)$ are constant, making $p(\theta|m,s) \propto p(m|s,\theta)$, and this together with A2 gives us the result that in $A$'s opinion

Theorem 3.1

$\sqrt{c^2\nu/sm^2}\ \theta \sim \chi\ ((\nu/s)+1)$, where the notation "$X \sim \chi(\nu)$" denotes the fact that X has a chi distribution with $\nu$ degrees of freedom.

Assumption 1, which states that s on its own gives no information about $\theta$, may not be true. The following alternate assumption may be a more appropriate one to consider.

Assumption 1' (A1'). For some $\nu^* > 0$

$$\frac{\nu^* \theta^2}{s^2} \sim \chi^2(\nu^*).$$

A1' implies that were $\theta$ to be known to $E$, then in the opinion of $A$, $E$ would specify s in such a manner that with probability 95%, $s \in [w_1^{-1}\theta, w_1\theta]$, where $\log w_1 = \sqrt{2/\nu^*}$. Thus $\nu^* > 0$ captures $A$'s opinion about $E$'s uncertainty in specifying $\theta$. For example, if $A$ feels that given $\theta$, $E$ has a wide (narrow) range of values for s, then $\nu^*$ will be small (large). Thus $\nu^* = 1$ implies that $s \in [\theta e^{-\sqrt{2}}, \theta e^{\sqrt{2}}]$, $\nu^* = 2$ implies that $s \in [\theta e^{-1}, \theta e]$, and $\nu^* = .5$ implies that $s \in [\theta e^{-2}, \theta e^{2}]$. Such a scheme allows us to model the proportional effect of $\theta$; for large $\theta$ we may want to choose small $\nu^*$ and vice versa.

If $A$ does have some prior information about $\theta$, then A3 may not be meaningful and the following alternate assumption may be plausible. Assumption 3' (A3'). $p(\theta)$ is Gaussian with mean $\mu$ and variance $\tau^2$, where $\tau$ is chosen in such a manner that $P(\theta < 0)$ is very small.

The new assumptions A1' and A3' would lead to results analogous to Theorem 3.1. These are:

Theorem 3.2. Under the assumptions A1', A2, and A3, $A$'s opinion about $\theta$ can be described as

$$\sqrt{\frac{c^2\nu}{sm^2} + \frac{\nu^*}{s^2}}\,\theta \sim \chi\left(\frac{\nu}{s} + \nu^* + 1\right).$$

Theorem 3.3. Under the assumptions A1', A2, and A3', $A$'s opinion about $\theta$ can be described as

$$p(\theta|m,s,\nu,\nu^*,c,\mu,\tau) \propto \theta^{\frac{\nu}{s}+\nu^*} \exp\{-\frac{A}{2}(\theta-B)^2\},\ \theta > 0,$$

where

$$A = \frac{c^2\nu}{sm^2} + \frac{\nu^*}{s^2} + \frac{1}{\tau^2}, \text{ and}$$

$$B = \mu/\tau^2 A.$$

In Theorems 3.1 - 3.3 we see how $A$ has modulated $E$'s testimony on $\theta$ to reflect $A$'s judgment on the expertise of $E$. Extensions to cover the case of several experts is straightforward except that now we have to incorporate into our mathematics $A$'s perceived correlations between the experts declared values.

## 3.2 The Codification of Expert Opinion on the Weibull Shape Parameter

As is well known, the shape parameter $\beta$ of the Weibull distribution characterizes ageing, in the sense that $\beta >(<) 1$ implies that $\frac{\beta}{\alpha}\left(\frac{x}{\alpha}\right)^{\beta-1}$, the failure rate increases (decreases) in x; it is a constant for $\beta$=1. Thus if expert opinion suggests that the item degrades (improves) with age, then $\beta$ is likely to be greater (less) than 1, and if the item neither degrades nor improves with age, then $\beta$ is more likely to be in the vicinity of 1. In principle, we should use a methodology analogous to that described in Section 3.1, for eliciting and modulating expert opinion on $\beta$. However, in the interest of simplicity we choose to model the uncertainty about $\beta$ via a gamma density. We have

Assumption 4 (A4). For some $\lambda > 0$ and $p > 0$

$$p(\beta|\lambda,p) \propto \beta^{p-1} e^{-\lambda\beta}, \quad \beta > 0.$$

The parameters $\lambda$ and p are specified by $\mathcal{A}$ and are based upon $\mathcal{E}$'s view and conviction of the ageing characteristics of the item. A convenient way for $\mathcal{A}$ to specify $\lambda$ and p is to use the fact that the mean, mode, and variance of the kernel in A4 are given by $p/\lambda$, $(p-1)/\lambda$, and $p/\lambda^2$, respectively.

## 3.3 The Distribution Induced by Expert Opinion on the Weibull Parameters

The elicitation (and modulation) of expert opinion discussed in the previous two sections has resulted in the distributions determined by Theorems 3.1 - 3.3 and Assumption 4. If we assume that $\beta$ is independent of $\theta$, then the joint densities at $\theta$ and $\beta$, corresponding to Theorems 3.1 - 3.3 are given by:

$$\pi_1(\theta,\beta|m,s,\nu,c,\lambda,p) \propto \theta^{\nu/s}\, \beta^{p-1} \exp(-\lambda\beta - \frac{c^2\nu\theta^2}{2sm^2}) = \tilde{\pi}_1(\theta,\beta|\cdot).$$

$$\pi_2(\theta,\beta|m,s,\nu,\nu^*,c,\lambda,p) \propto \theta^{\frac{\nu}{s}+\nu^*}\, \beta^{p-1} \exp(-\lambda\beta - \frac{\theta^2}{2}(\frac{c^2\nu}{sm^2} + \frac{\nu^*}{s^2}))$$

$$= \tilde{\pi}_2(\theta,\beta|\cdot),$$

and

$$\pi_3(\theta,\beta|m,s,\nu,\nu^*,c,\mu,\tau,\lambda,p) \propto \theta^{\frac{\nu}{s}+\nu^*}\, \beta^{p-1} \exp(-\lambda\beta - \frac{A}{2}(\theta-B)^2)$$

$$= \tilde{\pi}_3(\theta,\beta|\cdot),$$

respectively.

The above *joint prior densities* at $\theta$ and $\beta$, based on expert opinion alone. When the data $\underset{\sim}{d}$ becomes available, we will obtain the joint posterior density using (2.3). Suppressing the conditioning arguments in $\pi_1$, $\pi_2$, and $\pi_3$ the *joint posterior densities* at $\theta$ and $\beta$ corresponding to $\pi_1$, $\pi_2$, and $\pi_3$ are given by:

$$\pi_1(\theta,\beta|\underset{\sim}{d},\cdot) \propto \theta^{\frac{\nu}{s}-n\beta}\, \beta^{n+p-1}\exp\{-\lambda\beta - \frac{c^2\nu\theta^2}{2sm^2} - a\sum_{i=1}^{n}(\frac{t_i}{\theta})^{\beta} + (\beta-1)\sum_{i=1}^{n}\ell n t_i\}$$

$$= \tilde{\pi}_1(\theta,\beta|\underset{\sim}{d},\cdot),$$

$$\pi_2(\theta,\beta|\underset{\sim}{d},\cdot) \propto \theta^{\frac{\nu}{s}+\nu^*-n\beta}\, \beta^{n+p-1}\exp\{-\lambda\beta - \frac{\theta^2}{2}(\frac{c^2\nu}{sm^2} + \frac{\nu^*}{s^2}) - a\sum_{i=1}^{n}(\frac{t_i}{\theta})^{\beta} + (\beta-1)\sum_{i=1}^{n}\ell n t_i\}$$

$$= \tilde{\pi}_2(\theta,\beta|\underset{\sim}{d},\cdot), \text{ and}$$

$$\pi_3(\theta,\beta|\underset{\sim}{d},\cdot) \propto \theta^{\frac{\nu}{s}+\nu^*-n\beta}\beta^{n+p-1}\exp\{-\lambda\beta-\frac{A}{2}(\theta-B)^2 - a\sum_{i=1}^{n}(\frac{t_i}{\theta})^{\beta}+(\beta-1)\sum_{i=1}^{n}\ell n t_i\}$$

$$= \tilde{\pi}_3(\theta,\beta|\underset{\sim}{d},\cdot), \text{ respectively.}$$

The use of $\pi_i(\theta,\beta|\cdot)$ and $\pi_i(\theta,\beta|\underset{\sim}{d},\cdot)$, i=1,2,3, for estimating reliabilities is discussed in Section 4.

### 3.4 Assessing the Impact of Data on Expert Opinion

In order to rate and score the expert and also A's assessment of the expertise of the expert, or otherwise to enhance the expertise of the expert (for future use), it is of interest to provide *feedback to the expert* based on the observed $\underset{\sim}{d}$. For the parameter β this is done via a comparison of $p(\beta|\lambda,p)$ with

$$\pi_i(\beta|\underset{\sim}{d},\cdot) = \int_\theta \pi_i(\theta,\beta|\underset{\sim}{d},\cdot)d\theta, \quad i=1,2,3. \tag{3.3}$$

For the median θ, the effect of $\underset{\sim}{d}$ is assessed via a comparison of the results of Theorems 3.1 through 3.3 with

$$\pi_i(\theta|\underset{\sim}{d},\cdot) = \int_\beta \pi_i(\theta,\beta|\underset{\sim}{d},\cdot)d\beta, \quad i=1,2,3, \text{ respectively.} \tag{3.4}$$

The computation of (3.3) and (3.4) is discussed in Section 5.

## 4. BAYESIAN ESTIMATION OF THE RELIABILITY FUNCTION

As outlined in Section 1, an estimation of the reliability function can be undertaken using expert opinion alone or using both, expert opinion and failure data. To see how the above can be done, let $t_0$ denote the mission time, and recall that $R(t_0|\theta,\beta) = \exp(-a(t_0|\theta)^\beta)$, where $a = \ell n2$. Then, based on expert opinion alone, we have from (2.1)

$$R_i(t_0|\cdot) = \int_\theta\int_\beta R(t_0|\theta,\beta)\ \pi_i(\theta,\beta|\cdot)d\beta d\theta, \tag{4.1}$$

and based on both failure data and expert opinion, we have from (2.5)

$$R_i(t_0|\underset{\sim}{d},\cdot) = \int_\theta\int_\beta R(t_0|\theta,\beta)\pi_i(\theta,\beta|\underset{\sim}{d},\cdot)d\beta d\theta, \quad i=1,2,3. \tag{4.2}$$

To discuss an evaluation of the above quantities, let us focus attention on (4.1). We first note that for i=1,2,3,

$$R_i(t_0|\cdot) = \frac{\int_\theta\int_\beta R(t_0|\theta,\beta)\ \tilde{\pi}_i(\theta,\beta|\cdot)d\beta d\theta}{\int_\theta\int_\beta \tilde{\pi}_i(\theta,\beta|\cdot)d\beta d\theta},$$

and write

$$R_i(t_0|\cdot) = \frac{\int_\theta\int_\beta e^{N L_i^*(\theta,\beta)}d\beta d\theta}{\int_\theta\int_\beta e^{N L_i(\theta,\beta)}d\beta d\theta}, \text{ where} \tag{4.3}$$

$$N L_i^*(\theta,\beta) \overset{\text{def}}{=} \log\ [R(t_0|\theta,\beta)\ \tilde{\pi}_i(\theta,\beta|\cdot)],$$

$N\,L_i(\theta,\beta) \overset{\text{def}}{=} \log[\tilde{\pi}_i(\theta,\beta|\cdot)]$, and N is an integer which reflects A's guess about the number of previous observations* upon which E has based the declared values m and s.

The ratio of the two integrals in (4.3) is difficult to evaluate analytically. An approximation due to Laplace, described in De Bruijen (1961), and recently studied by Tierney and Kadane (1986) works very well for large values of N. Following the material in Section 2 of Tierney and Kadane (1986), we are able to say, that for large N

$$R_i(t_0|\cdot) \approx \left[\frac{L_{i,11}L_{i,22} - L_{i,12}^2}{L_{i,11}^* L_{i,22}^* - L_{i,12}^{*2}}\right]^{1/2} e^{N\left[L_i^*(\hat{\theta}_i^*,\hat{\beta}_i^*) - L_i(\hat{\theta}_i,\hat{\beta}_i)\right]}, \quad (4.4)$$

where $\hat{\theta}_i$ and $\hat{\beta}_i$ ($\hat{\theta}_i^*$ and $\hat{\beta}_i^*$) maximize the function $L_i(\theta,\beta)$ ($L_i^*(\theta,\beta)$).

Since the functions $L_i$ and $L_i^*$ are continuously differentiable, $\hat{\theta},\hat{\beta}$, $\hat{\theta}^*$ and $\hat{\beta}^*$ are solutions to the equations $\frac{\partial L_i}{\partial\theta} = 0$, $\frac{\partial L_i}{\partial\beta} = 0$, $\frac{\partial L_i^*}{\partial\theta} = 0$, and $\frac{\partial L_i^*}{\partial\beta} = 0$, respectively. The quantities $L_{i,11} = \frac{\partial^2}{\partial\theta^2}L_i$, $L_{i,12} = \frac{\partial^2}{\partial\theta\partial\beta}L_i$, $L_{i,22} = \frac{\partial^2}{\partial\beta^2}L_i$, $L_{i,11}^* = \frac{\partial^2}{\partial\theta^2}L_i^*$, $L_{i,12}^* = \frac{\partial^2}{\partial\theta\partial\beta}L_i^*$ and $L_{i,22}^* = \frac{\partial^2}{\partial\beta^2}L_i^*$ evaluated at $\hat{\theta}$, $\hat{\beta}$ and $\hat{\theta}^*,\hat{\beta}^*$, respectively, expressions for these are given in Singpurwalla and Song (1986).

The strategy for evaluating (4.2) is analogous to the above except that now $L_i(\theta,\beta)$ and $L_i^*(\theta,\beta)$ are replaced by $\mathcal{L}_i(\theta,\beta|\underset{\sim}{d})$ and $\mathcal{L}_i^*(\theta,\beta|\underset{\sim}{d})$, respectively, where $n\mathcal{L}_i^*(\theta,\beta|\underset{\sim}{d}) \overset{\text{def}}{=} \log[R(t_0|\theta,\beta)\tilde{\pi}_i(\theta,\beta|\underset{\sim}{d},\cdot)]$, and $n\mathcal{L}_i(\theta,\beta|\underset{\sim}{d}) \overset{\text{def}}{=} \log[\tilde{\pi}_i(\theta,\beta|\underset{\sim}{d}\cdot)]$. The quantities $\mathcal{L}_{i,jk}$ and $\mathcal{L}_{i,jk}^*$, $i=1,2,3$, $j,k=1,2$, are defined like the $L_{i,jk}$ and $L_{i,jk}^*$, *mutatis-mutandis*. Here the n denote the number of failure times which go into making up the likelihood, and the ensuing approximation to be used is valid for large n.

Having evaluated

$$R_i(t_0|\underset{\sim}{d},\cdot) \approx \left[\frac{\mathcal{L}_{i,11}\mathcal{L}_{i,22} - \mathcal{L}_{i,12}^2}{\mathcal{L}_{i,11}^*\mathcal{L}_{i,22}^* - \mathcal{L}_{i,12}^{*2}}\right]^{1/2} e^{n[\mathcal{L}_i^*(\tilde{\theta}_i^*,\tilde{\beta}_i^*) - \mathcal{L}_i(\tilde{\theta}_i,\tilde{\beta}_i)]} \quad (4.5)$$

we may compare $R_i(t_0|\cdot)$ and $R_i(t_0|\underset{\sim}{d},\cdot)$, $i=1,2,3$, to assess the impact of $\underset{\sim}{d}$ on expert opinion. Such a comparison can also be used as a basis for calibrating expert opinion and its codification by A. Note that $\tilde{\theta}_i$ and $\tilde{\beta}_i$ ($\tilde{\theta}_i^*$ and $\tilde{\beta}_i^*$) maximize the function $\mathcal{L}_i(\theta,\beta)$ ($\mathcal{L}_i^*(\theta,\beta)$).

There is no simple approximation which enables us to evaluate (2.2) and (2.6) in closed form; the only recourse available to us is via numerical integration.

---

*In the case of multiple experts, N could represent the number of experts consulted.

## 5. POSTERIOR DISTRIBUTIONS OF MEDIAN LIFE AND THE SHAPE PARAMETER

In order to evaluate $\pi_i(\theta|\underset{\sim}{d},\cdot)$, the marginal posterior distribution of $\theta$ - see (3.4), we first note that for i=1,2,3,

$$\pi_i(\theta|\underset{\sim}{d},\cdot) = \frac{\int_\beta \tilde{\pi}_i(\theta,\beta|\underset{\sim}{d},\cdot)d\beta}{\int_\theta\int_\beta \tilde{\pi}_i(\theta,\beta|\underset{\sim}{d},\cdot)d\beta d\theta},$$

and write

$$\pi_i(\theta|\underset{\sim}{d},\cdot) = \frac{\int_\beta e^{nL_i(\theta,\beta)}\,d\beta}{\int_\theta\int_\beta e^{nL_i(\theta,\beta)}d\beta d\theta}, \text{ where} \tag{5.1}$$

$$NL_i(\theta,\beta) \overset{\text{def}}{=} \log(\tilde{\pi}_i(\theta,\beta|\underset{\sim}{d},\cdot)).$$

The ratio of the integrals in (5.1) is again difficult to evaluate analytically. However, following the material in Section 2 of Tierney and Kadane (1986), we are able to say, that for large n

$$\pi_i(\theta|\underset{\sim}{d},\cdot) \approx \left[\frac{L_{i,11}\,L_{i,22} - L^2_{i,12}}{(2\pi/n)\,L_{i,22}(\theta)}\right]^{1/2} e^{n[L_i(\theta,\tilde{\tilde{\beta}}_i) - L_i(\tilde{\theta}_i,\tilde{\beta}_i)]} \tag{5.2}$$

where $\tilde{\theta}_i$ and $\tilde{\beta}_i$ maximize the function $L_i(\theta,\beta)$ and $\tilde{\tilde{\beta}}_i$ maximizes the function $L_i(\theta,\beta)$ for a fixed value of $\theta$. The quantities $L_{i,jk}$, i=1,2,3, j,k = 1,2 have been defined in Section 4. The quantity $L_{i,22}(\theta) = \frac{\partial^2}{\partial\beta^2} L_i(\theta,\beta)\Big|_{\beta=\tilde{\tilde{\beta}}_i}$.

The marginal posterior distribution of the shape parameter $\beta$ is obtained via the approximation

$$\pi_i(\beta|\underset{\sim}{d},\cdot) \approx \left[\frac{L_{i,11}\,L_{i,22} - L^2_{i,12}}{(2\pi/n)\,L_{i,11}(\beta)}\right]^{1/2} e^{n[L_i(\tilde{\tilde{\theta}}_i,\beta) - L_i(\tilde{\theta}_i,\tilde{\beta}_i)]}, \tag{5.3}$$

where $\tilde{\tilde{\theta}}_i$ maximizes the function $L_i(\theta,\beta)$ for a fixed value of $\beta$, and

$$L_{i,11}(\beta) = \frac{\partial^2}{\partial\theta^2} L_i(\theta,\beta)\Big|_{\theta=\tilde{\tilde{\theta}}_i}.$$

## 6. ILLUSTRATIVE EXAMPLE

We illustrate the approach of this paper via an example involving simulated data from a Weibull distribution. Suppose that $\mathcal{A}$ elicits expert opinion on $\theta$, the median life of a component whose life length can be meaningfully described by a Weibull distribution. Suppose that $\mathcal{E}$ gives two numbers m=500 and s=200 as measures of location and scale, respectively, of the distribution describing $\mathcal{E}$'s uncertainty about $\theta$. Suppose that $\mathcal{A}$ uses A2, and assuming no bias in $\mathcal{E}$'s assessment of $\theta$, chooses c=1, $\nu$=1/2 and arrives at the result (via Theorem 3.1) that $10^{-4}\,\theta \sim \chi(\frac{1}{400} + 1)$; this implies that

$$p(\theta|m,s,\nu,c) = .5516\ \theta^{\frac{1}{400}}\ e^{-10^{-8}\cdot\theta^2/2}$$

A feature of the above distribution for $\theta$ is that it is nearly a constant, taking a value of approximately $.55 \times 10^{-4}$, for values of $\theta$ in the range 0 to 1000.

Suppose that expert opinion about $\beta$ is described via A4 with $p=6.25$ and $\lambda=12.5$. Then

$$\pi_1(\theta,\beta|m,s,\nu,c,\lambda,p) \propto \theta^{1/400}\ \beta^{5.25}\ e^{-(12.5\beta+10^{-8}\theta^2/2)} = \tilde{\pi}_1(\theta,\beta|\cdot),$$

and thus

$$NL_1(\theta,\beta) = \log[\tilde{\pi}_1(\theta,\beta|\cdot)] = \frac{1}{400}\log\theta + 5.25\log\beta - 12.5\beta - 10^{-8}\theta^2/2.$$

It is a straightforward matter to verify, that the maximum of the above function occurs at $\hat{\theta}_1 = 500$ and $\hat{\beta}_1 = .42$. Thus $NL_1(\hat{\theta}_1,\hat{\beta}_1) = -9.79009$, and $L_{1,11} = \frac{2\times10^{-8}}{N}$, $L_{1,12} = 0$, and $L_{1,22} = -\frac{29.7619}{N}$, so that

$$L_{1,11}\,L_{1,22} - L_{1,12}^2 = \frac{59.5238 \times 10^{-8}}{N^2}.$$

If the mission time $t_0$ is 100 (hrs), then $NL_1^*(\theta,\beta) = NL_1(\theta,\beta) - (\log 2)\ (\frac{100}{\theta})^{\beta}$, and the maximum of this function occurs at $\hat{\theta}_1^* = 2719$ and $\hat{\beta}_1^* = .439$. Thus $NL_1^*(\theta,\beta) = -9.9894$, $L_{1,11}^* = -\frac{2.432 \times 10^{-8}}{N}$, $L_{1,12}^* = \frac{.269 \times 10^{-4}}{N}$, and $L_{1,22}^* = -\frac{29.015}{N}$, so that $L_{1,11}^*\,L_{1,22}^* - L_{1,12}^{*2} = \frac{70.2368 \times 10^{-8}}{N}$.

Using the above, we can obtain as an approximation to the Bayes estimator of the reliability for a mission of 100 hours, using on expert opinion alone - see (4.4) - as

$$R_1(100|\cdot) \approx \left(\frac{59.5238 \times 10^{-8}}{70.2368 \times 10^{-8}}\right)^{1/2} e^{-9.9894+9.7901} = .6943.$$

Suppose now that a sample of size $n=50$ life-lengths is simulated from a Weibull distribution with $\theta = 10{,}000$ and $\beta = .8$. These data are shown in Table 1.

Undertaking the computations analogous to the ones described above we obtain (following the notation of Section 4) $\tilde{\theta}_1 = 13304$, $\tilde{\beta}_1 = .869$, $\tilde{\theta}_1^* = 13303$, and $\tilde{\beta}_1^* = .870$. Also, $n\mathcal{L}_1(\tilde{\theta}_1,\tilde{\beta}_1) = -545.73$, $n\mathcal{L}_1^*(\tilde{\theta}_1^*, \tilde{\beta}_1^*) = -545.74$, $\mathcal{L}_{1,11} = \frac{-2.4189 \times 10^{-7}}{n}$, $\mathcal{L}_{1,11}^* = \frac{-2.4233 \times 10^{-7}}{n}$, $\mathcal{L}_{1,12} = \frac{.0032}{n}$, $\mathcal{L}_{1,12}^* = \frac{.00322}{n}$, $\mathcal{L}_{1,22} = \frac{-157.99}{n}$, and $\mathcal{L}_{1,22}^* = \frac{-158.22}{n}$. Thus

$$\mathcal{L}_{1,11}\,\mathcal{L}_{1,22} - \mathcal{L}_{1,12}^2 = \frac{.2792 \times 10^{-4}}{n^2} \text{ and } \mathcal{L}_{1,11}^*\,\mathcal{L}_{1,22}^* - \mathcal{L}_{1,12}^{*2} = \frac{.27995 \times 10^{-4}}{n^2}$$

Thus a Bayes estimator of the reliability for a mission of 100 hours, based on expert opinion and failure data, is, from (4.5) approximated as

$$R_1(100|\underset{\sim}{d},\cdot) \approx \left(\frac{0.27922 \times 10^{-4}}{0.27995 \times 10^{-4}}\right)^{1/2} e^{-545.74 + 545.73} = .9875.$$

Compare the above to the "true" reliability

$$R(100) = e^{-a\left(\frac{100}{10,000}\right)^{.8}} = .9827.$$

TABLE I

Simulated Values of 50 Life Times from a Weibull Distribution with Median 10000 and Shape .8

| No | Life Time | No | Life Time | No | Life Time | No | Life Time | No | Life Time |
|---|---|---|---|---|---|---|---|---|---|
| 1 | 9533 | 11 | 16242 | 21 | 6708 | 31 | 15598 | 41 | 3046 |
| 2 | 15338 | 12 | 14464 | 22 | 17219 | 32 | 7987 | 42 | 35347 |
| 3 | 32897 | 13 | 116212 | 23 | 9645 | 33 | 1958 | 43 | 80528 |
| 4 | 3956 | 14 | 39 | 24 | 36082 | 34 | 22188 | 44 | 29150 |
| 5 | 8909 | 15 | 8316 | 25 | 48546 | 35 | 10315 | 45 | 995 |
| 6 | 1371 | 16 | 8281 | 26 | 27563 | 36 | 23081 | 46 | 2742 |
| 7 | 3954 | 17 | 48547 | 27 | 85619 | 37 | .6001 | 47 | 13728 |
| 8 | 4554 | 18 | 4969 | 28 | 6536 | 38 | 6653 | 48 | 26194 |
| 9 | 14222 | 19 | 16270 | 29 | 46673 | 39 | 18208 | 49 | 67621 |
| 10 | 16519 | 20 | 18426 | 30 | 8495 | 40 | 30311 | 50 | 73331 |

ACKNOWLEDGMENTS

We would like to thank Professor Jay Kadane for making available to us a pre-print of his paper with Professor Tierney. The work of the first author has been supported by Contract N00014-85-K-0202 Project NR 042-372, Office of Naval Research and Grant DAAG 29-84-K-0160, the U.S. Army Research Office.

REFERENCES

Apostolakis, G., and Mosleh, A. (1979). "Expert Opinion and Statistical Evidence: An Application to Reactor Core Melt Frequency," *Nuclear Science Engineering, No. 70.*

Barlow, R. E. and Proschan, F. (1981). Life Distribution Models and Incomplete Data. FSU Statistics Report M-568, The Florida State University, Department of Statistics, Tallahassee, Florida 32306.

De Bruijen, N. G. (1961). *Asymptotic Methods in Analysis*, Amsterdam: North-Holland.

Erto, P. (1982). New Practical Bayes Estimators for the 2- Parameter Weibull Distribution. *IEEE Transactions on Reliability, Vol. R-31, No. 2*, pp. 194-197.

Lindley, D. V. (1983). Reconciliation of Probability Distributions. *Operations Research, Vol. 31, No. 5*, pp. 866-880.

Lindley, D. V. and Singpurwalla, N. D. (1986). Reliability (and Fault Tree) Analysis Using Expert Opinion. *Journal of the American Statistical Association, Vol. 81, No. 393,* pp. 87-90.

Martz, H. and Bryson, M. (1984). A Statistical Model for Combining Biased Expert Opinions. *IEEE Transactions on Reliability, Vol. R-33, No. 3,* August 1984.

Martz, H. F., Bryson, C., and Waller, R. A. (1984). Eliciting and Aggregating Subjective Judgments - Some Experimental Results. *Proceedings of the 1984 Statistical Symposium on National Energy Issues,* NUREG/CP0063, pp. 63-82.

Mosleh, A., and Apostolakis, G. (1986). "The Assessment of Probability Distributions from Expert Opinions with an Application to Seismic Fragility Curves," accepted for publication, *Risk Analysis,* to appear in 1986.

Okrent, D. (1975). "A Survey of Expert Opinion on Low Probability Earthquakes," *Annals of Nuclear Energy, Vol. 2,* pp. 601-614.

Singpurwalla, N. D. (1986). An Interactive PC-Based Procedure for Reliability Assessment Incorporating Expert Opinion and Survival Data. The George Washington University Technical Report Serial TR-86/1.

Singpurwalla, N. D. and Song, M. S. (1986). The Analysis of Weibull Lifetime Data Incorporating Expert Opinion. Technical Report GWU/IRRA/Serial TR-86/9, The George Washington University, Washington, DC.

Soland, R. M. (1969). Bayesian Analysis of the Weibull Process with Unknown Scale and Shape Parameters. *IEEE Transactions in Reliability, Vol. R-18,* pp. 181-184.

Tierney, L. and Kadane, J. B. (1986). Accurate Approximations for Posterior Moments and Marginal Densities. *Journal of the American Statistical Association. Vol. 81, No. 393,* pp. 82-86.

Tsokos, C. P. (1972). A Bayesian Approach to Reliability: Theory and Simulation. *Proceedings 1972 Annual Reliability and Maintainability Symposium,* San Francisco, pp. 78-87.

Key words Reliability Analysis, Expert Opinion, Laplace's Method of Approximation

# ROBUST TESTS FOR TREND IN BINOMIAL PROPORTIONS

R. T. Smythe[1], D. Krewski[2,3] and A. Dewanji[2]

[1]Department of Statistics, George Washington University
Washington, D.C., U.S.A. 20052
[2]Health Protection Branch, Health and Welfare Canada
Ottawa, Ontario, Canada K1A 0L2
[3]Department of Mathematics and Statistics, Carleton University
Ottawa, Ontario, Canada K1S 5B6

Recent modifications to the Cochran-Armitage statistic used to test for trend in binomial proportions in carcinogenicity bioassays for which a series of historical control data is available employ a beta distribution for the between study variation in the binomial response rate in the control group. In this paper, the use of robust distributions with heavier tails than the beta is proposed as a means of accommodating the uncertainty as to the actual historical distribution of the binomial response rate. The robust distributions are selected from within a class of mixed distributions using a Γ-minimax criterion to select the most appropriate value of the mixing proportion. These tests are shown to be more robust than the existing tests with respect to inclusion or exclusion of individual historical control data points.

## 1. Introduction

Bioassay of small rodents is an important step in identifying chemicals with carcinogenic potential. These studies involve groups of animals exposed to different levels of the test agent as well as unexposed controls (Bickis and Krewski, 1985). Although such concurrent controls constitute the most appropriate reference group against which to compare the exposed groups, historical controls obtained from other studies contain some information regarding the spontaneous rate of occurrence of the lesion of interest and may therefore aid in evaluating results in the exposed groups in the experiment at hand (Haseman et al., 1984). In particular, historical controls may be useful in assessing rare tumours or interpreting a marginally significant result relative to the concurrent controls.

The first formal statistical procedure for utilizing historical control data in testing for carcinogenic effects in rodent bioassays was proposed by Tarone (1982). Extensions to this procedure have subsequently been proposed by Krewski et al. (1985) and Yanagawa and Hoel (1985). In this paper, we consider several robust alternatives to the existing tests based on the concept of Γ-minimax estimation (Albert, 1983).

## 2. Tests for Trend in Binomial Proportions

Consider an experiment with k+1 dose levels $0 = d_0 < d_1 < \ldots < d_k$

in which $x_i$ of the $n_i$ animals at dose $d_i$ respond ($i=0,1,\ldots,k$). We assume that $x_i$ follows a binomial distribution where the response probability $p_i = P(d_i)$ is given by the logistic dose response model.

$$P(d) = [1 + \exp\{-(a+bd)\}]^{-1} \tag{2.1}$$

($-\infty < a,b < +\infty$) for $d \geq 0$. Treating $a$ as a nuisance parameter, the score statistic for testing the null hypothesis $H_0 : b = 0$ against the one-sided alternative $H_1 : b > 0$ is given by

$$T_{CA} = \Sigma x_i d_i - \hat{p}\, \Sigma n_i d_i , \tag{2.2}$$

(Tarone and Gart, 1980), where $\hat{p} = x/n$ with $x = \Sigma x_i$ and $n = \Sigma n_i$. The variance of this statistic is

$$n^{-1} V(T_{CA}) = n^{-1} p(1-p) \{\Sigma n_i d_i^2 - (\Sigma n_i d_i)^2/n\} \sim p(1-p)\sigma_d^2 , \tag{2.3}$$

where $\sigma_d^2 = \Sigma\lambda_i (d_i - \bar{d})^2$, $\bar{d} = \Sigma\lambda_i d_i$, and $\sim$ denotes asymptotic equivalence as $n \to \infty$ with $n_i/n \to \lambda_i > 0$. The standardized test statistic $S_{CA} = T_{CA}/[V(T_{CA})]^{1/2}$, commonly called the Cochran-Armitage statistic, converges in distribution to the standard normal as $n \to \infty$ under the null hypothesis.

To incorporate information from historical controls, we regard $p = [1+\exp(-a)]^{-1}$ as a random variable, following the beta density

$$f(p|\alpha,\beta) = \frac{\Gamma(\alpha+\beta)}{\Gamma(\alpha)\Gamma(\beta)} p^{\alpha-1}(1-p)^{\beta-1} \tag{2.4}$$

($0 < p < 1$; $\alpha,\beta > 0$). For later applications, it will also be convenient to parametrize this distribution in terms of $\theta = \alpha/(\alpha+\beta)$ and $\rho = (\alpha+\beta)^{-1}$. The former quantity represents the mean of the distribution while the latter provides a measure of dispersion with $\rho = 0$ representing the limiting case of a degenerate distribution concentrated at $\theta$.

The score statistic based on the marginal likelihood obtained after integrating out $p$ is now given by

$$T_{HC}(\alpha,\beta) = \Sigma x_i d_i - \tilde{p}\, \Sigma n_i d_i , \tag{2.5}$$

where $\tilde{p} = (x+\alpha)/(n+\alpha+\beta)$. Note that $\tilde{p}$ coincides with the Bayes estimator of the binomial response probability $p$ in the concurrent control group under the beta prior in (2.4). This statistic has mean zero and variance

$$n^{-1} V(T_{HC}) = n^{-1} \frac{\alpha\beta}{(\alpha+\beta)(\alpha+\beta+1)} \left\{\Sigma n_i d_i^2 - \frac{1}{n+\alpha+\beta} (\Sigma n_i d_i)^2\right\}$$

$$\sim \frac{\alpha\beta}{(\alpha+\beta)(\alpha+\beta+1)} \sigma_d^2 . \tag{2.6}$$

Although the asymptotic null distribution of $S_{HC} = T_{HC}/[V(T_{HC})]^{1/2}$ is a mixture of normal distributions with mean zero and variance one, this misture is well approximated by a standard normal distribution (Krewski et al., 1985).

## 3. Robust Distributions

The methods outlined in section 2 employ a beta distribution for the binomial response probability $p$ in the concurrent control group. In order to allow for some uncertainty in the specification of this distribution, we consider three classes of modified distributions having heavier tails than the original beta distribution. These modified distributions reflect our relative ignorance of the distribution of p in these regions, and may be robust against misspecification of the distribution as a beta distribution.

Class I

Consider the class of distributions given by

$$\Gamma_1 = \{f_\varepsilon(p|\theta,\rho) = (1-\varepsilon)f(p|\theta,\rho)+\varepsilon\cdot 1 : 0 \leq \varepsilon \leq 1\}. \tag{3.1}$$

The densities $f_\varepsilon \in \Gamma_1$ are thus mixtures of beta and uniform densities defined on the interval (0,1). For $\varepsilon > 0$, these mixed distributions will have heavier tails than a pure beta distribution, reflecting our uncertainty as to the tail behaviour of the distribution of p.

Consider the marginal distribution of $x_0$ under the mixture $f_\varepsilon$ given by

$$m_\varepsilon(x_0) = \binom{n_0}{x_0} \int_0^1 p^{x_0}(1-p)^{n_0-x_0} f_\varepsilon(p)\,dp. \tag{3.2}$$

The special cases $\varepsilon = 0$ and 1 correspond to the beta-binomial and discrete uniform distributions respectively, with

$$m_0(x_0) = \binom{n_0}{x_0} \frac{\Gamma(\alpha+\beta)}{\Gamma(\alpha)\Gamma(\beta)} \frac{\Gamma(x_0+\alpha)\Gamma(n_0-x_0+\beta)}{\Gamma(n_0+\alpha+\beta)} \tag{3.3}$$

and

$$m_1(x_0) = (n_0+1)^{-1}. \tag{3.4}$$

The Bayes estimator of the value of p in the concurrent control group under the prior $f_\varepsilon$ is then given by

$$\delta_\varepsilon(x_0) = \lambda_1(x_0)\tilde{p}_0(\alpha,\beta)+(1-\lambda_1(x_0))\tilde{p}_0(1,1) \tag{3.5}$$

where

$$\lambda_1(x_0) = (1-\varepsilon)m_0(x_0)/[(1-\varepsilon)m_0(x_0)+\varepsilon m_1(x_0)] \tag{3.6}$$

and $\tilde{p}_0(\alpha,\beta) = (x_0+\alpha)/(n_0+\alpha+\beta)$ is the Bayes estimator of p under a pure beta prior based on the data from the concurrent control group only.

The Bayes risk of any estimator $\delta(x_0)$ of p is given by

$$r(f_\varepsilon,\delta) = \sum_{x_0=0}^{n_0} (\delta(x_0)-p)^2 m_\varepsilon(x_0), \tag{3.7}$$

where the Bayes risk of $\delta_\varepsilon$ satisfies

$$r_\varepsilon = r(f_\varepsilon,\delta_\varepsilon) = \inf_\delta r(f_\varepsilon,\delta) . \tag{3.8}$$

In order to find the best distribution in the class $\Gamma_1$, the $\Gamma$-minimax criterion may be used to determine the most suitable value of the mixing proportion $\varepsilon$. Thus, we seek $\varepsilon = \varepsilon^*$ such that

$$R = \inf_{0\leq\varepsilon\leq 1} \sup_{0\leq\varepsilon'\leq 1} \{r(f_{\varepsilon'},\delta_\varepsilon)-r_{\varepsilon'}\}$$

$$= \inf_{0\leq\varepsilon\leq 1} \max\{r(f_0,\delta_\varepsilon)-r_0, r(f_1,\delta_\varepsilon)-r_1\} \tag{3.9}$$

is attained at $\varepsilon = \varepsilon^*$. The mixing proportion can be found iteratively by solving the equation

$$r(f_0,\delta_{\varepsilon^*})-r_0 = r(f_1,\delta_{\varepsilon^*})-r_1 \tag{3.10}$$

for $\varepsilon^*$. (Although we have not examined the uniqueness of $\varepsilon^*$ analytically, $\varepsilon^*$ has been found to be unique in all of the examples we have considered to date.) We note that once $\varepsilon^*$ is determined by minimizing the maximum Bayes risk of $\delta(x_0)$, no further use of Bayesian ideas is required in the subsequent analysis.

Using the mixed density $f_{\varepsilon^*}\in\Gamma_1$ in place of $f_0$ leads to the score statistic

$$T_1 = \lambda_1(\underset{\sim}{x})T_{HC}(\alpha,\beta)+(1-\lambda_1(\underset{\sim}{x}))T_{HC}(1,1), \tag{3.11}$$

where

$$\lambda_1(\underset{\sim}{x}) = (1-\varepsilon^*)m_0(\underset{\sim}{x})/[(1-\varepsilon^*)m_0(\underset{\sim}{x})+\varepsilon^* m_1(\underset{\sim}{x})], \tag{3.12}$$

with the joint marginal null distribution of $\underset{\sim}{x}$ under $f_\varepsilon$ given by

$$m_\varepsilon(\underset{\sim}{x}) = \prod_{i=0}^{k}\binom{n_i}{x_i}\int_0^1 p^x(1-p)^{n-x}f_\varepsilon(p)\,dp. \tag{3.13}$$

In the special cases $\varepsilon = 0$ and $1$ we have

$$m_0(\underset{\sim}{x}) = \prod_{i=0}^{k}\binom{n_i}{x_i}\frac{\Gamma(\alpha+\beta)}{\Gamma(\alpha)\Gamma(\beta)}\frac{\Gamma(x+\alpha)\Gamma(n-x+\beta)}{\Gamma(n+\alpha+\beta)} \tag{3.14}$$

and

$$m_1(\underset{\sim}{x}) = (n+1)^{-1}\prod_{i=0}^{k}\binom{n_i}{x_i}/\binom{n}{x} \tag{3.15}$$

respectively. It follows from (3.12), (3.14) and (3.15) that $\lambda_1(\underset{\sim}{x})$ in fact depends on the data $\underset{\sim}{x}$ only through $x$. Since $T_{HC}(1,1) \approx T_{CA}$, $T_1$ may be essentially viewed as a linear combination of Tarone's statistic $T_{HC}(\alpha,\beta)$ and the Cochran-Armitage statistic $T_{CA}$.

An exact expression for the variance of $T_1$ is given in the Appendix. Asymptotically, we also have

$$n^{-1}V(T_1) \to [(1-\varepsilon^*)\frac{\alpha\beta}{(\alpha+\beta)(\alpha+\beta+1)} + \varepsilon^*/6]\sigma_d^2 \tag{3.16}$$

in probability, given that $f_{\varepsilon^*}$ represents the underlying distribution of $p$.

<u>Class II</u>

When $\varepsilon^*$ is large, $f_{\varepsilon^*}$ will have relatively heavy tails. In order not to alter the mass in the central portion of the distribution, we consider the restricted class

$$\Gamma_2 = \{f_\varepsilon(p) = (1-\varepsilon)f(p|\theta,\rho')+\varepsilon\cdot 1:\rho'\le\rho,\ \int_{c_1}^{c_2} f_\varepsilon(p)\,dp = 1-\gamma\}, \tag{3.17}$$

where $c_1$ and $c_2$ are chosen so that

$$\int_0^{c_1} f(p|\theta,\rho)\,dp = \int_{c_2}^{1} f(p|\theta,\rho)\,dp = \frac{\gamma}{2} \tag{3.18}$$

$(0 < \gamma < 1)$. Thus, all densities $f_\varepsilon \in \Gamma_2$ assign mass $(1-\gamma)$ to the interval $(e_1, e_2)$.

It follows from (3.17) and (3.18) that

$$\varepsilon = \frac{(1-\gamma) - \int_{c_1}^{c_2} f(p|\theta,\rho')dp}{(c_2-c_1) - \int_{c_1}^{c_2} f(p|\theta,\rho')dp} \tag{3.19}$$

so that $\varepsilon$ determines $\rho'$ and vice versa. If $\rho' = \rho$, then $\varepsilon = 0$; as $\rho' \to 0$, $\varepsilon \to \gamma[1-(c_2-c_1)]^{-1} = \varepsilon_0$. Thus, we have $0 \leq \varepsilon \leq \varepsilon_0$. The priors $f_0$ and $f_{\varepsilon_0}$ are the extremes in the class $\Gamma_2$ having the lightest and heaviest tails respectively.

The Bayes estimator of $p$ is now given by

$$\delta_\varepsilon(x_0) = \lambda_2(x_0)\tilde{p}_0(\alpha',\beta')+(1-\lambda_2(x_0))\tilde{p}_0(1,1), \tag{3.20}$$

where $\alpha' = \theta/\rho'$, $\beta' = (1-\theta)/\rho'$ and

$$\lambda_2(x_0) = (1-\varepsilon)m_0(x_0)/[(1-\varepsilon)m_0(x_0)+\varepsilon m_1(x_0)] \tag{3.21}$$

with $m_0(x_0)$ now defined as in (3.3), but $(\alpha',\beta')$ replacing $(\alpha,\beta)$. In analogy with (3.10), the value of $\varepsilon = \varepsilon^*$ using the $\Gamma$-minimax criterion is obtained by solving the equation

$$r(f_0,\delta_{\varepsilon^*})-r_0 = r(f_{\varepsilon_0},\delta_{\varepsilon^*})-r_{\varepsilon_0} \tag{3.22}$$

for $\varepsilon^*$. The score statistic is then

$$T_2 = \lambda_2(\underset{\sim}{x})T_{HC}(\alpha^*,\beta^*)+(1-\lambda_2(\underset{\sim}{x}))T_{HC}(1,1), \tag{3.23}$$

where $\lambda_2(\underset{\sim}{x})$ has the same form as (3.21) with $\varepsilon = \varepsilon^*$ and $(\alpha^*,\beta^*)$ corresponds to $(\theta^*,\rho^*)$ with $\rho^*$ being the value of $\rho'$ obtained from (3.19) with $\varepsilon = \varepsilon^*$.

<u>Class III</u>

Although the mixed densities in $\Gamma_2$ maintain the same mass in the central part of the distribution as the original beta density, the mean of the mixed distribution will be greater than the mean of a pure beta distribution whenever $\theta < 1/2$. Thus, we consider a third class of priors defined by

$$\Gamma_3 = \left\{f_\varepsilon(p) = (1-\varepsilon)f(p|\theta,\rho')+\varepsilon f(p|\theta,\theta):\rho'\leq\rho,\ \int_0^c f_\varepsilon(p)dp=1-\gamma\right\}, \tag{3.24}$$

where $c$ is chosen so that

$$\int_0^c f(p|\theta,\rho)dp = 1-\gamma . \tag{3.25}$$

As in (3.19), we have

$$\varepsilon = \frac{(1-\gamma)-\int_0^c f(p|\theta,\rho')dp}{[1-(1-c)^{\beta^{**}}]-\int_0^c f(p|\theta,\rho')dp}, \tag{3.26}$$

where $\beta^{**} = (1-\theta)/\theta$.

In this case, $\varepsilon \to \gamma(1-c)^{-\beta^{**}} = \varepsilon_0$ as $\rho' \to 0$ so that $0 \le \varepsilon \le \varepsilon_0$ with $f_0$ and $f_{\varepsilon_0}$ being the extreme elements in $\Gamma_3$.

The Bayes estimator of $p$ is given by

$$\delta_\varepsilon(x_0) = \lambda_3(x_0)\tilde{p}_0(\alpha',\beta') + (1-\lambda_3(x_0))\tilde{p}_0(1,\beta^{**}), \qquad (3.27)$$

Here, $\lambda_3(x_0)$ is defined as in (3.21) with $m_1(x_0)$ now being a beta-binomial distribution as in (3.3) with parameters $\alpha = 1$ and $\beta = \beta^{**}$.

After finding the value of $\varepsilon^*$ as in (3.22), the score statistic is

$$T_3 = \lambda_3(\underset{\sim}{x})T_{HC}(\alpha^*,\beta^*) + (1-\lambda_3(\underset{\sim}{x}))T_{HC}(1,\beta^{**}). \qquad (3.28)$$

## 4. Applications

In order to illustrate the use of the methods developed in section 3, consider the two examples in Table 1 previously analyzed by Smythe et al. (1986) and Dempster et al. (1983) respectively. The values of $\theta$ and $\rho$ used in these two examples were estimated by maximum likelihood using actual historical control data. Although these estimates are subject to sampling error, we will assume for purposes of illustration that they are known constants which characterize the distribution of $p$. Note that while the means of the prior distributions are comparable in these two examples, the dispersion of the prior distribution as measured by $\rho$ is much greater in example 2.

The results of applying the tests for trend discussed in section 2 and 3 to these data are summarized in Table 2. In the absence of prior information on the control response rate $p$, no strong evidence of an increasing trend in tumour occurrence with dose is provided by the Cochran-Armitage statistic $T_{CA}$ in example 1. With an informative beta prior for $p$, however, Tarone's statistic $T_{HC}$ is indicative of a significant trend.

Because this latter test may be expected to perform well only when the assumed beta prior is correct, we reanalyzed these data using robust mixed priors selected from classes I, II and III discussed in section 3.

Table 1. Two Examples of Experimental Data

| Example | Parameters of Beta Prior | | Doses: $d_0,\ldots,d_k = 1$ | | | |
|---|---|---|---|---|---|---|
| | $\theta$ | $\rho$ | Response Rates: $x_0/n_0,\ldots,x_k/n_k$ | | | |
| 1. Smythe et al. (1986) | 0.085 | 0.004 | 0<br>2/20 | 0.5<br>6/49 | | 1<br>10/49 |
| 2. Dempster et al. (1983) | 0.094 | 0.024 | 0<br>3/55 | 0.003<br>3/57 | 0.1<br>5/60 | 1<br>10/55 |

Table 2. Tests for Trend[a] in Examples 1 and 2

| Example | Test Statistic | Prior Mean | Prior Variance | $\varepsilon^*$ | $\lambda$ | p-value |
|---|---|---|---|---|---|---|
| 1 | $T_{CA}$ | - | - | - | - | 0.103 |
| | $T_{HC}$ | 0.085 | 0.0003 | - | - | 0.003 |
| | $T_1$ | 0.347 | 0.0929 | 0.62 | 0.43 | 0.030 |
| | $T_2$[b] | 0.128 | 0.0252 | 0.11 | 0.86 | 0.002 |
| | $T_3$[c] | 0.085 | 0.0020 | 0.34 | 0.40 | 0.009 |
| 2 | $T_{CA}$ | - | - | - | - | 0.004 |
| | $T_{HC}$ | 0.094 | 0.002 | - | - | 0.005 |
| | $T_1$ | 0.321 | 0.088 | 0.56 | 0.86 | 0.017 |
| | $T_2$[d] | 0.142 | 0.027 | 0.12 | 0.99 | 0.011 |
| | $T_3$[e] | 0.094 | 0.004 | 0.50 | 0.81 | 0.005 |

a $\gamma = 0.10$ in $T_2$ and $T_3$

b $(c_1,c_2) = (0.0592, 0.1131)$

c $(0,c) = (0, 0.1061)$

d $(c_1,c_2) = (0.0333, 0.1769)$

e $(0,c) = (0, 0.1543)$

With $T_1$, the p-value in example 1 is greatly increased due both to the prior mean being shifted to the right (thereby reducing the linear trend estimated by $T_1$) and the larger prior variance.

With $T_2$, which maintains the same mass in the central part of the mixed prior distribution as the original beta prior, the prior mean is again shifted slightly to the right, but far less so than with $T_1$. Although the mixed prior has larger variance than the original pure beta prior, the constraint on the mass in the central part of the mixture implies $\rho^* < \rho$. Because of the larger weight assigned to the first component of this statistic ($\lambda_2 = 0.86$), $T_2$ leads to a slightly more significant result than $T_{HC}$.

With $T_3$, mixing in a light tailed beta distribution highly skewed to the right rather than a uniform distribution leads to a relatively small prior variance. Because most of the weight ($1-\lambda_3 = 0.60$) is assigned to the component of $T_3$ with the larger prior variance, the significance level is somewhat greater than that for $T_2$.

The prior distributions for example 1 are illustrated graphically in Figure 1. Note that the prior for $T_1$ is much more diffuse than the prior for $T_{HC}$ due to the lack of any constraints on the mass in the central portion of the distribution. The priors for $T_2$ and $T_3$ are more peaked than that for $T_{HC}$ in the central portion of the distribution, but have heavier tails. The left tail for $T_3$ is notably heavier than that for $T_2$ due to the

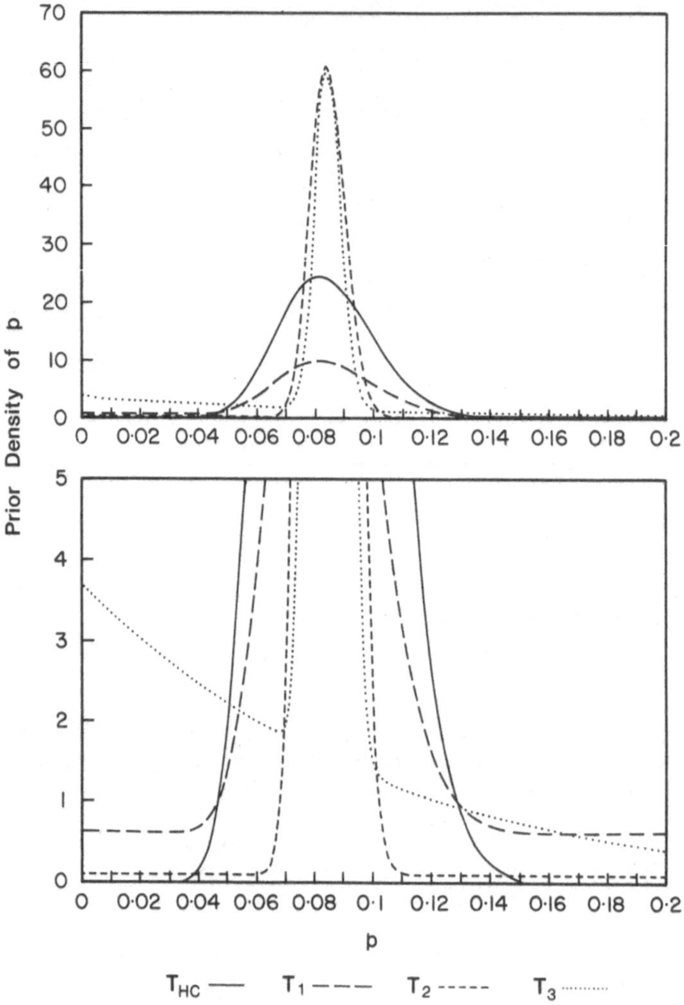


Figure 1. Prior Densities for p in Example 1.

greater weight placed on small values of p in the former case.

In example 2, the mean and variance of the three mixed priors demonstrate similar effects as observed in example 1. Although both $T_{CA}$ and $T_{HC}$ lead to similar significance because of the larger dispersion of the original beta prior, the results provided by the robust test statistics $T_1$ and $T_2$ are somewhat less significant. With $T_1$, this is again largely due to the notably larger prior mean.

5. Robustness Properties

One problem with the use of a pure beta distribution to model the variability in the historical control data is that the estimates of the parameters $\alpha$ and $\beta$ in (2.4) can be strongly influenced by the inclusion or exclusion of a single data point. This was pointed out by Tamura and Young (1986) in an early draft of their paper, although direct discussion of this issue was deleted from the published version in the interests of brevity.

To illustrate this point, consider the hypothetical sequence of historical controls given in Table 3. Note that the estimated values of $\alpha$ and $\beta$ are somewhat sensitive to the deletion of individual data points, as previously noted by Tamura and Young (1986). Although the estimated values of $\theta$ are more stable, the estimates of $\rho$ remain somewhat variable.

Consider now the hypothetical bioassay data in Table 4, which, by themselves, provide little evidence of a dose response relationship. Using $T_{CA}$ without historical controls, the p-value is 0.20. With the full set of historical control data from Table 3, however, the p-value based on $T_{HC}$ is 0.046.

Table 3. Estimates of the Parameters in the Beta Prior Based on a Hypothetical Sequence of Historical Controls Deleting Individual Data Points

| Historical Control Response Rate | Frequency of Occurrence | Parameter Estimates Deleting One Observation | | | |
|---|---|---|---|---|---|
| | | $\hat{\alpha}$ | $\hat{\beta}$ | $\hat{\theta}$ | $\hat{\rho}$ |
| 0/50 | 7 | 2.9 | 89.8 | 0.032 | 0.011 |
| 1/50 | 4 | 2.0 | 62.2 | 0.030 | 0.016 |
| 2/50 | 4 | 1.8 | 59.6 | 0.029 | 0.016 |
| 3/50 | 3 | 1.9 | 66.4 | 0.028 | 0.015 |
| 4/50 | 1 | 2.4 | 84.0 | 0.027 | 0.012 |
| 5/50 | 1 | 3.7 | 136.1 | 0.026 | 0.007 |

Table 4. A Hypothetical Example of Experimental Data

| Example | Parameters of Beta Prior | | Doses: $d_0,\ldots,d_k = 1$ | | |
|---|---|---|---|---|---|
| | $\theta$ | $\rho$ | Response Rates: $x_0/n_0,\ldots,x_k/n_k$ | | |
| 3 | 0.030 | 0.013 | 0 | 0.5 | 1 |
| | | | 2/50 | 3/50 | 4/50 |

In order to explore the robustness of the latter statistic, we recomputed the p-values after deleting in turn each of the six distinct historical control data points (Table 5). Depending on the point excluded, the resulting p-values for $T_{HC}$ ranged from 0.017 to 0.053, casting some doubt on whether or not the original result with $T_{HC}$ should be considered significant at the nominal 5% level of significance. An examination of the corresponding results for $T_1$, $T_2$ and $T_3$ reveals less variation in the p-values obtained after deleting a point with $T_2$ and $T_3$.

## 6. Summary and Conclusions

The tests for trend in carcinogen bioassay data for use with historical controls proposed to date have been based largely on the assumption that the binomial response rate in the control group varies among studies in accordance with a beta distribution. Because this distribution has relatively light tails, we considered several classes of mixed distributions having heavier tails than the beta. The optimum value of the mixing parameter is obtained using the $\Gamma$-minimax criterion. Two of these statistics ($T_2$ and $T_3$) were also shown to be more robust than the traditional statistic ($T_{HC}$) with respect to deletion of an individual historical control data point.

## Appendix A: Variance of the Score Statistic with Mixed Historical Control Distributions

Consider the general form of the mixed distribution for the response probability in the control group given by

$$f_\varepsilon(p) = (1-\varepsilon)f(p|\alpha,\beta) + \varepsilon f(p|\alpha^*,\beta^*). \tag{A.1}$$

Note that all of the mixed distributions in the classes $\Gamma_1$, $\Gamma_2$ and $\Gamma_3$ in section 3 are special cases of (A.1). Note also that under (A.1) the test statistic can be written as

$$T = \lambda(\underset{\sim}{x})T_{HC}(\alpha,\beta)+(1-\lambda(\underset{\sim}{x}))T_{HC}(\alpha^*,\beta^*), \tag{A.2}$$

where $\lambda(\underset{\sim}{x})$ is defined as in (3.12). This statistic can be expressed as

Table 5. Tests for Trend in Example 3 After Deleting One Historical Control Observation[a]

| Historical Control Response Rate | p-values | | | |
|---|---|---|---|---|
| | $T_{HC}$ | $T_1$ | $T_2$ | $T_3$ |
| 0/50 | 0.048 | 0.119 | 0.025 | 0.029 |
| 1/50 | 0.053 | 0.139 | 0.024 | 0.035 |
| 2/50 | 0.049 | 0.144 | 0.021 | 0.035 |
| 3/50 | 0.041 | 0.132 | 0.017 | 0.031 |
| 4/50 | 0.030 | 0.110 | 0.013 | 0.025 |
| 5/50 | 0.017 | 0.079 | 0.009 | 0.019 |

[a] $\gamma = 0.10$ in $T_2$ and $T_3$

$$T = \sum x_j d_j - n\bar{d}_n H(\underset{\sim}{x}) \tag{A.3}$$

where $\bar{d}_n = \sum n_i d_i / n$ and

$$H(\underset{\sim}{x}) = \lambda(\underset{\sim}{x}) \frac{x+\alpha}{n+\alpha+\beta} + (1-\lambda(\underset{\sim}{x})) \frac{x+\alpha^*}{n+\alpha^*+\beta^*}. \tag{A.4}$$

Note that $H(\underset{\sim}{x})$ depends on the data only through $x$.

From (A.3), we have

$$V(T) = V(\sum x_j d_j) - 2(n\bar{d}_n)\mathrm{Cov}(\sum x_j d_j, H(\underset{\sim}{x})) + (n\bar{d}_n)^2 V(H(\underset{\sim}{x})), \tag{A.5}$$

where

$$\begin{aligned} V(\sum x_j d_j) &= E[V(\sum x_j d_j | p)] + V[E(\sum x_j d_j | p)] \\ &= (\sum n_j d_j^2) E[p(1-p)] + (n\bar{d}_n)^2 V[p]. \end{aligned} \tag{A.6}$$

Under (A.1),

$$E(p) = (1-\varepsilon) \frac{\alpha}{\alpha+\beta} + \varepsilon \frac{\alpha^*}{\alpha^*+\beta^*} \tag{A.7}$$

and

$$E(p^2) = (1-\varepsilon) \frac{(\alpha+1)}{(\alpha+\beta)(\alpha+\beta+1)} + \varepsilon \frac{\alpha^*(\alpha^*+1)}{(\alpha^*+\beta^*)(\alpha^*+\beta^*+1)}, \tag{A.8}$$

from which $E[p(1-p)]$ and $V[p]$ can be calculated easily. Further,

$$\mathrm{Cov}(\sum x_j d_j, H(\underset{\sim}{x})) = \sum d_j \ \mathrm{Cov}(x_j, H(\underset{\sim}{x})) \tag{A.9}$$

with

$$\mathrm{Cov}(x_j, H(\underset{\sim}{x})) = \sum_{\underset{\sim}{x}} x_j (H(\underset{\sim}{x}) - E(H(\underset{\sim}{x}))) m_\varepsilon(\underset{\sim}{x}), \tag{A.10}$$

where

$$E[H(\underset{\sim}{x})] = (1-\varepsilon) \frac{\alpha}{\alpha+\beta} + \varepsilon \frac{\alpha^*}{\alpha^*+\beta^*}, \tag{A.11}$$

and $m_\varepsilon(\underset{\sim}{x})$ is the marginal null distribution of $\underset{\sim}{x}$ under (A.1) defined as in (3.13). Note that $m_\varepsilon(\underset{\sim}{x})$ can be written as

$$m_\varepsilon(\underset{\sim}{x}) = \left[\prod_j \binom{n_j}{x_j}\right] M(x),$$

where

$$M(x) = \int_0^1 p^x (1-p)^{n-x} f_\varepsilon(p)\, dp \tag{A.12}$$

depends only on $x$. Noting that (A.4) also depends on $x$ only, it follows from (A.10) that

$$\begin{aligned} \mathrm{Cov}(x_j, H(\underset{\sim}{x})) &= \sum_{x=0}^{n} (H(\underset{\sim}{x}) - E(H(\underset{\sim}{x})))M(x) \sum_{\underset{\sim}{x}:\sum x_j = x} x_j \left[\prod_i \binom{n_i}{x_i}\right] \\ &= n_j \sum_{x=0}^{n} (H(\underset{\sim}{x}) - E(H(\underset{\sim}{x}))M(x) \binom{n}{x} \frac{x}{n}. \end{aligned} \tag{A.13}$$

Thus, (A.9) can be easily evaluated using (A.13). Finally, it can be easily shown that

$$V(H(\underset{\sim}{x})) = \sum_{x=0}^{n} (H(\underset{\sim}{x})-E(H(\underset{\sim}{x})))^2 M(x)\binom{n}{x}. \tag{A.14}$$

V(T) can now be calculated using (A.5), (A.6), (A.9) and (A.14).

Acknowledgements

This research was supported in part by grant no. A8664 from the Natural Sciences and Engineering Research Council of Canada to D. Krewski, and by grant no. DMS-8503774 from the U.S. National Science Foundation to R.T. Smythe.

## References

Alberta, J.H., 1983, A Γ-minimax approach for estimating a binomial probability. Technical Report, Department of Mathematics & Statistics, Bowling Green State University, Bowling Green, Ohio.

Bickis, M. and Krewski, D., 1985, Statistical design and analysis of the long-term carcinogenicity bioassay. In: Toxicological Risk Assessment, Vol. I, Biological and Statistical Criteria (D. Clayson, D. Krewski & I.C. Munro, eds.). CRC Press, Boca Raton, Florida, pp. 125-147.

Dempster, A.P., Selwyn, M.D. and Weeks, B.J. (1983). Combining historical and randomized controls for assessing trends in proportions. Journal of the American Statistical Association 78, 221-227.

Haseman, J.K., Huff, J. and Boorman, G.A., 1984, Use of historical control data in carcinogenicity studies in rodents. Toxicologic Pathology 12, 126-135.

Krewski, D., Smythe, R.T., and Burnett, R., 1985, The use of historical control information in testing for trend in quantal response carcinogenicity data. In: Proceedings of the Symposium on Long-Term Animal Carcinogenicity Studies: A Statistical Perspective. American Statistical Association, Washington, D.C., pp. 56-62.

Smythe, R.T., Krewski, D. and Murdoch, D., 1986, The use of historical control information in modelling dose response relationships in carcinogenesis. Statistics & Probability Letters 4, 87-93.

Tamura, R. and Young, S., 1986, The incorporation of historical control information in tests of proportions: simulation study of Tarone's procedure. Biometrics 42, 343-349.

Tarone, R.E., 1982, The use of historical control information in testing for a trend in proportions. Biometrics 38, 215-220.

Tarone, R.E. and Gart, J.J., 1980, On the robustness of combined tests for trend in proportions. Journal of the American Statistical Association 75, 110-116.

Yanagawa, T. and Hoel, D.G., 1985, Use of historical controls for animal experiments. Environmental Health Perspectives 63, 217-224.

DECOMPOSITION OF WEIBULL MIXTURE-DISTRIBUTIONS

IN ACCELERATED LIFE TESTING BY BAYESIAN METHODS

Harald Strelec

Institute of Statistics
Technical University of Vienna
Vienna, Austria

## INTRODUCTION

Practical reliability analysis shows that in most cases a bathtub like hazard rate function best fits real life time data. This fact can be explained by a succession of time intervals in which early failures dominate in the first one, random failures in the second one and wear-out failures in the last part (see fig.1). But whereas there are many probability distributions which can describe monotonic hazard rate functions unfortunately there are only few ones which have bathtub like hazard rate functions. Perhaps the best one among the latter ones is the model of Hjorth (1980). He used three parameters to treat the problem. But the applicability of this model seems to be constrained.

Roughly speaking there are two simple models for describing bathtub like hazard rate functions which are using compositions of simple probability distributions. In one case this is done using an own probability distribution for each of the three time intervals where the starting points of these distributions equal the break points of the time area. This model is of certain mathematical simplicity but can only describe the fact of bathtub like hazard rate functions and not explain it.

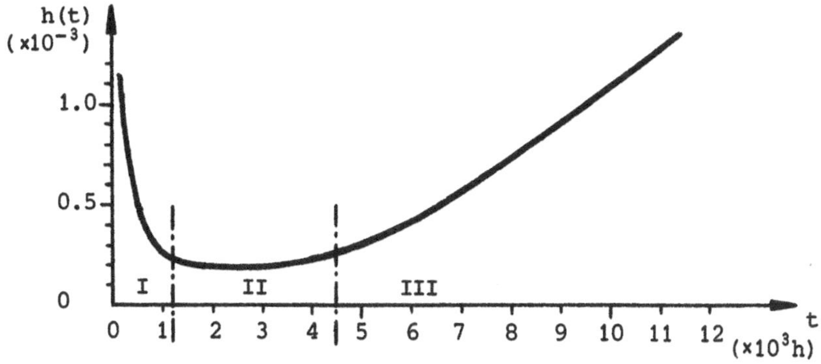


Fig. 1. Bathtub like hazard rate function h(t) (mean life time: 2 300 h). Region I corresponds to early type failures, region II to completely random and region III to wear-out failures.

On the other hand the life time distribution may be given by a superposition of three simple probability distributions (for instance Weibull distributions) where each of the latter ones describes one of the above mentioned failure types. Using this model we are able to explain the phenomenon of bathtub like hazard rate functions in an almost exact way. The definition and analysis of this model is given in the following chapters.

## THE MIXTURE MODEL

As mentioned before a mixture model seems to be an appropriate way of describing and explaining life time distributions with bathtub like hazard rate functions. Corresponding to the three different failure modes a mixture of three life time distributions seems to fit best where the first one has decreasing failure rate, the second one (nearly) constant and the last one increasing failure rate function. The cumulative distribution function for the global life time is now given by

$$F(t) = \sum_{i=1}^{3} p_i \cdot Pr(T \leq t \mid G_i) = \sum_{i=1}^{3} p_i \cdot F_i(t) \quad \text{for} \quad t > 0 \qquad (1)$$

where $G_1, G_2, G_3$ are the three subgroups of the total production corresponding to early, random or wear-out failing with portion $p_i$ $(i=1,2,3)$. The (conditional) life time distribution for each subgroup is given by the cumulative distribution function $F_i$ $(i=1,2,3)$.

The hazard rate function $h(t)$ in the mixture model (1) is now

$$h(t) = \frac{f(t)}{R(t)} = \sum_{i=1}^{3} h_i(t) \cdot p_i \cdot \frac{R_i(t)}{R(t)} \quad \text{for} \quad t > 0 \qquad (2)$$

with $R(t) = 1-F(t)$ (resp. $R_i(t) = 1-F_i(t)$) being the reliability function (survival function). It shows a weighted mean of single hazard rate functions $h_i(t)$ with (time dependent) weights $w_i(t) = p_i R_i(t)/R(t)$.

As it is possible to describe increasing, decreasing and constant failure rate functions by special Weibull-distributions the mixture (1) may be assumed to consist only of Weibull components. The cumulative distribution function is therefore given by

$$F(t) = 1 - \sum_{i=1}^{3} p_i \cdot \exp(-(t/\tau_i)^{\eta_i}) \quad \text{for} \quad t > 0 \qquad (3)$$

with scale parameters $0 < \tau_1 < \tau_2 < \tau_3$ and shape parameters $0 < \eta_1 < 1 < \eta_3$ and $\eta_2 = 1$ (or close to 1). The hazard rate function is therefore of the form

$$h(t) = \sum_{i=1}^{3} \left(\frac{\eta_i}{\tau_i}\right)\left(\frac{t}{\tau_i}\right)^{\eta_i - 1} \cdot p_i \cdot \frac{\exp(-(t/\tau_i)^{\eta_i})}{\sum_{j=1}^{3} p_j \cdot \exp(-(t/\tau_j)^{\eta_j})} \quad \text{for} \quad t > 0 \qquad (4)$$

corresponding to relation (2).

Unfortunately using this Weibull-mixture the resulting hazard rate function is not quite of the form shown in fig.1 but is decreasing after

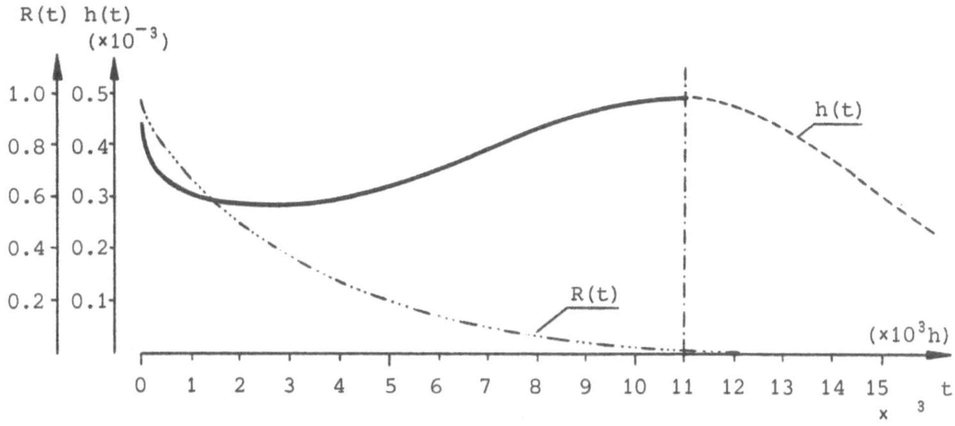


Fig. 2. Hazard rate function and reliablity function of a Weibull-mixture with weights $p_1=0.1$, $p_2=0.6$, $p_3=0.3$, scale parameters $\tau_1=1000\,h$, $\tau_2=2000\,h$, $\tau_3=6000\,h$ and shape parameters $\eta_1=0.5$, $\eta_2=1.0$, $\eta_3=2.0$ .

the bathtub like period and tends to zero for increasing t (see fig.2). This fact is caused by the dominance of $R_1(t)$ over $R_2(t)$ and of $R_2(t)$ over $R_3(t)$ for large t so that

$$\lim_{t\to\infty} w_2(t)/w_1(t) = \lim_{t\to\infty} w_3(t)/w_2(t) = 0 \,. \tag{5}$$

But in spite of that it does not matter from a practical point of view as long as the bathtub like period $0 < t \leq t'$ is important enough in the sense that

$$R(t') = \exp(-\int_0^{t'} h(t)\,dt)$$

is close to zero. For instance if $R(t') \leq 0.01$ should be fulfilled there must hold $\int_0^{t'} h(t)\,dt \geq 4.6$ .

Therefore to ensure the above requirement some analytical conditions on the parameters of the mixture model must be imposed: these are

- the three subpopulations should be evidently seperated by the scale parameters (characteristic life times) where $\tau_3$ should be at least twice $\tau_2$;
- the portion $p_2$ must not exceed $p_3$ for a greater amount and it is all the better the more $p_3$ exceeds $p_2$;
- the portion $p_1$ should be small.

Simulation studies show that the influence of changes of the mixture parameters on the hazard rate function of the mixture is moderate which is important for the Bayesian analysis of this model.

## ACCELERATED LIFE TESTING AND DECOMPOSITION

For products of high reliability (e.g. semiconductor technology) often mean life times of $10^6$ hours and more can be found. For these items life testing under environmental conditions is unsatisfactory. Because either it takes too long time and is therefore senseless from an economical or technical point of view on the one hand or life tests have to be performed highly censored on the other hand. In the latter case the estimations of the life time distribution are then comparatively bad because of the great influence of the early failures observed at censored life testing at an obviously higher rate.

Accelerated life testing is almost the only efficient possibility to treat the above mentioned problem. If it is possible to find the dependence between life time distribution and the stress level under which the items are tested and have to work, results found for high stress levels within a comparatively short time of test may be used to describe the life time distribution under a low usual stress. A rough survey is given by Strelec (1984), more details are found in a monography by Viertl (1987).

Of course life testing and especially accelerated life testing is performed to get information about the whole life time distribution. But when considering the decomposition problem at life time distributions with bathtub like hazard rate function most attention is put to that part of life time distribution which corresponds to early failures. Questions are of great interest concerning the portion of early failures and life time characteristics of that set of items. But it is almost impossible to get estimates for these parameters at usual stress and highly censored sampling because almost only early failures could be observed during this starting phase of sampling and these would then influence any estimation of life time characteristics very strongly and make them senseless. On the other hand sensibly organized accelerated life testing can reduce observed life times to a great amount so that it is possible to get information not only by the early part of the life time distribution but instead of that information is obtained from (almost) the whole range of the life time distribution. Therefore it is possible to find good estimations for the distribution of early failures (i.e. the first component of the mixture) for high stress levels which may be evaluated for usual stress by known relationships.

In order to describe the above mentioned relationship of life time characteristics and the stress level under which a certain device has to work parametric models for all life time and mixture characteristics of the underlying mixture distribution (3) are assumed in the following way:

(i) The portions $p_i$ are stress independent for i=1,2,3, that means that there is (almost) no influence to the proportion of failure types caused by the underlying stress level.

(ii) For the scale parameter $\tau_i = \tau_i(\underset{\sim}{s})$ generally a model of the form

$$\tau_i(s) = \tau_i(c_{i1},\dots,c_{ir_i};\underset{\sim}{s}) \qquad (6)$$

with some (stress independent) parameters $c_{i1},\dots,c_{ir_i}$ (i=1,2,3) is given. Examples are given by some of the classical parametric models like Arrhenius model, Eyring model or power rule.

(iii) The shape parameters $\eta_i$ are described by

$$\eta_i(\underset{\sim}{s}) = \exp(b_i \|\underset{\sim}{s}\|) \qquad (7)$$

with $b_1 < 0$, $b_2 = 0$, $b_3 > 0$ so that $\eta_i$ may either increase or decrease with raising stress level. Of course this simple model is admissible only in a reasonably (that means practically) constrained area of stress levels which is no real constraint because in life time analysis generally acceleration models of course are not appropriate for the whole area of possible but not senseful stress levels.

Within the above relations $\underset{\sim}{s} = (s_1,\dots,s_k)'$ is the vector of stress components (like voltage, temperature, humidity or the like) and $\|\underset{\sim}{s}\|$ stands for some appropriate norm for this stress vector (e.g. Euclidian, weighted sum of abolute components or similar). A slight simplification of the above model may be achieved when (iii) is substituted by

(iii') The shape parameters $\eta_1$ and $\eta_3$ are stress independent.

A further great advantage of the above model especially in connection with Bayesian analysis is given by the fact that it is not necessary to consider an own prior distribution for any of the parameters $\tau_i(\underset{\sim}{s})$ and probably $\eta_i(\underset{\sim}{s})$ (i=1,2,3) for each stress level $\underset{\sim}{s}$ what is tedious if there are many different stress levels at which life time data are observed. Contrary to that the above model reduces the number of unknown parameters to $c_{ij}$ and $b_i$ for which prior distributions are necessary anylonger. So life data under different stress levels may be analysed easily within one single step.

## BAYESIAN ANALYSIS IN ACCELERATED LIFE TESTING

For the following analysis the power rule model

$$\tau_i(\underset{\sim}{s}) = c_{i1}\cdot\|\underset{\sim}{s}\|^{c_{i2}} \qquad c_{i1}, c_{i2} > 0 \tag{8}$$

is assumed for the scale parameter $\tau_i$ (i=1,2,3) so that the number m of parameters to be estimated is at most 10, namely

2 for the mixture portions $p_1$ and $p_2$,
6 for the $c_{i1}, c_{i2}$ (i=1,2,3),
2 for the $b_1$ and $b_3$ for the shape parameters $\eta_1$ and $\eta_3$.

This analysis is based on a set of r ($\geq$m) samples $\underset{\sim}{t}_l = (t_{l1},\ldots,t_{ln_l})'$ of life data observed at life tests under stress level $\underset{\sim}{s}_l$. The choice of these stress levels depends on economic and precision arguments and is a problem of experimental design.

The likelihood function for such a set of life data received from uncensored experiments is then given by

$$l(\underset{\sim}{t}_1,\ldots,\underset{\sim}{t}_r | p_1,p_2,p_3;b_1,b_3;c_{11},c_{12},c_{21},c_{22},c_{31},c_{32}) = \tag{9}$$
$$= \prod_{l=1}^{r}\prod_{\nu=1}^{n_l}\sum_{i=1}^{3} p_i\left(\frac{\eta_i(\underset{\sim}{s}_l)}{\tau_i(\underset{\sim}{s}_l)}\right)\left(\frac{t_{l\nu}}{\tau_i(\underset{\sim}{s}_l)}\right)^{\eta_i(\underset{\sim}{s}_l)-1} .\exp(-(t_{l\nu}/\tau_i(\underset{\sim}{s}_l))^{\eta_i(\underset{\sim}{s}_l)})$$

with

$$\eta_i(\underset{\sim}{s}_l) = e^{b_i\|\underset{\sim}{s}_l\|}$$

and

$$\tau_i(\underset{\sim}{s}_l) = c_{i1}\cdot\|\underset{\sim}{s}_l\|^{c_{i2}}$$

for i=1,2,3 . This typical form of likelihood for mixture models makes sufficiency completely impossible so that only numerical solutions can be achieved.

A set of assumptions is put to possible prior distributions for the parameters of the considered model (9) which should be taken into account at the first step of analysis. These are

(P1) A Dirichlet distribution $D(\underset{\sim}{\alpha})$ for the portions $p_i$ (i=1,2,3) with $\underset{\sim}{\alpha} = (\alpha_1,\alpha_2,\alpha_3)'$ is assumed so that

$$f(p_1,p_2,p_3) = \Gamma(\alpha_o) \prod_{i=1}^{3} \frac{p_i^{\alpha_i - 1}}{\Gamma(\alpha_i)} \qquad \text{with } \alpha_o := \alpha_1+\alpha_2+\alpha_3 .$$

(P2) The domain of $\eta_1$ is restricted to $0 < \eta_1 < 1$ so that $b_1 < 0$ should hold. In the same way $\eta_3 > 1$ resp. $b_3 > 0$ has to be fulfilled. Gamma distributions $\gamma(\alpha_1,\beta_1)$ and $\gamma(\alpha_3,\beta_3)$ for $|b_1|$ and $b_3$ seem to be a good approximation for prior distributions. For the first step $\alpha_1 = \alpha_3 = \beta_1 = \beta_3 = 1$ is suggested.

(P3) The shape parameters $\eta_1$ and $\eta_3$ are mutually independent and independent from all the other parameters.

(P4) The prior distributions for the parameters $c_{ij}$ are independent for different failure modes and restricted to positive real numbers. For the first step uniform priors for suitable domains could be assumed.

Considering the above noted assumptions on the prior distributions for the 10 parameters of the model (9) the posterior distribution of these parameters given the set $(\underset{\sim}{t}_1,\ldots,\underset{\sim}{t}_r)$ of samples at different stress levels is of the form

$$f(\underset{\sim}{p},\underset{\sim}{b},\underset{\sim}{c}|\underset{\sim}{t}_1,\ldots,\underset{\sim}{t}_r) \propto l(\underset{\sim}{t}_1,\ldots,\underset{\sim}{t}_r|\underset{\sim}{p},\underset{\sim}{b},\underset{\sim}{c}) \times$$
$$\times f(p_1,p_2,p_3).f(b_1).f(b_3). \prod_{i=1}^{3} f(c_{i1},c_{i2}) \qquad (10)$$

with $\underset{\sim}{p} = (p_1,p_2,p_3)'$, $\underset{\sim}{b} = (b_1,b_3)'$, $\underset{\sim}{c} = (c_{11},c_{12},c_{21},c_{22},c_{31},c_{32})'$. This relation is of a form where only numerical methods are possible for the solution. Of course the above mentioned assumptions are not valid for the posterior distribution anylonger. Because of the complexity of the model the assumptions about independence of some parameters are at least slightly disturbed and the posterior density cannot be factorized sensefully anylonger. Then numerical analysis would become slower because of the given number of function calls a computer program would need and therefore Bayesian analysis would lose much of attractivity.

One possibility to treat the last mentioned problem is to make use of the fact that generally speaking dependencies caused by evaluating the posterior distribution (10) are comparatively small. Doing so the assumptions (P1) - (P4) have only to be updated by the information given by the posterior distribution. For instance updating of the assumed distribution for $\underset{\sim}{p}$ is possible by finding a good approximation of the posterior distribution by an appropriate Dirichlet distribution. If equal mean values and variances for $\underset{\sim}{p}$ are required new values for $\underset{\sim}{\alpha}$ are given by

$$\alpha_{new,i} = \bar{p}_i . k \qquad i=1,2,3$$

where

$$k = \sum_{i=1}^{3} \bar{p}_i^{-2}(1-\bar{p}_i)^2 \Big/ \sum_{i=1}^{3} \sigma_{p_i}^2 . \bar{p}_i(1-\bar{p}_i) - 1$$

with

$$\bar{p}_i = E(p_i|\underset{\sim}{t}_1,\ldots,\underset{\sim}{t}_r) \quad \text{and} \quad \sigma_{p_i}^2 = Var(p_i|\underset{\sim}{t}_1,\ldots,\underset{\sim}{t}_r) .$$

In a similar way the posterior distribution for $b_1$ and $b_3$ may be redefined. If the same requirements are put to the approximations as above $\alpha$ and $\beta$ of the Gamma-distributions are given by

$$\alpha_{new,i} = \sigma_{b_i}^2 / \bar{b}_i$$

and $i=1,3$ .

$$\beta_{new,i} = \bar{b}_i^2 / \sigma_{b_i}^2$$

If needed the distribution of c may be redefined too in the same way. Any further analysis like Bayesian point or interval estimation or the determination of prediction intervals (e.g. for early failures) may now be continued.

The integration procedure for the evaluation of posterior distributions and posterior characteristics may be of a simple type because of the smooth form of prior distributions and likelihood without any singularities or other difficulties. An appropriate procedure is given in the following way:

- Define a multidimensional rectangle (at most 10 dimensions) so that the posterior distribution (10) vanishes (at least almost) outside this area.
- Take a sequence of grids where each of them is created from the preceding one by doubling the number of points of support.
- Evaluate an approximation to the integral for any grid using rectangular rule.
- Perform a modified Rhomberg-integration by interpolating those approximate integral values by rational functions using Stoer-algorithm (see Stoer, 1972).

This method may be applied within a broad spectrum of computer configurations. If memory is great enough so that a great part of function values may be stored computer time can be reduced significantly especially when using the redefinitions of the posterior distribution. On the other hand such a program works also at an AT-compatible personal computer without any memory demands but with a comparatively great amount of computer time. In any case senseful results can be achieved by a senseful effort.

## REFERENCES

Berger, J. O., 1985, "Statistical Decision Theory and Bayesian Analysis," 2nd ed., Springer-Verlag, New York/Heidelberg/Tokyo.

Hjorth, U., 1980, A Reliability Distribution With Increasing, Decreasing, Constant and Bathtub-Shaped Failure Rates, Technometrics, 22:99.

Stoer, J., 1972, "Einführung in die Numerische Mathematik," Springer-Verlag, Berlin/Heidelberg/New York.

Strelec, H., 1984, Zeitraffende Zuverlässigkeitsanalyse, Res. Report TS-1984-2, Institut für Statistik und Wahrscheinlichkeitstheorie, Technische Universität Wien.

Strelec, H. and Viertl, R., 1983, Über die Schätzung von Koeffizienten bei Beschleunigungsfunktionen, Res. Report TS-1983-1, Institut für Statistik und Wahrscheinlichkeitstheorie, Technische Universität Wien.

Viertl, R., 1987, "Statistical Methods in Accelerated Life Testing," Vandenhoeck & Ruprecht, Göttingen.

# ROBUST BAYESIAN METHODS

Daniel Thorburn

Department of Statistics
University of Stockholm
Stockholm, Sweden

## 1. INTRODUCTION

With robust statistics we mean methods that work well, if a chosen model is true and that are acceptable if the model is only an approximation. But if the model is far from the true one robust methods may be very bad (Huber 1980, Hampel & al 1986). Thus robust statistics should be used whenever we know that the chosen model is only an approximation to the true model.

A true Bayesian should describe his beliefs in his prior. He should then use his prior to compute his posterior. The uncertainty about a probability distribution (density function) could be modelled in the following way. The true distribution, $\Psi(x,\theta)$, is the product of a chosen model $f_0(x|\theta)$ and a small random multiplicative noise $\exp(\varepsilon(x))$. The noise $\varepsilon(x)$ is a priori a stochastic process, which fluctuates around zero. Note that the distribution of the noise may be modelled to depend on the unknown parameter $\theta$, and that both $x$ and $\theta$ may be multidimensional.

In a normal Bayesian model the parameter $\theta$ has a prior, $\pi(\theta)$, and the objective of the calculations is to compute the posterior $\pi(\theta|x_1,x_2,\ldots,x_n)$. The posterior is proportional to

$$\begin{aligned}&\pi(\theta)\ E\{f(x_1,x_2,\ldots,x_n|\theta)|x_1,x_2,\ldots,x_n,\theta\} = \\ &\pi(\theta)\ \Pi_1^n\ f_0(x_i|\theta)\quad E\{\exp(\Sigma_1^n\varepsilon(x_i))|x_1,x_2,\ldots,x_n,\theta\}\end{aligned} \tag{1.1}$$

We have here assumed that the observations are exchangeable or, in other words, that they are independent given $\varepsilon$ and $\theta$. In this paper we do not consider robustness against the exchangeability assumption. Nor do we consider robustness against misspecifying the prior.

The expression (1.1) is often difficult to compute. In this paper we shall derive a first order approximation, which is good as long as the noise $\varepsilon(x)$ is small. We shall also give some examples of its performance in the location parameter case.

## 2. THE GENERAL CASE

If $\varepsilon$ had been a Gaussian process with mean $m(x)$ and covariance $\sigma(x,y)$, the expected value in (1.1) would have been exactly

$$\exp\{E(\Sigma_1^n\varepsilon(x_i)) + Var(\Sigma_1^n\varepsilon(x_i))/2\} = \exp\{\Sigma_1^n m(x_i) + \Sigma_1^n\Sigma_1^n\sigma(x_i,x_j)/2\}. \quad (2.1)$$

Unfortunately, $\varepsilon$ cannot be exactly Gaussian, since

$$\int f_0(x|\theta)\exp(\varepsilon(x))dx = 1 \quad (2.2)$$

is a non-linear restriction on the sample space. But if $\varepsilon$ is small, this restriction can approximately be replaced by the linear restriction

$$\int f_0(x|\theta)\varepsilon(x)dx = 0. \quad (2.3)$$

In the following we will assume that $\varepsilon$ is a Gaussian process subject to the condition (2.2). Thorburn (1986) showed that this is a valid prior distribution. We also assume that the covariance function is so small that (2.2) can be replaced by (2.3). Under that condition we get from standard theory of conditional normal distributions

$$m'(x) = E(\varepsilon(x)|\cdot) = m(x) - \frac{\int\sigma(x,s)f_0(s|\theta)ds\int f_0(s|\theta)m(s)ds}{\int\int\sigma(s,t)f_0(s|\theta)f_0(t|\theta)dsdt} \quad (2.4a)$$

$$\sigma'(x,y) = Cov(\varepsilon(x),\varepsilon(y)|\cdot) = \sigma(x,y) - \frac{\int\sigma(x,s)f_0(s|\theta)ds\int\sigma(y,s)f_0(s|\theta)ds}{\int\int\sigma(s,t)f_0(s|\theta)f_0(t|\theta)dsdt}. \quad (2.4b)$$

These approximations and (2.1) give the following result.

Result 2.1 Let $x_1,x_2,\ldots,x_n$ be independent and identically distributed random variables given the density $\Psi(x,\theta)=f_0(x|\theta)\exp(\varepsilon(x))$. Further $\theta$ has a prior $\pi(\theta)$ and $\varepsilon$ is a small Gaussian process with mean value function $m(x)$ and covariance $\sigma(x,y)$ conditioned by $\int\Psi(x,\theta)dx=1$. The posterior of $\theta$ is then approximately proportional to

$$\pi(\theta)\ \Pi_1^n f_0(x_i|\theta)\exp\ (\Sigma_1^n m'(x_i)+\Sigma_1^n\Sigma_1^n\sigma'(x_i,x_j)),$$

where $m'$ and $\sigma'$ are given by (2.4).

With this formulation $\theta$ is not a unique function of the true density $\Psi(x,\theta)$, since the random fluctuations may take different $f_0(x|\theta)$ into the same $\Psi(x,\theta)$. In many applications, however, $\theta$ should be uniquely determined by the true distribution, e.g. be its mean or median. This can easily be solved by adding further conditions to the distribution of the noise, e.g.

$$\int sf_0(s|\theta)\exp(\varepsilon(s))ds = \theta \quad (2.5)$$

or

$$\int^{\theta} f_0(s|\theta)\exp(\varepsilon(s))ds = 0.5. \quad (2.6)$$

Let $C$ denote all the conditions imposed on $\varepsilon$, (e.g. (2.2) and (2.5)), and $C'$ be the corresponding linearized versions (e.g. (2.3) and $\int sf_0(s|\theta)\varepsilon(s)ds = 0$). Correspondingly we let $m_C'$ and $\sigma_C'$ be the conditional mean and covariance of $\varepsilon$ given $C'$. We then have the more general result that the posterior is approximately proportional to

$$\pi(\theta)\Pi_1^n f_0(x_i|\theta)\exp\ (\Sigma_1^n m_c'(x_i)+\Sigma_1^n\Sigma_1^n\sigma_c'(x_i,x_j)/2). \tag{2.7}$$

Remark 2.1 This is an approximation, which can be used only when the deviations from the model $f_0$ are small. What is meant by small may depend on the size of the sample. If $n$ is five the noise $\varepsilon$ may be bigger than if $n$ is one hundred. It should also be noted that the approximation may be very bad when $\theta$ is far from $\hat{\theta}$. As a consequence the approximated posterior may falsely become large far out in the tails.

## 3. THE GAUSSIAN LOCATION PARAMETER CASE

In this section we assume that

$$f_0(x|\theta) = \exp(-(x-\theta)^2/2)/\sqrt{2\pi},$$

and that the noise $\varepsilon(x+\theta)$ is independent of $\theta$ a priori. For simplicity we assume the diffuse prior for $\theta$ i.e. $\pi(\theta)\propto 1$.

The posterior distribution is thus approximately proportional to

$$\exp(-\tfrac{1}{2}\Sigma_1^n(x_i-\theta)^2+\Sigma_1^n m_c'(x_i-\theta)+\tfrac{1}{2}\Sigma_1^n\Sigma_1^n\sigma_c'(x_i-\theta,x_j-\theta)),$$

where $m_c'$ and $\sigma_c'$ are the conditional mean and covariance function of $\varepsilon$.

A natural choice of the mean value function, $m$, is zero, i.e. that the most likely log distribution is the model $\ln f_0$. Another choice is $m(x)=-\sigma^2(x)/2$ which implies that the expected value of the density is $f_0$. The second choice is probably not so good if the true distribution might have thicker tails than the normal one, since that choice implies that the most likely log density has thinner tails. We mostly use the first choice.

The choice of the covariance function requires a little more care. It should both reflect the prior opinion on possible departures from the model, $f_0(x,\theta)$, and be mathematically and computationally convenient. We will first give one simple example, that is unacceptable from many points of view but still gives some insight.

Example 3.1 "Stationary, differentiable noise":

$$\sigma(x,y)=\delta\exp(-k(x-y)^2).$$

If $C$ is the condition (2.2), simple calculations show that

$$\sigma_c'(x,y)=\delta\exp(-k(x-y)^2) - \frac{\delta\exp(-\frac{k}{2k+1}((x-\theta)^2+(y-\theta)^2))}{(2k+1)/(4k+1)^{0.5}}.$$

The exponent in the posterior is thus proportional to

$$\Sigma(-\tfrac{1}{2}(x_i-\theta)^2)-\frac{(4k+1)^{0.5}}{2(2k+1)}\delta(\sum_j\exp(-\tfrac{k}{2k+1}(x_j-\theta)^2))^2.$$

The maximum value is obtained for a $\hat{\theta}$, that is a weighted average of the observations $x_i$ with the following weights:

$$1-\frac{2k(4k+1)^{0.5}}{(2k+1)^2}\delta(\Sigma\exp(-\tfrac{k}{2k+1}(x_j-\hat{\theta})^2))\ \exp(-\tfrac{k}{2k+1}(x_i-\hat{\theta})^2).$$

The observations far from $\hat{\theta}$ have thus larger weights than those near $\hat{\theta}$. When one is uncertain about the central part of the density, but knows that the tails should be similar to the normal distribution this covariance function can be used.

The variance $\sigma(x,x)$ should increase faster than $x^2$ in order to give smaller weights to the observation far from $x$. On the other hand the covariance function should not increase so fast that the far tails might contain a large part of the probability mass. If $\sigma(x,x)>x^2n^2$ for large $x$, the posterior density gets false maxima in plus and minus infinity.

It is often believed that if the tail is thicker three standard deviations away, it is probably thicker than the model $f_0$ five or ten standard deviations away too. It is thus sensible to model larger correlations in the tails than in the centre of the distribution.

## 4. NUMERICAL EXAMPLES

The examples in this section are all computed numerically on a computer. We have not been able to find covariance functions which reflect all the properties we want to model, such that it is possible to do all the integrations exactly.

In the numerical examples below we have used the covariance function

$$\sigma(x,y)=a\left(\frac{(b+1)^2x'^2y'^2}{(b+x')(b+y')}\right)\exp\left(-c\left|\frac{(d+1)x}{d+x'}-\frac{(d+1)y}{d+y'}\right|\right),$$

where $x'=\max(1,|x|)$ and $y'=\max(1,|y|)$. The parameters of this function could be interpreted in the following way. The variance $\sigma(x,x)=a$ in the interval $(-1,1)$. The variance $\sigma(x,x)$ then increases roughly as $x^4$ immediately outside this interval. The rate of increase smoothly changes to $x^2$ at plus or minus infinity. The change takes place at about the points $b$ and $-b$. The correlation near origin decreases exponentially as $\exp(-c|x|)$. Finally the dependence of the extreme tails increases with $d$ and $c$. An approximate rule says that the tails start at $\pm cd^2$.

Our experience from the numerical computations is that the magnitude of the noise is the most important parameter, but that the posterior distribution is changed in the same direction for all levels of $a$. For large values of $a$ the approximations sometimes break down, particularly if $b$ is chosen large and $c$ small. However, it is very clear from the result, when the method works and when it does not. For most sample sizes and situations an $a$ below 0.01 seems to work. The other parameters can be chosen rather freely, without affecting the result too much. The best choice for robustness against outliers, seems to be a large $c$ and a small $cd^2$.

In the four examples below four different robust posteriors are computed. The first three assumes that $m=0$ and the fourth assumes that $m=-\sigma^2/2$. In the first and fourth case we only condition by (2.2). In the second and third case we also condition by (2.5) or (2.6) so that mean and median, respectively, is preserved.

Example 4.1 A correct model

One hundred normal random numbers were generated. Their sample mean and standard deviation were 4.90 and 0.94. The parameters of the prior covariance were chosen to be a=0.01, b=10, c=1 and d=5. The resulting posteriors are given in Fig. 4.1. The standard posterior and the robust ones, where the mean and median are fixed are impossible to distinguish from the figure. The two other robust posteriors are a little wider.

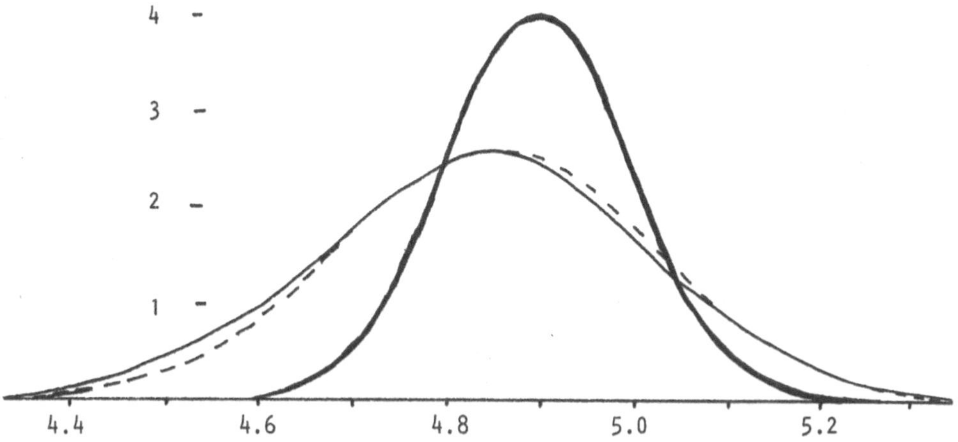


Figure 4.1 Posterior distributions of $\theta$ after 100 random normal numbers. Thick line: Standard normal and robust with median or mean fixed by the true density. Thin line: Robust with m=0. Broken line: Robust with $m=-\sigma^2/2$.

Example 4.2 Bimodal distribution

Twenty normal random numbers with mean zero and ten with mean two were generated. The sample mean and deviation became 0.71 and 1.44. The parameters of the prior mean and covariance were chosen to be a=0.01, b=5, c=2 and d=1. The resulting posteriors are given in Fig. 4.2. All the robust posteriors are flatter than the standard normal one, but they are still centered around the same point.

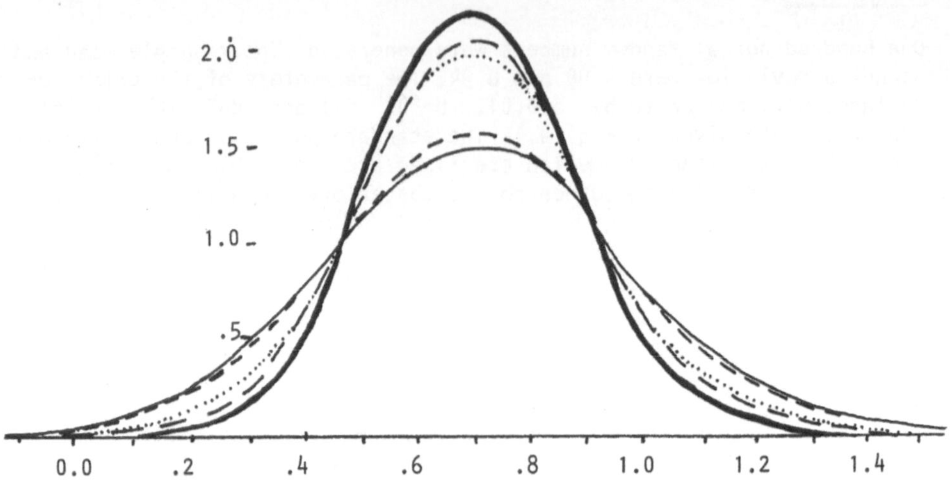


Figure 4.2 Posterior distributions after 30 random numbers from a mixed normal distribution with means 0 and 2. Thick line: Standard posterior. Thin line: Robust with m=0. Broken line (long dashes): Robust with fixed mean. Dotted line: Robust with fixed median. Broken line (short dashes): Robust with $m=-\sigma^2/2$.

Example 4.3 An outlier

We constructed a data set with one outlier: 0, 1, 1.5, 2, 2.5, 3, 3.5, 4, 5, 22.5. The sample mean is 4.5. The parameters of the prior covariance were chosen to be a=0.005, b=10, c=5 and d=0.8. The resulting posteriors are given in Fig. 4.3. In all the cases with m=0, the robust procedures moved the posterior to the left, i.e. they gave less weight to the outlier. The robust procedure with $m=\sigma^2/2$ did not give a sensible answer.

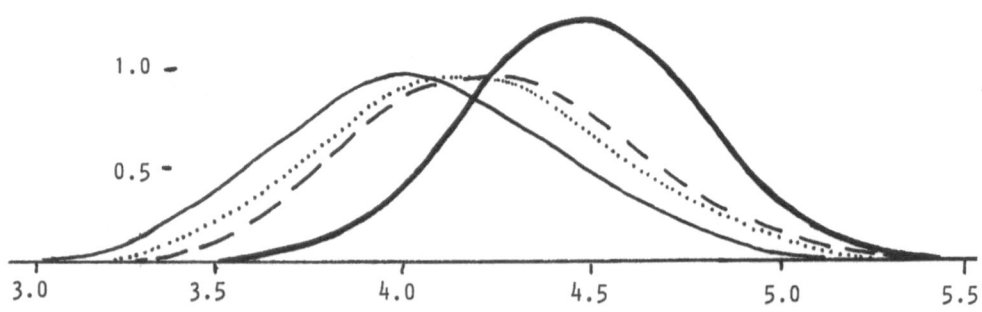

Figure 4.3 Posterior distributions after ten observations with one large outlier. Thick line: Standard posterior. Thin line: Robust with m=0. Broken line: Robust with fixed mean. Dotted line: Robust with fixed median.

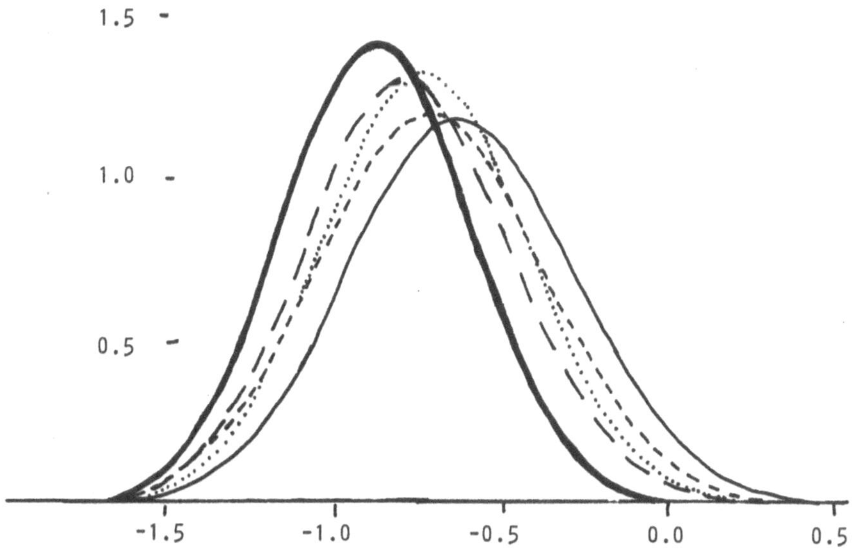


Figure 4.4 Posterior distributions after 50 Cauchy distributed random variables. Thick line: Standard normal posterior. Thin line: Robust with m=0. Broken line (long dashes): Robust with fixed mean. Dotted line: Robust with fixed median. Broken line (short dashes): Robust with $m=-\sigma^2/2$.

Example 4.4 Cauchy distribution

Fifty observations were generated from a standard Cauchy distribution. The extreme outliers were -23.10 and -18.67. The sample mean and deviation became -0.85 and 5.06. The parameters of the covariance function were chosen to be a=0.005, b=5, c=10 and d=0.5. The posteriors are given in Fig.4.4. All the robust procedures shift the posterior towards the median -0.20. An interval based on the 19th and 32nd ordered observation is (-0.83, 0.18).

These examples can be summarized in the following way. This method to obtain robust posterior distributions works quite well if the true model is in the neighbourhood of assumed model. However, in many situations where robust models are considered in the literature (Andrews & al 1972, Hampel & al 1985) the deviations are sometimes too large for the present method. In such cases the present method only indicates the direction in which the posterior should be shifted.

## 5. CONCLUDING REMARKS

In this paper only second order approximations were studied. It is possible to include fourth order corrections in Result 2.1. However, in order to get fully satisfactory results also for more distant alternatives other models must be considered.

In this paper numerical results were only given for the simple case with a standard normal distribution with known variance. The method works equally well for other one-parameter models, such as Poisson. The multi-dimensional case is numerically a little more tedious, since a multidimensional posterior must be computed, but no new theoretical problems are involved.

Consider a more general model e.g. the linear model where

$$E(Y_i|\theta) = \theta_0 + \theta_1 x_{1i} + \theta_2 x_{2i} + \ldots + \theta_k x_{ki}.$$

In that case it is not reasonable to assume that the distribution of $Y_i - E(Y_i|\theta)$ is dependent of $\theta$. This may be assumed as a first approximation in the underlying model, but not in the robust version. For that situation further developments must be made.

## REFERENCES


Andrews, D.F., Bickel, P.J., Hampel, F.R., Huber, P.J., Rogers, W.H. and Tukey, J.W., 1972, "Robust estimates of location: survey and advances". Princeton University Press, Princeton.

Hampel, F.R., Ronchetti, E.V., Rousseeuw, P.J. and Stahel, W.A., 1986, "Robust statistics - the approach based on influence functions". Wiley, New York.

Huber, P.J., 1981, "Robust statistics", Wiley, New York.

Thorburn, D., 1986, A Bayesian approach to density estimation. Biometrika 73:65.

# IS IT NECESSARY TO DEVELOP
# A FUZZY BAYESIAN INFERENCE ?

Reinhard Viertl

Institut für Statistik und Wahrscheinlichkeitstheorie
Technische Universität Wien
1040 Wien, Austria

## ABSTRACT

In applications data used for updating a-priori information are often fuzzy. These fuzzy data are usually not described by standard Bayesian inference. Statistical analysis has to take care of this fuzzyness which can be described by fuzzy numbers. Therefore the resulting fuzzyness of a-posteriori distributions has to be modelled and an analogue of predictive distributions under fuzzyness must be developed. Moreover for a fuzzy observation it is not always possible to decide if it is a member of a certain event. This kind of uncertainty states the following question: Is additivity for the measurement of uncertainty in general valid or a generalization of probability, postulating superadditivity, necessary.

## 1. INTRODUCTION

The elements of standard Bayesian statistical analysis with stochastic model $X \sim f(x|\theta)$, $\theta \in \Theta$ are

a) subjective a-priori information
b) objective data

where the a-priori information is expressed by the a-priori distribution and the data are considered as fixed numbers or vectors.

But in reality for nondiscrete observations usually fuzzyness is observed. This fuzzyness can be modelled by fuzzy numbers $x^*$ which are generalizations $\varphi(x)$ of numbers x and indicator functions $I_A(x)$. Typical examples of fuzzy numbers $x^* = \varphi(x)$ are given in figure 1 on the next page.

Fuzzy data $D^* = (\varphi_1(x),\dots,\varphi_n(x))$ are consisting of n fuzzy observations $\varphi_1(x),\dots,\varphi_n(x)$. This data set has to be used for statistical inference.

One could think of using probability densities instead of fuzzy observations. The reason why this is not generally reasonable is the probably insufficiency of probability measures to model uncertainty in general as described in section 3.

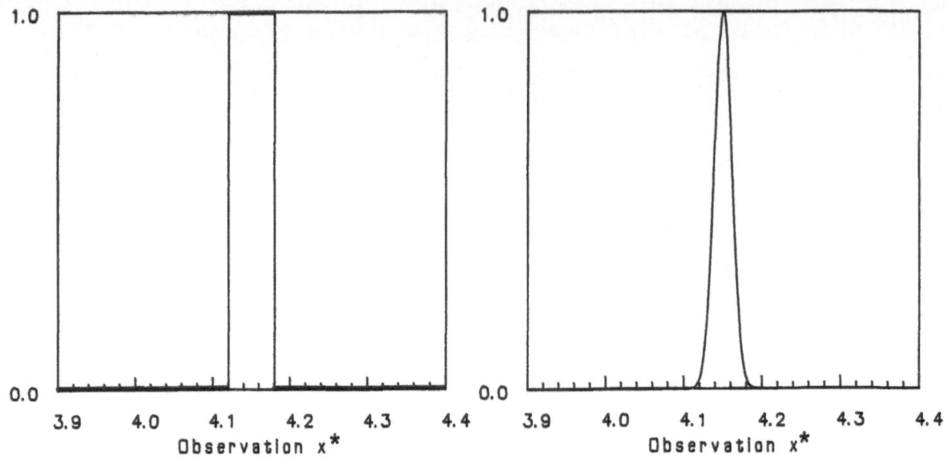


Figure 1. Examples of fuzzy numbers $x^* = \varphi(x)$ as model for fuzzy observations

## 2. BAYESIAN INFERENCE FOR FUZZY DATA

Starting with an a-priori distribution $\pi(\theta)$ for the parameter $\theta$ described by a stochastic quantity $\tilde{\theta}$ Bayes´ theorem

$$\pi(\theta|D) \propto \pi(\theta).1(\theta;D)$$

for precise data D can be used for fuzzy data $D^* = (\varphi_1(x),\ldots,\varphi_n(x))$ with $x \in \mathbb{R}$ in the univariate case.

One possibility is to use the combined fuzzy sample

$$\underline{x}^* = \varphi(\underline{x}) = \prod_{i=1}^{n} \varphi_i(x_i) \quad \text{for} \quad \underline{x} = (x_1,\ldots,x_n) \in \mathbb{R}^n .$$

For fixed $\theta$ by variation of $\underline{x}$ with corresponding grade of membership $\varphi(\underline{x})$ a fuzzy number $\psi_\theta(y)$, which forms the fuzzy value of $\pi^*(\theta|D^*)$ for the argument $\theta$, is obtained.

The fuzzy a-posteriori distribution $\pi^*(\theta|D^*)$ could be used for the construction of an analogue to HPD-regions. These regions will be fuzzy subsets of the parameter space $\Theta$. The construction and interpretation of fuzzy HPD*-regions would be an interesting problem.

Moreover it should be possible to use $\pi^*(\theta|D^*)$ for an analogue to predictive distributions. In order to do that an adaption of the equation

$$f(x|D) = \int_\Theta f(x|\theta)\pi(\theta|D)d\theta$$

for fuzzy a-posteriori distributions $\pi^*(\theta|D^*)$ is necessary.

## 3. FUZZY PROBABILITY MEASURES

Probability as a degree of believe that certain events occur is usually supposed to be additive. For fuzzy observations it is not deterministically decidable if an observation falls into an event or not. Therefore on the margins of an event A there may be uncertainty also after observation. In figure 2 on the next page this is depicted for one-dimensional observations.

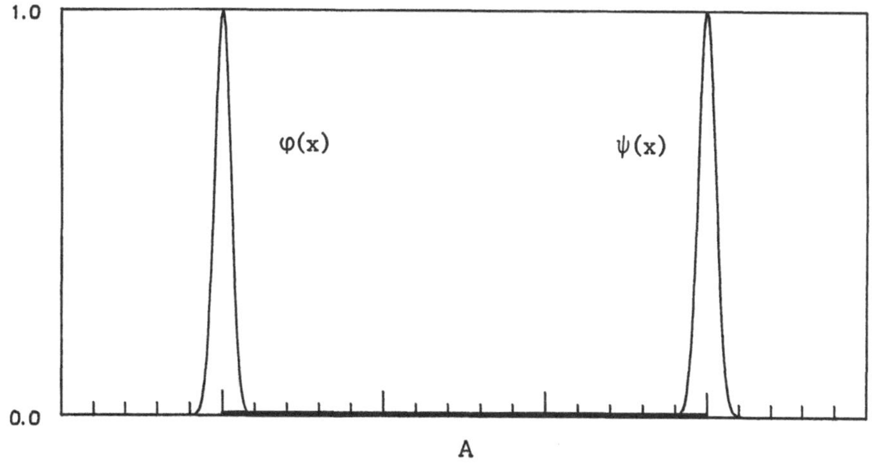


Figure 2. Uncertainty of membership for fuzzy observations φ(x) and ψ(x), and nonfuzzy event A

Using the example from figure 2 for two different events A and B it cannot be expected that probabilities as degrees of believe are always additive. This is explained in figure 3 and figure 4 for the union of the events A and B.

In figure 3 the uncertainty of membership of observations exists on all four endpoints of the two events A and B. For this situation additivity of degrees of believe seems to be an appropriate feature.

If the union of A and B becomes one interval, as in figure 4 on the next page, the uncertainty on the right end of A and on the left end of B vanishes. Therefore for a measure $\mu(.)$ of degree of believe that a fuzzy measurement is a member of AUB it is possible that

$$\mu(A \cup B) > \mu(A) + \mu(B) \ .$$

This is also supported by the superadditive nature of the relative frequency of observations which are certainly members of corresponding events.

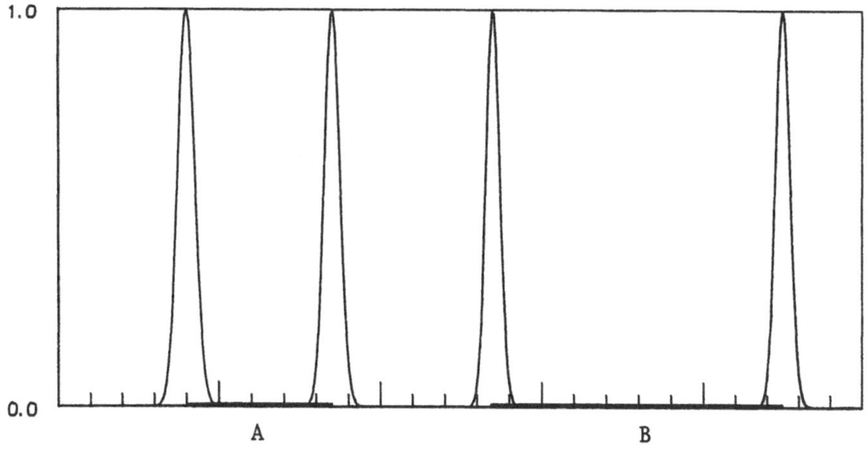


Figure 3. Uncertainty of membership after observation on four endpoints

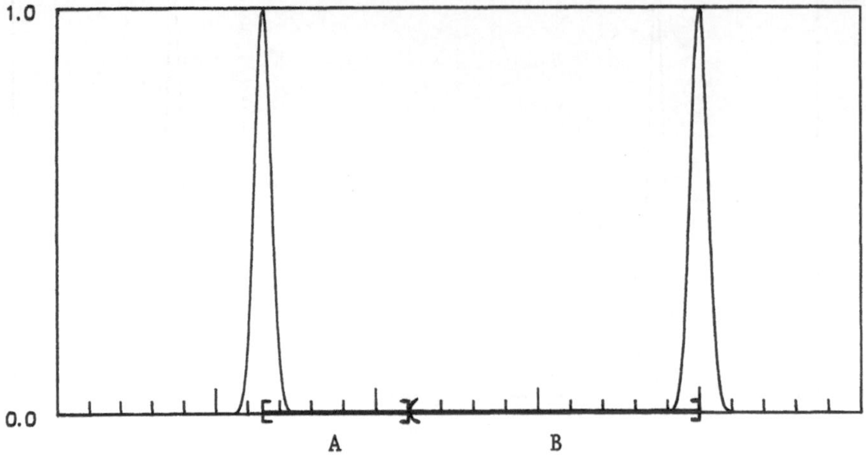


Figure 4. Argument against additivity of degrees of belief

Natural requirements for a measure of uncertainty in connection with fuzzy observations on a measurable space $(M,\mathfrak{A})$ are

(1) $\mu(\emptyset) = 0$ and $\mu(M) = 1$
(2) $A \subseteq B \Rightarrow \mu(A) \leq \mu(B)$
(3) $A_n \downarrow A \Rightarrow \mu(A_n) \rightarrow \mu(A)$.
(4) $A \cap B = \emptyset \Rightarrow \mu(A \cup B) \geq \mu(A) + \mu(B)$.

A set function $\mu: \mathfrak{A} \rightarrow [0,1]$ obeying the conditions (1) to (4) could be called a <u>fuzzy probability measure</u> . These fuzzy probability measures are special forms of so called fuzzy measures which are defined as set functions on a measurable space $(M,\mathfrak{A})$ fulfilling conditions (1) to (3) above (compare [2]).

Related to a dynamic interpretation of probability from the Bayesian view fuzzy probabilities should be subjective uncertainty judgements obeying the conditions (1), (2) and (4) above in conditional form and the coherence condition. Therefore <u>fuzzy subjective probabilites</u> $Pr(A|H)$ conditional on information H have to fulfill the following rules for general event systems A,B,E,....

($1^*$) $(B \Rightarrow A) \Rightarrow Pr(A|H) \leq Pr(B|H)$
($2^*$) (A disjunct B) $\Rightarrow P(A \wedge B|H) \geq Pr(A|H) + Pr(B|H)$
($3^*$) $Pr(A \wedge B|H) = Pr(A|B \wedge H) . Pr(B|H)$ .

In order to formalize the incorporation of new information to update fuzzy probability measures $\mu(.)$ a generalization of Bayes' theorem for fuzzy probability measures in form of an <u>information transformation formula</u> is necessary.

## 4. CONCLUSIONS

By the different problems in describing uncertainty it seems to be necessary to consider other measures than classical probability measures. Looking to the evolution of modelling real phenomena a path of development could be the following possible evolution diagram for modelling uncertainty which is depicted in figure 5 on the next page.

Deterministic Models
↓
True Stochastic Models
↓
Bayesian Models
↓
Fuzzy Bayesian Models

Figure 5. Evolution diagram for modelling uncertainty

REFERENCES


1. M. Eschbach, J. Cunningham: The logic of Fuzzy Bayesian Inference, Contributed paper at the international symposium on fuzzy information processing in artificial intelligence and operations research, Cambridge (1984).

2. T. Terano, M. Sugeno: Conditional Fuzzy Measures and Their Applications, in: "Fuzzy Sets and Their Applications to Cognitive and Decision Processes", L.A. Zadeh, K.S. Fu, K. Tanaka, M. Shimura, ed., Academic Press, New York (1975).

3. S.D. Unwin: A Fuzzy Set Theoretic Foundation for Vagueness in Uncertainty Analysis, Risk Analysis, Vol.6, No.1: 27 (1986).

# A PREDICTIVE DENSITY CRITERION FOR SELECTING NON-NESTED LINEAR MODELS AND COMPARISON WITH OTHER CRITERIA


Hajime Wago — Tsukuba University, Sakura, Ibaraki 305, Japan

Hiroki Tsurumi — Rutgers University, New Brunswick, NJ 08903, U.S.A.


## 1. INTRODUCTION

The mean squared-errors of forecasts (MSEF) is a statistic used to evaluate post-sample prediction performance. The MSEF has been used as a descriptive measure, but its exact distribution can be derived either from a sample theoretical or from a Bayesian perspective if the MSEF is computed from a linear regression model. In this paper, Bayesian and sampling distributions of the MSEF are derived, and it is suggested that the MSEF may be used as a statistic for linear model selection. Using sampling experiments, we compare the MSEF criterion with other model selection criteria. The organization of the paper is as follows. In section 2, we give the Bayesian and sampling distributions of the MSEF. In section 3, after presenting Akaike's information criterion, AIC, [Akaike (1974)], Efron's confidence interval for the mean squared errors, the N- and J- tests, we make sampling experiments to compare the Bayesian MSEF criterion with these other criteria.

## 2. BAYESIAN AND SAMPLING DISTRIBUTIONS OF THE MSEF

Let the linear model be given by

$$y = X\beta + u, \tag{1}$$

where y is an $(n\times 1)$ vector of observations on the dependent variable, X is an $(n\times k)$ matrix of observations on the explanatory variables with rank k, u is an $(n\times 1)$ vector of error terms, and $\beta$ is a $(k\times 1)$ vector of unknown regression coefficients. Assume that $u \sim N(0,\sigma^2 I_n)$ and that $\beta$ is estimated by $\hat{\beta} = (X'X)^{-1} X'y$.

The mean-squared-error for the post-sample period, n+1,...,n+m is computed using the post-sample actual observations on y and X. Let $y_*$ and $X_*$ be, respectively, an $(m\times 1)$ vector and an $(m\times k)$ matrix of post-sample observations and assume that the rank of $X_*$ is min(m,k). Then the MSEF is

$$\text{MSEF} = \frac{1}{m}(\hat{y}_*-y_*)'(\hat{y}_*-y_*), \tag{2}$$

where $\hat{y}_* = X_*\hat{\beta}$. Given equation (1) and $\hat{\beta} = \beta + (X'X)^{-1} X'u$, equation

(2) can be written as

$$\text{MSEF} = \frac{1}{m}\epsilon_*'B'B\epsilon_* = \frac{1}{m}\sum_{i=1}^{m}\mu_i\epsilon_i^2, \tag{3}$$

where: $\epsilon_*=(u',u_*')'$, $B=(A,-I_m)$, $A=X_*(X'X)^{-1}X_*'$, and the $\mu_i$'s are the nonzero characteristic roots of B'B. The $\varepsilon_i$ are elements of $\varepsilon=c'\varepsilon_*$, where c is the matrix of characteristic vectors of B'B. In passing, let us note that the $\mu_i$'s are given by $\mu_i=1+\lambda_i$, $i=1,\ldots,m$ for $m\leq k$, and $\mu_i=1+\lambda_i$, $i=1,\ldots,k$; $\mu_i=1$, $i=k+1,\ldots,m$, for $m>k$, where $\lambda_i$ is the ith nonzero characteristic root of AA'.

Since $\epsilon_i \sim NID(0,\sigma^2)$, m·MSEF is a quadratic form in normal variables. The distribution of quadratic forms or ratios of quadratic forms has been investigated by many; some of the earlier works are by McCarthy (1939), von Neumann (1941), and Bhattacharyya (1943). Bhattacharyya (1954) and Hotelling (1948) employed Laguerre expansion, and Gurland (1953) and Johnson and Kotz (1970) refined further the convergent Laguerre expansions. In this paper we use the degenerate hyperbolic function, which is convenient for computational purposes. Theorem 1 belows summarizes the derivation.

Theorem 1: Let $x = m\cdot\text{MSEF}/\sigma^2$. Then the distribution of x is given by

$$f(x) = \frac{e^{-x/2\mu_m}\, x^{1/2m-1}}{2^{1/2m}\,\pi^{1/2}\prod_{i=1}^{m}\mu_i^{1/2}}\sum_{p=0}^{\infty} c(m,p)\, x^p \tag{4}$$

where c(m,p) is the recursive coefficient given by

$$c(m,p) = \frac{\Gamma(p+\frac{m-1}{2})}{\Gamma(p+\frac{m}{2})}\sum_{j=0}^{p}\frac{c(m-1,j)\, a_m^{p-j}}{(p-j)!}, \text{ for } m \geq 2, \tag{5}$$

and $c(1,0)=1$, $c(1,j)=0$ for $j\geq 1$; $a_m = \frac{1}{2}(\frac{1}{\mu_m} - \frac{1}{\mu_{m-1}})$, $\mu_1 \geq \mu_2 \geq \cdots \geq \mu_m$, for $m \geq 2$, $a_1 = \frac{1}{2\mu_1}$, and $a_i^0 = 1$ for all $i=1,\ldots,m$.

Remark 1: If m > k, then equation (4) becomes

$$f(x) = \frac{e^{-x/2}\, x^{m/2-1}}{2^{m/2}\sqrt{\pi}\prod_{i=1}^{k}\mu_i^{1/2}}\sum_{p=0}^{\infty}\frac{\Gamma(\frac{k}{2}+p)}{\Gamma(\frac{m}{2}+p)}\sum_{j=0}^{p}\frac{c(k,j)\, a_{k+1}^{p-j}\, x^p}{(p-j)!} \tag{6}$$

where $a_{k+1} = \frac{1}{2}(1 - \frac{1}{\mu_k})$.

Remark 2: Equation (4) is the pdf of $x = m \cdot MSEF/\sigma^2$. The pdf of $z$=MSEF may be obtained by transforming $z = (\frac{\sigma^2}{m})x$, and this becomes

$$f(z|m,\sigma^2) = c_1\, m^{1/2m} z^{1/2m-1}\, \sigma^{-m} \exp(-\frac{m}{2\mu_m} z\, \sigma^{-2}) \sum_{p=0}^{\infty} c(m,p) m^p z^p \sigma^{-2p} \tag{7}$$

where $c_1$ is the constant given by $c_1 = 1/(2^{1/2m}\, \pi^{1/2} \prod_{i=1}^{m} \mu_i^{1/2})$.

Equations (4) and (7) have upper and lower bounds that are chisquare distributions, and this is stated in the following lemma.

Lemma 1: Let the pdf of z be denoted by $f(z|m,\sigma^2)$ as in equation (7). Then

$$\frac{\Gamma(1/2)}{\Gamma(\frac{m}{2})} c_1 m^{1/2m} z^{1/2m-1}\, \sigma^{-m} \exp(-\frac{mz\sigma^{-2}}{2\mu_m}) \leq f(z|m,\sigma^2)$$

$$\leq \frac{\Gamma(1/2)}{\Gamma(\frac{m}{2})} c_1\, m^{1/2m} z^{1/2m-1}\, \sigma^{-m} \exp(-\frac{mz\sigma^{-2}}{2\mu_1}). \tag{8}$$

A predictive density of the MSEF will be given by

$$p(z|\text{data}) = \int_0^{\infty} f(z|\sigma^2,m)\, p(\sigma^2|\text{data})\, d\sigma^2 \tag{9}$$

where $p(\sigma^2|\text{data})$ is the posterior pdf of $\sigma^2$, which may be given by

$$p(\sigma^2|\text{data}) \propto \sigma^{-(\nu+1)} \exp(-\frac{\nu s^2}{2\sigma^2}) \tag{10}$$

where $\nu = n-k$, and $\nu s^2 = y'(I - X(X'X)^{-1}X')y$. Carrying out the integration in (9) we obtain

$$p(z|s^2,\nu,m) \propto \frac{z^{1/2m-1}}{(\nu s^2 + mz/\mu_m)^{(m+\nu)/2}} \cdot \sum_{p=0}^{\infty} \Gamma(\frac{m+\nu}{2} + p)\, 2^p\, c(m,p) \left[\frac{mz}{\nu s^2 + mz/\mu_m}\right]^p \tag{11}$$

Using equation (8) we can show that the predictive pdf of z is bounded by two F distributions:

$$c_1 \frac{\Gamma(1/2)\Gamma(\frac{m+\nu}{2})}{\Gamma(\frac{m}{2})} 2^{(m+\nu)/2-1} \frac{m^{1/2m} z^{1/2m-1}}{(\nu s^2 + mz/\mu_m)^{(m+\nu)/2}}$$

$$\leq p(z|s^2,\nu,m) \leq c_1 \frac{\Gamma(1/2)\Gamma(\frac{m+\nu}{2})}{\Gamma(\frac{m}{2})} 2^{(m+\nu)/2-1} \frac{m^{1/2} z^{1/2m-1}}{(\nu s^2 + mz/\mu_i)^{(m+\nu)/2}} \tag{12}$$

The equalities hold if $\mu_m = \mu_i$. When m=1, the predictive density of the squared root of the MSEF is identical to the predictive density for one period ahead forecast [Zellner (1971, pp.72-73)], which in turn is equal to the predictive density in the sampling theory framework.

In Theorem 1 we used the degenerate hyperbolic function. The distribution of quadratic forms is often given by the Laguerre polynomials. If we rearrange the Laguerre expansions given in Johnson and Kotz (1970, pp.159-160) to fit more conveniently in our case, the distribution of $x = m \cdot MSEF/\sigma^2$ is given by

$$f(x) = \frac{1}{\beta} p_m(x/\beta,1)\Gamma(\frac{m}{2}) \sum_{p=0}^{\infty} \frac{(-2)^{-p}}{p!\Gamma(p+1/2m)} \cdot \left( \sum_{j=p}^{\infty} \frac{j!}{(j-p)!} c_j \beta^{-j} \right) (x/\beta)^p \tag{13}$$

## 3. COMPARISON OF CERTAIN MODEL SELECTION CRITERIA

The Bayesian predictive density of the MSEF that is given in (11) may be used as a criterion for selecting linear models. For each model we may draw the predictive density, and choose the model that has the mass of its density closest to zero. Or, we may choose the model that minimizes an expected loss. The choice of a quadratic loss function leads to the mean of the MSEF as the selection criterion.

The Bayesian criterion above belongs to the class of model selection criteria that are based on measures of how well each model explains data. Akaike's (1974) information criterion (AIC) and Efron's (1984) confidence interval for the mean squared errors also belong to this class.

Efron's confidence interval for the mean squared errors may be interpreted as an inferential procedure for the $C_p$ that is suggested by Mallows (1973). Let two linear regression models be given by

$$\begin{aligned} &\text{Model A:} \quad y = X_A\beta_A + \varepsilon \\ &\text{Model B:} \quad y = X_B\beta_B + \varepsilon \end{aligned} \tag{14}$$

where y is an (nx1) vector of observations on the dependent variable; $X_i$ is an $(nxk_i)$ matrix of observations on the $k_i$ explanatory variables of model i (i=A,B) and $\beta_i$ is a $(k_i x1)$ vector of regression coefficients of model i (i=A,B), and $\varepsilon$ is an (nx1) vector of error terms, The unbiased estimator of the difference of the mean squared errors (MSE) of models A and B, $\Delta = MSE_B - MSE_A$, is given by

$$\hat{\Delta} = (|y_{Bo}|^2 - |y_{Ao}|^2) + 2(d_B - d_A)\bar{\sigma}^2 \quad (15)$$

where $|y_{Bo}|^2 = \hat{\beta}_A' X_A' M_B X_A \hat{\beta}_A$, $|y_{Ao}|^2 = \hat{\beta}_B' X_B' M_A X_B \hat{\beta}_B$, $M_i = I - X_i(X_i'X_i)^{-1}X_i'$,

$i = A,B$, $\bar{\sigma}^2 = y'[I-XX^+]y$, $X = [X_A, X_B]$, and $d_i$ is the dimension of model i. Efron decomposes the MSE ($\Delta$) and its estimate ($\hat{\Delta}$) in a symmetric coordinate system and proposes to compute confidence intervals in the symmetric coordinate system. The computation of confidence intervals is suggested either by parametric bootstrapping or by non-parametric bootstrapping.

In contrast to the class of model selection criteria based on measures of 'goodness of fit', Cox's tests of separate families [Cox (1962)] are based on the translation of non-nested models into hypothesis testing on parameters. Pesaran (1974, 1982) proposes the N-tests. Davidson and MacKinnon (1981) suggest the J-tests. The N-tests are given by

$$N_0 = \frac{n}{2} \log (\hat{\sigma}_B^2/\hat{\sigma}_{BA}^2)/\{\frac{\hat{\alpha}_A^2}{\hat{\sigma}_{BA}^4} \hat{\beta}_A' X_A' M_B M_A M_B X_A \hat{\beta}_A\}^{1/2} \quad (16)$$

where $\hat{\sigma}_i^2 = y'M_i y/n$ (i=A,B), and $\hat{\sigma}_{BA}^2 = \hat{\sigma}_A^2 + (\hat{\beta}_A' X_A' M_B X_A \hat{\beta}_A)/n$. The $N_0$ test is computed using model A in (17) as the null hypothesis. By using model B as the null hypothesis, one obtains the $N_1$ test the formula of which is given by interchanging subscripts A and B in (19). The J test by Davidson and MacKinnon is the t-test on parameter $\lambda$ in

$$y = X_A b_A + \lambda(X_B \hat{\beta}_B) + u \quad (17)$$

where $b_A = (1-\lambda)\beta_A$. Again, a symmetric test can arise by interchanging subscripts A and B in (20).

The N- and J- tests give rise to cases where one either rejects or accepts both models. Table 1 gives four possible cases.

Table 1. Four Cases of N- and J- Tests

| | $N_0$ $(J_0)$ - Test | |
|---|---|---|
| $N_1$ $(J_1)$- Test | Case 1<br>Accept Model A<br>Reject Model B<br>$(p_1)$ | Case 2<br>Reject Model A<br>Accept Model B<br>$(p_2)$ |
| | Case 3<br>Reject Model A<br>Reject Model B<br>$(p_3)$ | Case 4<br>Accept Model A<br>Accept Model B<br>$(p_4)$ |

Note: $p_i$ is the probability of case i,(i=1,...,4).

As is obvious from equation (16), the N-tests can only be defined if $X_A'M_BM_AM_BX_A \neq 0$. Sufficient conditions for making this quantity zero are $M_BX_A = 0$ or $M_BX_A = X_A$. $M_BX_A = 0$ occurs if the columns of $X_A$ are linear combination of the columns of $X_B$, and $M_BX_A = X_A$ occurs if $X_B'X_A = 0$ (i.e. when the explanatory variables of the two models are orthogonal.) As for the J-tests, they cannot be defined if linear dependence exists between $X_A$ and $X_B$.

Let us make sampling experiments to compare the powers of the Bayesian MSEF criterion, the AIC, Efron's confidence interval, N-tests, and J-tests. In evaluating these tests we need to develop a measure of nearness of competing two models. Pesaran (1982) introduces a sequence of local alternatives

$$X_B = X_AC + n^{-1/2} D + o(n^{-1/2}) \tag{18}$$

where C and D are $k_A x k_B$ and $nxk_B$ nonzero matrices of constants, and $D'M_AD/n$ exists. Pesaran uses the local alternatives (18) so that he can derive asymptotic non-null distributions of the test statistics.

Instead of (18), the measure of nearness of two models may be given by the measure of correlation among non-overlapping explanatory variables of the two models. Let models A and B be written as

$$y = X_1\beta_{A1} + X_2\beta_{A2} + \varepsilon$$

$$y = X_1\beta_{B1} + Z\beta_{B2} + \varepsilon$$

so that $X_A = [X_1, X_2]$ and $X_B = [X_1, Z]$. The non-overlapping explanatory variables of the two models are $X_2$ and Z, and the measure of nearness of the two models may be given by

$$\rho_{AB}^2 = \text{Min}(\lambda_i^2)$$

where $\lambda_i^2$ is the square of the i-th nonzero cannonical correlation coefficient between $X_2$ and Z. $\rho_{AB}^2$ is bounded between 0 and 1, and if $\rho_{AB}^2 = 1$, the models A and B can be thought to be identical, whereas $\rho_{AB}^2 = 0$ indicates that the two models are farthest apart.

Sampling experiments are made by specifying the two models as

$$\text{Model A:}\quad y_t = \beta_0 + \beta_1 x_{t1} + \beta_2 x_{t2} + \varepsilon_t$$

$$\text{Model B:}\quad y_t = \gamma_0 + \gamma_1 x_{t1} + \gamma_2 z_{t2} + \varepsilon_t\ .$$

Hence, the models A and B have $(1, x_{t1})$ as the common variables, whereas $x_{t2}$ and $z_{t2}$ are uncommon variables. As in Pesaran's (1982) experiments, $x_{ti}$'s are drawn from N(0, 1), and $z_{t2}$ is generated by

$$z_{t2} = \lambda_2 x_{t2} + v_{t2}\ ,\ v_{t2} \sim N(0,1).$$

$\lambda_2$ is controlled by the correlation between $z_{t2}$ and $x_{t2}$:

$$\lambda_2 = \rho_2/(1-\rho_2^2)^{1/2}$$

where $\rho_2 = \mathrm{Corr}(x_{t2}, z_{t2})$.

The model selection criteria are also influenced by the 'fit' of the true model as measured by the coefficient of determination of the true model (model A in our experiments), $R^2$, and by the relative sizes of $\beta_1$ and $\beta_2$. In our experiments, we set $R^2$ at .5 ($R^2$=.5), and in Table 2 we set $\beta_1$ and $\beta_2$ to be (1.0, .5), respectively, whereas in Table 3 the values of $\beta_1$ and $\beta_2$ are switched: (.5, 1.0). The constant term $\beta_0$ is set at 1.0 in both tables. The number of replications for each value of $\rho_2^2$ is 500.

The following observations can be made from Table 2 and 3:

(1) As the sample sizes increase the powers of all the criteria tend to increase for given values of $\rho_2^2$.

(2) Comparing Table 3 with Table 2, we see that the powers in Table 3 are larger than those in Table 2.

(3) The N-test tends to perform better than the J-test. For $\rho_2^2$ = 0.1 or 0, the powers of the N-test decline. This is due to the fact that for low values of $\rho_2^2$, the nonoverlapping variables $x_{t2}$ and $z_{t2}$ tend to be orthogonal, and this brings the N-test closer to the case in which it is not defined (i.e. $X_A'X_B = 0$).

(4) The powers of the Bayesian MSEF criterion tend to dominate those of the other criteria, especially for the cases of sample size 20. For larger sample sizes, the AIC performs as good as the Bayesian MSEF criterion.

(5) Efron's 90% confidence interval (CI) appears to be too conservative, and for small sample sizes, the powers are substantially lower than the other criteria.

For the N- and J- tests we presented two measures of power, $p_1$ and 1-$\beta$ , respectively. The probability of Type II error is $\beta$, and 1-$\beta$ is the conventional concept of power in a nested hypothesis. As Pesaran (1974) states, however, for a non-nested hypothesis, a suitable concept of power is the probability of making correct decision, which is $p_1$. Pesaran (1982) uses 1-$\beta$ as the measure of power in his experiments.

Efron (1984) suggests a non-parametric bootstrapping procedure in addition to a parametric bootstrapping procedure. Since the non-parametric bootstrapping procedure requires considerable computational time in generating empirical powers, we did not carry it out in our experiments. Confidence intervals that are generated by non-parametric bootstrapping tend to be larger than those by parametric bootstrapping, and their powers are in general lower than those by parametric bootstrapping.

In our sampling experiments, we set the prediction period, m, at 10. We varied m at different values, and the results are comparable to those of m = 10.

Complete paper is available from authors upon request.

Table 2. Empirical Powers of Model Selection Criteria [$R^2$ = .5, $(\beta_1,\beta_2)$=(1.0, .5)]

| | $\delta$ | N-Test $p_1^{(2)}$ | N-Test $1-\beta^{(3)}$ | J-Test $p_1^{(2)}$ | J-Test $1-\beta^{(3)}$ | Difference of Predictive Means (4) | Difference of AIC's (5) | Efron's 90% CI (6) |
|---|---|---|---|---|---|---|---|---|
| | | | | n=20 | | | | |
| $\rho_2^2$=1.0 | 0 | ND(7) | ND(7) | ND(7) | ND(7) | 1(8) | 1(8) | 0(8) |
| .9 | 2 | .216 | .218 | .054 | .084 | .84 | .662 | 0 |
| .7 | 6 | .402 | .402 | .15 | .17 | .932 | .782 | .002 |
| .5 | 10 | .466 | .470 | .138 | .156 | .964 | .828 | .024 |
| .3 | 14 | .668 | .702 | .318 | .336 | .978 | .859 | .028 |
| .1 | 18 | .680 | .724 | .308 | .334 | .988 | .872 | .078 |
| 0 | 20 | .722 | .898 | .418 | .452 | .982 | .848 | .154 |
| | | | | n=60 | | | | |
| $\rho_2^2$=1.0 | 0 | ND | ND | ND | ND | 1 | 1 | 0 |
| .9 | 6 | .262 | .272 | .182 | .200 | .710 | .752 | 0 |
| .7 | 18 | .600 | .604 | .392 | .414 | .876 | .884 | .004 |
| .5 | 30 | .770 | .778 | .638 | .652 | .936 | .936 | .114 |
| .3 | 42 | .880 | .946 | .772 | .824 | .972 | .982 | .182 |
| .1 | 54 | .548 | 1 | .902 | .928 | .990 | .990 | .574 |
| 0 | 60 | .812 | 1 | .888 | .936 | .994 | .990 | .642 |
| | | | | n=100 | | | | |
| $\rho_2^2$=1.0 | 0 | ND | ND | ND | ND | 1.0 | 1.0 | 0 |
| .9 | 10 | .406 | .418 | .288 | .310 | .790 | .800 | 0 |
| .7 | 30 | .700 | .712 | .570 | .592 | .920 | .916 | .040 |
| .5 | 50 | .906 | .946 | .834 | .866 | .972 | .962 | .450 |
| .3 | 70 | .940 | 1.0 | .928 | .974 | .992 | .964 | .702 |
| .1 | 90 | .906 | 1.0 | .946 | .996 | 1.0 | 1.0 | .812 |
| 0 | 100 | .566 | 1.0 | .952 | .998 | 1.0 | 1.0 | .962 |

Table 3. Empirical Powers of Model Selection Criteria [$R^2 = .5$, $(\beta_1,\beta_2)=(.5, 1.0)$]

| | $\delta$ | N-Test $p_1$ | N-Test $1-\beta$ | J-Test $p_1$ | J-Test $1-\beta$ | Difference of Predictive Means | Difference of AIC's | Efron's 90% CI |
|---|---|---|---|---|---|---|---|---|
| | | | | n=20 | | | | |
| $\rho_2^2=1.0$ | 0 | ND | ND | ND | ND | 1 | 1 | 0 |
| .9 | 2 | .308 | .412 | .110 | .128 | .888 | .822 | 0 |
| .7 | 6 | .784 | .808 | .474 | .504 | .972 | .930 | .076 |
| .5 | 10 | .862 | .914 | .694 | .718 | .996 | .976 | .248 |
| .3 | 14 | .884 | .962 | .746 | .876 | 1.0 | .986 | .316 |
| .1 | 18 | .872 | .966 | .812 | .852 | 1.0 | .996 | .456 |
| 0 | 20 | .846 | .998 | .906 | .962 | 1.0 | 1.0 | .696 |
| | | | | n=60 | | | | |
| $\rho_2^2=1.0$ | 0 | ND | ND | ND | ND | 1.0 | 1.0 | 0 |
| .9 | 6 | .600 | .608 | .502 | .516 | .920 | .896 | .002 |
| .7 | 18 | .946 | .978 | .926 | .958 | .992 | .992 | .428 |
| .5 | 30 | .906 | .998 | .914 | .996 | .998 | .998 | .808 |
| .3 | 42 | .964 | 1.0 | .960 | .998 | 1.0 | 1.0 | .900 |
| .1 | 54 | .698 | 1.0 | .970 | 1.0 | 1.0 | 1.0 | .998 |
| 0 | 60 | .876 | 1.0 | .948 | 1.0 | 1.0 | 1.0 | 1.0 |
| | | | | n=100 | | | | |
| $\rho_2^2=1.0$ | 0 | ND | ND | ND | ND | 1 | 1 | 0 |
| .9 | 10 | .858 | .874 | .814 | .836 | .922 | .942 | .174 |
| .7 | 30 | .954 | .998 | .952 | .998 | .998 | .998 | .774 |
| .5 | 50 | .956 | 1.0 | .948 | 1.0 | 1.0 | 1.0 | .978 |
| .3 | 70 | .952 | 1.0 | .966 | 1.0 | 1.0 | 1.0 | .998 |
| .1 | 90 | .936 | 1.0 | .950 | 1.0 | 1.0 | 1.0 | 1.0 |
| 0 | 100 | .714 | 1.0 | .954 | 1.0 | 1.0 | 1.0 | 1.0 |

Notes: For each vale of $\rho_2$, the number of replications in 500.

(1) $\rho_2=\text{Corr}(x_{t2}, z_{t2})$, and $\delta$ is the measure of the distance of two models, D, in equation (21), and it is given by
$\delta = \lim_{n\to\infty} D'M_A D/\lim_{n\to\infty}(X_B'X_B/n)$. In our experimental design becomes
$\delta = n(1-\rho_2^2)$.

(2) $p_1$ is the probability of accepting model A and rejecting model B.

(3) $\beta$ is the probability of Type II errors, and it is given by $\beta = p_2+p_3$ in Table 1.

(4) The predictive mean is computed by $E(MSEF|\cdot) = \int zp(z|\text{data})dz$ for each model, and the difference is $E(MSEF_A|\cdot) - E(MSEF_B|\cdot)$.

(5) The difference of the AIC's is $AIC_A - AIC_B$.

(6) Efron's 90% confidence interval (CI) is computed by assuming that the sample estimate, $\bar{\sigma}^2$ is true (hence $d_E = \infty$ in Efron's notation), and by the Edgeworth expansions for the parametric bootstrap distribution without resorting to Monte Carlo.

(7) For $\rho_2^2 = 1.0$ the N- and J- tests are not defined.

(8) For $\rho_2^2 = 1.0$, the difference of the predictive means and the difference of the AIC's both become 1 by construction. Efron's CI becomes zero by construction.

(9) For all the sample sizes, the period of prediction, m, is set at 10.

REFERENCES


Akaike, H., 1974, A New Look at the statistical model identification, IEEE Transactions in Automatic Control, AC-19, 716-723.

Bhattacharyya, A., 1945, A Note on the distribution of the sum of chisquares, Sankhya, 7, 27-28.

Bhattacharyya, B.C., 1943, On an aspect of the Pearson system of curves, and a few analogies, Sankhya, 6, 415-448.

Davidson, R. and MacKinnon, J.G., 1981, Several tests for model specification in the presence of alternative hypotheses, Econometrica, 49, 781-793.

Efron, B., 1984, Comparing non-nested linear models, Journal of the American Statistical Association, 79, 791-803.

Gurland, J., 1953, Distribution of quadratic forms and ratios of quadratic forms, Annals of Mathematical Statistics, 24, 416-427.

Hotelling, H., 1940, The selection of variates for use in prediction with some comments on the general problem of nuisance parameters, Annals of Mathematical Statistics, 11, 271-283.

Hotelling, H., 1948, Some new methods for distributions of quadratic forms, (Abstract), Annals of Mathematical Statistics, 19, 119.

Johnson, N.L. and Kotz, S., 1970, Continuous univariate distributions - 2, John Wiley and Sons, New York.

Mallows, C., 1973, Some commets on $C_p$, Technometrics, 15, 661-675.

McCarthy, M.D., 1939, On the application of the z-test to randomized blocks, Annals of Mathematical Statistics, 10, 337-359.

Neumann, J. von., 1941, Distribution of the ratio of the mean square successive difference to variance, Annals of Mathematical Statistics, 12, 367-395.

Pesaran, M.H., 1974, On the general problem of model selection, Review of Economic Studies, 41, 153-171.

Pesaran, M.H., 1982, Comparison of local power of alternative tests of non-nested regression models, Econometrica, 50, 1287-1304.

Zellner, A., 1971, An introduction to Bayesian Inference in Econometrics, John Wiley and Sons, New York.

BAYESIAN MODELS AND METHODS FOR BINARY TIME SERIES

Mike West

Department of Statistics
University of Warwick
Coventry CV4 7AL
England

Julia Mortera

Dept. Statistics and Probability
Università degli Studi di Roma
00100 Roma
Italy

## 1. INTRODUCTION

From the perspective of applied statistical modelling, binary time series analysis and forecasting are relatively undeveloped areas. This paper reports on preliminary investigations of the use of some Bayesian models, discussing a variety of mathematical and practical modelling issues. A flexible class of models is that based on logistic linear regressions which, with an emphasis on sequential forecasting, are provided as a subset of the class of dynamic generalised linear models (West, Harrison and Migon, 1985). Special cases are Markov chains, considered here in detail, and non-stationary Markov chains with time evolving transition probabilities. In Sections 2 and 3 we discuss the use of low order Markov chains to model the non-Markov structure of binary series derived as qualitative summaries of underlying quantitative processes. An example concerns binary data indicating when a real valued process exceeds a specified threshold level. Such *clipped* processes arise naturally in monitoring problems in, for example, river flow and dam water level management; pollution emission regulation; clinical measurements such as blood pressure, in patient care; financial and economic time series forecasting; and so forth. In the context of an underlying gaussian process generated by a simple, yet widely used, dynamic linear model we show how simple Markov models can approximate derived binary processes.

In section 4 we extend the autoregressive Markov chain model to include independent variable information using logistic linear models. An example concerns forecasting the rise/fall (i.e. turning point) behaviour of a financial exchange rate series using external probability forecasts from an advisor as regressor information. The approach follows West (1986 a,b) and provides: (a) probability forecasting using external probabilities as independent variables; (b) data-based assessment of predictive accuracy of such external forecasts; and (c) *recalibration* of such forecasts to correct for systematic biases, optimism/pessimism, and deficiencies in autocorrelation structure.

## 2. STOCHASTIC STRUCTURE OF SIMPLE CLIPPED PROCESSES

To obtain insight into the structure of clipped processes we consider a real valued series following the simplest, yet most widely used and applicable, dynamic linear model, namely the first order polynomial. The series $Z_t$, $(t = 1,2,\ldots)$, is given by

$$Z_t = \theta_t + v_t, \qquad v_t \sim N[0, V];$$

$$\theta_t = \theta_{t-1} + w_t, \qquad w_t \sim N[0, VW],$$

with $V, W > 0$. Suppose we observe only the rise/fall behaviour of $Z_t$ via the indicator series

$$X_t = \begin{cases} 1, & \text{if } Y_t \geq 0; \\ 0, & \text{if } Y_t < 0, \end{cases}$$

where $Y_t = Z_t - Z_{t-1}$. $X_t$ is said to be obtained by clipping $Y_t$ at level 0; clearly $X_t = 1$ if and only if the Z series rises at time t. Two basic questions of interest are (A) can we model $X_t$ using simply estimated Markov models, and (B) if so what do these models say about the original model for $Z_t$?

A: Probability structure and Markov approximations

From the model for $Z_t$ it follows that the differenced series $Y_t$ is a stationary gaussian process whose autocorrelation structure is MA(1). In fact the lag-1 correlation is just $p = \text{corr}(Y_t, Y_{t-1}) = -1/(2 + W)$ and higher lag correlations are zero. Note that $-0.5 < p < 0$ whereas more general MA(1) processes have $-0.5 < p < 0.5$. Results in Kedem (1980) can be used to show that $X_t$ is stationary though not Markov with first and second order transition probabilities defined as follows. For $X_{t-1}$ and $X_{t-2}$ taking values 0 or 1, define quantities $P_i = P[X_t = 1|X_{t-1} = i]$ and $P_{ij} = P[X_t = 1|X_{t-1} = i, X_{t-2} = j]$, and let $\alpha = 0.5 + \arcsin(p)/\pi$, for $-1 < p < 1$. Then we can deduce, using the considerable symmetry in the model for $X_t$, that

$$P_1 = 1 - P_0 = \alpha, \quad P_{11} = 1 - P_{00} = 1-1/(4\alpha), \quad P_{10} = 1-P_{01} = 1/[4(1-\alpha)].$$

Note that these, and higher order, probabilities depend only on p, and also that this dependency is via $\alpha$ alone. Figure 1 displays $P_1$, $P_{10}$ and $P_{11}$ as functions of p (recall $|p|<0.5$ in our model); $P_0$, $P_{01}$ and $P_{00}$ are simply reflections of these. Apparently the second order dependence is small away from the extremes of p, with, for example, $P_{10}$ and $P_{11}$ being close to $P_1$. This is supported analytically by way of Taylor series expansions about $\alpha = 0.5$, corresponding to $p = 0$, which give $P_{ij} = P_i + o(p^2)$. This suggests that first order dependencies alone may in practice provide adequate approximations to the second order Markov model, the latter generally being sufficient to capture the structure of the series. This is supported empirically be experience with simulations, some of which is reported in Section 3 below. As an aside, note that similar features are found when $Y_t$ follows an AR(1) process with correlation p.

B: Inference about p from binary series

In applications concerning many similar $Y_t$ series and/or fast data rates, binary indicators are economic and easily processed summaries. Questions then arise about the information content of the indicator series relative to the underlying process. Kedem (1980) discusses this in the context of AR(1) models for $Y_t$, demonstrating the usefulness of first order Markov approximations to the $X_t$ process in making inferences about p. In our MA(1) model, an observed series of length n + 1 provides a log likelihood

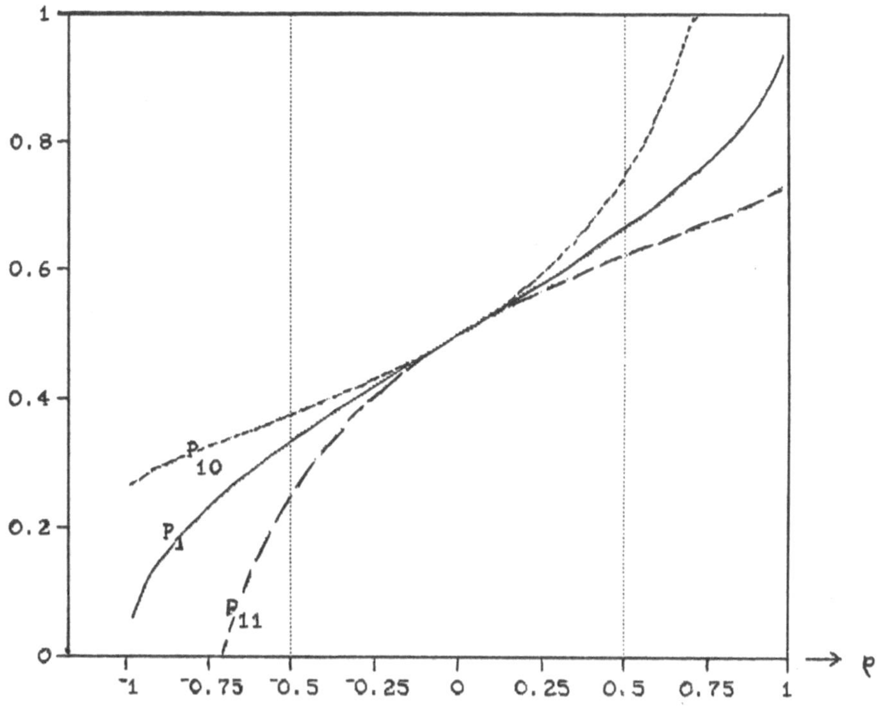


Fig. 1. Probabilities from the MA model

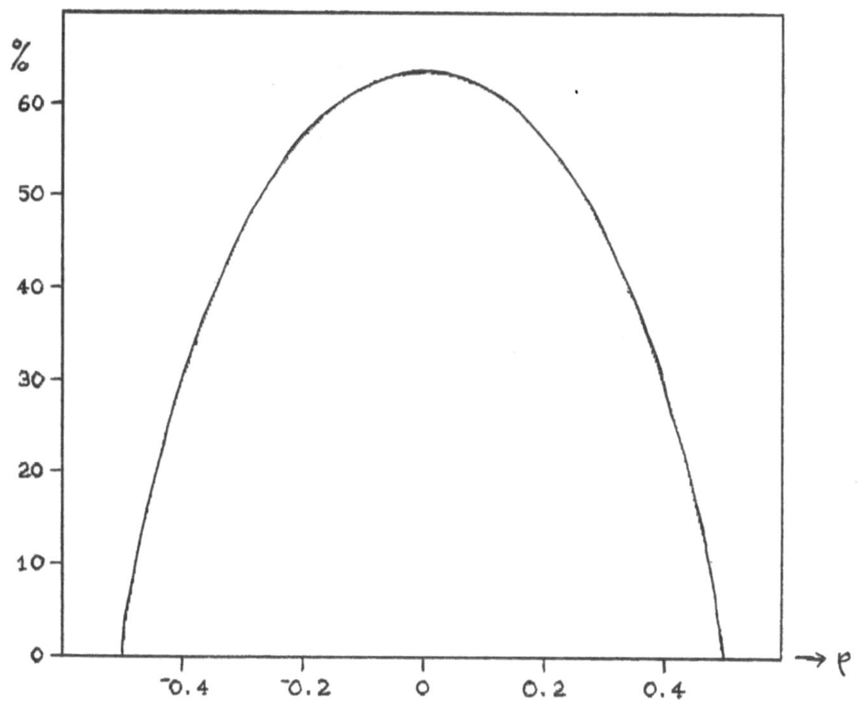


Fig. 2. Relative information for p

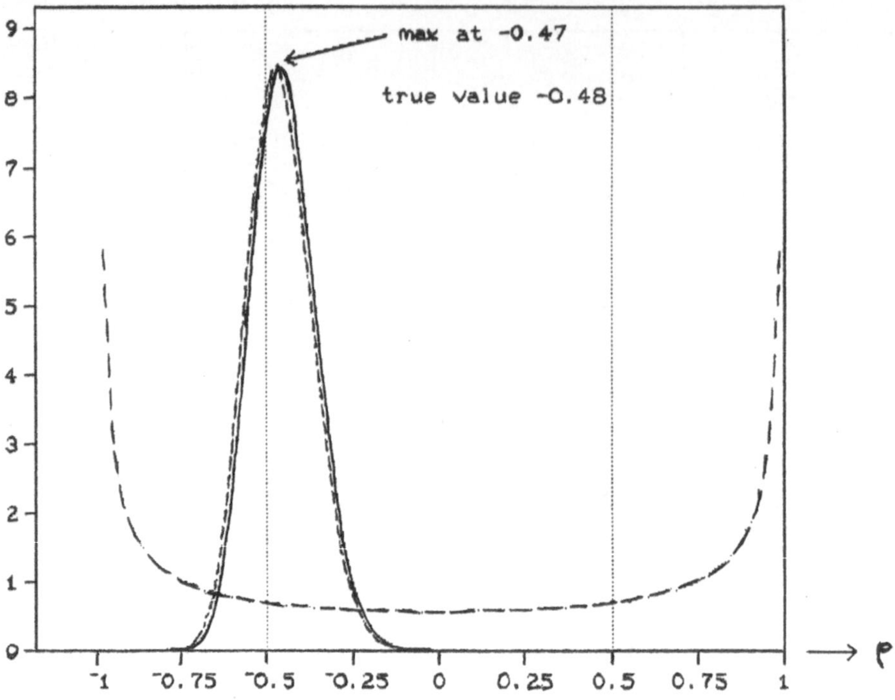


Fig. 3. Posteriors for p from binary data

$T \log(\alpha) + (n - T) \log(1 - \alpha)$ where $T = \Sigma[X_t X_{t-1} + (1 - X_t)(1 - X_{t-1})]$ is the number of runs of length 2. Substituting $\alpha$ as a function of p, we deduce a likelihood for p whose expected (Fisher) information is given by $I(p)$ where $I(p)^{-1} = \pi^2 \alpha(1 - \alpha)(1 - p^2)$. Relative to the information expected from the actual $Y_t$ data, measured by the square root of Fisher information, the binary series is roughly 60% as informative when $|p| \leq 0.2$, decaying to less than 30% when $|p| > 0.4$. Figure 2 provides a plot of the relative information as a function of p; clearly there is always some loss of precision. Experience with simulations, however, suggest that the location of the likelihood is often close to that from the full data. A typical example appears in Figure 3. Here 200 observations were generated from an MA(1) model with $p = -0.48$, near the boundary where the Markov model is expected to do worst. The dashed line represents the reference prior proportional to $\sqrt{I(p)}$, (proper over the range $|p| < 0.5$ of interest), the full line is the posterior from this example, with mode at -0.47. A posterior based on a second order Markov model is similar, being slightly more peaked at the mode. We return to this data in the next section.

## 3. SEQUENTIAL AUTOREGRESSION MODELS

A subset of dynamic generalised linear models (DGLM's) provides suitable models for binary time series. These are used in West (1986a and b), where further technical details appear, and here in the context of second order Markov chains. We define the quantities $D_t = \{X_t, X_{t-1}, \ldots , X_1\}$; $\underline{\theta}^T = [\mu, \beta_1, \beta_2]$; $\underline{F}_t^T = [1, X_{t-1}, X_{t-2}]$; $\eta_t = \underline{F}_t^T \underline{\theta} = \mu + \beta_1 X_{t-1} + \beta_2 X_{t-2}$, and $\pi_t = 1/[1 + \exp(-\eta_t)]$ so that $\eta_t = \log[\pi_t/(1-\pi_t)]$. We can express a second order Markov chain via the logistic linear model $P[X_t = 1 \mid \theta, D_{t-1}] = \pi_t$. Note the emphasis on sequential prediction here, $\pi_t$ is conditional on past data $D_{t-1}$.

The DGLM technique applies, as in the above references, to give the following ingredients: (a) at any time t - 1, the prior information about $\underline{\theta}$ is summarised in terms of a mean $m_{t-1}$ and variance matrix $C_{t-1}$, denoted $(\underline{\theta}|D_{t-1}) \sim [\underline{m}_{t-1}, C_{t-1}]$; (b) $(\pi_t|D_{t-1})$ is constrained to be a Beta prior with parameters appropriately matching the moments of $\eta_t$ implied in (a); (c) forecasts are based on $P[X_t = 1|D_{t-1}] = E[\pi_t|D_{t-1}]$; and (d) linear Bayes' techniques are used to update to $(\underline{\theta}|D_t)$ on observing $X_t$. The plots in Figures 4, 5 and 6 are based on this model applied to the simulated data of Section 2. With a relatively vague prior ($\underline{m}_0 = 0, C_0 = I$) we sequentially compute $\underline{m}_t$ and $C_t$ for t = 1,2, ..., 200. For each of the three elements of $\underline{\theta}$, the figures display posterior means, and two standard deviation intervals about the means, for each t; $\mu$ and $\beta_1$ are significantly non-zero, but $\beta_2$ is not. Final values at t = 200, with standard deviations, are $\mu$ : 0.61 (0.26); $\beta_1$ : -1.27 (0.30); $\beta_2$ : -0.38 (0.30). The sign of $\beta_1$ is appropriately negative and the insignificance of $\beta_2$ further supports a first order Markov model for the non-Markov process $X_t$.

As a follow up consider inference for p. We have a final posterior $(\underline{\theta}|D_{200}) \sim [m_{200}, C_{200}]$ to which we apply two constraints. First, we impose $\beta_2 = 0$ via a linear constraint $(0, 0, 1)\theta = 0$. Secondly, if the first order transition probabilities are to agree with the true values $P_1$ and $P_0$ of Section 2, then, logically, $\log[P_1/(1 - P_1)] = \mu + \beta_1$ and $\log[P_0/(1 - P_0)] = \mu$. However, since $P_1 = 1 - P_0 = \alpha$, we have a second constraint $2\mu + \beta_1 = 0$ or $(2, 1, 0)\theta = 0$. These two linear constraints are used to condition the posterior moments, revising them to $m^*_{200}$ and $C^*_{200}$, the latter now of rank 1 rather than 3. This conditioning uses linear Bayes theory, as in the DGLM, and is thus similar to standard normal theory, applying as if $\underline{\theta}$ were normal. Finally, the DGLM analysis leads to a Beta posterior for $\alpha = 1/[1 + \exp(\mu)]$ with parameters determined by the conditioned mean and variance of $\mu$. Transforming to $p = \sin[\pi(\alpha - 0.5)]$ leads to the posterior

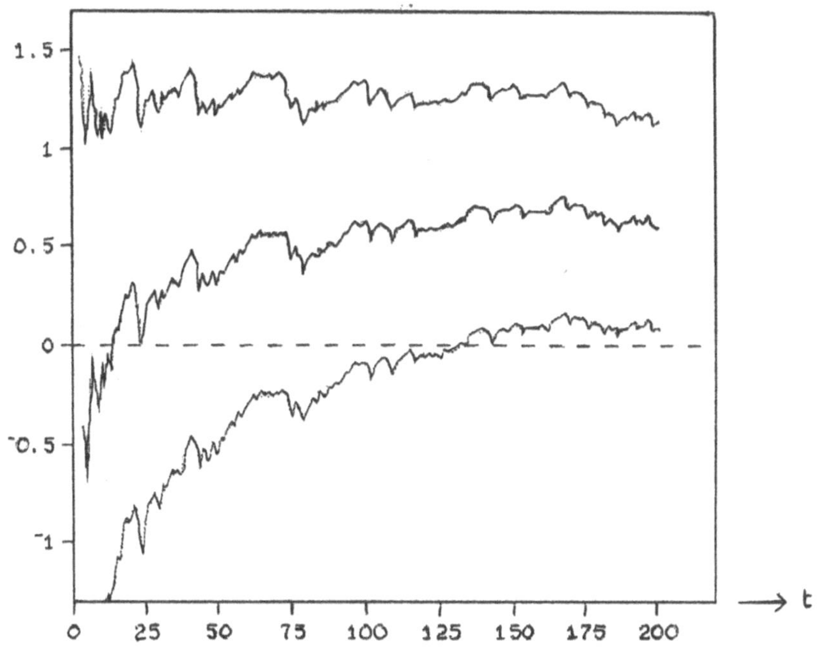


Fig. 4. Posterior intervals for $\mu$

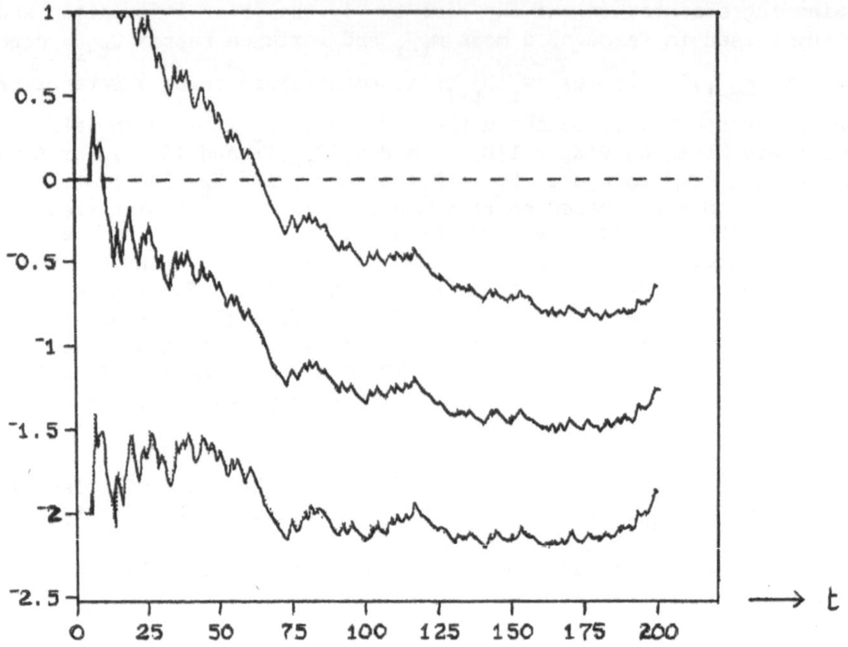


Fig. 5. Posterior intervals for $\beta_1$

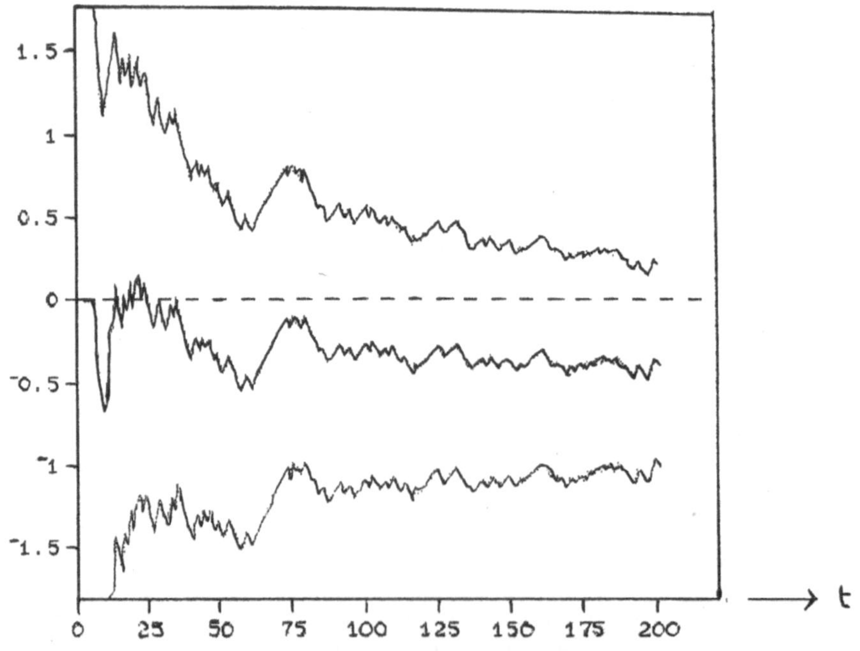


Fig. 6. Posterior intervals for $\beta_2$

for p plotted as a dotted line in Figure 3 closely agreeing with that calculated in Section 2. This agreement validates the sequential DGLM procedure, and restresses our conclusion about the utility of first order Markov approximations to the non-Markov process producing the binary series.

## 4. REGRESSION ON EXTERNAL FORECASTS

The logistic linear model of Section 3 allows obvious extension to include further regression terms. One such extension is considered in West (1986b) in the context of assessing and recalibrating probability forecasts of individuals or models. Suppose we have a series of probability forecasts $q_t$ for $X_t$ generated sequentially over time. We can simply extend the regression vector $F_t$ to include a further independent variable $\log[q_t/(1 - q_t)]$, similarly extending $\underline{\theta}$ to include an associated coefficient $\gamma$, say. Sequential analysis within the DGLM framework provides assessment of the accuracy of the $q_t$ forecasts via inference about $\theta$; $\mu$ allows for systematic *location* bias in the $q_t$, $\gamma$ for *scale* bias, and the autoregressive coefficients $\beta_1$ and $\beta_2$ allow for deficiencies in correlation structure. The sequential forecasts from this model, $P[X_t = 1 \mid D_{t-1}]$, now represent *data-based recalibrations* of the $q_t$. Such an approach provides a formal, model based alternative to empirical recalibration methods such as in Dawid (1984). Additionally, these models can be given Bayesian foundation using the framework of Lindley (1985). Specifically, a forecaster may model the way in which the $q_t$ sequence is generated such that $\pi_t$ is his/her own posterior probability for $X_t = 1$, conditional on $\underline{\theta}$, $D_{t-1}$ and $q_t$.

As an illustration, the model was applied to a series of 114 observations based on the monthly British Pound/Italian Lira exchange rate (January 1975 - August 1984). Here $X_t = 1$ indicates, as in Section 2, a rise in the rate in month t. An analyst provides the naive forecasts

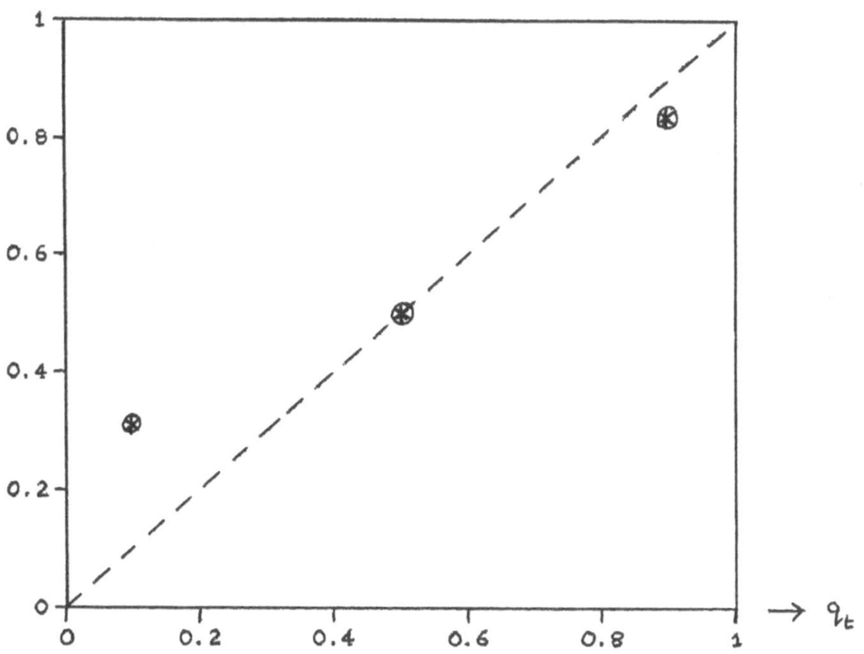


Fig. 7. Calibration curve for $q_t$

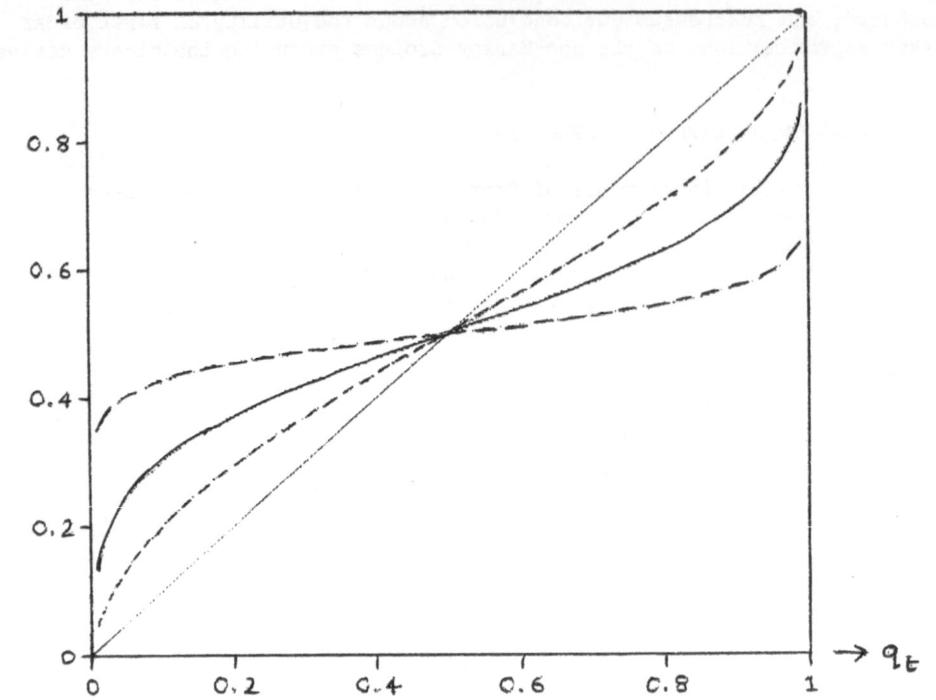


Fig. 8. Recalibration of $q_t$ : $X_{t-1} = 0$

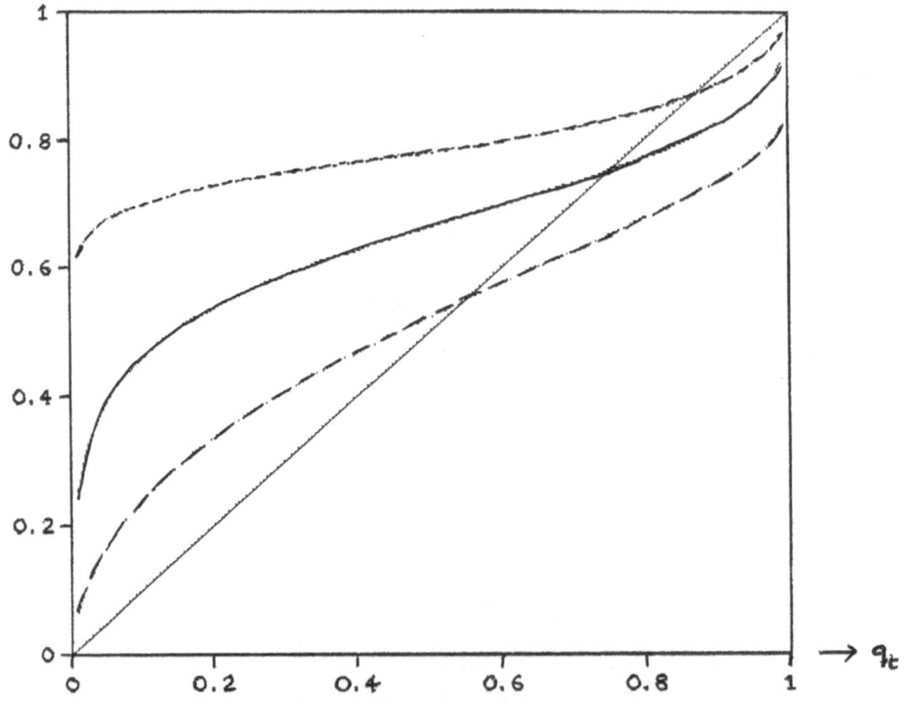


Fig. 9. Recalibration of $q_t$ : $X_{t-1} = 1$

$$q_t = \begin{cases} 0.9, & \text{if } X_{t-1} = X_{t-2} = 1; \\ 0.1, & \text{if } X_{t-1} = X_{t-2} = 0; \\ 0.5, & \textit{otherwise.} \end{cases}$$

The asterisks in Figure 7 are the corresponding points on the analyst's empirical calibration curve (the relative frequency of occurrences of $X_t = 1$ when the forecast is $q_t$) indicating that the forecasts 0.1 and 0.9 are rather extreme. We report some features of a regression model in which $\mu = \beta_2 = 0$ so that, redefining $\beta = \beta_1$, we have $\eta_t = \log[\pi_t/(1-\pi_t)] = \beta X_{t-1} + \gamma\log[q_t/(1-q_t)]$. This is chosen as the best predictive model in terms of aggregate predictive probability over the series, and clearly out-performs the raw $q_t$ with a log-Bayes' factor of more than 3. Final means and standard deviations at t = 114 are $\beta$ : 0.68(0.36) and $\gamma$ : 0.38(0.15), with correlation 0.59. The posterior mean of 0.38 for $\gamma$ indicates that, whilst in positive accord with the data ($\gamma > 0$), the naive forecasts tend to be overly extreme, confirming the message from the calibration plot. Figure 8 is a *recalibration curve* for the $q_t$, with an associated interval. The full line is the predictive probability from the DGLM model for $(X_t = 1 \mid X_{t-1}, q_t)$, simply the posterior mean of $\eta_t$ converted to the probability scale. The dotted and dashed lines provide an interval about this probability calculated similarly from 1.65 standard deviations limits for $\eta_t$. The shrinkage of extreme values towards 0.5 is clear from this curve, correcting the scale bias (over precision) in the naive forecasts. For the AR term, the posterior for $\beta$ indicates residual positive dependence in the series over and above the structure predicted by $q_t$. A similar recalibration curve with $X_{t-1} = 1$ appears in Figure 9.

The utility of our models is evident here. Further discussion and illustration appear in West(1986b), where parameters such as $\beta$ and $\gamma$ are modelled as dynamic. This allows for the possibility of time-varying biases and relationships, providing in particular for non-stationary Markov transition probabilities, and more fully exploits the time series modelling concepts underlying the DGLM framework. Other extensions appear in West (1986a) where more than one $q_t$ forecast sequence are available, providing an approach to comparison and aggregation of probability forecasts from several sources.

ACKNOWLEDGEMENTS

This research was completed during the S.E.R.C. supported Bayesian Statistics Study Year at the University of Warwick. We acknowledge financial support for the second author from the Italian National Research Council (C.N.R.) under grant No. 20/2-7.

REFERENCES

DAWID, A.P., 1984, Statistical theory: the prequential approach. J.R. Statist. Soc., A, 2, 278-292.

LINDLEY, D.V., 1985, Reconciliation of discrete probability distributions, in "Bayesian Statistics 2" eds: J.M. Bernardo, M.H. deGroot, D.V. Lindley and A.F.M. Smith, North-Holland, Amsterdam.

KEDEM, B., 1980, "Binary Time Series", Marcel Dekker, New York.

WEST, M., HARRISON, P.J. and MIGON, H.S., 1985, Dynamic generalised linear models and Bayesian forecasting (with Discussion), J. Amer. Statist. Ass., 80, 389, 73-97.

WEST, M., 1986a, Combining probability forecasts, J.R. Statist. Soc., B. (to appear)

WEST, M., 1986b, Assessment and control of probability forecasts, Warwick Research Report #69.

# SEMI-SUFFICIENCY IN ACCELERATED LIFE TESTING

Rudolf Willing

Institut für Statistik und Wahrscheinlichkeitstheorie
Technische Universität Wien
1040 Wien, Austria

## INTRODUCTION

Extremely long testing time causes a well known problem in reliability analysis. The most successful method of treating it is the application of accelerating stress. S. S is a vector of physical effects like temperature, voltage or pressure. The lifetime of an object put under stress S is described by a stochastic quantity $T_S$. The distribution belongs to a family parameterized by $\theta(S)$, the vector of statistical parameters given stress S. In practice the relation between these parameters and the stress components is often known to be of some functional form $\psi$ with unknown physical parameters $c=(c_1,\dots,c_n)$

$$\theta(S) = \psi(S,c). \tag{1}$$

All knowledge about these physical parameters c before the experiment is put into the prior density $\pi(c)$. Now we can make stochastically independent observations on m different stress levels $S_i$ $i=1(1)m$. On each of these we get a sample of size $k_i$. Using the life time densities f(.) we get the likelihood function $l(c;D)$ given the data $D=(t_{ij},\ i=1(1)m,\ j=1(1)k_i)$

$$l(c;D) = \prod_{i=1}^{m} \prod_{j=1}^{k_i} f(t_{ij}|S_i,c). \tag{2}$$

Calculating the posterior density $\pi(c|D)$ via Bayes' theorem

$$\pi(c|D) \propto \pi(c)\ l(c;D) \tag{3}$$

we can make estimates about the physical parameters c, the statistical parameters given the usual stress $S_u$

$$\theta(S_u) = \psi(S_u,c) \tag{4}$$

or the predictive density $f(t|S_u)$ of an object under usual stress $S_u$. If we think of quadratic loss the best way to do that, is to calculate the posterior expectations of these quantities. For instance the statistical parameters at usual stress $S_u$ can be estimated by

$$\hat{\theta}(S_u) = E_{\pi(c|D)}\, \psi(S_u, \tilde{c}), \tag{5}$$

where the swung dash denotes the stochastic quantity that describes the unknown.

Only in the case of using exponential families as life time models and their natural conjugate priors these integrations can be done analytically. But exponential families are very restrictive. To apply Bayesian analysis we need a broader class of distributions. These distributions should be easy to handle but allow for a variety of different models. Therefore the author wants to introduce a new set of distributions.

## THE CONCEPT OF SEMISUFFICIENCY

Definition: Let X be a stochastic quantity, the distribution of which is depending on a parametervector $\theta=(\theta_1,\theta_2)$, where $\theta_1$ and $\theta_2$ are two subvectors of $\theta$. A statistic $S(X|\theta_2)$ is called semisufficient for $\theta$, iff the posterior distribution of $\tilde{\theta}_1$ given $\tilde{\theta}_2=\theta_2$ depends on X only through $S(X|\theta_2)$ no matter what prior distribution of $\tilde{\theta}$ is used.

$$\pi(\theta_1|\theta_2,x) = g(\theta_1|\theta_2,S(x|\theta_2)) \tag{6}$$

The subvector $\theta_2$ is not considered to be a nuisance parameter like in Dawid (1979) who defined other concepts of relaxed sufficiency.

Theorem: $S(X|\theta_2)$ is a semisufficient statistic for $\theta$ iff the condional density $f(x|\theta)$ factors like

$$f(x|\theta) = h(S(x|\theta_2)|\theta_1,\theta_2)\; i(x|\theta_2), \tag{7}$$

where h depends on both subparametervectors $\theta_1$ and $\theta_2$ but only through $S(x|\theta_2)$ on x and i depends on x but only on $\theta_2$.

Proof: Using Bayes´ theorem we obtain

$$\pi(\theta_1,\theta_2|x) = \pi(\theta_1,\theta_2)\; h(S(x|\theta_2)|\theta_1,\theta_2)\; i(x|\theta_2)/K$$

where

$$K = \int \pi(\theta_1,\theta_2)\; h(S(x|\theta_2)|\theta_1,\theta_2)\; i(x|\theta_2)\; d(\theta_1,\theta_2).$$

The marginal density of $\tilde{\theta}_2$ given the data x equals to

$$\pi(\theta_2|x) = \int \pi(\theta_1,\theta_2)\; h(S(x|\theta_2)|\theta_1,\theta_2)\; d\theta_1\; \frac{i(x|\theta_2)}{K}$$

so that the conditional density of $\tilde{\theta}_1$ given $\tilde{\theta}_2=\theta_2$ after the data x is observed equals to

$$\pi(\theta_1|\theta_2,x) = \frac{\pi(\theta_1,\theta_2|x)}{\pi(\theta_2|x)} = \frac{\pi(\theta_1,\theta_2)\; h(S(x|\theta_2)|\theta_1,\theta_2)}{\int \pi(\theta_1,\theta_2)\; h(S(x|\theta_2)|\theta_1,\theta_2)\; d\theta_1}\,.$$

On the other hand given

$$f(x|\theta_1,\theta_2) = \frac{\pi(\theta_1|\theta_2,x)\ \pi(\theta_2|x)\ f(x)}{\pi(\theta_1,\theta_2)} = \frac{g(\theta_1|\theta_2,S(x|\theta_2))\pi(\theta_2|x)\ f(x)}{\pi(\theta_1,\theta_2)}$$

so we can choose

$$h(S(x|\theta_2)|\theta_1,\theta_2) = \frac{g(\theta_1|\theta_2,S(x|\theta_2))}{\pi(\theta_1,\theta_2)}$$

and

$$i(x|\theta_2) = \pi(\theta_2|x)\ f(x),$$

where $f(x)$ is the prior predictive density.

## SEMI-EXPONENTIAL FAMILIES

Now we want to use this result. If $f(x|\theta)$ does not belong to an exponential family but $\theta$ consists of two subvectors $\theta_1$ and $\theta_2$ such that (7) is valid and additionally if $f(x|\theta_1,\theta_2)$ belongs to an exponential family in $\theta_1$ given $\theta_2$, then $S(x|\theta_2)$ is a semisufficient statistic of a fixed dimensionality for $\theta$. That means the dimension of the statistic does not increase when we switch from the stochastic quantity to a random sample.

Definition: A family of distributions of a stochastic quantity X forms a semi-exponential family of dimension k with parametervector $\theta$ iff $\theta$ consists of $\theta_1$ and $\theta_2$ so that the density can be written in the form

$$f(x|\theta_1,\theta_2) = G_1(\theta_1|\theta_2)\ G_2(\theta_2)\ H(x|\theta_2)\ \exp(-\sum_{j=1}^{k}\psi_j(\theta_1|\theta_2)T_j(x|\theta_2)). \quad (8)$$

Theorem: Let $X=(X_1,\ldots,X_n)$ be a sample of iid. stochastic quantities belonging to the semi-exponential family in (8), then we can immediately write down a natural conjugate family of prior distributions for the parametervector $\theta=(\theta_1,\theta_2)$

$$\pi(\theta_1,\theta_2|\alpha_0,\alpha_1,\ldots,\alpha_k,p) = p(\theta_2)\ K(\alpha_0,\alpha_1(\theta_2),\ldots,\alpha_k(\theta_2),\theta_2)\times$$
$$\times\ G_1(\theta_1|\theta_2)^{\alpha_0}\ \exp(-\sum_{j=1}^{k}\ \psi_j(\theta_1|\theta_2)\ \alpha_j(\theta_2)). \quad (9)$$

That is a fairly extensive family because $p(\theta_2)$ can be any marginal density for $\theta_2$. $K(\alpha_0,\alpha_1(\theta_2),\ldots,\alpha_k(\theta_2),\theta_2)$ is the normalizing factor for $\pi(\theta_1|\theta_2)$, where $\alpha_0,\alpha_1(\theta_2),\ldots,\alpha_k(\theta_2)$ are the hyperparameterfunctions depending on $\theta_2$. After the observation of $D=(x_1,\ldots,x_n)$ we get the following posterior density for $\bar{\theta}$

$$\pi(\theta_1,\theta_2|D) = \pi(\theta_1,\theta_2|\bar{\alpha}_0,\bar{\alpha}_1,\ldots,\bar{\alpha}_k,\bar{p}) \quad (10)$$

with the new hyperparameterfunctions

$$\bar{\alpha}_0 = \alpha_0 + n$$

$$\bar{\alpha}_j(\theta_2) = \alpha_j(\theta_2) + \sum_{i=1}^{n} T_j(x_i|\theta_2) \qquad j=1(1)k$$

$$p(\theta_2) = \frac{p(\theta_2)\, G_2(\theta_2)^n \sum_{i=1}^{n} H(x_i|\theta_2)\, \dfrac{K(\alpha_0,\alpha_1(\theta_2),\ldots,\alpha_k(\theta_2),\theta_2)}{K(\bar{\alpha}_0,\bar{\alpha}_1(\theta_2),\ldots,\bar{\alpha}_k(\theta_2),\theta_2)}}{\int p(\theta_2)\, G_2(\theta_2)^n \sum_{i=1}^{n} H(x_i|\theta_2)\, \dfrac{K(\alpha_0,\alpha_1(\theta_2),\ldots,\alpha_k(\theta_2),\theta_2)}{K(\bar{\alpha}_0,\bar{\alpha}_1(\theta_2),\ldots,\bar{\alpha}_k(\theta_2),\theta_2)}\, d\theta_2}$$

This is a generalized way of updating hyperparameters.

Proof: Straight forward calculation.

All this expense pays only profit if $K(\alpha_0,\alpha_1(\theta_2),\ldots,\alpha_k(\theta_2),\theta_2)$ can be evaluated analytically.

Example: Let $X=(X_1,\ldots,X_n)$ be a sample of Weibull-distributed stochastic quantities with density function

$$f(x_1,\ldots,x_n|\lambda,\beta) = \lambda^n\, \beta^n \Big(\prod_{i=1}^{n} x_i\Big)^{\beta-1} \exp\Big(-\lambda \sum_{i=1}^{n} x_i^{\beta}\Big). \tag{11}$$

Now let us take

$\theta_1 = \lambda$

$\theta_2 = \beta$

and we obtain the factorization (7) with the functions

$$h(S(x|\theta_2)\theta_1,\theta_2) = \lambda^{S_1} \exp(-\lambda S_2)$$

$$i(x|\theta_2) = \beta^n \Big(\prod_{i=1}^{n} x_i\Big)^{\beta-1}$$

where $S_1$ and $S_2$ are the two components of the semisufficient statistic of dimension two

$$S(X|\theta_2) = \Big(n, \sum_{i=1}^{n} X_i^{\beta}\Big) = (S_1,S_2)\ .$$

The Weibull-distribution does not belong to an exponential family but it belongs to a semi-exponential family as can be seen by putting

$k = 1$

$G_1(\theta_1|\theta_2) = \lambda$ $\qquad G_2(\theta_2) = \beta$

$H(x|\theta_2) = \prod_{i=1}^{n} x_i^{\beta-1}$

$\psi_1(\theta_1|\theta_2) = \lambda$ $\qquad T_1(x|\theta_2) = \sum_{i=1}^{n} x_i^{\beta}.$

For the Weibull-distribution a choice of a natural conjugate family for $\tilde{\lambda}$ and $\tilde{\beta}$ is

$$\pi(\lambda,\beta) = \frac{a^r}{\Gamma(r)}\, \beta^{r-1}\, e^{-a\beta}\, \frac{\alpha_1^{\alpha_0+1}}{\Gamma(\alpha_0+1)}\, \lambda^{\alpha_0}\, e^{-\lambda\alpha_1}. \tag{12}$$

The connexion with the former definitions (9) can be seen as

$$p(\theta_2) = \frac{a^r}{\Gamma(r)}\ \beta^{r-1}\, e^{-a\beta}$$

which is a Gamma-distribution with hyperparameters a and r,

$$\alpha_0 = \alpha_0$$

$$\alpha_1(\theta_2) = \alpha_1$$

where $\alpha_1$ does not a-priori depend on $\theta_2=\beta$,

$$K(\alpha_0,\alpha_1(\theta_2),\ldots,\alpha_k(\theta_2),\theta_2) = \frac{\alpha_1^{\alpha_0+1}}{\Gamma(\alpha_0+1)}$$

$$G_1(\theta_1|\theta_2)^{\alpha_0} = \lambda^{\alpha_0} .$$

The posterior distribution of $\tilde{\lambda}$ and $\tilde{\beta}$ given the data $D=(x_1,\ldots,x_n)$ of life time observations is given by

$$\pi(\lambda,\beta|D) = \frac{\beta^{r+n-1}\, e^{-(a-\sum_{i=1}^{n}\ln x_i)\beta}\,(\alpha_1+\sum_{i=1}^{n}x_i^{\beta})^{-\alpha_0-n-1}}{\int \beta^{r+n-1}\, e^{-(a-\sum_{i=1}^{n}\ln x_i)\beta}\,(\alpha_1+\sum_{i=1}^{n}x_i^{\beta})^{-\alpha_0-n-1}\,d\beta} \times$$

$$\times \frac{(\alpha_1+\sum_{i=1}^{n}x_i^{\beta})^{\alpha_0+n+1}}{\Gamma(\alpha_0+n+1)}\,\lambda^{\alpha_0+n}\,e^{-\lambda(\alpha_1+\sum_{i=1}^{n}x_i^{\beta})} \tag{13}$$

with the new hyperparameters

$$\bar{\alpha}_0 = \alpha_0 + n$$

$$\bar{\alpha}_1(\theta_2) = \alpha_1 + \sum_{i=1}^{n}x_i^{\beta}$$

$$p(\theta_2) = \frac{\beta^{r+n-1}\, e^{-(a-\sum_{i=1}^{n}\ln x_i)\beta}\,(\alpha_1+\sum_{i=1}^{n}x_i^{\beta})^{-\alpha_0-n-1}}{\int \beta^{r+n-1}\, e^{-(a-\sum_{i=1}^{n}\ln x_i)\beta}\,(\alpha_1+\sum_{i=1}^{n}x_i^{\beta})^{-\alpha_0-n-1}\,d\beta} .$$

And so the conditional distribution of $\tilde{\lambda}$ given $\tilde{\beta}=\beta$ is a Gamma with the updated hyperparameters $\bar{\alpha}_0$ and $\bar{\alpha}_1(\theta_2)$.

## APPLICATION IN THE ANALYSIS OF PRESSURE BEARINGS

The computational simplifications of this method had been put into praxis for the problem investigated by Viertl and Willing(1985). The abrasion of bearings put under higher pressures was measured to get their life times. Their distribution was assumed to be Weibull with parameters $\lambda$ and $\beta$. These depend on the accelerating stress the pressure S in the following way

$$\lambda(S) = \lambda\, e^{\delta S^3}$$

$$\beta(S) = \beta . \tag{14}$$

This corresponds to the relation (1) with the vector $c=(\lambda,\beta,\delta)$. Collecting data $D=(t_{ij}, i=1(1)m, j=1(1)k_i)$ on m different stress levels $S_i$. We get the likelihood-function (2)

$$l(c;D) = \lambda^n \beta^n e^{\sum_{i=1}^{m} k_i \delta S_i^3} \prod_{i=1}^{m} \prod_{j=1}^{k_i} t_{ij}^{\beta-1} e^{-\lambda \sum_{i,j} t_{ij}^{\beta} e^{\delta S_i^3}} \qquad (15)$$

with

$$n = \sum_{i=1}^{m} k_i$$

the total number of objects tested. Now this belongs to a semi-exponential family (8) of dimension k=1 with parameters $\theta$=c and

$$\theta_1 = \lambda$$

$$\theta_2 = \begin{pmatrix} \beta \\ \delta \end{pmatrix}.$$

The semisufficient statistic S a function of D depending on $\beta$ and $\delta$ equals to

$$S(D|\theta_2) = (n, \sum_{i,j} t_{ij}^{\beta} e^{\delta S_i^3}).$$

The other functions in (8) become

$$G_1(\theta_1|\theta_2) = \lambda^n \qquad G_2(\theta_2) = \beta^n e^{\sum_{i=1}^{m} k_i \delta S_i^3}$$

$$H(X|\theta_2) = \prod_{i,j} t_{ij}^{\beta-1}$$

$$\psi_1(\theta_1|\theta_2) = \lambda \qquad T_1(X|\theta_2) = \prod_{i,j} t_{ij}^{\beta} e^{\delta S_i^3} .$$

Taking a non informative prior distribution of the natural conjugate family (9) with

$$p(\theta_2) = \frac{1}{\beta \delta}$$

$$\alpha_0 = -1$$

$$\alpha_1(\theta_2) = 0$$

we get an improper prior density

$$\pi(\lambda,\beta,\delta) \propto \frac{1}{\lambda \beta \delta} . \qquad (16)$$

This leads to the posterior density

$$\pi(\lambda,\beta,\delta|D) \propto \lambda^{n-1} e^{-\lambda \sum_{i,j} t_{ij}^{\beta} e^{\delta S_i^3}} \beta^{n-1} \prod_{i,j} t_{ij}^{\beta-1} e^{\sum_i k_i \delta S_i^3} , \qquad (17)$$

where the conditional density of $\lambda$ given $\beta$ and $\delta$ is a Gamma-distribution with parameters n and

$$\sum_{i,j} t_{ij}^{\beta} e^{\sum_i \delta S_i^3}$$

and the marginal density of $\tilde{\beta}$ and $\tilde{\delta}$

$$\pi(\beta,\delta|D) \propto \frac{\beta^{n-1} \prod_{i,j} t_{ij}^{\beta-1} e^{\sum_i k_i \delta S_i^3}}{\delta\left(\sum_{i,j} t_{ij}^{\beta} e^{\delta S_i^3}\right)^n} . \qquad (18)$$

Now it is easier to calculate the expectations in (5) because only a two dimensional integration is necessary. The computation time for the analysis in Willing(1985) has been reduced by 95%. This allows for a more detailed study of acceleration models and higher dimensional models become treatable.

ACKNOWLEDGEMENT

The author wants to thank the Austrian Science Foundation for support of this work.

REFERENCES

Dawid, A.P., 1979, A Bayesian Look at Nuisance Parameters, in: "Bayesian Statistics", Proceedings of the First International Meeting held in Valencia, J.M. Bernardo et al., eds., Univ. Press of Valencia.

Viertl, R., Willing, R., 1985, Bayesian Analysis for Weibull-distributed Life Times in Accelerated Life Testing, in: "Proceedings of the 45th Session of the Int. Stat. Institute", Contributed Papers, Book 1, Amsterdam.

Willing, R., 1985, Bayes´sche Analyse von beschleunigten Lebensdauerversuchen bei Weibull Verteilungsannahme, Research Rep. RIS-1985-13, Institut f. Statistik u. Wahrscheinlichkeitstheorie, Technische Universität Wien, Wien.

INDEX